Cryptography and Embedded Systems Security

Xiaolu Hou · Jakub Breier

Cryptography and Embedded Systems Security

Springer

Xiaolu Hou
Faculty of Informatics and Information
Technologies
Slovak University of Technology
Bratislava, Slovakia

Jakub Breier
TTControl GmbH
Vienna, Austria

ISBN 978-3-031-62204-5   ISBN 978-3-031-62205-2   (eBook)
https://doi.org/10.1007/978-3-031-62205-2

This Springer imprint is published by the registered company Springer Nature Switzerland AG
The registered company address is: Gewerbestrasse 11, 6330 Cham, Switzerland

If disposing of this product, please recycle the paper.

For our Aurel

Foreword

In an era defined by interconnectedness, the importance of security is undeniable. Across billions of devices and computing systems, cryptographic algorithms and protocols stand as sentinels, safeguarding the confidentiality, integrity, and non-repudiation of transactions. However, even with the remarkable capabilities of cryptographic algorithms, the systems they safeguard are not necessarily immune to vulnerabilities. These vulnerabilities frequently emerge during the transition from theory to practical implementations, underscoring the pivotal role of cryptographic engineering in achieving comprehensive security measures. The present book serves to nicely bridge this gap and provide practitioners and researchers interested in the world of embedded security a wide perspective of secure implementations of cryptographic algorithms.

While strong cryptographic algorithms are an important starting point in the design of secured systems, they also need to be efficiently implemented for real-life practical applications. While in the early days they were implemented largely on general-purpose computers, it was gradually felt necessary to realize them on hardware and embedded platforms. This shift was an outcome of multiple factors. The complexity of cryptographic algorithms and their real-time requirements to ensure practical applications motivated researchers to implement the ciphers on hardware and embedded platforms. Moreover, because of the various attacks on software platforms, designing security systems relying on hardware root-of-trusts became a popular design choice. Further, the growth of embedded applications, and thereof the advent of Cyber-Physical Systems (CPS) and Internet-of-Things (IoT), obviated the integration of cryptographic algorithms into special-purpose devices. However, great care needs to be taken in such implementations, as apart from the classic design objectives, like power, energy, throughput, and area, designers also need to tackle side-channel information leakages which can be exploited by attackers with physical access to the devices. Common side-channel attacks based on power/electromagnetic analysis and fault analysis have become one of the biggest threats in deploying crypto algorithms on embedded devices. The ubiquitousness of such devices and easy physical access by adversaries offer novel attack surfaces which can cripple the best of crypto-algorithms if suitable countermeasures are not implemented along with.

The contribution of this book is to address these aspects of secured crypto-design and provide a vivid description to develop an end-to-end understanding. The designs of cryptographic algorithms and their analysis are often based on mathematical and statistical tools. The book starts with a nice summary of important mathematical principles, which are needed to comprehend the cipher constructions and their attack analysis. Subsequently, the book provides a summary of both classical and modern cryptosystems. The following chapters also stress on implementations of these modern cryptosystems, before delving into various forms of physical attacks on the implementations. The book discusses techniques for side-channel analysis of both symmetric-key and public-key cryptosystems, along with suitable countermeasures. The book then presents a contemporary summary of various forms of fault attacks on cryptosystems, and countermeasures against them. The book concludes with practical aspects of physical attacks, providing much-needed details of physical setups, useful to develop practical setups for hardware security research.

Engaging and informative, this book is fine reading for anyone fascinated by the intricate realm of embedded security and cryptographic engineering. It offers a compelling glimpse into the workings of attacks on cryptosystems in embedded devices and provides actionable strategies for mitigation. Enjoy the journey into the captivating world of security engineering!

Kharagpur, India   Debdeep Mukopadhyay
April 2024

Starting my doctoral studies several decades ago, I found myself immensely interested in the area of physical side channels and the resulting attacks, which at the time disrupted the way in which cryptographers approached designing and analyzing ciphers. This was a fortunate encounter for me: today my research is still driven by the challenge of efficiently detecting, quantifying, and as far as possible mitigating physical side channels.

Research in the area of side channels has developed and grown, not only in volume but also in maturity. In the early days, researchers playfully discovered how to tap into side channels, as well as how to extract more information from available side channels, and to make side channels harder to exploit. There was little emphasis on the development of a methodology. Countermeasures were (re)invented and applied to different types of cryptosystems, acknowledging, but not systematizing, that different discoveries were in fact related.

Only when, together with two colleagues, I wrote the first comprehensive research book on power analysis attacks, a clearer picture emerged of the factors that contribute to the success of attacks and how we can mitigate leakage. Other researchers pushed our initial attempts further, and today, we have sound theories for many aspects of side-channel attacks and countermeasures. Similarly, the area of fault attacks has seen significant progress over the past two decades.

This book here provides a contemporary summary of techniques for attacks and countermeasures. There are many good examples provided: I encourage all readers of this book to pay particular attention to these and implement and extend as many as possible. The best way to understand the foundational aspects of any field is by active learning: do as much as you can yourself!

Birmingham, UK

Elisabeth Oswald

April 2024

Preface

Cryptography is an indispensable tool used to protect information in computing systems. Billions of people all over the world use it in their daily lives without even noticing there is some cryptographic algorithm running behind the scenes. Cryptographic computations can be found in any form of electronic communication, electronic passports, security tokens, payment systems, etc.

Cryptographic algorithms in use nowadays are considered secure in theory. But in the real world, these algorithms are implemented on physical devices in the form of integrated circuits. These circuits have their physical properties, such as power consumption dependent on the processed data, emanation of electromagnetic waves, and susceptibility to computational errors due to environmental influences. To evaluate the security level of cryptographic implementations, it is necessary to include the physical security assessment.

There are various physical attack methods, e.g., fault attacks, side-channel attacks, hardware trojans, etc. Side-channel attacks can be divided into different specific attacks, depending on the exploited information, e.g., electromagnetic/power analysis, timing analysis, cache attacks. In this book, we focus on fault attacks and electromagnetic/power analysis attacks on cryptographic implementations.

We assume the readers have basic knowledge of real numbers, rational numbers, integers, and complex numbers, which will be denoted by $\mathbb{R}$, $\mathbb{Q}$, $\mathbb{Z}$, and $\mathbb{C}$ respectively in this book. We also assume the readers have completed a course in linear algebra.

This book is primarily aimed at graduate students who take a course on hardware security and/or cryptography. However, it provides useful resources for anyone willing to explore the exciting world of physical attacks—designers, implementers, evaluators, as well as academic scholars.

Bratislava, Slovakia   Xiaolu Hou
Vienna, Austria   Jakub Breier
April 2024

Acknowledgment

We would like to thank Debdeep Mukhopadhyay and Elisabeth Oswald for writing a nice and motivating foreword for this book.

Our thanks also go to Mladen Kovačević, Romain Poussier, and Dirmanto Jap for proofreading an earlier version of the book and for their detailed and constructive comments.

We would also like to acknowledge the editorial team of Springer Nature, especially Bakiyalakshmi R M, and Charles Glaser.

For the unwavering support and encouragement from our parents and especially our son, Aurel, who turned our writing process into a wild adventure.

This project has received funding from the European Union's Horizon 2020 Research and Innovation Programme under the Programme SASPRO 2 COFUND Marie Sklodowska-Curie grant agreement No. 945478.

Contents

List of Figures

List of Tables

List of Algorithms

Chapter 1
Mathematical and Statistical Background

To study attacks on cryptographic algorithms, we need to first understand the computations that are carried out in each step of those algorithms. To achieve this, we need knowledge of certain math concepts. In this chapter, we will introduce the necessary mathematical background for the rest of the book, including abstract algebra, linear algebra, coding theory, and probability theory. In Sect. 1.8, we will also provide statistical tools that will be useful for Chap. 4.

1.1 Preliminaries

Before we start with math, let us first introduce the basic notations.

1.1.1 Sets

By a *set*, we refer to a collection of objects without repetition. We will normally use a capital letter to denote a set. For example, $A = \{0, 1, 2\}$ is a set consisting of three numbers, and $B = \{\circ, \triangle, \square\}$ is a set consisting of three shapes. The objects in a set S are called *elements* of S. If an element a is in a set S, we write $a \in S$. If an element a is not in S, we write $a \notin S$. When there is no element in a set, we call it an *empty set* and denote it by $\emptyset$. The total number of elements in a set S is called the *cardinality* of S, denoted by $|S|$.

Now let us look at two sets, S and T. We say S is a *subset* of T, denoted by $S \subseteq T$, if any element of S is also an element of T. Namely, $S \subseteq T$ if for any $s \in S$, $s \in T$. Two sets are said to be equal if they contain the same elements. In other words, $S = T$ if $S \subseteq T$ and $T \subseteq S$. The *power set* of a set S, denoted by 2^S, is the

X. Hou, J. Breier, *Cryptography and Embedded Systems Security*,
https://doi.org/10.1007/978-3-031-62205-2_1

set of all subsets of S. We note that by definition, $S \in 2^S$, $\emptyset \in 2^S$, and $\emptyset \subseteq S$ for any set S.

Example 1.1.1 Let $T = \{0, 1, 2, 3\}$ and $S = \{2, 3\}$, then

- $S \subseteq T$ and $T \not\subseteq S$.
- $2 \in S, 0 \notin S$.
- $|S| = 2, |T| = 4$.
- $2^S = \{\emptyset, S, \{2\}, \{3\}\}$.

The *union* of two sets A and B, denoted $A \cup B$, is the set that contains all elements from A or B.

$$A \cup B := \{x \mid x \in A \text{ or } x \in B\}.$$

The *intersection* of A and B, denoted $A \cap B$, is the set that contains elements in both A and B.

$$A \cap B := \{x \mid x \in A \text{ and } x \in B\}.$$

Example 1.1.2 Let $A = \{0, 1, 2\}$ and $B = \{2, 3, 4\}$, then $A \cup B = \{0, 1, 2, 3, 4\}$ and $A \cap B = \{2\}$.

Similarly, the union and the intersection of n sets $A_1, A_2, \ldots, A_n$ are defined as follows:

$$\bigcup_{i=1}^{n} A_i := \{a \mid a \in A_i \text{ for some } i\}, \quad \bigcap_{i=1}^{n} A_i := \{a \mid a \in A_i \text{ for all } i\}.$$

The *difference* between set A and set B is the set of all elements of A that are not in B:

$$A - B := \{a \mid a \in A, a \notin B\}. \tag{1.1}$$

The *complement* of a set A in a set S is the difference between S and A,

$$A^c := S - A = \{s \mid s \in S, s \notin A\}.$$

The *Cartesian product* of A and B is the set of *ordered pairs* (a, b) such that $a \in A$ and $b \in B$,

$$A \times B := \{(a, b) \mid a \in A, b \in B\}.$$

The Cartesian product of n sets can be defined similarly,

$$\prod_{i=1}^{n} A_i := \{(a_1, a_2, \ldots, a_n) \mid a_i \in A_i \text{ for all } i\}.$$

Example 1.1.3 Let $A = \{2, 4, 6\}$, $B = \{1, 3, 5\}$, and $S = A \cup B$. Then $A - B = A$; the complement of A in S is B, and

$$A \times B = \{(2, 1), (2, 3), (2, 5), (4, 1), (4, 3), (4, 5), (6, 1), (6, 3), (6, 5)\}.$$

We note that, in general, $A \times B \neq B \times A$. In Example 1.1.3,

$$B \times A = \{(1, 2), (3, 2), (5, 2), (1, 4), (3, 4), (5, 4), (1, 6), (3, 6), (5, 6)\} \neq A \times B.$$

1.1.2 Functions

Functions (also called maps) will be used a lot in the rest of the book. Here we provide the formal definition.

Definition 1.1.1 A *function/map* $f : S \rightarrow T$ is a rule that assigns each element $s \in S$ a unique element $t \in T$.

- S is called the *domain* of f.
- T is called the *codomain* of f.
- If $f(s) = t$, then t is called the *image* of s, and s is called a *preimage* of t.
- For any $A \subseteq T$,

$$f^{-1}(A) := \{s \in S \mid f(s) \in A\}$$

is called the *preimage of A under f*.

Example 1.1.4 Define

$$f : \mathbb{R} \rightarrow \mathbb{R}$$
$$x \mapsto x^2,$$

where $\mathbb{R}$ is the set of real numbers. Then f has domain $\mathbb{R}$ and codomain $\mathbb{R}$.

Let $A = \{1\} \subseteq \mathbb{R}$, the preimage of A under f is given by

$$f^{-1}(A) = \{-1, 1\}.$$

1 is the image of -1 and -1 is a preimage of 1. 1 is another preimage of 1.

Let $B = \{-1\} \subseteq \mathbb{R}$, then $f^{-1}(B) = \emptyset$.

We note that the image of an element $s \in S$ is unique, and preimages of $t \in T$ may not exist. Even if a preimage of $t \in T$ exists, it may not be unique. In case every $t \in T$ has a preimage, we say that f is *surjective*. In case such a preimage is also unique, we say that f is *bijective*.

Definition 1.1.2

- A function $f : S \to T$ is called *onto* or *surjective*, if given any $t \in T$, there exists $s \in S$, such that $t = f(s)$.
- A function $f : S \to T$ is said to be *one-to-one* (written 1-1) or *injective* if for any $s_1, s_2 \in S$ such that $s_1 \neq s_2$, we have $f(s_1) \neq f(s_2)$.
- f is called *1-1 correspondence* or *bijective* if f is 1-1 and onto.

Example 1.1.5

- Define f

$$f : \mathbb{R} \to \mathbb{R}_{\geq 0}$$
$$x \mapsto x^2,$$

  then f is surjective as for any $y \in \mathbb{R}_{\geq 0}$, we can find a preimage x of y by calculating $x = \sqrt{y}$. But f is not injective, since $f(-1) = f(1) = 1$.
- Define g

$$g : \mathbb{R} \to \mathbb{R}$$
$$x \mapsto x.$$

It can be easily seen that g is bijective.

As mentioned above, if $f : S \to T$ is not surjective, there exists $t \in T$ such that $f^{-1}(t) = \emptyset$. If f is not injective, there are at least two $s_1, s_2 \in S$ such that $s_1 \neq s_2$ and $f(s_1) = f(s_2) = t$, which means $f^{-1}(t)$ is not a unique element. However, when f is bijective, $f^{-1} : T \to S$ is a function—it assigns to each $t \in T$ a unique element $s \in S$. In such a case, f^{-1} is called the *inverse* of f.

Example 1.1.6 Define f

$$f : \mathbb{R} \to \mathbb{R}$$
$$x \mapsto x^3.$$

Then the inverse of f exists and is given by

$$f^{-1} : \mathbb{R} \to \mathbb{R}$$
$$x \mapsto \sqrt[3]{x}.$$

When the domain of one function is the codomain of another function, we can define the composition of those two functions.

Definition 1.1.3 For two functions $f : T \to U$, $g : S \to T$, the *composition* of f and g, denoted by $f \circ g$, is the function

$$f \circ g : S \to U$$
$$s \mapsto f(g(s)).$$

Example 1.1.7 Suppose we have f

$$f : \mathbb{R} \to \mathbb{R}$$
$$x \mapsto x^2$$

and g

$$g : \mathbb{R} \to \mathbb{R}$$
$$x \mapsto x^3.$$

Then the composition of f and g is given by

$$f \circ g : \mathbb{R} \to \mathbb{R}$$
$$x \mapsto (x^3)^2 = x^6.$$

For a function whose domain and codomain are the same, say $f : S \to S$, we can define $f \circ f \circ \cdots \circ f$ in a similar way. For simplicity, we write f^n for the composition of n copies of f. When $f : S \to S$ is bijective, f^{-1} is a function. And we write f^{-n} for the composition of n copies of f^{-1}.

Example 1.1.8 Define

$$f : \mathbb{R} \to \mathbb{R}$$
$$x \mapsto x^2,$$

then

$$f^n : \mathbb{R} \to \mathbb{R}$$
$$x \mapsto x^{2^n}.$$

1.1.3 Integers

We deal with integers every day. We would write one hundred and twenty-three as 123 because

$$123 = 1 \times 100 + 2 \times 10 + 3 \times 1.$$

Such a representation of an integer is called a base-10 representation. In general, for any integer $b \geq 2$, we can have a base$-b$ representation for a positive integer.

Theorem 1.1.1 *Let $b \geq 2$ be an integer. Then any $n \in \mathbb{Z}$, $n > 0$, can be expressed uniquely in the form*

$$n = \sum_{i=0}^{\ell-1} a_i b^i, \tag{1.2}$$

where $0 \leq a_i < b$ $(0 \leq i < \ell)$, $a_{\ell-1} \neq 0$, and $\ell \geq 1$. $a_{\ell-1} a_{\ell-2} \ldots a_1 a_0$ is called a base$-b$ representation *for n. ℓ is called the* length *of n in base$-b$ representation.*

The proof can be found in, e.g., [Kos02, page 81]. To emphasize the base b, we sometimes put b as a subscript for the representation. When $b = 2$, a base-2 representation is also called a *binary representation*, ℓ is also called the *bit length* of n, a_0 is said to be the *least significant bit* (LSB) of n, and $a_{\ell-1}$ is said to be the *most significant bit* of n. When $b = 16$, a base-16 representation is also called a *hexadecimal representation*.

   The correspondence between decimal numerals and hexadecimal (base $b = 16$) numerals is listed in Table 1.1.

Example 1.1.9

$$3_{10} = 11_2 = 3_{16}.$$
$$4_{10} = 100_2 = 4_{16}.$$
$$60_{10} = 111100_2 = 3C_{16}.$$

We have learned in primary school that when we divide 6 by 4 we get quotient 1 and remainder 2. Such a computation can be done thanks to the following theorem. The proof involves well-ordering principles of integers, which will not be covered in this book. Interested readers are referred to, e.g., [Her96, page 22].

Theorem 1.1.2 *If $m, n \in \mathbb{Z}$, $n > 0$, then there exist $q, r \in \mathbb{Z}$, such that $0 \leq r < n$ and $n = qm + r$.*

q is called the *quotient* and r is called the *remainder*.

Definition 1.1.4 Given $m, n \in \mathbb{Z}$, if $m \neq 0$ and $n = am$ for some integer a, we say that m *divides* n, written $m | n$. We call m a *divisor* of n and n a *multiple* of m. If m does not divide n, we write $m \nmid n$.

Table 1.1 Correspondence between decimal and hexadecimal (base $b = 16$) numerals

Base 10	0	1	2	3	4	5	6	7	8	9	10	11	12	13	14	15
Base 16	0	1	2	3	4	5	6	7	8	9	A	B	C	D	E	F

Example 1.1.10

- $3|6, -2|4, 1|8, 5|5$.
- $7 \nmid 9, 4 \nmid 6$.
- All the positive divisors of 4 are $1, 2, 4$.
- All the positive divisors of 6 are $1, 2, 3, 6$.

We can see that there are some common divisors between 4 and 6. The largest of them will be of importance to us. Formally, we can define the greatest common divisor between two integers that are not both zero.

Definition 1.1.5 Take $m, n \in \mathbb{Z}$, $m \neq 0$ or $n \neq 0$, and the *greatest common divisor* of m and n, denoted $\gcd(m, n)$, is given by $d \in \mathbb{Z}$ such that

- $d > 0$.
- $d|m, d|n$.
- If $c|m$ and $c|n$, then $c|d$.

Example 1.1.11

- Continuing Example 1.1.10, common divisors of 4 and 6 are 1 and 2. So $\gcd(4, 6) = 2$.
- All the positive divisors of 2 are 1 and 2. All the positive divisors of 3 are 1 and 3. So $\gcd(2, 3) = 1$.

It can be proven that the greatest common divisor of two integers (not both zero) always exists and it is unique. The proof of the theorem can be found in, e.g., [Her96, page 23].

Theorem 1.1.3 (Bézout's identity) *For any $m, n \in \mathbb{Z}$, such that $m \neq 0$ or $n \neq 0$, $\gcd(m, n)$ exists and is unique. Moreover, there exist $s, t \in \mathbb{Z}$ such that $\gcd(m, n) = sm + tn$.*

The equation $\gcd(m, n) = sm + tn$ is usually called *Bézout's identity*. We note that the choices of s, t are not unique. Indeed, if $\gcd(m, n) = sm + tn$, then $\gcd(m, n) = (s + n)m + (t - m)n$.

Example 1.1.12

$$\gcd(4, 6) = 2 = (-1) \times 4 + 1 \times 6.$$

$$\gcd(2, 3) = 1 = (-4) \times 2 + 3 \times 3.$$

Next, we prove some simple but useful results.

Lemma 1.1.1 *For any $m, n, a \in \mathbb{Z}$, we have*

(1) $1|n$ for all n.
(2) If $m \neq 0$, then $m|0$.
(3) If $m|n$ and $n|a$, then $m|a$.
(4) If $m|1$, then $m = \pm 1$.

(5) If $m|n$ and $n|m$, then $m = \pm n$.
(6) If $m|n$ and $m|a$, then $m|(un + va)$, $\forall u, v \in \mathbb{Z}$.[1]
(7) If $a|mn$ and $\gcd(a, m) = 1$, then $a|n$.
(8) If $m|a$, $n|a$ and $\gcd(m, n) = 1$, then $mn|a$.

Proof Proofs of (1)–(4) easily follow from the definitions.

To prove (5), as $m|n$ and $n|m$, by Definition 1.1.4, there are integers c_1, c_2 such that $n = mc_1$ and $m = c_2 n$. This gives $n = nc_1 c_2$ and we have $c_1 c_2 = 1$. Since all the divisors of 1 are ± 1, we have $c_1 = c_2 = 1$ or $c_1 = c_2 = -1$.

To prove (6), since $m|n, m|q$, there are integers c_1, c_2 such that $n = mc_1$ and $q = mc_2$. Then

$$un + vq = uc_1 m + vc_2 m = (uc_1 + vc_2)m$$

is a multiple of m.

To prove (7), we note that by Bézout's identity, there exist $s, t \in \mathbb{Z}$ such that $as + mt = 1$. Multiplying both sides by n, we get $asn + mnt = n$. Since $a|asn$ and $a|mnt$, we have $a|n$.

Finally, we prove (8). Since $m|a$, $a = mk$ for some $k \in \mathbb{Z}$. We have $n|mk$. Now because $\gcd(m, n) = 1$, by (7), $n|k$ and so $k = nk'$ for some $k' \in \mathbb{Z}$. Thus $a = mnk'$ is divisible by mn. $\square$

In general, to find $\gcd(m, n)$, it would be too time-consuming to list all the divisors of m and n. The following theorem allows us to simplify the computation.

Theorem 1.1.4 (Euclid's division) *Given $m, n \in \mathbb{Z}$, take q, r such that $n = qm + r$. Then $\gcd(m, n) = \gcd(m, r)$.*

Proof We first note that we can find q, r by Theorem 1.1.2. By Lemma 1.1.1 (6), $\gcd(m, n)|n - qm$, i.e., $\gcd(m, n)|r$. Similarly we have $\gcd(m, r)|qm + r$, i.e., $\gcd(m, r)|n$.

By Definition 1.1.5, $\gcd(m, n)|\gcd(m, r)$ and $\gcd(m, r)|\gcd(m, n)$. By Lemma 1.1.1 (5), $\gcd(m, r) = \pm\gcd(m, n)$. By Definition 1.1.5, $\gcd(m, r) > 0$ and $\gcd(m, n) > 0$. We have $\gcd(m, n) = \gcd(m, r)$. $\square$

Thus, to find $\gcd(m, n)$, we can compute Euclid's division repeatedly until we get $r = 0$.

Example 1.1.13 We can calculate $\gcd(120, 35)$ as follows:

$$
\begin{aligned}
120 &= 35 \times 3 + 15 \quad & \gcd(120, 35) &= \gcd(35, 15), \\
35 &= 15 \times 2 + 5 \quad & \gcd(35, 15) &= \gcd(15, 5), \\
15 &= 5 \times 3 \quad & \gcd(15, 5) &= 5 \implies \gcd(120, 35) = 5.
\end{aligned}
$$

---

[1] The notation $\forall$ stands for "for all."

The procedure is called the *Euclidean algorithm,* and the details are provided in Algorithm 1.1. By Theorem 1.1.4, $\gcd(m, n) = \gcd(m, r)$ after each loop from line 1. In the end, we get $\gcd(m, n)$.

Algorithm 1.1: Euclidean algorithm

Input: m, n // $m, n \in \mathbb{Z}$, $m \neq 0$
Output: $\gcd(m, n)$

```
1  while m ≠ 0 do
2  │   r = n%m // remainder of n divided by m
3  │   n = m
4  └   m = r
5  return r
```

Furthermore, with the intermediate results we have from the Euclidean algorithm, we can also find a pair of s, t such that $\gcd(m, n) = sm + tn$ (Bézout's identity).

Example 1.1.14 Continuing Example 1.1.13, we can find integers s, t such that $\gcd(120, 35) = 120s + 35t$ as follows:

$$5 = 35 - 15 \times 2, \quad 15 = 120 - 35 \times 3,$$
$$\implies 5 = 35 - (120 - 35 \times 3) \times 2 = 120 \times (-2) + 35 \times 7.$$

Such a procedure is called the *extended Euclidean algorithm.*

Example 1.1.15 We can calculate $\gcd(160, 21)$ using the Euclidean algorithm

$$
\begin{aligned}
160 &= 21 \times 7 + 13 & \gcd(160, 21) &= \gcd(21, 13), \\
21 &= 13 \times 1 + 8 & \gcd(21, 13) &= \gcd(13, 8), \\
13 &= 8 \times 1 + 5 & \gcd(13, 8) &= \gcd(8, 5), \\
8 &= 5 \times 1 + 3 & \gcd(8, 5) &= \gcd(5, 3), \\
5 &= 3 \times 1 + 2 & \gcd(5, 3) &= \gcd(3, 2), \\
3 &= 2 \times 1 + 1 & \gcd(3, 2) &= \gcd(2, 1), \\
2 &= 1 \times 2 & \gcd(2, 1) &= 1 \implies \gcd(160, 21) = 1.
\end{aligned}
$$

By the extended Euclidean algorithm, we can also find integers s, t such that $\gcd(160, 21) = s160 + t35$

$$
\begin{aligned}
1 &= 3 - 2, & 2 &= 5 - 3, \\
3 &= 8 - 5, & 5 &= 13 - 8, \\
8 &= 21 - 13, & 13 &= 160 - 21 \times 7.
\end{aligned}
$$

We have

$$1 = 3 - (5 - 3) = 3 \times 2 - 5 = 8 \times 2 - 5 \times 3 = 8 \times 2 - (13 - 8) \times 3$$
$$= 8 \times 5 - 13 \times 3 = 21 \times 5 - 13 \times 8 = 21 \times 5 - (160 - 21 \times 7) \times 8$$
$$= (-8) \times 160 + 61 \times 21.$$

An algorithmic description of the extended Euclidean algorithm is shown in Algorithm 1.2. By Definition 1.1.5, $m \neq 0$ or $n \neq 0$. If $m = 0$, $\gcd(m, n) = n$. If $n = 0$, $\gcd(m, n) = m$. Both cases are trivial; hence in the algorithm, we assume $n \neq 0$ and $m \neq 0$. We also note that we can just compute the coefficient s and then compute t using s.

Algorithm 1.2: Extended Euclidean algorithm

Input: m, n // $m, n \in \mathbb{Z}$, $n \neq 0$, $m \neq 0$
Output: s, t such that $\gcd(m, n) = sm + tn$
1   $s = 0, ss = 1, r = m, rr = n$
2   **while** $r \neq 0$ **do**
     // quotient of rr divided by r
3      $q = rr/r$
4      $tmp = r$
     // remainder of rr divided by r
5      $r = rr\%r$
6      $rr = tmp$
7      $tmp = s$
8      $s = ss - q * s$
9      $ss = tmp$
     // $rr = \gcd(m, n)$
10   $t = (rr - ss * n)/m$
11   **return** ss, t

Definition 1.1.6

- For $m, n \in \mathbb{Z}$ such that $m \neq 0$ or $n \neq 0$, m and n are said to be *relatively prime/coprime* if $\gcd(m, n) = 1$.
- Given $p \in \mathbb{Z}$, $p > 1$. p is said to be *prime* (or a *prime number*) if for any $m \in \mathbb{Z}$, either m is a multiple of p (i.e., $p|m$) or m and p are coprime (i.e., $\gcd(p, m) = 1$).
- Given $n \in \mathbb{Z}$, $n > 1$. If n is not prime, it is said to be *composite* (or a *composite number*).

Example 1.1.16

- 4 and 9 are relatively prime.
- 8 and 6 are not coprime.
- $2, 3, 5, 7$ are prime numbers.

- 6, 9, 21 are not prime numbers.

We have the following lemma concerning prime numbers.

Lemma 1.1.2 *For $p \in \mathbb{Z}$ a prime number, if $p | \prod_{i=1}^{n} a_i$, where $a_i \in \mathbb{Z}$, then $p | a_i$ for some i $(1 \leq i \leq n)$.*

Proof If $p | a_1$, then we are done. Otherwise, $\gcd(p, a_1) = 1$, and by Lemma 1.1.1 (7), we have $p | \prod_{i=2}^{n} a_i$. We can repeat the argument and conclude that $p | a_i$ for some i. $\qquad\qquad\square$

It can be proven that an integer $n > 1$ is either a prime number or a product of prime numbers (see, e.g., [Her96, page 26]). Then, we have the Fundamental Theorem of Arithmetic which says that this product is unique up to permutation.

Theorem 1.1.5 (The Fundamental Theorem of Arithmetic) *For any $n \in \mathbb{Z}$, $n > 1$, n can be written in the form*

$$n = \prod_{i=1}^{k} p_i^{e_i},$$

where the exponents e_i are positive integers, and $p_1, p_2, \ldots, p_k$ are prime numbers that are pairwise distinct and unique up to permutation.

Proof We prove by contradiction. Assume the theorem is false. Let $n \in \mathbb{Z}$ $(n > 1)$ be the smallest integer with two distinct factorizations. We can write

$$n = \prod_{i=1}^{k} p_i^{e_i} = \prod_{j=1}^{\ell} q_j^{d_j}.$$

Since $p_1 | \prod_{j=1}^{\ell} q_j^{d_j}$, by Lemma 1.1.2, $p_1 | q_j$ for some j. Without loss of generality, we assume $p_1 | q_1$. Since p_1 and q_1 are prime numbers, we have $p_1 = q_1$. Then the integer $n' = \prod_{i=2}^{k} p_i^{e_i} = \prod_{j=2}^{\ell} q_j^{d_j}$ has two distinct factorizations and $n' < n$, which contradicts the minimality of n. $\qquad\qquad\square$

Example 1.1.17 $20 = 2^2 \times 5$, $135 = 3^3 \times 5$.

1.2 Abstract Algebra

In this section, we discuss the basics of abstract algebra and get to know a few abstract structures. Most of us are already familiar with examples of such structures, probably just not by the name. Those structures will become useful when we discuss modern cryptographic algorithms.

1.2.1 Groups

First, we define a group.

Definition 1.2.1 A *group* $(G, \cdot)$ is a nonempty set G with a binary operation $\cdot$ satisfying the following conditions:

- G is closed under $\cdot$ (closure property), $\forall g_1, g_2 \in G$, $g_1 \cdot g_2 \in G$.
- $\cdot$ is associative, $\forall g_1, g_2, g_3 \in G$, $g_1 \cdot (g_2 \cdot g_3) = (g_1 \cdot g_2) \cdot g_3$.
- $\exists e \in G$, an identity element, such that $\forall g \in G$, $e \cdot g = g \cdot e = g$.[2]
- Every $g \in G$ has an inverse $g^{-1} \in G$ such that $g \cdot g^{-1} = g^{-1} \cdot g = e$.

When it is clear from the context, we omit $\cdot$ and say that G is a group.

Example 1.2.1 There are many examples of groups that we are familiar with.

- $(\mathbb{Z}, +)$, the set of integers with addition, is a group. The identity element is 0.
- Similarly, $(\mathbb{Q}, +)$ and $(\mathbb{C}, +)$ are groups.
- $(\mathbb{Q}, \times)$ is not a group. Because $0 \in \mathbb{Q}$ does not have an inverse with respect to multiplication.
- But $(\mathbb{Q} \setminus \{0\}, \times)$ is a group. The identity element is 1.

Next, we give an example of formally proving that a set with a binary operation is a group. Let $G = \mathbb{R}^+$ be the set of positive real numbers, and let $\cdot$ be the multiplication of real numbers, denoted $\times$. We will show that $(\mathbb{R}^+, \times)$ is a group.

1. $\mathbb{R}^+$ is closed under $\times$: for any $a_1, a_2 \in \mathbb{R}^+$, $a_1 \times a_2 \in \mathbb{R}$ and $a_1 \times a_2 > 0$, hence $a_1 \times a_2 \in \mathbb{R}^+$.
2. $\times$ is associative: $\forall a_1, a_2, a_3 \in \mathbb{R}^+$, $a_1 \times (a_2 \times a_3) = (a_1 \times a_2) \times a_3$.
3. 1 is the identity element in $\mathbb{R}^+$: $\forall a \in \mathbb{R}^+$, $1 \times a = a \times 1 = a$.
4. Take any $a \in \mathbb{R}^+$, $\frac{1}{a} \in \mathbb{R}$, and $\frac{1}{a} > 0$, so $\frac{1}{a} \in \mathbb{R}^+$. Moreover,

$$a \times \frac{1}{a} = \frac{1}{a} \times a = 1,$$

hence $a^{-1} = \frac{1}{a} \in \mathbb{R}^+$.

By definition, we have proved that $(\mathbb{R}^+, \times)$ is a group.

Definition 1.2.2 Let $(G, \cdot)$ be a group. If $\cdot$ is commutative, i.e.,

$$\forall g_1, g_2 \in G, \; g_1 \cdot g_2 = g_2 \cdot g_1,$$

then the group is called *abelian*.

---

[2] The notation $\exists$ stands for "there exist."

The name abelian is in honor of the great mathematician Niels Henrik Abel (1802–1829).

Example 1.2.2 The groups we have seen so far $(\mathbb{Z}, +)$, $(\mathbb{R}^+, \times)$, $(\mathbb{Q} \backslash \{0\}, \times)$, $(\mathbb{Q}, +)$, and $(\mathbb{C}, +)$ are all abelian groups.

Example 1.2.3 Let us consider the set of 2×2 matrices with coefficients in $\mathbb{R}$. We denote this set by $\mathcal{M}_{2 \times 2}(\mathbb{R})$. Recall that matrix addition, denoted by $+$, is defined componentwise. For any $\begin{pmatrix} a_{00} & a_{10} \\ a_{01} & a_{11} \end{pmatrix}$, $\begin{pmatrix} b_{00} & b_{10} \\ b_{01} & b_{11} \end{pmatrix}$ in $\mathcal{M}_{2 \times 2}(\mathbb{R})$,

$$\begin{pmatrix} a_{00} & a_{10} \\ a_{01} & a_{11} \end{pmatrix} + \begin{pmatrix} b_{00} & b_{10} \\ b_{01} & b_{11} \end{pmatrix} = \begin{pmatrix} a_{00} + b_{00} & a_{10} + b_{10} \\ a_{01} + b_{01} & a_{11} + b_{11} \end{pmatrix}.$$

$(\mathcal{M}_{2 \times 2}(\mathbb{R}), +)$ is an abelian group: Closure, associativity, and commutativity of $+$ are easy to show. The identity element is the zero matrix $\begin{pmatrix} 0 & 0 \\ 0 & 0 \end{pmatrix}$. The inverse of any matrix $\begin{pmatrix} a_{00} & a_{10} \\ a_{01} & a_{11} \end{pmatrix}$ is $\begin{pmatrix} -a_{00} & -a_{10} \\ -a_{01} & -a_{11} \end{pmatrix}$, which is also in $\mathcal{M}_{2 \times 2}(\mathbb{R})$. Section 1.3.1 presents a more general discussion on matrices.

Example 1.2.4 Let $\mathbb{F}_2 := \{0, 1\}$. We define *logical XOR*, denoted $\oplus$, in $\mathbb{F}_2$ as follows:

$$0 \oplus 0 = 0, \quad 0 \oplus 1 = 1 \oplus 0 = 1, \quad 1 \oplus 1 = 0.$$

Closure, associativity, and commutativity can be directly seen from the definition. The identity element is 0, and the inverse of 1 is 1. Hence $(\mathbb{F}_2, \oplus)$ is an abelian group.

Example 1.2.5 Let $E = \{a, b\}$, $a \neq b$. Define addition in E as follows:

$$a + a = a, \quad a + b = b + a = b, \quad b + b = a.$$

Closure, associativity, and commutativity can be directly seen from the definition. The identity element is a, and the inverse of b is b. Hence $(E, +)$ is an abelian group.

Next, we will see a group that is not abelian. To introduce this group, we start by defining permutations.

Definition 1.2.3 A *permutation* of a set S is a bijective function $\sigma : S \to S$.

Example 1.2.6

- Let $S = \{0, 1, 2\}$. Define $\sigma : S \to S$ as follows:

$$0 \mapsto 1, \quad 1 \mapsto 2, \quad 2 \mapsto 0.$$

Then σ is a permutation of S.

- Let $S = \{\circ, \triangle, \square\}$. Define $\tau : S \to S$ as follows:

$$\circ \mapsto \triangle, \quad \triangle \mapsto \square, \quad \square \mapsto \circ.$$

Then τ is a permutation of S.

We note that what matters for a permutation is how many objects we have, not the objects' nature. We can label a set of n objects with $1, 2, \ldots, n$. In Example 1.2.6, we can label $\circ$ as 0, $\triangle$ as 1, and $\square$ as 2. Then σ and τ are the same permutation.

Now, we take a set S of n elements. Labeling the elements allows us to consider $S = \{1, \ldots, n\}$. Let S_n denote the set of all permutations of S. And let $\circ$ denote the composition of functions (see Definition 1.1.3). Then we have the following lemma:

Lemma 1.2.1 $(S_n, \circ)$ *is a group.*

The proof is easy. We leave it as an exercise for the readers.

We note that the identity element in the group is the identity function $\sigma : S \to S$, $\sigma(s) = s \ \forall s \in S$. Any $\sigma \in S_n$ is bijective (see Definition 1.1.2), and the inverse of σ in S_n is then given by σ^{-1}.

Definition 1.2.4 $(S_n, \circ)$ is called the *symmetric group of degree n*.

Example 1.2.7 Let $n = 2$ and $S = \{1, 2\}$. There are only two ways to permute two elements. So $S_2 = \{\sigma_1, \sigma_2\}$, where $\sigma_1 : S \to S, 1 \mapsto 1, 2 \mapsto 2$ is the identity, and $\sigma_2 : S \to S, 1 \mapsto 2, 2 \mapsto 1$.

Example 1.2.8 (A group that is not abelian) Let $n = 3$ and $S = \{1, 2, 3\}$. There are $3! = 6$ ways of permuting three elements. In particular, we have the following two permutations:

$$\sigma_1 : S \to S, 1 \mapsto 2, 2 \mapsto 3, 3 \mapsto 1; \quad \sigma_2 : S \to S, 1 \mapsto 3, 2 \mapsto 2, 3 \mapsto 1.$$

We note that $\sigma_1 \circ \sigma_2 \neq \sigma_2 \circ \sigma_1$ since

$$\sigma_1 \circ \sigma_2(1) = 1, \text{ but } \sigma_2 \circ \sigma_1(1) = 2.$$

Hence, S_3 is not abelian.

We can extend σ_1 and σ_2 in Example 1.2.8 to permuting n elements by keeping the other $n - 3$ elements unchanged. Thus S_n is not abelian for any $n \geq 3$.

Definition 1.2.5 The *order* of a group $(G, \cdot)$ is the number of elements in G or the cardinality of the set G, $|G|$. A group G is said to be *finite* if $|G| < \infty$ and *infinite* if $|G| = \infty$.

Example 1.2.9

- We have seen a few infinite groups, for example, $(\mathbb{Z}, +)$ and $(\mathbb{R}^+, \times)$.
- We have also seen two finite groups, $|S_2| = 2$ and $|S_3| = 6$.

- Let $S = \{1, 2, \ldots, n\}$. To permute the elements in S, there are n choices for the image of 1, $n - 1$ choices for the image of 2, etc. Thus $|S_n| = n!$, and S_n is a finite group.

Definition 1.2.6 Let $(G, \cdot)$ be a group with identity element e. The *order* of an element $g \in G$, denoted $\mathrm{ord}\,(g)$, is the smallest positive integer k such that

$$\underbrace{g \cdot g \cdots g}_{k \text{ times}} = g^k = e.$$

When such a k does not exist, we define $\mathrm{ord}\,(g) = \infty$.

Example 1.2.10

- In $(\mathbb{Z}, +)$, the identity element is 0, $\mathrm{ord}\,(1) = \infty$.
- Continuing Example 1.2.7, σ_1 is the identity. And $\sigma_2^2 : S \to S, 1 \mapsto 1, 2 \mapsto 2$. Hence $\mathrm{ord}\,(\sigma_2) = 2$.

Definition 1.2.7 A group G is called *cyclic* if it is generated by one element, i.e., if there exists an element $g \in G$ such that

$$G = \left\{ g^k \mid k \in \mathbb{Z} \right\}.$$

Example 1.2.11 We have seen in Example 1.2.7, $S_2 = \{\sigma_1, \sigma_2\}$, where σ_1 is the identity element. In Example 1.2.10, we discussed that $\sigma_2^2 = \sigma_1$. Hence $S_2 = \left\{\sigma_2, \sigma_2^2\right\}$ is a cyclic group.

We now state a very useful theorem about the order of a group and the order of an element in the group. The proof follows from a famous theorem (Lagrange Theorem) named after Joseph-Louis Lagrange (1736–1813). Details can be found in, e.g., [Her96, page 59].

Theorem 1.2.1 *Let $(G, \cdot)$ be a finite group with identity element e. For any $g \in G$, $\mathrm{ord}\,(g)$ divides $|G|$, in particular, $g^{|G|} = e$.*

A direct corollary is as follows.

Corollary 1.2.1 *Let G be a group. If $|G|$ is a prime number, then G is cyclic.*

Proof Let e denote the identity element in G. Take any element $g \in G$ such that $g \neq e$. By Theorem 1.2.1, $\mathrm{ord}\,(g)$ divides $|G|$. Since $|G|$ is prime and g is not the identity element, $\mathrm{ord}\,(g) = |G|$.

We claim that

$$G = \left\{ g, g^2, g^3, \ldots, g^{|G|} \right\}.$$

Otherwise, we would have $g^i = g^j$ for some $1 \leq i, j \leq |G|$, where $i \neq j$.

Without loss of generality, we assume $i \geq j$. Multiplying both sides of $g^i = g^j$ by g^{-j}, we get $g^{i-j} = e$.

By Definition 1.2.6, since $0 \leq i - j < \mathrm{ord}\,(g)$, we must have $i = j$, a contradiction. Hence $G = \left\{ g, g^2, g^3, \ldots, g^{|G|} \right\}$. $\square$

1.2.2 Rings

Next, we move to another abstract structure, rings.

Definition 1.2.8 A set R together with two binary operations $+$ and $\cdot$, $(R, +, \cdot)$, is a *ring* if $(R, +)$ is an abelian group, and for any $a, b, c \in R$, the following conditions are satisfied:

- R is closed under $\cdot$ (closure), $a \cdot b \in R$.
- $\cdot$ is associative, $(a \cdot b) \cdot c = a \cdot (b \cdot c)$.
- The distributive laws holds: $a \cdot (b+c) = a \cdot b + a \cdot c$ and $(b+c) \cdot a = b \cdot a + c \cdot a$.
- The identity element for $\cdot$ exists, which is different from the identity element for $+$.

Definition 1.2.9 If $a \cdot b = b \cdot a$ for all $a, b \in R$, R is a *commutative ring*.

Remark 1.2.1

- For most cases, we will denote the identity element for $+$ as 0 and the identity element for $\cdot$ as 1.
- We normally refer to the operation $+$ as addition and 0 as the *additive identity*. Similarly, we refer to the operation $\cdot$ as multiplication and 1 as the *multiplicative identity*.
- The inverse of an element $a \in R$ with respect to $+$ is called the *additive inverse* of a, usually denoted by $-a$.
- The last condition in Definition 1.2.8 implies that a set consisting of only 0 is not a ring.
- For simplicity, we sometimes write ab instead of $a \cdot b$.
- When the operations in $(R, +, \cdot)$ are clear from the context, we omit them and write R.

Example 1.2.12 We have seen that $(\mathbb{Z}, +)$ is an abelian group, and the identity element is 0. It can be easily shown that $(\mathbb{Z}, +, \times)$ is a commutative ring. The identity element for $\times$ is 1.

Similarly $(\mathbb{Q}, +, \times)$, $(\mathbb{R}, +, \times)$, and $(\mathbb{C}, +, \times)$ are all commutative rings with 0 as the additive identity and 1 as the multiplicative identity.

Example 1.2.13 In Example 1.2.3, we have shown that $(\mathcal{M}_{2\times 2}(\mathbb{R}), +)$ is an abelian group. We recall matrix multiplication, denoted by $\times$, for 2×2 matrices: For any $\begin{pmatrix} a_{00} & a_{10} \\ a_{01} & a_{11} \end{pmatrix}, \begin{pmatrix} b_{00} & b_{10} \\ b_{01} & b_{11} \end{pmatrix}$ in $\mathcal{M}_{2\times 2}(\mathbb{R})$,

$$\begin{pmatrix} a_{00} & a_{10} \\ a_{01} & a_{11} \end{pmatrix} \times \begin{pmatrix} b_{00} & b_{10} \\ b_{01} & b_{11} \end{pmatrix} = \begin{pmatrix} a_{00}b_{00} + a_{10}b_{01} & a_{00}b_{10} + a_{10}b_{11} \\ a_{01}b_{00} + a_{11}b_{01} & a_{01}b_{10} + a_{11}b_{11} \end{pmatrix}.$$

$(\mathcal{M}_{2\times 2}(\mathbb{R}), +, \times)$ is a ring: Associativity and distributive laws are easy to show. The identity element for $\times$ is the 2×2 identity matrix $\begin{pmatrix} 1 & 0 \\ 0 & 1 \end{pmatrix}$. We note that $(\mathcal{M}_{2\times 2}(\mathbb{R}), +, \times)$ is not a commutative ring. For example,

$$\begin{pmatrix} 1 & 0 \\ 0 & 0 \end{pmatrix}\begin{pmatrix} 0 & 0 \\ 1 & 0 \end{pmatrix} = \begin{pmatrix} 0 & 0 \\ 0 & 0 \end{pmatrix}, \text{ but } \begin{pmatrix} 0 & 0 \\ 1 & 0 \end{pmatrix}\begin{pmatrix} 1 & 0 \\ 0 & 0 \end{pmatrix} = \begin{pmatrix} 0 & 0 \\ 1 & 0 \end{pmatrix}.$$

Example 1.2.14 In Example 1.2.4 we have shown that $(\mathbb{F}_2, \oplus)$ is an abelian group. Let us define *logical AND*, denoted $\&$, in $\mathbb{F}_2$ as follows:

$$0 \,\&\, 0 = 0, \quad 1 \,\&\, 0 = 0 \,\&\, 1 = 0, \quad 1 \,\&\, 1 = 1.$$

Closure of $\mathbb{F}_2$ with respect to $\&$, associativity and commutativity of $\&$, and the distributive laws are easy to see from the definitions. The identity element for $\&$ is 1. $(\mathbb{F}_2, \oplus, \&)$ is a commutative ring.

Example 1.2.15 In Example 1.2.5 we showed that $(E, +)$ is an abelian group. Define multiplication in E as follows:

$$a \cdot a = a, \quad a \cdot b = b \cdot a = a, \quad b \cdot b = b.$$

Closure of E with respect to $\cdot$, associativity of $\cdot$, commutativity of $\cdot$, and the distributive laws are easy to see from the definitions. The identity element for $\cdot$ is b. Thus $(E, +, \cdot)$ is a commutative ring.

Definition 1.2.10 Let $(R, +, \cdot)$ be a ring with additive identity 0 and multiplicative identity 1. Let $a, b \in R$. If $a \neq 0$ and $b \neq 0$ but $a \cdot b = 0$, then a and b are called *zero divisors*. If $a \cdot b = b \cdot a = 1$, a (also b) is said to be *invertible*, and it is called a *unit*.

Example 1.2.16

- There are no zero divisors in $(\mathbb{Z}, +, \times)$, $(\mathbb{Q}, +, \times)$, $(\mathbb{R}, +, \times)$, or $(\mathbb{C}, +, \times)$.
- Any nonzero element in $(\mathbb{Z}, +, \times)$, $(\mathbb{Q}, +, \times)$, $(\mathbb{R}, +, \times)$, or $(\mathbb{C}, +, \times)$ is a unit.

Example 1.2.17 As shown in Examples 1.2.3 and 1.2.13, $(\mathcal{M}_{2\times 2}(\mathbb{R}), +, \times)$ is a ring. The additive identity is $\begin{pmatrix} 0 & 0 \\ 0 & 0 \end{pmatrix}$. Since

$$\begin{pmatrix} 1 & 0 \\ 0 & 0 \end{pmatrix}\begin{pmatrix} 0 & 0 \\ 1 & 0 \end{pmatrix} = \begin{pmatrix} 0 & 0 \\ 0 & 0 \end{pmatrix},$$

by Definition 1.2.10, $\begin{pmatrix} 1 & 0 \\ 0 & 0 \end{pmatrix}$ and $\begin{pmatrix} 0 & 0 \\ 1 & 0 \end{pmatrix}$ are zero divisors.

Definition 1.2.11 An *integral domain* is a commutative ring with no zero divisors.

Example 1.2.18 $(\mathbb{Z}, +, \times)$, $(\mathbb{Q}, +, \times)$, $(\mathbb{R}, +, \times)$, and $(\mathbb{C}, +, \times)$ are all integral domains.

1.2.3 Fields

Definition 1.2.12 A *field* is a commutative ring in which every nonzero element is invertible.

By definition, for any $a \in F$, there exists $b \in F$ such that $a \cdot b = b \cdot a = 1$. Then b is called the *multiplicative inverse* of a. It is easy to show that the multiplicative inverse of an element a is unique: Let $b, c \in F$ be such that

$$ab = ac = 1.$$

Multiplying by b on the left, we get

$$bab = bac = b \implies b = c = b.$$

We will denote the multiplicative inverse of a nonzero element $a \in F$ by a^{-1}.

Lemma 1.2.2 *A field is an integral domain.*

Proof Let F be a field. Suppose there are zero divisors in F. By Definition 1.2.10, there exist $a, b \in F$ such that $a \neq 0$, $b \neq 0$, and $a \cdot b = 0$. Since F is a field, by the above discussion, $a^{-1} \in F$. Multiplying both sides of $a \cdot b = 0$ by a^{-1}, we get

$$a^{-1} \cdot a \cdot b = 1 \cdot b = 0 \implies b = 0,$$

a contradiction. $\square$

Example 1.2.19

- $(\mathbb{Q}, +, \times)$, $(\mathbb{R}, +, \times)$, and $(\mathbb{C}, +, \times)$ are all fields.
- $(\mathbb{Z}, +, \times)$ is not a field. For example, $2 \in \mathbb{Z}$ is not invertible and $2 \neq 0$.

  For the rest of this subsection, let F be a field with addition $+$ and multiplication $\cdot$.

Definition 1.2.13 A field with finitely many elements is called a *finite field*.

Example 1.2.20 In Example 1.2.14 we have shown that $(\mathbb{F}_2, \oplus, \&)$ is a commutative ring. The only nonzero element is 1, which has inverse 1 with respect to $\&$. Thus $(\mathbb{F}_2, \oplus, \&)$ is a finite field.

Example 1.2.21 In Example 1.2.15 we have shown that $(E, +, \cdot)$ is a commutative ring with additive identity a and multiplicative identity b. The only nonzero element, i.e., the element not equal to the additive identity, is b. b has multiplicative inverse b since $b \cdot b = b$. Hence $(E, +, \cdot)$ is a finite field.

For an element $a \in F$ and an integer p, we define

$$p \odot a = \sum_{i=1}^{p} a.$$

Definition 1.2.14 The *characteristic* of a field F is the smallest positive integer p such that $p \odot 1 = 0$, where 1 is the multiplicative identity of F. If no such p exists, we define the characteristic of the field to be 0.

Example 1.2.22

- The characteristics of $\mathbb{R}$, $\mathbb{Q}$, and $\mathbb{C}$ are 0.
- The characteristic of the field $\mathbb{F}_2$ in Example 1.2.20 is 2 since

$$2 \odot 1 = 1 \oplus 1 = 0.$$

- The characteristic of the field E in Example 1.2.21 is 2 since

$$2 \odot b = b + b = a.$$

Theorem 1.2.2 *The characteristic of a field is either* 0 *or a prime number.*

Proof First, we note that the characteristic of a field is not equal to 1 since $1 \odot 1 = 1 \neq 0$.
Suppose the characteristic $p = mn$ is not a prime, where $m, n \in \mathbb{Z}$ and $1 < m, n < p$. Let $a = n \odot 1, b = m \odot 1$. Then

$$a \cdot b = (n \odot 1) \cdot (m \odot 1) = \left(\sum_{i=1}^{n} 1 \right) \cdot \left(\sum_{j=1}^{m} 1 \right) = (mn) \odot 1 = 0 \implies n \odot 1 = 0 \text{ or } m \odot 1 = 0,$$

where the last part follows from Lemma 1.2.2. As n, m are both strictly smaller than p, we have a contradiction. $\qquad \square$

Definition 1.2.15 Let E, F be two fields with $F \subset E$. F is called a *subfield* of E if the addition and multiplication of E, when restricted to F, are the same as those in F.

Example 1.2.23 $\mathbb{Q}$ is a subfield of $\mathbb{R}$, and $\mathbb{R}$ is a subfield of $\mathbb{C}$.

Definition 1.2.16 Let $(F, +_F, \cdot_F), (E, +_E, \cdot_E)$ be two fields. F is said to be *isomorphic* to E, written $F \cong E$, if there is a bijective function $f : F \to E$ such that for any $a, b \in F$,

(1) $f(a +_F b) = f(a) +_E f(b)$.
(2) $f(a \cdot_F b) = f(a) \cdot_E f(b)$.

The function f is called a *field isomorphism*.

A function $f : F \to E$ that satisfies condition (1) in Definition 1.2.16 is said to *preserve the addition*. Similarly, a function $g : F \to E$ that satisfies condition (2) in Definition 1.2.16 is said to *preserve the multiplication*.

Example 1.2.24 Let us consider the fields $(\mathbb{F}_2, \oplus, \&)$ from Example 1.2.20 and $(E, +, \cdot)$ from Example 1.2.21. Define $f : F \to E$, such that

$$f(0) = a, \quad f(1) = b.$$

It is easy to see that f is bijective. Also, it can be shown that f preserves both addition and multiplication. For example,

$$f(1 \oplus 0) = f(1) = a, \ f(1) + f(0) = a + b = a \implies f(1 \oplus 0) = f(1) + f(0).$$

Thus f is a field isomorphism and $\mathbb{F}_2 \cong E$.

In fact, it can be shown that any finite field with two elements is always isomorphic to $\mathbb{F}_2$. The next theorem says that, in general, there is only one finite field up to isomorphism. The proof can be found in, e.g., [Her96, page 224].

Theorem 1.2.3

- *Let K be a finite field of characteristic p. Then K contains p^n elements.*
- *For any prime p and any positive integer n, there exists, up to isomorphism, a unique field with p^n elements.*

Remark 1.2.2

- We will use $\mathbb{F}_{p^n}$ to denote the unique finite field with p^n elements.
- Let K be a finite field with characteristic p and multiplicative identity 1. Then K contains $1, 2, \ldots, p - 1, 0$, the p multiples of 1. Thus, K contains a subfield isomorphic to $\mathbb{F}_p$.

Furthermore, we define the notion of bit formally.

Definition 1.2.17

- Variables that range over $\mathbb{F}_2$ are called *Boolean variables* or *bits*.
- Addition of two bits is defined to be logical **XOR** , also called *exclusive OR*.
- Multiplication of two bits is defined to be logical **AND**.
- When the value of a bit is changed, we say the bit is *flipped*.

1.3 Linear Algebra

The most readers are probably very familiar with linear algebra. However, when we learned about matrices in high school, we focused on the case when the underlying abstract structure is a field. In Sect. 1.3.1 we will see the general case when the underlying abstract structure is a commutative ring. Then in Sect. 1.3.2 we recap concepts for vector spaces.

1.3.1 Matrices

Let R be a commutative ring with additive identity 0 and multiplicative identity 1 throughout this subsection.

Definition 1.3.1 A *matrix with coefficients in R* is a rectangular array where each entry is an element of R.

Matrix A as shown in Eq. 1.3 is said to have m *rows* and n *columns* and is of size $m \times n$. The *transpose* of A, denoted $A^\top$, is the $n \times m$ matrix obtained by interchanging the rows and columns of A.

$$
A = \begin{pmatrix} a_{00} & \cdots & a_{0(n-1)} \\ a_{10} & \cdots & a_{1(n-1)} \\ & \vdots & \\ a_{(m-1)0} & \cdots & a_{(m-1)(n-1)} \end{pmatrix}, \quad A^\top = \begin{pmatrix} a_{00} & \cdots & a_{(m-1)0} \\ a_{01} & \cdots & a_{(m-1)1} \\ & \vdots & \\ a_{0(n-1)} & \cdots & a_{(m-1)(n-1)} \end{pmatrix}. \tag{1.3}
$$

The ith row of A is

$$
\begin{pmatrix} a_{i0} & a_{i1} & \ldots & a_{i(n-1)} \end{pmatrix},
$$

and the jth column of A is

$$
\begin{pmatrix} a_{0j} \\ a_{1j} \\ \vdots \\ a_{(m-1)j} \end{pmatrix},
$$

where a_{ij} denotes the entry in the ith row and jth column. If $a_{ij} = 0$ for $i \neq j$, A is said to be a *diagonal matrix*. An *n-dimensional identity matrix*, denoted I_n, is a diagonal matrix whose diagonal entries are 1, i.e., $a_{ii} = 1$ for $i = 0, 1, \ldots, n - 1$. A $1 \times n$ matrix is called a *row vector*. An $n \times 1$ matrix is called a *column vector*. An $n \times n$ matrix is called a *square matrix* (i.e., a matrix with the same number of rows and columns).

Example 1.3.1 Let $R = \mathbb{Z}$.

- $A = \begin{pmatrix} 9 & 1 \\ 0 & -2 \end{pmatrix}$ is a 2×2 matrix with coefficients in $\mathbb{Z}$. $a_{00} = 9$ and $a_{01} = 1$.

- $I_2 = \begin{pmatrix} 1 & 0 \\ 0 & 1 \end{pmatrix}$ and $I_3 = \begin{pmatrix} 1 & 0 & 0 \\ 0 & 1 & 0 \\ 0 & 0 & 1 \end{pmatrix}$.

- $\begin{pmatrix} 5 & 0 \\ 0 & -1 \end{pmatrix}$ is a diagonal matrix.

We define the *addition* of two $m \times n$ matrices componentwise:

$$
\begin{pmatrix}
a_{00} & \cdots & a_{0(n-1)} \\
a_{10} & \cdots & a_{1(n-1)} \\
& \vdots & \\
a_{(m-1)0} & \cdots & a_{(m-1)(n-1)}
\end{pmatrix}
+
\begin{pmatrix}
b_{00} & \cdots & b_{0(n-1)} \\
b_{10} & \cdots & b_{1(n-1)} \\
& \vdots & \\
b_{(m-1)0} & \cdots & b_{(m-1)(n-1)}
\end{pmatrix}
$$

$$
=
\begin{pmatrix}
a_{00} + b_{00} & \cdots & a_{0(n-1)} + b_{0(n-1)} \\
a_{10} + b_{10} & \cdots & a_{1(n-1)} + b_{1(n-1)} \\
& \vdots & \\
a_{(m-1)0} + b_{(m-1)0} & \cdots & a_{(m-1)(n-1)} + b_{(m-1)(n-1)}
\end{pmatrix}.
$$

$$(1.4)$$

Example 1.3.2 Let $R = \mathbb{Z}$. Below is an example of addition between two 2×2 matrices with coefficients in $\mathbb{Z}$:

$$
\begin{pmatrix} 2 & 3 \\ 1 & -1 \end{pmatrix} + \begin{pmatrix} 4 & -2 \\ 0 & -5 \end{pmatrix} = \begin{pmatrix} 6 & 1 \\ 1 & -6 \end{pmatrix}.
$$

Definition 1.3.2 The *scalar product* of a $1 \times n$ row vector $\mathbf{v} = (v_0, v_1, \ldots, v_{n-1})$ with an $n \times 1$ column vector $\mathbf{w} = (w_0, w_1, \ldots, w_{n-1})^\top$ is given by

$$
\mathbf{v} \cdot \mathbf{w} = \begin{pmatrix} v_0 & v_1 & \ldots & v_{n-1} \end{pmatrix} \begin{pmatrix} w_0 \\ w_1 \\ \vdots \\ w_{n-1} \end{pmatrix} = \sum_{i=0}^{n-1} v_i w_i.
$$

Example 1.3.3 Let $R = \mathbb{Z}$. The scalar product of $\begin{pmatrix} 2 & 3 \end{pmatrix}$ and $\begin{pmatrix} 4 & 0 \end{pmatrix}^\top$ is

$$
\begin{pmatrix} 2 & 3 \end{pmatrix} \begin{pmatrix} 4 \\ 0 \end{pmatrix} = 2 \times 4 + 3 \times 0 = 8 + 0 = 8.
$$

We define the *multiplication* of an $m \times n$ matrix A with an $n \times r$ matrix B as follows:

$$
AB = \begin{pmatrix} a_{00} & \cdots & a_{0(n-1)} \\ a_{10} & \cdots & a_{1(n-1)} \\ & \vdots & \\ a_{(m-1)0} & \cdots & a_{(m-1)(n-1)} \end{pmatrix} \begin{pmatrix} b_{00} & \cdots & b_{0(r-1)} \\ b_{10} & \cdots & b_{1(r-1)} \\ & \vdots & \\ b_{(n-1)0} & \cdots & b_{(n-1)(r-1)} \end{pmatrix}
$$
$$
= \begin{pmatrix} c_{00} & \cdots & c_{0(r-1)} \\ c_{10} & \cdots & c_{1(r-1)} \\ & \vdots & \\ c_{(m-1)0} & \cdots & c_{(m-1)(r-1)} \end{pmatrix},
$$

(1.5)

where c_{ij} is the scalar product of the ith row of A and the jth column of B:

$$
c_{ij} = \sum_{k=0}^{n-1} a_{ik}b_{kj}, \quad i = 0, 1, \ldots, m-1, \; j = 0, 1, \ldots, r-1.
$$

Example 1.3.4 Let $R = \mathbb{Z}$. Below is an example for multiplication of two 2×2 matrices with coefficients in $\mathbb{Z}$:

$$
\begin{pmatrix} 2 & 3 \\ 1 & -1 \end{pmatrix} \begin{pmatrix} 4 & -2 \\ 0 & -5 \end{pmatrix} = \begin{pmatrix} 8 & -19 \\ 4 & 3 \end{pmatrix}.
$$

Definition 1.3.3 An $n \times n$ square matrix A is said to be *invertible* if there exists an $n \times n$ matrix B such that

$$
AB = BA = I_n.
$$

B is called the *inverse* of A. We will use A^{-1} to denote this matrix.

Example 1.3.5 Let $R = \mathbb{Z}$. We have

$$
\begin{pmatrix} 2 & 1 \\ 1 & 1 \end{pmatrix} \begin{pmatrix} 1 & -1 \\ -1 & 2 \end{pmatrix} = \begin{pmatrix} 1 & -1 \\ -1 & 2 \end{pmatrix} \begin{pmatrix} 2 & 1 \\ 1 & 1 \end{pmatrix} = \begin{pmatrix} 1 & 0 \\ 0 & 1 \end{pmatrix}.
$$

Hence, the 2×2 matrix $A = \begin{pmatrix} 2 & 1 \\ 1 & 1 \end{pmatrix}$ is invertible and its inverse $A^{-1} = \begin{pmatrix} 1 & -1 \\ -1 & 2 \end{pmatrix}$.

Theorem 1.3.1 *Let n be a positive integer. We define $\mathcal{M}_{n \times n}(R)$ to be the set of $n \times n$ square matrices with coefficients in R. Then $\mathcal{M}_{n \times n}(R)$ together with addition and multiplication defined in Eqs. 1.4 and 1.5 is a ring. It is not a commutative ring when $n \geq 2$.*

Proof In Examples 1.2.3 and 1.2.13 we have shown that $\mathcal{M}_{2 \times 2}(\mathbb{R})$ is a ring. The proof for the general case is similar.

The closure of $\mathcal{M}_{n\times n}(R)$ with respect to both operations is easy to see. Associativity and distributive laws for addition and multiplication follow from the corresponding properties of R.

The additive identity is the zero matrix of size $n \times n$. The additive inverse of a matrix A with coefficients a_{ij} $(0 \le i, j \le n - 1)$ is given by $-A$ with coefficients $-a_{ij}$, $(0 \le i, j \le n - 1)$. The multiplicative identity is I_n.

When $n = 1$, $\mathcal{M}_{1\times 1}(R)$ is a commutative ring because R is commutative.

When $n \ge 2$, let

$$
A = \begin{pmatrix} 1 & 0 & \dots & 0 \\ 0 & 0 & \dots & 0 \\ \vdots & \vdots & \ddots & \vdots \\ 0 & 0 & \dots & 0 \end{pmatrix}, \quad
B = \begin{pmatrix} 0 & 0 & \dots & 0 \\ 0 & 0 & \dots & 0 \\ \vdots & \vdots & \ddots & \vdots \\ 1 & 0 & \dots & 0 \end{pmatrix}.
$$

Then

$$
AB = \begin{pmatrix} 0 & 0 & \dots & 0 \\ 0 & 0 & \dots & 0 \\ \vdots & \vdots & \ddots & \vdots \\ 0 & 0 & \dots & 0 \end{pmatrix}, \quad
BA = \begin{pmatrix} 0 & 0 & \dots & 0 \\ 0 & 0 & \dots & 0 \\ \vdots & \vdots & \ddots & \vdots \\ 1 & 0 & \dots & 0 \end{pmatrix}.
$$

Hence $AB \ne BA$ and $\mathcal{M}_{n\times n}(R)$ is not commutative for $n \ge 2$ $\square$

In general, not every matrix is invertible. To find the inverse of an invertible matrix, we will need the following definition.

Definition 1.3.4 Let n be a positive integer. For any $A \in \mathcal{M}_{n\times n}(R)$, the *determinant* of A, denoted $\det(A)$, is defined as follows:

- If $n = 1$, $A = (a)$, $\det(A) := a$.
- If $n > 1$, let A_{ij} denote the matrix obtained from A by deleting the ith row and the jth column. Fix an i_0,

$$
\det(A) := \sum_{j=0}^{n-1} (-1)^{i_0+j} a_{i_0 j} \det(A_{i_0 j}). \tag{1.6}
$$

We note that the value of $\det(A)$ is independent of the choice of i_0 in Eq. 1.6 (see Appendix A.1). Similarly, $\det(A)$ can also be found by fixing a j_0 and computing

$$
\det(A) = \sum_{i=0}^{n-1} (-1)^{i+j_0} a_{i j_0} \det(A_{i j_0}).
$$

Example 1.3.6 Let $n = 2$; for any $A \in \mathcal{M}_{2 \times 2}(R)$, we can write $A = \begin{pmatrix} a_{00} & a_{01} \\ a_{10} & a_{11} \end{pmatrix}$. Take $i_0 = 0$,

$$\det(A) = \sum_{j=0}^{n-1} (-1)^{i_0 + j} a_{i_0 j} \det(A_{i_0 j}) = \sum_{j=0}^{1} (-1)^j a_{0j} \det(A_{0j}) = a_{00}a_{11} - a_{01}a_{10}.$$

Theorem 1.3.2 *A matrix $A \in \mathcal{M}_{n \times n}(R)$ is invertible in $\mathcal{M}_{n \times n}(R)$ if and only if* $\det(A)$ *is a unit in R.*

When $\det(A)$ *is a unit in R, if $n = 1$ and $A = (a)$, then $A^{-1} = (a^{-1})$. If $n > 1$, we define the* adjoint matrix *of A as follows:*

$$adjA := \begin{pmatrix} (-1)^{0+0} \det(A_{00}) & (-1)^{0+1} \det(A_{10}) & \cdots & (-1)^{0+(n-1)} \det(A_{(n-1)0}) \\ \vdots & \vdots & \ddots & \vdots \\ (-1)^{(n-1)+0} \det(A_{0(n-1)}) & (-1)^{(n-1)+1} \det(A_{1(n-1)}) & \cdots & (-1)^{(n-1)+(n-1)} \det(A_{(n-1)(n-1)}) \end{pmatrix},$$

where the (i, j)-entry of $adjA$ is given by $(-1)^{i+j} \det(A_{ji})$. Then

$$A^{-1} = (\det(A))^{-1} adjA.$$

The proof can be found in, e.g., [Hun12, page 353].

Example 1.3.7 Let $n = 2$; by Example 1.3.6 and Theorem 1.3.2, a matrix $A = \begin{pmatrix} a_{00} & a_{01} \\ a_{10} & a_{11} \end{pmatrix}$ from $\mathcal{M}_{2 \times 2}(R)$ is invertible if and only if $a_{00}a_{11} - a_{01}a_{10}$ is a unit in R. When $a_{00}a_{11} - a_{01}a_{10}$ is a unit in R, the adjoint matrix of A is given by

$$adjA = \begin{pmatrix} a_{11} & -a_{01} \\ -a_{10} & a_{00} \end{pmatrix}.$$

And the inverse of matrix A is given by

$$A^{-1} = (a_{00}a_{11} - a_{01}a_{10})^{-1} \begin{pmatrix} a_{11} & -a_{01} \\ -a_{10} & a_{00} \end{pmatrix}. \tag{1.7}$$

Example 1.3.8 Let $R = \mathbb{Z}$. By Example 1.3.6, $A = \begin{pmatrix} 2 & 3 \\ 4 & 7 \end{pmatrix}$ has determinant $14 - 12 = 2$. 2 is not a unit in $\mathbb{Z}$. By Theorem 1.3.2, A is not invertible in $\mathcal{M}_{2 \times 2}(\mathbb{Z})$. However, if we consider $R = \mathbb{Q}$, 2 is a unit in $\mathbb{Q}$. By Theorem 1.3.2, A is invertible in $\mathcal{M}_{2 \times 2}(\mathbb{Q})$, and we can compute A^{-1} using Eq. 1.7:

$$A^{-1} = \frac{1}{2} \begin{pmatrix} 7 & -3 \\ -4 & 2 \end{pmatrix} = \begin{pmatrix} 3.5 & -1.5 \\ -2 & 1 \end{pmatrix}.$$

1.3.2   Vector Spaces

Let F be a field with additive identity 0 and multiplicative identity 1.

Definition 1.3.5 (Vector space) A nonempty set V, together with two binary operations—*vector addition* (denoted by $+$) and *scalar multiplication by elements of F*, which is a map with domain $V \times F$ and codomain V, is called a *vector space* over F if $(V, +)$ is an abelian group, and for any $v, w \in V$ and any $a, b \in F$, we have:

1. $a(v + w) = av + aw$.
2. $(a + b)v = av + bv$.
3. $a(bv) = (ab)v$.
4. $1v = v$, where 1 is the multiplicative identity of F.

Elements of V are called *vectors*, and elements of F are called *scalars*.

Remark 1.3.1 It is easy to see that if 0 is the additive identity in F and v any vector in V, then $0v = \mathbf{0}$ is the additive identity in V (or the identity for vector addition).

Example 1.3.9 The set of complex numbers $\mathbb{C} = \{x + iy \mid x, y \in \mathbb{R}\}$ is a vector space over $\mathbb{R}$. Note that for any $a_1 + b_1 i, a_2 + b_2 i \in \mathbb{C}$, vector addition is defined as

$$(a_1 + b_1 i) + (a_2 + b_2 i) = (a_1 + a_2) + (b_1 + b_2)i.$$

And for any $a \in \mathbb{R}$, scalar multiplication by elements of $\mathbb{R}$ is defined as

$$a(a_1 + b_1 i) = aa_1 + ab_1 i.$$

The identity element for vector addition is 0. Furthermore, for any $a + bi \in \mathbb{C}$, its inverse with respect to vector addition is given by $-a - bi$.

Let $F^n = \{(v_0, v_1, \ldots, v_{n-1}) \mid v_i \in F\ \forall i\}$ be the set of n-tuples over F. Define vector addition and scalar multiplication by elements of F componentwise: For any $v = (v_0, v_1, \ldots, v_{n-1}) \in F^n$, $w = (w_0, w_1, \ldots, w_{n-1}) \in F^n$, and any $a \in F$,

$$v + w := (v_0 + w_0, v_1 + w_1, \ldots, v_{n-1} + w_{n-1}), \tag{1.8}$$

$$av := (av_0, av_1, \ldots, av_{n-1}). \tag{1.9}$$

Theorem 1.3.3 *Together with vector addition and scalar multiplication by elements of F defined in Eqs. 1.8 and 1.9, respectively, $F^n = \{(v_0, v_1, \ldots, v_{n-1}) \mid v_i \in F\ \forall i\}$ is a vector space over F.*

Proof Take any $v = (v_0, v_1, \ldots, v_{n-1})$, $w = (w_0, w_1, \ldots, w_{n-1})$ from F^n and any $a, b \in F$.

By Eq. 1.8, it is easy to see that F^n is closed under vector addition. The associativity and commutativity of vector addition follow from that for addition in

F. The identity element for vector addition is $(0, 0, \ldots, 0)$, where 0 is the additive identity in F. The inverse of $v \in F^n$ is $(-v_0, -v_1, \ldots, -v_{n-1})$, where $-v_i$ is the additive inverse of v_i in F. Thus F^n with vector addition is an abelian group.

By the definition of scalar multiplication by elements of F (Eq. 1.9), $av \in F^n$. Properties 1 and 2 in Definition 1.3.5 follow from distributive law in F. Property 3 follows from the associativity of multiplication in F. Property 4 follows from the definition of multiplicative identity in F. $\qquad \square$

Example 1.3.10 Let $F = \mathbb{F}_2$, the unique finite field with two elements (see Example 1.2.20 and Theorem 1.2.3). Let n be a positive integer, and it follows from Theorem 1.3.3 that $\mathbb{F}_2^n$ is a vector space over $\mathbb{F}_2$.

The identity element for vector addition is $(0, 0, \ldots, 0)$. For any $v = (v_0, v_1, \ldots, v_{n-1}) \in \mathbb{F}_2^n$, the inverse of v with respect to vector addition is $(-v_0, -v_1, \ldots, -v_{n-1}) = v$.

Recall that variables ranging over $\mathbb{F}_2$ are called bits (see Definition 1.2.17). We have shown that $(\mathbb{F}_2, \oplus, \&)$ is a finite field (see Example 1.2.20), where $\oplus$ is logical XOR (see Example 1.2.4), and $\&$ is logical AND (see Example 1.2.14).

Definition 1.3.6 Vector addition in $\mathbb{F}_2^n$ is called *bitwise XOR*, also denoted $\oplus$. Similarly, we define *bitwise AND* between any two vectors $v = (v_0, v_1, \ldots, v_{n-1})$, $w = (w_0, w_1, \ldots, w_{n-1})$ from $\mathbb{F}_2^n$ as follows:

$$v \& w := (v_0 \& w_0, v_1 \& w_1, \ldots, v_{n-1} \& w_{n-1}).$$

Remark 1.3.2 Another useful binary operation, logical OR, denoted $\vee$, on $\mathbb{F}_2$ is defined as follows:

$$0 \vee 0 = 0, \quad 1 \vee 0 = 1, \quad 0 \vee 1 = 1, \quad 1 \vee 1 = 1.$$

It can also be extended to $\mathbb{F}_2^n$ in a bitwise manner, and we get *bitwise OR*.

For simplicity, we sometimes write $v_0 v_1 \ldots v_{n-1}$ instead of $(v_0, v_1, \ldots, v_{n-1})$.

Example 1.3.11 Let $n = 3$, and take $111, 101 \in \mathbb{F}_2^3$, $111 \oplus 101 = 010$, $111 \& 101 = 101$, $111 \vee 101 = 111$.

Definition 1.3.7 A vector in $\mathbb{F}_2^n$ is called an *n-bit binary string*. A 4-bit binary string is called a *nibble*. An 8-bit binary string is called a *byte*.

Example 1.3.12

- $1010, 0011 \in \mathbb{F}_2^4$ are two nibbles. Furthermore,

$$1010 \oplus 0011 = 1001, \quad 1010 \& 0011 = 0010.$$

- 00101100 is a byte.

Remark 1.3.3 By Theorem 1.1.1, a byte can be considered as a base-2 representation/binary representation of an integer (see Theorem 1.1.1). By Eq. 1.2, the value of this integer is between 0 and 255 or between $\mathbf{00}_{16}$ and $\mathbf{FF}_{16}$ with base-16 representation/hexadecimal representation.

For the rest of this section, let V be a vector space over F.

Definition 1.3.8 A nonempty subset $U \subseteq V$ is called a *subspace* of V if U is a vector space over F under the same operations (vector addition and scalar multiplication by elements of F) in V.

Remark 1.3.4 To show $U \subset V$ is a subspace of V, by Definitions 1.3.5, 1.2.1, and 1.2.2, we need to prove the following:

1. $(U, +)$ is an abelian group.

   (a) U is closed under $+$ (closure property): $\forall u, v \in U, u + v \in U$.
   (b) $+$ is associative: $\forall u, v, w \in U, u + (v + w) = (u + v) + w$.
   (c) The identity element for vector addition in V is also in U.
   (d) For $v \in U$, its additive inverse in V is also in U.

2. Scalar multiplication by elements of F is a function with domain $U \times F$ and codomain U.
3. For any $v, w \in U$ and any $a, b \in F$, we have

   (a) $a(v + w) = av + aw$.
   (b) $(a + b)v = av + bv$.
   (c) $a(bv) = (ab)v$.
   (d) $1v = v$, where 1 is the multiplicative identity in F.

We note that 1-(b) and 3 follow from the corresponding properties of V. Thus, to prove U is a subspace of V, we need to prove 1-(a), 1-(c), 1-(d), and 2.

In case $F = \mathbb{F}_2$, by Example 1.3.10, 1-(d) is true by default. Furthermore, 2 is also true as there are only two elements in $\mathbb{F}_2$: 0 and 1. To show U is a subspace when $F = \mathbb{F}_2$, it suffices to prove 1-(a) and 1-(c).

Definition 1.3.9 A *linear combination* of $v_1, v_2, \ldots, v_r \in V$ is a vector of the form $a_1 v_1 + a_2 v_2 + \cdots + a_r v_r$, where $a_i \in F$ $\forall i$.

Lemma 1.3.1 *For any $v_1, v_2, \ldots, v_r \in V$ $(r \geq 1)$, $U := \{a_1 v_1 + a_2 v_2 + \cdots + a_r v_r \mid a_i \in F\}$ is a subspace of V.*

Proof By Remark 1.3.4, we will prove 1-(a), 1-(c), 1-(d), and 2.

Take any $v = \displaystyle\sum_{i=1}^{r} a_i v_i \in U.$

1-(a). For any $u = \displaystyle\sum_{i=1}^{r} b_i v_i \in U,$

$$v + u = \sum_{i=1}^{r} a_i v_i + \sum_{i=1}^{r} b_i v_i = \sum_{i=1}^{r} (a_i + b_i) v_i \in U.$$

1-(c). Let $a_i = 0 \in F$, then (see Remark 1.3.1)

$$0 = \sum_{i=1}^{r} a_i v_i \in U.$$

1-(d). The inverse of v with respect to vector addition is given by

$$u := \sum_{i=1}^{r} (-a_i) v_i$$

because $v + u = 0$. Furthermore, since $-a_i \in F$, we have $u \in U$.

2. For any $\alpha \in F$,

$$\alpha \sum_{i=1}^{r} a_i v_i = \sum_{i=1}^{r} (\alpha a_i) v_i \in U.$$

$\square$

Definition 1.3.10 Let $S = \{ v_1, v_2, \ldots, v_r \} \subseteq V$,

$$\langle S \rangle := \{ a_1 v_1 + a_2 v_2 + \cdots + a_r v_r \mid a_i \in F \}$$

is called the *(linear) span* of S over F. For any subspace $U \subseteq V$ and a subset S of U, if $U = \langle S \rangle$, S is called a *generating set* for U.

We note that if S is a subspace of V, then $\langle S \rangle = S$.

Example 1.3.13 Let $V = \mathbb{F}_2^3$ and $S = \{ 001, 100 \}$, then $\langle S \rangle = \{ 000, 001, 100, 101 \}$

Definition 1.3.11 A set of vectors $\{ v_1, v_2, \ldots, v_r \} \subseteq V$ are *linearly independent over F* if

$$\sum_{i=1}^{r} a_i v_i = 0 \implies a_i = 0 \ \forall i.$$

Otherwise, they are said to be *linearly dependent over F*.

Example 1.3.14

- Let $F = \mathbb{F}$, $V = \mathbb{F}_2^3$. 001 and 100 are linearly independent.
- For any $S \subseteq V$, if $0 \in S$, then the vectors in S are linearly dependent.
- Let $F = \mathbb{R}$, $V = \mathbb{R}^3$, and $(0, 1, 0)$ and $(0, 0, 1)$ are linearly independent.

$(0, 1, 0), (2, 3, 0), (1, 0, 0)$ are linearly dependent since, for example, we have

$$3 \cdot (0, 1, 0) + (-1) \cdot (2, 3, 0) + 2 \cdot (1, 0, 0) = (0, 0, 0).$$

Definition 1.3.12 Let B be a nonempty subset of V. If $V = \langle B \rangle$ and vectors in B are linearly independent, then B is called a *basis* for V over F.

Remark 1.3.5 Suppose B is a basis for V and $B = \{v_1, v_2, \ldots, v_r\}$. Then any element $v \in V$ has a unique representation as a linear combination of vectors in B:

$$v = \sum_{i=1}^{r} a_i v_i = \sum_{i=1}^{r} b_i v_r \implies \sum_{i=1}^{r} (a_i - b_i) v_i = 0 \implies a_i = b_i.$$

Example 1.3.15

- Let $F = \mathbb{R}$, $V = \mathbb{R}^3$, and $B = \{(1, 0, 0), (0, 1, 0), (0, 0, 1)\}$. It is easy to see that vectors in B are linearly independent. For any $v = (v_0, v_1, v_2) \in \mathbb{R}^3$, we have

$$v = v_0(1, 0, 0) + v_1(0, 1, 0) + v_2(0, 0, 1).$$

  Thus, B is a generating set of V. By definition, B is a basis for V over $\mathbb{R}$.
- Let $F = \mathbb{F}_2$ and $V = \mathbb{F}_2^3$; similarly, we can show $\{(1, 0, 0), (0, 1, 0), (0, 0, 1)\}$ is a basis for V over $\mathbb{F}_2$.

Example 1.3.16 Let $V = F^n$ and $B = \{v_0, v_1, \ldots, v_{n-1}\}$, where

$$v_i = (v_{i0}, v_{i1}, \ldots, v_{i(n-1)}), \quad v_{ii} = 1 \text{ and } v_{ij} = 0 \text{ for } i \neq j.$$

It is easy to see that vectors in B are linearly independent. For any $u = (u_0, u_1, \ldots, u_{n-1}) \in V$, we can write

$$u = \sum_{\ell=0}^{n-1} u_\ell v_\ell.$$

Thus, B is a generating set of V. By definition, B is a basis for V over F.

Lemma 1.3.2 *Let B_1, B_2 be subsets of V. If $V = \langle B_1 \rangle$ and vectors in B_2 are linearly independent, then $|B_1| \geq |B_2|$.*

Proof Suppose $B_1 = \{v_1, v_2, \ldots, v_{r_1}\}$ and $B_2 = \{w_1, w_2, \ldots, w_{r_2}\}$. Since $V = \langle B_1 \rangle$,

$$w_1 = \sum_{j=1}^{r_1} a_j v_j$$

for some $a_j \in F$. Moreover, at least one of $a_j \neq 0$ as vectors in B_2 are linearly independent. Without loss of generality, let us assume $a_1 \neq 0$, then

$$\boldsymbol{v}_1 = -\sum_{j=2}^{r_1} \frac{a_j}{a_1} \boldsymbol{v}_j + \frac{1}{a_1} \boldsymbol{w}_1,$$

and we have $\{\, \boldsymbol{w}_1, \boldsymbol{v}_2, \ldots, \boldsymbol{v}_{r_1} \,\}$ spans V. Then, we can write

$$\boldsymbol{w}_2 = b_1 \boldsymbol{w}_1 + \sum_{j=2}^{r_1} b_j \boldsymbol{v}_j,$$

where $b_j \in F$, and at least one of $b_j \neq 0$ for $2 \leq j \leq r_1$, otherwise $\boldsymbol{w}_2$ is a linear combination of $\boldsymbol{w}_1$. Suppose $b_2 \neq 0$. We have

$$\boldsymbol{v}_2 = -\frac{b_1}{b_2} \boldsymbol{w}_1 - \sum_{j=3}^{r_1} \frac{b_j}{b_2} \boldsymbol{v}_j + \frac{1}{b_2} \boldsymbol{w}_2,$$

which means $\{\, \boldsymbol{w}_1, \boldsymbol{w}_2, \boldsymbol{v}_3, \ldots, \boldsymbol{v}_{r_1} \,\}$ spans V.

We can continue in this manner, if $r_1 < r_2$, we will deduce that $\{\, \boldsymbol{w}_1, \boldsymbol{w}_2 \ldots, \boldsymbol{w}_{r_1} \,\}$ spans V, and $\boldsymbol{w}_{r_1+1}$ can be written as a linear combination of $\{\, \boldsymbol{w}_1, \boldsymbol{w}_2 \ldots, \boldsymbol{w}_{r_1} \,\}$, a contradiction. $\qquad\square$

We have the following direct corollary.

Corollary 1.3.1 *If B_1 and B_2 are bases of V, then $|B_1| = |B_2|$.*

Proof By Lemma 1.3.2, $|B_1| \leq |B_2|$ and $|B_2| \leq |B_1|$. $\qquad\square$

Definition 1.3.13 The *dimension* of V over F, denoted $\dim(V)_F$, is given by the cardinality of B, $|B|$, where B is a basis of V over F.

Example 1.3.17 Continuing Example 1.3.16, $\dim(F^n)_F = n$.

Lemma 1.3.3 *Let $F = \mathbb{F}_2$; if $\dim(V)_{\mathbb{F}_2} = k$, then $|V| = 2^k$.*

Proof Let $B = \{\, \boldsymbol{v}_1, \boldsymbol{v}_2, \ldots, \boldsymbol{v}_k \,\}$ be a basis for V. We have discussed in Remark 1.3.5 that every $\boldsymbol{w} \in V$ has a unique representation as a linear combination of vectors in B. In other words,

$$V = \left\{\, \sum_{i=1}^{k} a_i \boldsymbol{v}_i \;\middle|\; a_i \in \mathbb{F}_2,\ 1 \leq i \leq k \,\right\},$$

where there are two choices for each a_i. $\qquad\square$

Example 1.3.18 Let $F = \mathbb{F}_2$, $S = \{0010, 1000\}$, and $V = \langle S \rangle$. It is easy to see that vectors in S are linearly independent. By Definition 1.3.13, $\dim(V)_{\mathbb{F}_2} = 2$. By Lemma 1.3.3, $|V| = 4$. We can verify that $V = \{0000, 0010, 1000, 1010\}$.

For any $v = (v_0, v_2, \ldots, v_{n-1}) \in \mathbb{F}_2^n$ and $w = (w_0, w_2, \ldots, w_{n-1}) \in \mathbb{F}_2^n$, we can consider v as a row vector and w as a column vector and compute the *scalar product* (see Definition 1.3.2) between v and w:

$$v \cdot w = \sum_{i=0}^{n-1} v_i w_i.$$

We note for any $u = (u_0, u_1, \ldots, u_{n-1}) \in \mathbb{F}_2^n$

$$(v + w) \cdot u = \sum_{i=0}^{n-1} (v_i + w_i) u_i = \sum_{i=0}^{n-1} v_i u_i + \sum_{i=0}^{n-1} w_i u_i = v \cdot u + w \cdot u. \tag{1.10}$$

Definition 1.3.14

- For any $v, w \in \mathbb{F}_2^n$, v and w are said to be *orthogonal* if $v \cdot w = 0$.
- Let $S \subseteq \mathbb{F}_2^n$ be nonempty. The *orthogonal complement*, denoted $S^\perp$, of S is given by

$$S^\perp = \left\{ v \mid v \in \mathbb{F}_2^n, v \cdot s = 0 \; \forall s \in S \right\}.$$

- If $S = \emptyset$, we define $S^\perp = \mathbb{F}_2^n$.

By definition, it is easy to see that $\langle S \rangle^\perp = S^\perp$.

Lemma 1.3.4 *For any $S \subseteq V$, $S^\perp$ is a subspace of $\mathbb{F}_2^n$.*

Proof By Remark 1.3.4, we will prove 1-(a) and 1-(c).
1-(a). Take any $v, u \in S^\perp$ and any $s \in S$, by Eq. 1.10, we have

$$(v + w) \cdot s = v \cdot s + u \cdot s = 0,$$

hence $v + w \in S^\perp$.
1-(c). $\mathbf{0} \cdot s = 0$ for any $s \in S$. Hence $\mathbf{0} \in S^\perp$. $\qquad\qquad \square$

1.4   Modular Arithmetic

In this section, let $n > 1$ be an integer.

We are interested in the set $\{0, 1, 2 \ldots, n - 1\}$. It can be considered as the set of possible remainders when dividing by n (see Theorem 1.1.2). We will also associate each integer with one element in the set—namely the remainder of this integer divided by n. Here we would like to provide a rigorous definition for this association. First, we introduce the notion of equivalence relations.

Definition 1.4.1 A relation $\sim$ on a set S is called an *equivalence relation* if $\forall a, b, c \in S$, and the following conditions are satisfied.

- $a \sim a$ (reflexivity).
- If $a \sim b$, then $b \sim a$ (symmetry).
- If $a \sim b$ and $b \sim c$, then $a \sim c$ (transitivity)

Let us define a relation $\sim$ on the set $\mathbb{Z}$ as follows:

$$a \sim b \quad \text{if and only if} \quad n|(b - a). \tag{1.11}$$

We can see that this is an equivalence relation on $\mathbb{Z}$.

- $\forall a \in \mathbb{Z}, 0 = a - a$ and $n|0$; hence $a \sim a$ (reflexivity).
- If $n|(a - b)$, then $n|(b - a)$, and we have $a \sim b$ implies $b \sim a$ (symmetry).
- If $n|(a - b)$ and $n|(b - c)$, then

$$n|((a - b) + (b - c)) \implies n|(a - c).$$

Thus $a \sim b$ and $b \sim c$ imply $a \sim c$ (transitivity).

Definition 1.4.2 Take $a, b \in \mathbb{Z}$. If $a \sim b$, i.e., $n|(b - a)$, then we say that a is *congruent to b modulo n*, written $a \equiv b \bmod n$. n is called the *modulus*.

By the above definitions, saying a is congruent to b modulo n is equivalent to saying that the remainder of a divided by n is the same as the remainder of b divided by n.

Definition 1.4.3 If $\sim$ is an equivalence relation on a set S, then the *equivalence class* of an element $a \in S$, denoted $\overline{a}$, is defined by

$$\overline{a} := \{b \mid b \in S, b \sim a\}.$$

Theorem 1.4.1 *If $\sim$ is an equivalence relation on a set S, then $\sim$ partitions S into disjoint equivalence classes. That is,*

$$S = \bigcup \overline{a}, \quad \text{and} \quad \overline{a} \bigcap \overline{b} = \emptyset \text{ if } \overline{a} \neq \overline{b}.$$

Proof It is easy to see that $S = \bigcup \overline{a}$.

To prove the second part, we show that the following equivalent claim is true:

$$\text{if } \overline{a} \bigcap \overline{b} \neq \emptyset, \text{ then } \overline{a} = \overline{b}.$$

Let c be an element of $\overline{a} \bigcap \overline{b}$. By Definition 1.4.3, $c \sim a$ and $c \sim b$. By symmetry (Definition 1.4.1), $a \sim c$. By transitivity (Definition 1.4.1), $a \sim b$. Hence $a \in \overline{b}$.

Now for any $d \in \bar{a}$, $d \sim a$. By transitivity (Definition 1.4.1), $d \sim b$. Then by Definition 1.4.3, $d \in \bar{b}$. We have $\bar{a} \subset \bar{b}$.

Similarly, we can prove $\bar{b} \subset \bar{a}$. Hence $\bar{a} = \bar{b}$. $\qquad\square$

Definition 1.4.4 For any $a \in \mathbb{Z}$, the *congruence class of a modulo n*, denoted $\bar{a}$, is defined to be the equivalence class of a with respect to the equivalence relation $\sim$ defined in Eq. 1.11.

We note that the set $\bar{a}$ consists of all integers of the form $a + nk$ for some $k \in \mathbb{Z}$.

Lemma 1.4.1 *Let $\mathbb{Z}_n$ denote the set of all congruence classes of $a \in \mathbb{Z}$ modulo n. Then*

$$\mathbb{Z}_n = \left\{ \bar{0}, \bar{1}, \ldots, \overline{n-1} \right\}.$$

Proof By Theorem 1.1.2, given any $b \in \mathbb{Z}$, we can find $q, r \in \mathbb{Z}$ such that

$$0 \le r < n \text{ and } b = qn + r \implies b \sim r.$$

By Theorem 1.4.1, we have $\bar{b} = \bar{r}$. Hence the set $\left\{ \bar{0}, \bar{1}, \ldots, \overline{n-1} \right\}$ contains all the congruence classes of integers modulo n, possibly with some repetitions.

If $\bar{r}_1 = \bar{r}_2$ for some $0 \le r_1, r_2 < n$, then $n | (r_1 - r_2)$. Since $0 \le r_1, r_2 < n$, we have $r_1 = r_2$. Thus $\bar{0}, \bar{1}, \ldots, \overline{n-1}$ are all distinct. $\qquad\square$

Remark 1.4.1 $\bar{a} = \bar{b}$ if and only if $a \equiv b \bmod n$.

Example 1.4.1 Let $n = 5$. We have $\bar{1} = \bar{6} = \overline{-4}$. By Lemma 1.4.1, $\mathbb{Z}_5 = \left\{ \bar{0}, \bar{1}, \bar{2}, \bar{3}, \bar{4} \right\}$.

We define the addition operation on the set $\mathbb{Z}_n$ as follows:

$$\bar{a} + \bar{b} = \overline{a + b}. \tag{1.12}$$

If $\bar{a} = \bar{a}'$ and $\bar{b} = \bar{b}'$, we have $n | (a' - a)$ and $n | (b' - b)$, therefore

$$n | ((a' - a) + (b' - b)) \implies n | ((a' + b') - (a + b)) \implies (a + b) \sim (a' + b')$$

$$\implies \overline{a + b} = \overline{a' + b'}.$$

Thus the addition in Eq. 1.12 is well-defined.

Example 1.4.2

- Let $n = 7$, $\bar{3} + \bar{2} = \bar{5}$.
- Let $n = 4$, $\bar{2} + \bar{2} = \bar{4} = \bar{0}$.

Proposition 1.4.1 $(\mathbb{Z}_n, +)$, *the set $\mathbb{Z}_n$ together with addition defined in Eq. 1.12, is an abelian group.*

Proof For any $\bar{a}, \bar{b} \in \mathbb{Z}_n$, $\overline{a+b} \in \mathbb{Z}_n$. Hence $\mathbb{Z}_n$ is closed under $+$. The associativity follows from the associativity of the addition of integers. The identity element is $\bar{0}$, and the inverse of $\bar{a}$ is $\overline{n-a}$:

$$\bar{a} + \overline{n-a} = \overline{n-a} + \bar{a} = \bar{n} = \bar{0}.$$

The commutative property follows from that for integer addition. $\qquad\square$

Remark 1.4.2 The proof also shows that the additive inverse of an element $\bar{a} \in \mathbb{Z}_n$ is $\overline{n-a} = \overline{-a}$, and the identity element is $\bar{0}$.

Example 1.4.3

- Let $n = 5$; the inverse of $\bar{1}$ in $(\mathbb{Z}_5, +)$ is $\overline{5-1} = \bar{4}$.
- Let $n = 8$; the inverse of $\bar{2}$ in $(\mathbb{Z}_8, +)$ is $\overline{8-2} = \bar{6}$.

Lemma 1.4.2 $(\mathbb{Z}_n, +)$ *is a cyclic group.*

Proof Recall that the identity element in $(\mathbb{Z}_n, +)$ is $\bar{0}$. It is easy to see that $\bar{1}$ has order n (see Definition 1.2.6):

$$\bar{1} + \bar{1} = \bar{2}$$
$$\bar{1} + \bar{1} + \bar{1} = \bar{3}$$
$$\vdots$$
$$\underbrace{\bar{1} + \bar{1} + \ldots \bar{1}}_{n-1 \text{ times}} = \overline{n-1}$$
$$\underbrace{\bar{1} + \bar{1} + \ldots \bar{1}}_{n \text{ times}} = \bar{n} = \bar{0}.$$

$\qquad\square$

We define the multiplication on $\mathbb{Z}_n$ as follows:

$$\bar{a} \cdot \bar{b} = \overline{ab}. \tag{1.13}$$

If $\overline{a'} = \bar{a}$ and $\overline{b'} = \bar{b}$, then we can write $a' = a + sn$, $b' = b + tn$ for some integers s, t. We have

$$a'b' = ab + n(at + sb + st) \implies a'b' \sim ab.$$

Hence $\overline{a'b'} = \overline{ab}$, and the multiplication in Eq. 1.13 is well-defined.

Example 1.4.4 Let $n = 5$,

$$\overline{-2} \cdot \overline{13} = \bar{3} \cdot \bar{3} = \bar{9} = \bar{4}.$$

Theorem 1.4.2 $(\mathbb{Z}_n, +, \cdot)$, *the set $\mathbb{Z}_n$ together with addition defined in Eq. 1.12 and multiplication defined in Eq. 1.13, is a commutative ring. It is an integral domain if and only if n is prime.*

Proof In Proposition 1.4.1 we have shown that $(\mathbb{Z}_n, +)$ is an abelian group.

Take any $\overline{a}, \overline{b} \in \mathbb{Z}_n$, $\overline{ab} \in \mathbb{Z}_n$. Hence $\mathbb{Z}_n$ is closed under $\cdot$. Associativity, commutativity of multiplication, and distributive laws follow from that for the integers. The identity element for multiplication is $\overline{1}$. We have proved that $(\mathbb{Z}_n, +, \cdot)$ is a commutative ring.

If n is not a prime, let m be a prime that divides n. Then $d = n/m$ is an integer and $d \neq 0$. We have

$$\overline{m} \cdot \overline{d} = \overline{n} = \overline{0}.$$

By Definition 1.2.10, $\overline{d}, \overline{m}$ are zero divisors in $\mathbb{Z}_n$. By Definition 1.2.11, $\mathbb{Z}_n$ is not an integral domain.

Let n be a prime. Suppose there are $\overline{a}, \overline{b} \in \mathbb{Z}_n$, such that $\overline{a} \neq \overline{0}, \overline{b} \neq \overline{0}$, and $\overline{a} \cdot \overline{b} = \overline{0}$. By definition, we have $n|ab$. By Lemma 1.1.2, $n|a$ or $n|b$, which gives $\overline{a} = \overline{0}$ or $\overline{b} = \overline{0}$, a contradiction. $\qquad\square$

For simplicity, we write a instead of $\overline{a}$, and to make sure there is no confusion with $a \in \mathbb{Z}$, we would specify that $a \in \mathbb{Z}_n$. In particular, $\mathbb{Z}_n = \{0, 1, 2, \ldots, n-1\}$. Furthermore, to emphasize that multiplication or addition is done in $\mathbb{Z}_n$, we write $ab \bmod n$ or $a + b \bmod n$.

Example 1.4.5 Let $n = 5$, and we write

$$4 \times 2 \bmod 5 = 8 \bmod 5 = 3, \quad \text{or} \quad 4 \times 2 \equiv 8 \equiv 3 \bmod 5.$$

Lemma 1.4.3 *For any $a \in \mathbb{Z}_n$, $a \neq 0$, a has a multiplicative inverse, denoted $a^{-1} \bmod n$, if and only if $\gcd(a, n) = 1$.*

Proof By Bézout's identity (Theorem 1.1.3), $\gcd(a, n) = sa + tn$ for some $s, t \in \mathbb{Z}$.

$\Longleftarrow$ If $\gcd(a, n) = 1$, then $sa + tn = 1$, i.e., $n|(1 - sa)$. By definition, $sa \equiv 1 \bmod n$; thus $a^{-1} \bmod n = s$.

$\Longrightarrow$ On the other hand, if a has a multiplicative inverse, then there exists $s \in \mathbb{Z}_n$ such that $as \bmod n = 1$, which gives $n|(as - 1)$. Hence there is some $t \in \mathbb{Z}$ such that $1 = as + tn$. By Lemma 1.1.1 (6), $\gcd(a, n)|1$. As $\gcd(a, n) > 0$, we have $\gcd(a, n) = 1$. $\qquad\square$

Remark 1.4.3 Recall that by the extended Euclidean algorithm (Algorithm 1.2), we can find integers s, t such that $\gcd(a, n) = sa + tn$ for any $a, n \in \mathbb{Z}$. In particular, when $\gcd(a, n) = 1$, we can find s, t such that $1 = as + tn$, which gives $as \bmod n = 1$. Thus, we can find $a^{-1} \bmod n = s \bmod n$ by the extended Euclidean algorithm.

Example 1.4.6 We calculated in Example 1.1.15 that $\gcd(160, 21) = 1$ and $1 = (-8) \times 160 + 61 \times 21$. We have $21^{-1} \bmod 160 = 61$.

Example 1.4.7 Let

$$p = 5, \quad q = 7.$$

By the extended Euclidean algorithm,

$$7 = 5 \times 1 + 2, \quad 5 = 2 \times 2 + 1,$$

$$1 = 5 - 2 \times 2 = 5 - (7 - 5) \times 2 = 5 \times 3 - 7 \times 2.$$

We have

$$p^{-1} \bmod q = 5^{-1} \bmod 7 = 3, \quad q^{-1} \bmod p = 7^{-1} \bmod 5 = -2 \bmod 5 = 3.$$

Example 1.4.8 Let

$$p = 7, \quad q = 47.$$

By the extended Euclidean algorithm,

$$47 = 7 \times 6 + 5, \quad 7 = 5 \times 1 + 2, \quad 5 = 2 \times 2 + 1,$$

$$1 = 5 - 2 \times 2 = 5 - (7 - 5) \times 2 = 5 \times 3 - 7 \times 2 = (47 - 7 \times 6) \times 3 - 7 \times 2$$

$$= 47 \times 3 - 7 \times 20.$$

We have

$$p^{-1} \bmod q = 7^{-1} \bmod 47 = -20 \bmod 47 = 27, \quad q^{-1} \bmod p = 47^{-1} \bmod 7 = 3.$$

Corollary 1.4.1 *$\mathbb{Z}_n$ is a field if and only if n is prime.*

Proof By Theorem 1.4.2, $\mathbb{Z}_n$ is a commutative ring. By Definition 1.2.12 and Lemma 1.4.3, $\mathbb{Z}_n$ is a field if and only if for any $a \in \mathbb{Z}_n$ such that $a \neq 0$, we have $\gcd(a, n) = 1$, which is true if and only if n is a prime. $\square$

Corollary 1.4.2 *For any $a \in \mathbb{Z}_n$, if $\gcd(a, n) = 1$, then the set $\{ ab \mid b \in \mathbb{Z}_n \} = \mathbb{Z}_n$.*

Proof It is clear from the definition that $\{ ab \mid b \in \mathbb{Z}_n \} \subseteq \mathbb{Z}_n$. As there are n distinct values for b, it suffices to prove that $ab_1 \not\equiv ab_2 \bmod n$ for $b_1, b_2 \in \mathbb{Z}_n$ with $b_1 \neq b_2$. We will prove the claim by contradiction.

Assume

$$ab_1 \equiv ab_2 \bmod n \tag{1.14}$$

and $b_1 \neq b_2$. By Lemma 1.4.3, a^{-1} exists. Multiply both sides of Eq. 1.14 by a^{-1}, we get $b_1 \equiv b_2 \bmod n$, a contradiction. $\square$

We note that when p is prime, $\mathbb{Z}_p$ is the unique finite field $\mathbb{F}_p$ up to isomorphism (see Theorem 1.2.3 and Remark 1.2.2).

Lemma 1.4.3 leads us to the following definition.

Definition 1.4.5 Let $\mathbb{Z}_n^*$ denote the set of congruence classes in $\mathbb{Z}_n$ which have multiplicative inverses:

$$\mathbb{Z}_n^* := \{\, a \mid a \in \mathbb{Z}_n, \gcd(a, n) = 1 \,\}.$$

Euler's totient function, φ, is a function defined on the set of integers bigger than 1 such that $\varphi(n)$ gives the cardinality of $\mathbb{Z}_n^*$:

$$\varphi(n) = |\mathbb{Z}_n^*|.$$

Example 1.4.9

- Let $n = 3$, $\mathbb{Z}_3^* = \{1, 2\}$, $\varphi(3) = 2$.
- Let $n = 4$, $\mathbb{Z}_4^* = \{1, 3\}$, $\varphi(4) = 2$.
- Let $n = p$ be a prime number, $\mathbb{Z}_p^* = \mathbb{Z}_p - \{0\} = \{1, 2, \ldots, p-1\}$,[3] $\varphi(p) = p-1$.

Lemma 1.4.4 $(\mathbb{Z}_n^*, \cdot)$, *the set $\mathbb{Z}_n^*$ together with the multiplication defined in $\mathbb{Z}_n$ (Eq. 1.13), is an abelian group.*

Proof For any $a, b \in \mathbb{Z}_n^*$, $a^{-1}, b^{-1} \in \mathbb{Z}_n^*$. We note that $(ab)(b^{-1}a^{-1}) = 1$; hence ab has an inverse in $\mathbb{Z}_n^*$ and $ab \in \mathbb{Z}_n^*$ (closure). The associativity follows from that for multiplications in $\mathbb{Z}$. The identity element is 1, and Lemma 1.4.3 proves that every element has an inverse in $\mathbb{Z}_n^*$. $\square$

Recall by the Fundamental Theorem of Arithmetic (Theorem 1.1.5), every integer $n > 1$ is either a prime or can be written as a product of primes in a unique way. We have the following result concerning Euler's totient function. The proof can be found in, e.g., [Sie88, page 247].

Theorem 1.4.3 *For any $n \in \mathbb{Z}$, $n > 1$,*

$$if \quad n = \prod_{i=1}^{k} p_i^{e_i}, \quad then \quad \varphi(n) = n \prod_{i=1}^{k} \left(1 - \frac{1}{p_i}\right), \tag{1.15}$$

where p_i are distinct primes.

---

[3] Recall the difference between sets defined in Eq. 1.1.

Example 1.4.10

- Let $n = 10$. $10 = 2 \times 5$. We can count the elements in $\mathbb{Z}_{10}$ that are coprime to 10 (labeled in red color):

$$\mathbb{Z}_{10} = \{0, 1, 2, 3, 4, 5, 6, 7, 8, 9\}.$$

There are four of them. By Eq. 1.15, we also have

$$\varphi(10) = 10 \times \left(1 - \frac{1}{2}\right) \times \left(1 - \frac{1}{5}\right) = 4.$$

- Let $n = 120$. $120 = 2^3 \times 3 \times 5$. We have

$$\varphi(120) = 120 \times \left(1 - \frac{1}{2}\right) \times \left(1 - \frac{1}{3}\right) \times \left(1 - \frac{1}{5}\right) = 32.$$

- Let $n = pq$, where p and q are two distinct primes. Then

$$\varphi(n) = pq \left(1 - \frac{1}{p}\right) \left(1 - \frac{1}{q}\right) = (p-1)(q-1).$$

- Let $n = p^k$, where p is a prime and $k \in \mathbb{Z}$, $k \geq 1$. Then

$$\varphi(p^k) = p^k \left(1 - \frac{1}{p}\right) = p^{k-1}(p-1).$$

- In particular, if $p = 2$,

$$\varphi(2^k) = 2^{k-1}.$$

Theorem 1.4.4 (Euler's Theorem)   *For any* $a \in \mathbb{Z}$, $a^{\varphi(n)} \equiv 1 \mod n$ *if* $\gcd(a, n) = 1$.

Proof By definition, $|\mathbb{Z}_n^*| = \varphi(n)$. If $\gcd(a, n) = 1$, then $a \in \mathbb{Z}_n^*$. The result follows from Theorem 1.2.1. $\qquad\qquad\qquad\qquad\qquad\qquad\qquad\qquad\qquad\qquad\qquad\quad\square$

Example 1.4.11 Let $n = 4$. We have calculated that $\varphi(4) = 2$ in Example 1.4.9. And

$$3^2 = 9 \equiv 1 \mod 4.$$

Let $n = 10$. We have calculated that $\varphi(10) = 4$ in Example 1.4.10. And

$$3^4 = 81 \equiv 1 \mod 10.$$

Since $\varphi(p) = p - 1$ (Example 1.4.9), a direct corollary of Euler's Theorem is Fermat's Little Theorem.

Theorem 1.4.5 (Fermat's Little Theorem) *Let p be a prime. For any $a \in \mathbb{Z}$, if $p \nmid a$, then $a^{p-1} \equiv 1 \bmod p$.*

Example 1.4.12

- Let $p = 3$. $2^2 = 4 \equiv 1 \bmod 3$.
- Let $p = 5$. $2^4 = 16 \equiv 1 \bmod 5$.

Corollary 1.4.3 *Let p be a prime. Then for any $a, b, c \in \mathbb{Z}$ such that $b \equiv c \bmod (p - 1)$, we have*

$$a^b \equiv a^c \bmod p.$$

In particular,

$$a^b \equiv a^{b \bmod (p-1)} \bmod p.$$

Proof By Fermat's Little Theorem (Theorem 1.4.5),

$$a^{p-1} \equiv \begin{cases} 1 \bmod p & \text{if } p \nmid a \\ 0 \bmod p & \text{otherwise.} \end{cases}$$

Since $b \equiv c \bmod (p - 1)$, $b - c = (p - 1)k$ for some $k \in \mathbb{Z}$. And

$$a^b \equiv a^{c+(p-1)k} \equiv a^c a^{(p-1)k} \equiv \begin{cases} a^c \bmod p & \text{if } p \nmid a \\ 0 \bmod p & \text{otherwise} \end{cases} \equiv a^c \bmod p.$$

$\square$

Example 1.4.13 Let $p = 5$, $a = 2$, $b = 6$. Then

$$2^6 \equiv 2^{6 \bmod 4} \equiv 2^2 \equiv 4 \bmod 5.$$

We can verify that indeed

$$2^6 \equiv 64 \equiv 4 \bmod 5.$$

Corollary 1.4.4 *Let p be a prime and b be an integer coprime to $\varphi(p)$. For any $a_1, a_2 \in \mathbb{Z}_p$, if $a_1 \neq a_2$, then $a_1^b \not\equiv a_2^b \bmod p$.*

Proof Suppose $a_1 \neq a_2$ and $a_1^b \equiv a_2^b \bmod p$. Let $c = b^{-1} \bmod \varphi(p)$, then

$$a_1^{bc} \equiv a_2^{bc} \bmod p, \quad \text{and} \quad bc \equiv 1 \bmod \varphi(p).$$

By Corollary 1.4.3, $a_1 \equiv a_2 \bmod p$. Since $a_1, a_2 \in \mathbb{Z}_p$, we have $a_1 = a_2$, a contradiction. $\qquad\square$

Example 1.4.14 Let $p = 7$, then $\varphi(p) = 6$. Let $a_1 = 3$, $a_2 = 4$, $b = 5$. Then

$$a_1^b \equiv 3^5 \equiv 243 \equiv 5 \bmod 7, \quad a_2^b \equiv 4^5 \equiv 1024 \equiv 2 \bmod 7.$$

This agrees with Corollary 1.4.4. On the other hand, if we let $b = 2$, which is no coprime to $\varphi(p)$, we have

$$a_1^b \equiv 3^2 \equiv 9 \equiv 2 \bmod 7, \quad a_2^b \equiv 4^2 \equiv 16 \equiv 2 \bmod 7.$$

1.4.1 Solving Linear Congruences

In this part, we will discuss how to solve a system of linear congruences in $\mathbb{Z}_n$.

We first consider one linear congruence equation.

Lemma 1.4.5 *For any $a, b \in \mathbb{Z}$, the linear congruence*

$$ax \equiv b \bmod n$$

has at least one solution in $\mathbb{Z}$ if and only if $\gcd(a, n) | b$.

Proof By Definition 1.4.2, the linear congruence is equivalent to the following equation for some $k \in \mathbb{Z}$:

$$ax + kn = b. \tag{1.16}$$

$\Longrightarrow$ By Lemma 1.1.1 (6), $\gcd(a, n) | b$.

$\Longleftarrow$ Assume $\gcd(a, n) | b$, then $\frac{b}{\gcd(a,n)}$ is an integer. By Bézout's identity (Theorem 1.1.3), we can find integers s, t such that $as + tn = \gcd(a, n)$. Multiplying both sides by $\frac{b}{\gcd(a,n)}$, we have

$$a\frac{sb}{\gcd(a, n)} + n\frac{tb}{\gcd(a, n)} = b.$$

Thus $\frac{sb}{\gcd(a,n)}$ is a solution for Eq. 1.16. $\qquad\square$

Example 1.4.15 Let $n = 10$, $a = 4$. Then $\gcd(a, n) = 2$. By Lemma 1.4.5, the linear congruence $4x \equiv 1 \bmod 10$ has no solution. Indeed, if we try to multiply any integer by 4 and divide by 10, we will not get an odd remainder.

On the other hand, the linear congruence $4x \equiv 2 \bmod 10$ has at least one solution. For example, $x = 3$ is a solution ($4 \times 3 \equiv 12 \equiv 2 \bmod 10$).

Theorem 1.4.6 *For any $a, b \in \mathbb{Z}$, the linear congruence*

$$ax \equiv b \bmod n$$

has a unique solution $x \in \mathbb{Z}_n$ if and only if $\gcd(a, n) = 1$

Proof $\Longrightarrow$ Suppose $\gcd(a, n) > 1$, and $x_0 \in \mathbb{Z}_n$ is a solution for the linear congruence. Let $x_1 = x_0 + \frac{n}{\gcd(a,n)}$, then

$$ax_1 \equiv ax_0 + \left(\frac{a}{\gcd(a, n)} \right) n \equiv ax_0 \bmod n.$$

Since $\gcd(a, n) > 1$, $\frac{n}{\gcd(a,n)} \neq 0 \bmod n$, and we have $x_1 \not\equiv x_0 \bmod n$. Thus $x_1 \bmod n$ is another solution in $\mathbb{Z}_n$.

$\Longleftarrow$ Suppose $\gcd(a, n) = 1$. Take any two solutions $x_0, x_1 \in \mathbb{Z}_n$, and we have $ax_1 \equiv ax_0 \bmod n$. Then

$$a(x_0 - x_1) \equiv 0 \bmod n \Longrightarrow n | a(x_0 - x_1).$$

Since $\gcd(n, a) = 1$, $n \nmid a$. By Lemma 1.1.1 (7), $n | (x_0 - x_1)$. As $x_0, x_1 \in \mathbb{Z}_n$, $0 \leq x_0, x_1 < n$, we must have $x_0 - x_1 = 0$. $\qquad\square$

Example 1.4.16

- Let $n = 10$, $a = 3$. $3x \equiv 4 \bmod 10$ has a unique solution $x = 8 \in \mathbb{Z}_{10}$.
- Let $n = 10$, $a = 4$. $4x \equiv 4 \bmod 10$ has two solutions in $\mathbb{Z}_{10}$: $x = 1, 6$.

We now know when there are solutions for a linear congruence and when the solution is unique in $\mathbb{Z}_n$. Next, we will discuss the formulas to find the solution when it is unique. Also, instead of only looking at one equation, the method can find the solution for a few equations, which are called a *system of simultaneous congruences*, at the same time.

Such a problem was mentioned in an ancient Chinese math book called "Sun Zi Suan Jing." The question in the book asks: "There is something whose amount is unknown. If we count by threes, 2 are remaining; by fives, 3 are remaining; and by sevens, 2 are remaining. How many things are there?" Translating to our notations, the question is

$$x \equiv 2 \bmod 3$$

$$x \equiv 3 \bmod 5$$

$$x \equiv 2 \bmod 7$$

$$x \equiv ? \tag{1.17}$$

Before answering the question, we provide the solution for a more general case. Let us consider a system of simultaneous linear congruences

$$x \equiv a_1 \bmod m_1$$

$$x \equiv a_2 \bmod m_2$$

$$\vdots$$

$$x \equiv a_k \bmod m_k, \tag{1.18}$$

where m_i are pairwise coprime positive integers, i.e., $\gcd(m_i, m_j) = 1$ for $i \neq j$.
  Define

$$m = \prod_{i=1}^{k} m_i, \quad M_i = \frac{m}{m_i}, \quad 1 \leq i \leq k. \tag{1.19}$$

Since m_i are pairwise coprime, m_i and M_i are coprime. By Lemma 1.4.3, $y_i :=$
$M_i^{-1} \bmod m_i$ exists. It can be computed by the extended Euclidean algorithm (See
Remark 1.4.3). Let

$$x = \sum_{i=1}^{k} a_i y_i M_i \bmod m. \tag{1.20}$$

Since $y_i = M_i^{-1} \bmod m_i$ and $m_j | M_i$ for $j \neq i$, we have

$$a_i y_i M_i \equiv a_i \bmod m_i, \quad \text{and} \quad a_j y_j M_j \equiv 0 \bmod m_i \text{ if } j \neq i.$$

Then,

$$x \equiv a_i y_i M_i + \sum_{1 \leq j \leq n, j \neq i} a_j y_j M_j \equiv a_i \bmod m_i \quad \text{for all } i.$$

Thus, x is a solution to the system of simultaneous linear congruences in Eq. 1.18.
  Now, we can compute a solution to Eq. 1.17. We have

$$m_1 = 3, \quad m_2 = 5, \quad m_3 = 7, \quad a_1 = 2, \quad a_2 = 3, \quad a_3 = 2,$$

and

$$m = 3 \times 5 \times 7 = 105, \quad M_1 = 35, \quad M_2 = 21, \quad M_3 = 15.$$

By the extended Euclidean algorithm, we get

$$y_1 = M_1^{-1} \bmod 3 = 2^{-1} \bmod 3 = 2,$$

$$y_2 = M_2^{-1} \bmod 5 = 1^{-1} \bmod 5 = 1,$$

$$y_3 = M_3^{-1} \bmod 7 = 1^{-1} \bmod 7 = 1.$$

And a solution to Eq. 1.17 is given by

$$x = \sum_{i=1}^{3} a_i y_i M_i \bmod n = 2 \times 2 \times 35 + 3 \times 1 \times 21 + 2 \times 1 \times 15 \bmod 105$$

$$= 233 \bmod 105 = 23 \bmod 105.$$

Example 1.4.17 Let us solve the following system of simultaneous linear congruences:

$$x \equiv 2 \bmod 5$$

$$x \equiv 1 \bmod 7$$

$$x \equiv 5 \bmod 11$$

$$x \equiv ? \bmod 385.$$

Following the above procedures, we have

$$m_1 = 5, \quad m_2 = 7, \quad m_3 = 11, \quad a_1 = 2, \quad a_2 = 1, \quad a_3 = 5,$$

$$m = 5 \times 7 \times 11 = 385, \quad M_1 = 77, \quad M_2 = 55, \quad M_3 = 35.$$

Then

$$M_1 \equiv 77 \equiv 2 \bmod 5, \quad M_2 \equiv 55 \equiv 6 \bmod 7, \quad M_3 \equiv 35 \equiv 2 \bmod 11.$$

With the extended Euclidean algorithm, we have found

$$y_1 = M_1^{-1} \bmod 5 = 3, \quad y_2 = M_2^{-1} \bmod 7 = 6, \quad y_3 = M_3^{-1} \bmod 11 = 6.$$

And

$$x = \sum_{i=1}^{3} a_i y_i M_i \bmod m = 2 \times 3 \times 77 + 1 \times 6 \times 55 + 5 \times 6 \times 35 \bmod 385$$

$$= 1842 \bmod 385 = 302.$$

We have shown how to find a solution to a system of simultaneous linear congruences. The following theorem says that our solution is unique in $\mathbb{Z}_m$.

Theorem 1.4.7 (Chinese Remainder Theorem) *Let $m_1, m_2, \ldots, m_k$ be pairwise coprime integers. For any $a_1, a_2, \ldots, a_k \in \mathbb{Z}$, the system of simultaneous congruences*

$$x \equiv a_1 \bmod m_1, \quad x \equiv a_2 \bmod m_2, \quad \ldots \quad x \equiv a_k \bmod m_k$$

has a unique solution modulo $m = \prod_{i=1}^{k} m_i$.

Proof The discussions above have shown the existence of such a solution. To prove the uniqueness, let $x_1, x_2 \in \mathbb{Z}_m$ be two solutions for the system of simultaneous congruences. Then

$$x_1 \equiv x_2 \bmod m_1, \quad x_1 \equiv x_2 \bmod m_2, \quad \ldots \quad x_1 \equiv x_2 \bmod m_k.$$

By definition, we have

$$m_1 | (x_1 - x_2), \quad m_2 | (x_1 - x_2), \quad \ldots \quad m_k | (x_1 - x_2).$$

Since m_is are pairwise coprime, by Lemma 1.1.1 (8), we can conclude that $m = \prod_{i=1}^{k} m_i$ divides $x_1 - x_2$. As x_1 and x_2 are from $\mathbb{Z}_m$, we must have $x_1 = x_2$. $\square$

Example 1.4.18 Let $p = 3, q = 5, n = 15, a = 10$. We would like to find the unique solution $x \in \mathbb{Z}_{15}$ such that

$$x \equiv 10 \bmod 3, \quad x \equiv 10 \bmod 5.$$

We have

$$m_1 = p = 3, \quad m_2 = q = 5, \quad a_1 = a_2 = a = 10.$$

Hence

$$m = n = 15, \quad M_1 = 5, \quad M_2 = 3, \quad y_1 = 5^{-1} \bmod 3 = 2, \quad y_2 = 3^{-1} \bmod 5 = 2.$$

And

$$x = a_1 y_1 M_1 + a_2 y_2 M_2 \bmod n = 10 \times 2 \times 5 + 10 \times 2 \times 3 \bmod 15 = 160 \bmod 15 = 10.$$

Example 1.4.19 Take two distinct primes p, q, and let $n = pq$. By Theorem 1.4.7, for any $a \in \mathbb{Z}_n$, there is a unique solution $x \in \mathbb{Z}_n$ such that

$$x \equiv a \bmod p, \quad x \equiv a \bmod q. \tag{1.21}$$

Since $a \equiv a \bmod p$ and $a \equiv a \bmod q$, the unique solution is given by $x = a \in \mathbb{Z}_n$. In other words, there is no other element $b \in \mathbb{Z}_n$ different from a that satisfies Eq. 1.21.

On the other hand, following the above procedures for finding the solution, we have

$$m_1 = p, \quad m_2 = q, \quad a_1 = a_2 = a.$$

And

$$m = n = pq, \quad M_1 = q, \quad M_2 = p, \quad y_1 = q^{-1} \bmod p, \quad y_2 = p^{-1} \bmod q.$$

Then

$$x = a_1 y_1 M_1 + a_2 y_2 M_2 \bmod n = (a(q^{-1} \bmod p)q + a(p^{-1} \bmod q)p) \bmod n$$
$$= (a((q^{-1} \bmod p)q + (p^{-1} \bmod q)p)) \bmod n.$$

By definition,

$$(q^{-1} \bmod p)q = pk_1 + 1, \quad (p^{-1} \bmod q)p = qk_2 + 1,$$

for some integers k_1, k_2. Thus

$$p|((q^{-1} \bmod p)q + (p^{-1} \bmod q)p - 1),$$

and

$$q|((q^{-1} \bmod p)q + (p^{-1} \bmod q)p - 1).$$

By Lemma 1.1.1 (8), we have

$$n|((q^{-1} \bmod p)q + (p^{-1} \bmod q)p - 1) \implies (q^{-1} \bmod p)q + (p^{-1} \bmod q)p \equiv 1 \bmod n.$$

Thus

$$x = (a((q^{-1} \bmod p)q + (p^{-1} \bmod q)p)) \bmod n = a \bmod n.$$

Corollary 1.4.5 *Let p and q be two distinct primes and $n = pq$. For any $a, b \in \mathbb{Z}$, we have*

$$a^b \equiv a^{b \bmod \varphi(n)} \bmod n.$$

Proof Since $\varphi(n) = (p-1)(q-1)$,

$$b \bmod \varphi(n) \equiv b \bmod (p-1), \quad b \bmod \varphi(n) \equiv b \bmod (q-1).$$

By Corollary 1.4.3,

$$a^b \equiv a^{b \bmod \varphi(n)} \bmod p, \quad a^b \equiv a^{b \bmod \varphi(n)} \bmod q.$$

By Example 1.4.19,

$$a^b \equiv a^{b \bmod \varphi(n)} \bmod n.$$

$\square$

Example 1.4.20 Let $p = 3$, $q = 5$, $a = 2$, $b = 9$. Then $n = 15$ and $\varphi(n) = 2 \times 4 = 8$. And

$$2^9 \equiv 2^{9 \bmod 8} \equiv 2 \bmod 15.$$

We can check that

$$2^9 \equiv 512 \equiv 2 \bmod 15.$$

Corollary 1.4.6 *Let p and q be two distinct primes and $n = pq$. For any $a_1, a_2 \in \mathbb{Z}_n$ and $b \in \mathbb{Z}_{\varphi(n)}^*$, if $a_1 \neq a_2$, then $a_1^b \not\equiv a_2^b \bmod n$.*

Proof Suppose $a_1^b \equiv a_2^b \bmod n$. Let $c = b^{-1} \bmod \varphi(n)$, then

$$a_1^{bc} \equiv a_2^{bc} \bmod n, \quad \text{and} \quad bc \equiv 1 \bmod \varphi(n).$$

By Corollary 1.4.5, $a_1 \equiv a_2 \bmod n$. Since $a_1, a_2 \in \mathbb{Z}_n$, we have $a_1 = a_2$, a contradiction. $\square$

Example 1.4.21 Let $p = 5$, $q = 7$, $a_1 = 4$, $a_2 = 6$. Then $n = 35$ and $\varphi(n) = 4 \times 6 = 24$. Choose $b = 5$, and we have

$$a_1^b \equiv 4^5 \equiv 9 \bmod 35, \quad a_2^b \equiv 6^5 \equiv 6 \bmod 35.$$

1.5 Polynomial Rings

In this section, we introduce another example of commutative rings—polynomial rings. Throughout this section, let $(F, +, \cdot)$ be a field with additive identity 0 and multiplicative identity 1.

Definition 1.5.1

- Define

$$F[x] := \left\{ \sum_{i=0}^{n} a_i x^i \,\middle|\, a_i \in F, n \geq 0 \right\}.$$

An element $f(x) = a_n x^n + a_{n-1} x^{n-1} + \cdots + a_1 x + a_0 \in F[x]$ is called a *polynomial over F*.

- If $a_n \neq 0$, we define *degree of $f(x)$*, denoted $\deg(f(x))$, to be n. Following the convention, we define $\deg(0) = -\infty$.

Example 1.5.1 Let $F = \mathbb{R}$, then $f(x) = x + 1 \in \mathbb{R}[x]$ is a polynomial over $\mathbb{R}$ and $\deg(f(x)) = 1$.

Take $f(x) = a_n x^n + a_{n-1} x^{n-1} + \cdots + a_0$, $g(x) = b_m x^m + b_{m-1} x^{m-1} + \cdots + b_0$ from $F[x]$. Without loss of generality, let us assume $n \geq m$. Then we can write $g(x) = b_n x^n + b_{n-1} x^{n-1} + \cdots + b_0$, where $b_i = 0$ for $i > m$. We define addition $+_{F[x]}$ and multiplication $\times_{F[x]}$ as follows:

$$f(x) +_{F[x]} g(x) := c_n x^n + c_{n-1} x^{n-1} + \cdots + c_0, \quad \text{where } c_i = a_i + b_i \qquad (1.22)$$

and

$$f(x) \times_{F[x]} g(x) := d_n x^n + d_{n-1} x^{n-1} + \cdots + d_0, \quad \text{where } d_i = \sum_{j=0}^{i} a_j b_{i-j}. \qquad (1.23)$$

It is easy to show the following proposition.

Proposition 1.5.1 *With the addition $+_{F[x]}$ and multiplication $\times_{F[x]}$ defined in Eqs. 1.22 and 1.23, $(F[x], +_{F[x]}, \times_{F[x]})$ is a commutative ring. It is called the polynomial ring over F.*

The identity element for $+_{F[x]}$ is 0, the additive identity in F. The identity element for $\times_{F[x]}$ is 1, the multiplicative identity in F. The additive inverse of a polynomial

$$f(x) = a_n x^n + a_{n-1} x^{n-1} + \cdots + a_0$$

is given by

$$-f(x) = -a_n x^n - a_{n-1} x^{n-1} - \cdots - a_0,$$

where $-a_i$ is the additive inverse of a_i in F. For simplicity, we will write $f(x)g(x)$ and $f(x) + g(x)$ instead of $f(x) \times_{F[x]} g(x)$ and $f(x) +_{F[x]} g(x)$.

Example 1.5.2 Let $F = \mathbb{R}$, and $\mathbb{R}[x]$ is a ring. The identity element for multiplication is 1. The identity element for addition is 0. Take $f(x) = x + 1$, $g(x) = x$ in $\mathbb{R}[x]$,

$$f(x) + g(x) = 2x + 1, \quad f(x)g(x) = x^2 + x.$$

The additive inverse of $f(x)$ is

$$-x - 1.$$

Lemma 1.5.1 *For any $f(x), g(x) \in F[x]$, such that $f(x) \neq 0, g(x) \neq 0$, we have*

$$\deg(f(x)g(x)) = \deg(f(x)) + \deg(g(x)).$$

Proof Let $m = \deg(f(x))$ and $n = \deg(g(x))$. Then we can write

$$f(x) = \sum_{i=0}^{m} a_i x^i, \ g(x) = \sum_{j=0}^{n} b_j x^j, \ \text{where } a_m \neq 0, \ b_n \neq 0.$$

By Eq. 1.23, $f(x)g(x) = d(x)$, where the highest power of x in $d(x)$ is $m + n$, and its coefficient is $a_m b_n \neq 0$. We have $\deg(d(x)) = m + n$. $\square$

Lemma 1.5.2 *$F[x]$ is an integral domain.*

Proof For any $f(x), g(x) \in F[x]$, such that $f(x) \neq 0, g(x) \neq 0$, we have $\deg(f(x)) \geq 0, \deg(g(x)) \geq 0$. By Lemma 1.5.1, $\deg(f(x)g(x)) \geq 0$, and hence $f(x)g(x) \neq 0$. $\square$

Similar to Euclid's algorithm (Theorem 1.1.2), we have the following theorem. The proof can be found in, e.g., [Her96, page 155].

Theorem 1.5.1 (Division Algorithm) *For any $f(x), g(x) \in F[x]$, of $\deg(f(x)) \geq 1$, there exist $s(x), r(x) \in F[x]$ such that $\deg(r(x)) < \deg(f(x))$ and*

$$g(x) = s(x)f(x) + r(x).$$

$r(x)$ *is called the* remainder, *and $s(x)$ is called the* quotient.

Definition 1.5.2 Let $f(x), g(x) \in F[x]$; if $f(x) \neq 0$ and $g(x) = s(x)f(x)$ for some $s(x) \in F[x]$, then we say $f(x)$ *divides* $g(x)$, written $f(x)|g(x)$.

Example 1.5.3 Let $F = \mathbb{F}_5$. Take $g(x) = 4x^5 + x^3$, $f(x) = x^3 \in \mathbb{F}_5[x]$, then

$$g(x) = f(x)(4x^2 + 1),$$

and $f(x)|g(x)$.

Definition 1.5.3 A polynomial $f(x) \in F[x]$ of positive degree is said to be *reducible (over F)* if there exist $g(x), h(x) \in F[x]$ such that

$$\deg(g(x)) < \deg(f(x)), \ \deg(h(x)) < \deg(f(x)), \ \text{and } f(x) = g(x)h(x).$$

Otherwise, it is said to be *irreducible (over F)*.

It is easy to show the following lemma from the above definitions.

Lemma 1.5.3 *A polynomial $f(x) \in F[x]$ of degree n is reducible over F if and only if it is divisible by an irreducible polynomial of degree at most $\lfloor n/2 \rfloor$.*

Remark 1.5.1

- $f(x) \in F[x]$ of degree 2 or 3 is reducible over F if and only if it has a root in F.[4]
- Let $f(x) = \sum_{i=0}^{n} a_i x^i \in F[x]$. Then $f(0) = a_0$. Thus $f(x)$ is reducible if $a_0 = 0$.
- Let $f(x) = \sum_{i=0}^{n} a_i x^i \in \mathbb{F}_2[x]$. Then $f(1) = \sum_{i=0}^{n} a_i$. If $|\{a_i \mid a_i \neq 0\}|$ is even, then $f(1) = 0$ and $f(x)$ is reducible over $\mathbb{F}_2$. In other words, any $f(x) \in \mathbb{F}_2[x]$ with an even number of nonzero terms is reducible over $\mathbb{F}_2$.

Example 1.5.4

- $h(x) = 4x^5 + x^3 \in \mathbb{F}_3[x]$ has degree 5, and it is reducible since $h(x) = x^3(4x^2 + 1)$.
- $g(x) = x^2 \in \mathbb{F}_2[x]$ has degree 2, and it is reducible, $g(x) = x \cdot x$.

Example 1.5.5 Let $F = \mathbb{F}_2$.

- All the polynomials of degree 2 are $x^2, x^2+1, x^2+x+1, x^2+x$. By Remark 1.5.1, the only irreducible polynomial of degree 2 is $x^2 + x + 1$.
- All the degree 3 polynomials with an odd number of nonzero terms are $x^3, x^3 + x + 1, x^3 + x^2 + 1, x^3 + x^2 + x$. Among those, the polynomials with $a_0 \neq 0$ are the irreducible polynomials of degree 3:

$$x^3 + x + 1, \; x^3 + x^2 + 1.$$

- Degree 4 polynomials with $a_0 \neq 0$ and an odd number of nonzero terms are

$$x^4 + x + 1, \; x^4 + x^2 + 1, \; x^4 + x^3 + 1, \; x^4 + x^3 + x^2 + x + 1.$$

By our choice, they are not divisible by degree 1 polynomials. By Lemma 1.5.3, any of them is reducible if and only if it is divisible by $x^2 + x + 1$, which can be verified using the Division Algorithm (Theorem 1.5.1). For example,

$$x^4 + x + 1 = x^2(x^2 + x + 1) + (x^3 + x + x^2 + 1)$$

is not divisible by $x^2 + x + 1$. And

$$x^4 + x^2 + 1 = (x^2 + x + 1)(x^2 + x + 1)$$

is divisible by $x^2 + x + 1$.

Finally, we have all the degree 4 irreducible polynomials over $\mathbb{F}_2$:

$$x^4 + x + 1, \; x^4 + x^3 + 1, \; x^4 + x^3 + x^2 + x + 1.$$

---

[4] An element $a \in F$ is a *root* of $f(x)$ if $f(a) = 0$.

We note that there are many analogies between a polynomial ring $F[x]$ and the ring of integers $\mathbb{Z}$. For example, a polynomial $f(x)$ corresponds to an integer n. An irreducible polynomial $p(x)$ corresponds to a prime p.

For the rest of the section, let us fix a polynomial $f(x) \in F[x]$ such that $f(x) \neq 0$. The Same as in Eq. 1.11, we define a relation $\sim$ on $F[x]$ as follows:

$$g(x) \sim h(x) \text{ if } f(x) \mid (g(x) - h(x)).$$

We have shown that the relation in Eq. 1.11 is an equivalence relation on $\mathbb{Z}$, and a similar proof shows that $\sim$ is an equivalence relation on $F[x]$. We can also define congruence in $F[x]$ (cf. Definition 1.4.2).

Definition 1.5.4 For any $g(x), h(x) \in F[x]$, if $g(x) \sim h(x)$, i.e., $f(x)|(g(x) - h(x))$, we say $h(x)$ *is congruent to* $g(x)$ *modulo* $f(x)$, written $g(x) \equiv h(x) \bmod f(x)$.

The congruence class of $g(x)$ modulo $f(x)$ is given by

$$\{ h(x) | h(x) \equiv g(x) \bmod f(x) \}.$$

Similar proofs for Lemma 1.4.1 can be applied to prove the following lemma.

Lemma 1.5.4 *Suppose $f(x)$ has degree n, where $n \geq 1$. Let $F[x]/(f(x))$ denote the set of all congruence classes of $g(x) \in F[x]$ modulo $f(x)$. Then*

$$F[x]/(f(x)) = \left\{ \sum_{i=0}^{n-1} a_i x^i \,\middle|\, a_i \in F \text{ for } 0 \leq i < n \right\}$$

can be identified with the set of all polynomials of degree less than n.

Example 1.5.6 Let $f(x) = x^2 + x + 1 \in \mathbb{F}_2[x]$. By Lemma 1.5.4,

$$\mathbb{F}_2[x]/(f(x)) = \{ 1, x, x + 1 \}.$$

Similarly, let $g(x) = x^2 \in \mathbb{F}_2[x]$. Then

$$\mathbb{F}_2[x]/(g(x)) = \{ 1, x, x + 1 \}.$$

We can see that $\mathbb{F}_2[x]/(f(x))$ and $\mathbb{F}_2[x]/(g(x))$ contain equivalent classes generated by the same polynomials.

Naturally, for any $g(x), h(x) \in F[x]/(f(x))$, the same as in Eqs. 1.12 and 1.13, addition and multiplication in $F[x]/(f(x))$ are computed modulo $f(x)$.

Example 1.5.7 Let $f(x) \in \mathbb{F}_2[x]$ be a polynomial of degree n. For any

$$g(x) = \sum_{i=0}^{n-1} a_i x^i, \quad h(x) = \sum_{i=0}^{n-1} b_i x^i$$

from $\mathbb{F}_2[x]/(f(x))$, we have

$$g(x) + h(x) \bmod f(x) = \sum_{i=0}^{n-1} c_i x^i, \quad \text{where} \quad c_i = a_i + b_i \bmod 2.$$

Thus the addition computations in $\mathbb{F}_2[x]/(f(x))$ are the same for all $f(x)$ of the same degree.

Example 1.5.8 Let $F = \mathbb{F}_2$, $f(x) = x^2+x+1 \in \mathbb{F}_2[x]$, $g(x) = x \in \mathbb{F}_2[x]/(f(x))$, and $h(x) = x \in \mathbb{F}_2[x]/(f(x))$. We have

$$g(x) + h(x) \bmod f(x) = x + x \bmod f(x) = 0,$$
$$g(x)h(x) \bmod f(x) = x^2 \bmod f(x) = x + 1.$$

Example 1.5.9 Let $f(x) = x^2 + x + 1$, $g(x) = x^2 \in \mathbb{F}_2[x]$. The addition and multiplication computations in $\mathbb{F}_2[x]/(f(x))$ and $\mathbb{F}_2[x]/(g(x))$ are shown in Tables 1.2 and 1.3, respectively. We note that the addition computations for $\mathbb{F}_2[x]/(f(x))$ and $\mathbb{F}_2[x]/(g(x))$ are the same as discussed in Example 1.5.7.

We also have the notion of the *greatest common divisors* between two nonzero polynomials in $F[x]$ (cf. Definition 1.1.5). Then, for any $g(x) \in F[x]$, the modified version of the Euclidean algorithm (Algorithm 1.1) can be applied to find the greatest common divisor for $g(x)$ and $f(x)$, denoted $\gcd(g(x), f(x))$. Similarly the extended Euclidean algorithm (Algorithm 1.2) can be applied to find the inverse of $g(x)$ modulo $f(x)$ when $\gcd(f(x), g(x)) = 1$. More details are presented in [LX04, Section 3.2].

Example 1.5.10 Let $F = \mathbb{F}_2$ and $f(x) = x^2 + x + 1$, $g(x) = x \in \mathbb{F}_2[x]$. By the Euclidean algorithm, we have

$$f(x) = (x + 1)g(x) + 1, \quad \gcd(g(x), f(x)) = \gcd(g(x), 1) = 1.$$

Table 1.2 Addition and multiplication in $\mathbb{F}_2[x]/(f(x))$, where $f(x) = x^2 + x + 1$

$+$	0	1	x	$x+1$
0	0	1	x	$x+1$
1	1	0	$x+1$	x
x	x	$x+1$	0	1
$x+1$	$x+1$	x	1	0

$\times$	0	1	x	$x+1$
0	0	0	0	0
1	0	1	x	$x+1$
x	0	x	$x+1$	1
$x+1$	0	$x+1$	1	x

Table 1.3 Addition and multiplication in $\mathbb{F}_2[x]/(g(x))$, where $g(x) = x^2$

$+$	0	1	x	$x+1$
0	0	1	x	$x+1$
1	1	0	$x+1$	x
x	x	$x+1$	0	1
$x+1$	$x+1$	x	1	0

$\times$	0	1	x	$x+1$
0	0	0	0	0
1	0	1	x	$x+1$
x	0	x	0	x
$x+1$	0	$x+1$	x	1

By the extended Euclidean algorithm,

$$1 = g(x)(x+1) + f(x).$$

We have $g(x)^{-1} \bmod f(x) = x + 1$.

Example 1.5.11 Let $F = \mathbb{F}_2$ and $f(x) = x^2 + x + 1, g(x) = x^2 \in \mathbb{F}_2[x]$. By the Euclidean algorithm, we have

$$f(x) = g(x) + (x+1), \qquad \gcd(g(x), f(x)) = \gcd(g(x), (x+1)),$$
$$g(x) = (x+1)(x+1) + 1, \ \gcd(g(x), (x+1)) = 1.$$

By the extended Euclidean algorithm,

$$1 = g(x) + (x+1)(x+1) = g(x) + (x+1)(f(x)+g(x)) = g(x)x + (x+1)f(x).$$

And $g(x)^{-1} \bmod f(x) = x$.

Similar proofs for Theorem 1.4.2 and Corollary 1.4.1 can be applied to show the following theorem.

Theorem 1.5.2 *Together with addition and multiplication modulo $f(x)$, $F[x]/(f(x))$ is a commutative ring. It is a field if and only if $f(x)$ is irreducible.*

Example 1.5.12 Let $F = \mathbb{R}$. By Remark 1.5.1, $f(x) = x^2 + 1$ is irreducible over $\mathbb{R}$. By Theorem 1.5.2, $\mathbb{R}/(f(x))$ is a field. By Lemma 1.5.4,

$$\mathbb{R}/(f(x)) = \{a + bx \mid a, b \in \mathbb{R}\}.$$

Recall that

$$\mathbb{C} = \{a + bi \mid a, b \in \mathbb{R}\}.$$

It is easy to see that $\mathbb{R}/(f(x)) \cong \mathbb{C}$ by mapping x to i (see Definition 1.2.16).

Example 1.5.13 In Examples 1.5.4 and 1.5.5, we have shown that $g(x) = x^2$ is reducible and $f(x) = x^2 + x + 1$ is irreducible over $\mathbb{F}_2$.

By Theorem 1.5.2, $\mathbb{F}_2/(g(x))$ is not a field and $\mathbb{F}_2/(f(x))$ is a field. Indeed, in Examples 1.5.6 and 1.5.9, we have seen that even though $\mathbb{F}_2[x]/(f(x))$ and

$\mathbb{F}_2[x]/(g(x))$ contain equivalent classes generated by the same elements, the multiplication computations are different in those two rings. In particular, x is a zero divisor in $\mathbb{F}_2/(g(x))$ (Table 1.3) but has inverse $x + 1$ in $\mathbb{F}_2/(f(x))$ (Table 1.2).

We have discussed that there is only one finite field up to isomorphism (Theorem 1.2.3). The following theorem specifies the field structures for $F[x]/(f(x))$ when $F = \mathbb{F}_p$, where p is a prime.

Theorem 1.5.3 *Let p be a prime, and let $f(x) \in \mathbb{F}_p[x]$ be an irreducible polynomial of* $\deg(f(x)) = n$. *Then* $\mathbb{F}_p[x]/(f(x)) \cong \mathbb{F}_{p^n}$.

Proof By Lemma 1.5.4,

$$\mathbb{F}_p[x]/(f(x)) = \left\{ \sum_{i=0}^{n-1} a_i x^i \;\middle|\; a_i \in \mathbb{F}_p \text{ for } 0 \le i < n \right\}.$$

There are p choices for each of the n a_is. Hence the cardinality of $\mathbb{F}_p[x]/(f(x))$ is p^n. The result follows from Theorem 1.2.3. $\qquad\qquad\square$

Example 1.5.14 Let $f(x) = x^2 + x + 1 \in \mathbb{F}_2[x]$; by Theorem 1.5.3, $\mathbb{F}_2[x]/(f(x)) \cong \mathbb{F}_{2^2}$.

1.5.1 Bytes

Throughout this subsection, let $f(x) = x^8 + x^4 + x^3 + x + 1 \in \mathbb{F}_2[x]$.

It can be shown that $f(x)$ is irreducible over $\mathbb{F}_2$ using Lemma 1.5.3 and Example 1.5.5. Then by Lemma 1.5.4,

$$\mathbb{F}_2[x]/(f(x)) = \left\{ \sum_{i=0}^{7} b_i x^i \;\middle|\; b_i \in \mathbb{F}_2 \; \forall i \right\}.$$

By Theorem 1.5.3, $\mathbb{F}_2[x]/(f(x)) \cong \mathbb{F}_{2^8}$.

We note that any

$$b_7 x^7 + b_6 x^6 + b_5 x^5 + b_4 x^4 + b_3 x^3 + b_2 x^2 + b_1 x + b_0 \in \mathbb{F}_2[x]/(f(x))$$

can be stored as a byte $b_7 b_6 b_5 b_4 b_3 b_2 b_1 b_0 \in \mathbb{F}_2^8$ (see Definition 1.3.7), which represents an integer between 0 (00_{16}) and 255 (FF_{16}) (see Remark 1.3.3). There are 256 different values for a byte, and $|\mathbb{F}_{2^8}| = 2^8 = 256$. Then φ defined as follows

$$\varphi : \mathbb{F}_2[x]/(f(x)) \to \mathbb{F}_2^8$$

$$b_7 x^7 + b_6 x^6 + b_5 x^5 + b_4 x^4 + b_3 x^3 + b_2 x^2 + b_1 x + b_0 \mapsto b_7 b_6 b_5 b_4 b_3 b_2 b_1 b_0$$

is a bijective function. Thus, with addition and multiplication modulo $f(x)$ in $\mathbb{F}_2[x]/(f(x))$, we can define the corresponding addition and multiplication between bytes.

Definition 1.5.5 For any two bytes $v = v_7 v_6 \ldots v_1 v_0$ and $w = w_7 w_6 \ldots w_1 w_0$, let $g_v(x) = v_7 x^7 + v_6 x^6 + \cdots + v_1 x + v_0$ and $g_w(x) = w_7 x^7 + w_6 x^6 + \cdots + w_1 x + w_0$ be the corresponding polynomials in $\mathbb{F}_2[x]/(f(x))$. We define

$$v + w = g_v(x) + g_w(x) \bmod f(x), \quad v \times w = g_v(x) g_w(x) \bmod f(x).$$

In particular, by Example 1.5.7,

$$v + w = c_7 c_6 \ldots c_1 c_0, \quad \text{where } c_i = v_i + w_i \bmod 2.$$

Remark 1.5.2 Recall that a byte is also a vector in $\mathbb{F}_2^8$, we have defined vector addition as bitwise XOR (see Definition 1.3.6), and

$$v +_{\mathbb{F}_2^8} w = u_7 u_6 \ldots u_1 u_0, \quad \text{where } u_i = v_i \oplus w_i.$$

We note that $a + b \bmod 2 = a \oplus b$ for $a, b \in \mathbb{F}_2$. Thus, our definition of addition between two bytes (Definition 1.5.5) agrees with the vector addition between two vectors in $\mathbb{F}_2^8$.

Example 1.5.15 Take $x^6 + x^4 + x^2 + x + 1 \in \mathbb{F}_2[x]/(f(x))$, which corresponds to $01010111_2 = 57_{16}$. And $x^7 + x + 1 \in \mathbb{F}_2[x]/(f(x))$, which corresponds to $10000011_2 = 83_{16}$. We have

$$\begin{aligned}
57_{16} + 83_{16} &= (x^6 + x^4 + x^2 + x + 1) + (x^7 + x + 1) \bmod f(x) \\
&= x^7 + x^6 + x^4 + x^2 \bmod f(x) = 11010100_2 = D4_{16}.
\end{aligned}$$

We note that

$$01010111_2 \oplus 10000011_2 = 11010100_2.$$

For multiplication, we compute

$$(x^6 + x^4 + x^2 + x + 1)(x^7 + x + 1) = x^{13} + x^{11} + x^9 + x^8 + x^6 + x^5 + x^4 + x^3 + 1,$$

and

$$\begin{aligned}
x^8 &= x^4 + x^3 + x + 1 \bmod f(x), \\
x^9 &= x^5 + x^4 + x^2 + x \bmod f(x), \\
x^{11} &= x^7 + x^6 + x^4 + x^3 \bmod f(x),
\end{aligned}$$

$$x^{13} = x^9 + x^8 + x^6 + x^5 \bmod f(x).$$

Thus

$$x^{13}+x^{11}+x^9+x^8+x^6+x^5+x^4+x^3+1 = x^{11}+x^4+x^3+1 = x^7+x^6+1 \bmod f(x),$$

which gives

$$57_{16} \times 83_{16} = 11000001_2 = C1_{16}.$$

Example 1.5.16 In this example, we would like to compute the formula for a byte multiplied by $02_{16} = x$. Take any $g(x) = b_7 x^7 + b_6 x^6 + \cdots + b_1 x + b_0 \in \mathbb{F}_2[x]/(f(x))$,

$$g(x)x \bmod f(x)$$
$$= (b_7 x^7 + b_6 x^6 + b_5 x^5 + b_4 x^4 + b_3 x^3 + b_2 x^2 + b_1 x + b_0)x \bmod f(x)$$
$$= b_7 x^8 + b_6 x^7 + b_5 x^6 + b_4 x^5 + b_3 x^4 + b_2 x^3 + b_1 x^2 + b_0 x \bmod f(x)$$
$$= b_6 x^7 + b_5 x^6 + b_4 x^5 + b_3 x^4 + b_2 x^3 + b_1 x^2 + b_0 x + b_7 x^4 + b_7 x^3 + b_7 x$$
$$+ b_7 \bmod f(x)$$
$$= b_6 x^7 + b_5 x^6 + b_4 x^5 + (b_3 + b_7)x^4 + (b_2 + b_7)x^3 + b_1 x^2 + (b_0 + b_7)x$$
$$+ b_7 \bmod f(x).$$

Thus, for any byte $b_7 b_6 \ldots b_1 b_0$, multiplication by 02_{16} is equivalent to left shift by 1 and XOR with $00011011_2 = 1B_{16}$ if $b_7 = 1$.

Example 1.5.17

- $57_{16} = 01010111_2$, $02_{16} \times 57_{16} = 10101110 = AE_{16}$.
- $83_{16} = 10000011_2$, $02_{16} \times 83_{16} = 00000110_2 \oplus 00011011_2 = 00011101_2 = 1D_{16}$.
- $D4_{16} = 11010100_2$, $02_{16} \times D4_{16} = 10101000_2 \oplus 00011011_2 = 10110011_2 = B3_{16}$.

Example 1.5.18 Now, let us compute the multiplication of a byte by $03_{16} = x + 1$. Take any $h(x) = b_7 x^7 + b_6 x^6 + \cdots + b_1 x + b_0 \in \mathbb{F}_2[x]/(f(x))$,

$$h(x)(x + 1) \bmod f(x) = h(x)x + h(x) \bmod f(x).$$

Thus, for any byte $b_7 b_6 \ldots b_1 b_0$, multiplication by 03_{16} is equivalent to first multiplying by 02_{16} (left shift by 1 and XOR with $00011011_2 = 1B_{16}$ if $b_7 = 1$) and then XOR with the byte itself ($b_7 b_6 \ldots b_1 b_0$).

Example 1.5.19 Continuing Example 1.5.17,

- $03_{16} \times 57_{16} = AE_{16} \oplus 57_{16} = F9_{16}.$
- $03_{16} \times 83_{16} = 1D_{16} \oplus 83_{16} = 9E_{16}.$
- $03_{16} \times D4_{16} = B3_{16} \oplus D4_{16} = 67_{16}.$

Example 1.5.20 $03_{16} \times BF_{16} = 01111110_2 \oplus 00011011_2 \oplus 10111111_2 = 11011010_2 = DA_{16}.$

We can also compute the inverse of elements in $\mathbb{F}_2[x]/(f(x))$ using the extended Euclidean algorithm (Algorithm 1.2) as in Example 1.5.10, thus, enabling us to find the inverse of a byte as an element in $\mathbb{F}_2[x]/(f(x))$.

Example 1.5.21 $03_{16} = 00000011_2 = x + 1$. By the Euclidean algorithm (Algorithm 1.1),

$$f(x) = (x + 1)(x^7 + x^6 + x^5 + x^4 + x^2 + x) + 1 \implies \gcd(f(x), (x + 1)) = 1.$$

See also Appendix B for the computation.
By the extended Euclidean algorithm,

$$1 = f(x) - (x + 1)(x^7 + x^6 + x^5 + x^4 + x^2 + x).$$

We have

$$03_{16}^{-1} = (x + 1)^{-1} \bmod f(x) = x^7 + x^6 + x^5 + x^4 + x^2 + x = 11110110_2 = F6_{16}.$$

1.6 Coding Theory

In this section, we give a brief discussion on binary codes, which will be useful for the design of countermeasures against side-channel attacks (Sect. 4.5.1.1) and fault attacks (Sect. 5.2.1).

Let n be a positive integer in the rest of this section. To study binary codes, we look at the vector space $\mathbb{F}_2^n$, and we refer to vectors in $\mathbb{F}_2^n$ as *words* of length n.

Definition 1.6.1

- $w = w_0 w_1 \ldots w_{n-1} \in \mathbb{F}_2^n$ is called a *binary word of length n*.
- A nonempty set $C \subset \mathbb{F}_2^n$ is called a *binary code* of *length n*.
- An element of a binary code C is called a *codeword* of C.
- Cardinality of C is called the *size* of C.
- A code of length n and size M is called a *binary (n, M)-code*.

Example 1.6.1

- $C = \{00, 11\}$ is a binary $(2, 2)$-code.
- $C = \{010, 001, 110, 111\}$ is a binary $(3, 4)$-code.

Definition 1.6.2 For any v, $u \in \mathbb{F}_2^n$, the *Hamming distance* between v and u, denoted dis (v, u), is defined as follows:

$$\text{dis}\,(v, u) = \sum_{i=0}^{n-1} \text{dis}\,(v_i, u_i)\,, \quad \text{where dis}\,(v_i, u_i) = \begin{cases} 1 & \text{if } v_i \neq u_i \\ 0 & \text{if } v_i = u_i. \end{cases} \tag{1.24}$$

Example 1.6.2 dis $(001, 111) = 2$. dis $(00000, 10101) = 3$.

Lemma 1.6.1 *For any* v, u, $w \in \mathbb{F}_2^n$, *we have*

1. $0 \leq \text{dis}\,(v, u) \leq n$.
2. dis $(v, u) = 0$ *if and only if* $v = u$.
3. dis $(v, u) = \text{dis}\,(u, v)$.
4. dis $(v, w) \leq \text{dis}\,(v, u) + \text{dis}\,(u, w)$ *(triangle inequality)*.

Proof (1)–(3) are easy to see. We provide the proof for (4). By Eq. 1.24, it suffices to consider $n = 1$. Take any $v, w, u \in \mathbb{F}_2$.
    If $v = w$,

$$\text{dis}\,(w, w) = 0 \leq \text{dis}\,(v, u) + \text{dis}\,(u, w)\,.$$

If $v \neq w$, dis $(v, w) = 1$, and dis $(v, u) = 1$ or dis $(u, w) = 1$.   □

Definition 1.6.3 Let $C \subset \mathbb{F}_2^n$ be a binary code containing at least two codewords, and the *(minimum) distance*, denoted dis (C), is given by

$$\text{dis}\,(C) = \min\{\,\text{dis}\,(c_1, c_2) \mid c_1, c_2 \in C, c_1 \neq c_2\,\}\,.$$

Definition 1.6.4 A binary code of length n, size M, and distance d is called a *binary* (n, M, d)*-code*.

Example 1.6.3 Let $C = \{\,0011, 1101, 1000\,\}$; we can calculate that

$$\text{dis}\,(0011, 1101) = 3, \quad \text{dis}\,(0011, 1000) = 3, \quad \text{dis}\,(1101, 1000) = 2.$$

Thus C is a binary $(4, 3, 2)$-code

Recall that when the value of a bit is changed we say that the bit is flipped (Definition 1.2.17).

Definition 1.6.5 A binary code C is said to be *k-error detecting* for a positive integer k if for any $c \in C$, whenever at least 1 but at most k bits of c are flipped, the resulting word is not a codeword in C. If C is k-error detecting but not $(k+1)$−error detecting, then we say C is *exactly k-error detecting*.

Example 1.6.4 Let $C = \{\,0011, 1101, 1000\,\}$. Since

$$\text{dis}\,(0011, 1101) = \text{dis}\,(0011, 1000) = 3, \quad \text{dis}\,(1101, 1000) = 2,$$

with 1-bit flip from any codeword, we cannot get another codeword. But with 2-bit flips, we can change 1101 to 1000. Thus C is exactly 1-error detecting.

Theorem 1.6.1 *A binary (n, M, d)-code C is k-error detecting if and only if $d \geq k + 1$, i.e., C is an exactly $(d - 1)$-error detecting code.*

Proof $\Longleftarrow$ If $d \geq k + 1$, take $c \in C$ and $x \in \mathbb{F}_2^n$ such that $1 \leq \mathrm{dis}\,(c, x) \leq k$. Then $x \notin C$, and C is k-error detecting.

$\Longrightarrow$ If $d < k + 1$, take $c_1, c_2 \in C$ such that $\mathrm{dis}\,(c_1, c_2) = d$. Flipping d bits of c_1 we can get $c_2 \in C$. Hence C is not k-error detecting. $\square$

Let us consider binary (n, M, d)-codes with $M = 2^k$ for some positive integer k. When a binary code is used for transmitting information, every information word $u \in \mathbb{F}_2^k$ is assigned a unique codeword $c(u) \in C$. We say that u is *encoded* as $c(u)$. Suppose Alice would like to send information u to Bob using C. Alice sends codeword $c(u)$ to Bob. Due to transmission noise, Bob might receive a word $x \in \mathbb{F}_2^n$ not equal to $c(u)$. Thus we need to define a *decoding rule* for Bob that allows him to find u given x.

We are interested in a *minimum distance decoding rule*, which specifies that after receiving x, Bob computes

$$c_x = \arg\min_{c} \{\, \mathrm{dis}\,(x, c) \mid c \in C \,\}, \quad \text{i.e.,} \quad \mathrm{dis}\,(c_x, x) = \min_{c} \{\, \mathrm{dis}\,(x, c) \mid c \in C \,\}.$$

If more than one codeword is identified as c_x, there are two options. An *incomplete decoding rule* says that Bob should request Alice for another transmission. Following a *complete decoding rule*, Bob would then randomly select one codeword.

Example 1.6.5 Let $C = \{\, 0000, 0111, 1110, 1111 \,\}$. We use C to encode information words $u \in \mathbb{F}_2^2$ with encoding designed as follows:

$$c(00) = 0000, \quad c(01) = 0111, \quad c(10) = 1110, \quad c(11) = 1111.$$

Suppose Alice was sending information 00 with codeword 0000 to Bob. Due to an error during the transmission, Bob received 0001. By the minimum distance decoding rule, Bob computes the distances between 0001 and codewords in C.

$$\mathrm{dis}\,(0001, 0000) = 1, \quad \mathrm{dis}\,(0001, 0111) = 2, \quad \mathrm{dis}\,(0001, 1110) = 4,$$

$$\mathrm{dis}\,(0001, 1111) = 3.$$

Thus $c_{0001} = 0000$, and Bob gets the correct information 00.

Definition 1.6.6 A binary code C is said to be *k-error correcting* if the minimum distance decoding outputs the correct codeword when k or fewer bits are flipped. If C is k-error correcting but not $k + 1$-error correcting, then we say that C is *exactly k-error correcting*.

Example 1.6.6 Let $C = \{000, 111\}$.

- If 000 was sent and 1 bit flip occurred, the received word $\{001, 010, 100\}$ will be decoded to 000.
- If 111 was sent and 1 bit flip occurred, the received word $\{110, 011, 101\}$ will be decoded to 111.
- If 000 was sent and 011 was received, the decoding result will be 111.

Thus C is exactly 1-error correcting.

Theorem 1.6.2 *A binary* (n, M, d)*-code* C *is* k*-error correcting if and only if* $d \geq 2k + 1$*, i.e.,* C *is an exactly* $\lfloor (d - 1)/2 \rfloor$*-error correcting code.*

Proof $\Longleftarrow$ We assume $d \geq 2k + 1$. Suppose c was sent, v was received, and k or fewer bit flips occurred, i.e., dis $(c, v) \leq k$. For any codeword $c' \in C$ different from c,

$$\mathrm{dis}\left(v, c'\right) \geq \mathrm{dis}\left(c, c'\right) - \mathrm{dis}\left(v, c\right) \geq 2k + 1 - k = k + 1 > \mathrm{dis}\left(v, c\right).$$

Thus C is k-error correcting.

$\Longrightarrow$ Now suppose C is k-error correcting and $d < 2k + 1$. Take $c, c' \in C$ such that dis $\left(c, c'\right) = d$. By definition, C is also k-error detecting. By Theorem 1.6.1, dis $\left(c, c'\right) = d \geq k + 1$. Without loss of generality, assume c and c' differ in the first d bits.

Define $v \in \mathbb{F}_2^n$ as

$$v_i = \begin{cases} c_i & 0 \leq i < k \\ c_i' & k \leq i < d \\ c_i = c_i' & k \geq d. \end{cases}$$

Then

$$\mathrm{dis}\left(v, c'\right) = d - k \leq k = \mathrm{dis}\left(v, c\right).$$

If c is sent and v is received, the minimum distance decoding cannot uniquely decode v to c. $\qquad\square$

Definition 1.6.7 Let $C \subseteq \mathbb{F}_2^n$ be a binary code. C is said to be *linear* if it is a vector space over $\mathbb{F}_2$. Otherwise, it is said to be *nonlinear*.

In other words, a binary linear code C is a subspace of $\mathbb{F}_2^n$ (see Definitions 1.3.5 and 1.3.8).

Remark 1.6.1 By Remark 1.3.4, to show a binary code C is linear, we need to prove that $\mathbf{0} \in C$ and for any $c, c' \in C, c + c' \in C$.

We have defined dimensions for vector spaces in Definition 1.3.13.

Definition 1.6.8 The *dimension* of a binary linear code C is given by $\dim(C)_{\mathbb{F}_2}$, the dimension of C as a vector space over $\mathbb{F}_2$. A binary linear code C of length n and dimension k is called a *binary $[n, k]$-linear code*. If C has distance d, then it is called a *binary $[n, k, d]$-linear code*.

By Lemma 1.3.3, we can calculate the size of a linear code C using its dimension, $|C| = 2^{\dim(C)_{\mathbb{F}_2}}$. Thus a binary $[n, k]$-linear code is also a binary $(n, 2^k)$-code (see Definition 1.6.1).

Example 1.6.7

- Let $C = \{00, 11, 01, 10\} = \mathbb{F}_2^2$, then $\dim(C)_{\mathbb{F}_2} = 2$, and C is a binary $[2, 2, 1]$-linear code.
- Let $C = \langle 111 \rangle = \{000, 111\}$, then $\{111\}$ is a basis for C, and $\dim(C)_{\mathbb{F}_2} = 1$. C is a binary $[3, 1, 3]$-linear code.

Example 1.6.8 (Repetition code) Let

$$C = \langle 11 \ldots 11 \rangle = \{00 \ldots 00, 11 \ldots 11\} \subseteq \mathbb{F}_2^n.$$

Then $\{11 \ldots 11\}$ is a basis for C, and C is a binary $[n, 1, n]$-linear code. C is called the *binary n-repetition code*. By Theorems 1.6.1 and 1.6.2, C is exactly $(n-1)$-error detecting and exactly $\lfloor (n - 1)/2 \rfloor$-error correcting.

Example 1.6.9 (Parity-check code) Suppose we would like to encode information words

$$\boldsymbol{u} = (u_0, u_1, \ldots, u_{n-2}) \in \mathbb{F}_2^{n-1}.$$

We add one parity-check bit and encode $\boldsymbol{u}$ using

$$\boldsymbol{c} = (u_0, u_1, \ldots, u_{n-2}, c_{n-1}), \quad \text{where } c_{n-1} = \sum_{i=0}^{n-2} u_i.$$

The corresponding code C consists of codewords that have an even number of 1s.

$$C = \left\{ (c_0, c_1, \ldots, c_{n-2}, c_{n-1}) \;\middle|\; c_{n-1} = \sum_{i=0}^{n-2} c_i \right\} \subseteq \mathbb{F}_2^n. \tag{1.25}$$

It is easy to see that $\boldsymbol{0} \in C$. Take $\boldsymbol{v} = (v_0, v_1, \ldots, v_{n-1})$, $\boldsymbol{w} = (w_0, w_1, \ldots, w_{n-1})$ from C, then

$$\boldsymbol{v} + \boldsymbol{w} = (v_0 + w_0, v_1 + w_1, \ldots, v_{n-1} + w_{n-1}), \quad v_{n-1} + w_{n-1} = \sum_{i=0}^{n-2} v_i + \sum_{i=0}^{n-2} w_i$$

$$= \sum_{i=0}^{n-2} (v_i + w_i).$$

We have $v + w \in C$. By Remark 1.6.1, C is a linear code.

C is called the *binary parity-check code of length n*. By Eq. 1.25, the vectors v_i $(0 \leq i \leq n - 1)$, where $v_{ij} = 0$ for $j \neq i$ and $v_{ii} = 1$, form a basis for C. Thus, $\dim(C) = n - 1$. Furthermore, we note that the minimum distance between the first $n - 1$ bits of codewords in C is 1. The parity-check bit for two codewords will be different if they differ only at one position in the first $n - 1$ bits. Thus, the minimum distance of C is 2, and C is a binary $[n, n - 1, 2]$-linear code. By Theorems 1.6.1 and 1.6.2, C is exactly 1-error detecting and cannot correct errors.

Definition 1.6.9 The *dual code* of a binary linear code C is the orthogonal complement of C, $C^{\perp}$.

By Lemma 1.3.4, $C^{\perp}$ is a binary linear code. It is easy to see that $(C^{\perp})^{\perp} = C$.

Example 1.6.10 Let C be a binary parity-check code of length n (see Example 1.6.9). Then $c \in C^{\perp}$ if and only if $c \cdot v = 0 \ \forall v \in C$, i.e.,

$$\sum_{i=0}^{n-1} c_i v_i = 0 \iff \sum_{i=0}^{n-2} c_i v_i + \left(c_{n-1} \sum_{i=0}^{n-2} v_i \right) = 0 \iff \sum_{i=0}^{n-2} (c_i + c_{n-1}) v_i = 0$$

for all $v_i = 0, 1 (0 \leq i \leq n-2)$, which is equivalent to $c_i = c_{n-1}$ for all $0 \leq i \leq n - 2$. Thus $C^{\perp} = \{00 \ldots 00, 11 \ldots 11\}$ is the n-repetition code (see Example 1.6.8).

Example 1.6.11 Let $C = \{000, 111\}$ be the binary 3-repetition code, then

$$C^{\perp} = \{000, 011, 101, 110\}$$

is the binary parity-check code of length 3.

Definition 1.6.10 Let $v \in \mathbb{F}_2^n$ be a word, and the *Hamming weight* of v, denoted by $\mathrm{wt}\,(v)$, is given by the number of nonzero bits in v, or equivalently,

$$\mathrm{wt}\,(v) = \mathrm{dis}\,(v, \mathbf{0}) .$$

We note that when $n = 1$, $\mathrm{wt}\,(v) = 1$ if $v = 1$ and $\mathrm{wt}\,(v) = 0$ if $v = 0$. Then, for any $v = (v_0, v_1, \ldots, v_{n-1})$ from $\mathbb{F}_2^n$,

$$\mathrm{wt}\,(v) = \sum_{i=0}^{n-1} \mathrm{wt}\,(v_i) . \tag{1.26}$$

Lemma 1.6.2 *For any* $u, v \in \mathbb{F}_2^n$, $\mathrm{dis}\,(u, v) = \mathrm{wt}\,(u + v)$.

Proof Take any $u, v \in \mathbb{F}_2$,

$$\mathrm{dis}\,(u, v) = \begin{cases} 0 & \text{if } u = v \\ 1 & \text{if } u \neq v, \text{ i.e., } u + v = 0. \end{cases}$$

The lemma follows from Eq. 1.26. $\qquad\square$

Example 1.6.12 Let $u = (1, 0, 0, 1)$, $v = (0, 1, 1, 1)$, then dis $(u, v) = 3$ and

$$\text{wt}(u + v) = \text{wt}((1, 1, 1, 0)) = 3.$$

Theorem 1.6.3 *Let C be a binary linear code, define*

$$\text{wt}(C) := \min\{\,\text{wt}(c) \mid c \in C,\ c \neq 0\,\}.$$

Then dis $(C) = \text{wt}(C)$.

Proof Take $v, u \in C$, such that dis $(v, u) = $ dis (C). By Lemma 1.6.2, $\text{wt}(v + u) = $ dis (C). Since C is a vector space, $v + u \in C$. We have dis $(C) \geq \text{wt}(C)$.

Now, take $w \in C$ such that $\text{wt}(C) = \text{wt}(w)$. We have

$$\text{wt}(C) = \text{wt}(w) = \text{dis}(w, 0) \geq \text{dis}(C).$$

$\qquad\square$

Definition 1.6.11 Let C be a binary liner code. A *generator matrix* for C is a matrix whose rows form a basis for C. A *parity-check matrix* for C is a generator matrix for $C^\perp$.

Example 1.6.13 Let $C = \{\,000, 111\,\}$, and we know that $C^\perp = \{\,000, 011, 101, 110\,\}$ (see Example 1.6.11). Let

$$G = \begin{pmatrix} 1 & 1 & 1 \end{pmatrix}, \quad H = \begin{pmatrix} 0 & 1 & 1 \\ 1 & 0 & 1 \end{pmatrix}.$$

Then G is a generator matrix for C and a parity-check matrix for $C^\perp$. H is a generator matrix for $C^\perp$ and a parity-check matrix for C.

Let C be a binary $[n, k, d]$ linear code. If G is a generator matrix for C and H is a parity-check matrix for C, then $HG^\top = O$, where O denotes a matrix with all entries equal to zero. Also, the size of G is $k \times n$.

Let $\{v_1, \ldots, v_k\}$ be the rows of G. Then for any $u = (u_0, u_1, \ldots, u_{k-1})$ in $\mathbb{F}_2^k$,

$$uG = \sum_{i=0}^{k-1} u_i v_i \in C.$$

On the other hand, by Remark 1.3.5, any $c \in C$ has a unique representation of the form

$$c = \sum_{i=0}^{k-1} u_i v_i, \quad \text{where } u_i \in \mathbb{F}_2.$$

Thus, each $u \in \mathbb{F}_2^k$ can be encoded as uG.

Example 1.6.14 It follows from Example 1.6.9 that the binary parity-check code of length n has generator matrix $(I_{n-1} \mid \mathbf{1})$, where $\mathbf{1}$ represents a column vector of length $n - 1$ with each entry equal to 1.

- The binary parity-check code of length 2 is given by $\{00, 11\}$. It has a generator matrix $(1\ 1)$.
- The binary parity-check code of length 3 is given by $\{000, 011, 101, 110\}$. It has a generator matrix

$$\begin{pmatrix} 0 & 1 & 1 \\ 1 & 0 & 1 \end{pmatrix}.$$

Theorem 1.6.4 *Let C be a binary linear code with at least two codewords, and let H be a parity-check matrix for C. Then* $\mathrm{dis}\,(C)$ *is given by d such that any $d - 1$ columns of H are linearly independent and H has d columns that are linearly dependent.*

Proof Take $v \in C$ such that $v \neq \mathbf{0}$. By definition,

$$vH = \sum_{i,\, v_i \neq 0} v_i h_i = 0,$$

where h_i denotes the ith column of H. We can see that the columns h_i, where $v_i \neq 0$, are linearly dependent. Note that $\mathrm{wt}\,(v) = |\{v_i \mid v_i \neq 0\}|$.

Thus, there exists $v \in C$ such that $\mathrm{wt}\,(v) = d$ (i.e., $\mathrm{dis}\,(C) \leq d$) if and only if there are d columns of H that are linearly dependent.

$\mathrm{dis}\,(C) \geq d$ if and only if there is no $v \in C$ such that $\mathrm{wt}\,(v) < d$, which is equivalent to that any $d - 1$ columns of H are linearly independent. $\quad\square$

Example 1.6.15 Let C be the binary parity-check code of length n (see Example 1.6.9). We have discussed that $C^\perp$ is the n-repetition code (see Example 1.6.10). Since $C^\perp = \langle 11 \ldots 11 \rangle$, it has generator matrix

$$H = \begin{pmatrix} 1 & 1 & \ldots & 1 \end{pmatrix}.$$

By definition, H is a parity-check matrix for C.

Any single column of H is linearly independent. H has two columns that are linearly dependent, e.g., the first two columns. In fact, any two columns of H are linearly dependent. By Theorem 1.6.4, $\mathrm{dis}\,(C) = 2$, which agrees with our observation in Example 1.6.9.

Definition 1.6.12 Let C be a binary (n, M, d)-code. We define the *maximum distance* of C to be

$$\text{maxdis}(C) := \max \{ \text{dis} (c_1, c_2) \mid c_1, c_2 \in C \}.$$

If $\text{maxdis}(C) = \delta$, C is called a *binary (n, M, d, δ)-anticode*.

The notion of anticode was first defined in [Far70], where an anticode refers to a two-dimensional array of bits such that the maximum Hamming distance between any pair of rows is at most δ, for some integer $\delta > 0$. In this original definition, repeated rows are allowed. In Definition 1.6.12, an (n, M, d, δ)-anticode does not have repeated codewords.

We note that $\delta \geq d$. And any binary code is a binary anticode. However, the notion of binary anticode captures the maximum distance of a code.

Example 1.6.16

- $C = \{ 01, 10 \}$ is a binary $(2, 2, 2, 2)$-anticode.
- $C = \{ 001, 011, 111 \}$ is a binary $(3, 3, 1, 2)$-anticode.
- An n-repetition code is a binary $(n, 2, n, n)$-anticode.
- A binary parity-check code of length n is a binary $(n, 2^{n-1}, 2, n)$-anticode if n is even. And it is a binary $(n, 2^{n-1}, 2, n-1)$-anticode if n is odd.

1.7 Probability Theory

This section aims to provide a rigorous introduction to probabilities, random variables, and distributions.

Probability theory studies the mathematical theory behind random experiments. A random experiment is an experiment whose output cannot be predicted with certainty in advance. However, if the experiment is repeated many times, we can see "regularity" in the average output. For example, if we roll a die, we cannot predict the output of one roll. But if we roll it many times, we would expect to see the number 1 in $1/6$ of the outcomes assuming the die is fair.

For a given random experiment, we define *sample space*, denoted by Ω, to be the set of all possible outcomes. A subset A of Ω is called an *event*. If the outcome of the experiment is contained in A, then we say that A has *occurred*. The empty set $\emptyset$ denotes the event that consists of no outcomes. $\emptyset$ is also called the *impossible event*.

Example 1.7.1

- When the random experiment is rolling a die, the sample space is

$$\Omega = \{ 1, 2, 3, 4, 5, 6 \}.$$

$A = \{ 1, 2, 3 \} \subseteq \Omega$ is an event.

- When the random experiment is rolling two dice,

$$\Omega = \{ (i, j) \mid 1 \leq i, j \leq 6 \}.$$

One possible event is $A = \{ (1, 2), (1, 1) \}$.

Recall that we have defined complement, unions, and intersections of sets in Sect. 1.1.1. Fix a sample space Ω. Take two events, A and B. We say that $A \cup B$ occurs if either A or B occurs. Similarly, $\bigcup_{i=1}^{n} A_i$ occurs when at least one A_i occurs. $A \cap B$ occurs if both A and B occur, and $\bigcap_{i=1}^{m} A_i$ occurs if all of the events A_i occur. If $A \cap B = \emptyset$, then A and B cannot both occur, and they are called *mutually exclusive*. The complement of A, A^c, contains events in Ω that are not in A.

1.7.1 σ-Algebras

Let Ω be a set, and let $\mathcal{A}$ denote a set of subsets of Ω. $\mathcal{A}$ is called a *σ-algebra* if it has the following properties:

- $\Omega \in \mathcal{A}$.
- If $A \in \mathcal{A}$, then $A^c \in \mathcal{A}$.
- $\mathcal{A}$ is closed under finite unions and intersections: If $A_1, A_2, \ldots A_n \in \mathcal{A}$, then $\bigcup_{i=1}^{n} A_i \in \mathcal{A}$ and $\bigcap_{i=1}^{n} A_i \in \mathcal{A}$.
- $\mathcal{A}$ is closed under countable unions and intersections: If $A_1, A_2, \cdots \in \mathcal{A}$, then $\bigcup_{i=1} A_i \in \mathcal{A}$ and $\bigcap_{i=1} A_i \in \mathcal{A}$.

The pair $(\Omega, \mathcal{A})$ is called a *measurable space*, meaning that it is a space on which we can put a measure.

Example 1.7.2

- For any set Ω, $\mathcal{A} = \{ \emptyset, \Omega \}$ is a σ-algebra.
- For any set Ω, the power set $\mathcal{A} = 2^{\Omega}$ is a σ-algebra.
- Let us consider the random experiment to roll a die. We know $\Omega = \{ 1, 2, 3, 4, 5, 6 \}$. Then,

$$\mathcal{A} = \{ \emptyset, \Omega, \{ 1 \}, \{ 2, 3, 4, 5, 6 \} \}$$

  is a σ-algebra.
- If we toss a coin, $\Omega = \{ H, T \}$. And

$$\mathcal{A} = 2^{\Omega} = \{ \emptyset, \Omega, \{ H \}, \{ T \} \}$$

  is a σ-algebra.

Definition 1.7.1 Let d be a positive integer and $\Omega = \mathbb{R}^d$. Ω consists of vectors $(x_0, x_1, \ldots, x_{d-1})$, where $x_i \in \mathbb{R}$ (see Theorem 1.3.3). The smallest σ-algebra[5] containing open sets in Ω is called the *Borel σ-algebra*, denoted $\mathcal{R}^d$. When $d = 1$, we write $\mathcal{R}$. Any set $B \in \mathcal{R}^d$ is called a *Borel set*.

Example 1.7.3 Here we list some examples of Borel sets. Take any $a, b, c \in \mathbb{R}$ such that $a < c < b$.

- By definition, open sets (a, b) are Borel sets.
- Since a σ-algebra contains the complement of a set, closed sets $[a, b]$ are also Borel sets.
- As $(a, b] = (a, c) \cup [c, b]$ and $(a, c), [c, b] \in \mathcal{R}$, we have $(a, b] \in \mathcal{R}$.
- Take a singleton set $\{a\}$, and we have

$$\{a\} = \bigcap_{n=1}^{\infty} \left(a - \frac{1}{n}, a + \frac{1}{n} \right).$$

  Thus $\{a\}$ is a Borel set.
- By definition, $\mathcal{R}$ is closed under countable unions, and it follows that a set of integers is a Borel set.

1.7.2 Probabilities

Let Ω be a sample space, and let $(\Omega, \mathcal{A})$ be a measurable space in this subsection.

Definition 1.7.2 A *probability measure* defined on a measurable space $(\Omega, \mathcal{A})$ is a function $P : \mathcal{A} \to [0, 1]$ such that

- $P(\Omega) = 1$, $P(\emptyset) = 0$.
- For any $A_1, A_2, \ldots \in \mathcal{A}$ that are pairwise disjoint, i.e., $A_{i_1} \cap A_{i_2} = \emptyset$ for $i_1 \neq i_2$,

$$P \left(\bigcup_{i=1}^{\infty} A_i \right) = \sum_{i=1}^{\infty} P(A_i).$$

This property is also called *countable additivity*.

$P(A)$ is called the *probability* of A. $(\Omega, \mathcal{A}, P)$ is called a *probability space*.

Example 1.7.4 Let us consider the random experiment of tossing a coin, the sample space $\Omega = \{H, T\}$. Let $\mathcal{A} = 2^{\Omega} = \{\emptyset, \Omega, \{H\}, \{T\}\}$. Define

---

[5] It is easy to show that the intersection of σ-algebras is again a σ-algebra. Since 2^{Ω} is a σ-algebra, it follows that the smallest σ-algebra containing open sets exists.

$$P(\emptyset) = 0, \quad P(\Omega) = 1, \quad P(\{H\}) = \frac{1}{2}, \quad P(\{T\}) = \frac{1}{2}.$$

It is easy to see that P is a probability measure on $(\Omega, \mathcal{A})$.

Example 1.7.5 Let Ω be a countable set (finite or countably infinite). Let $\mathcal{A} = 2^{\Omega}$. Then, any probability measure on $(\Omega, \mathcal{A})$ is a function such that for any $A \in \mathcal{A}$,

$$P(A) = \sum_{\omega \in A} P(\{\omega\}), \text{ where } P(\{\omega\}) \geq 0 \text{ and } \sum_{\omega \in \Omega} P(\{\omega\}) = 1.$$

For the rest of this section, let $(\Omega, \mathcal{A}, P)$ be a probability space.

Lemma 1.7.1

- *For any $A_i \in \mathcal{A}$, $0 \leq i \leq m$, pairwise disjoint, we have*

$$P\left(\bigcup_{i=1}^{m} A_i\right) = \sum_{i=1}^{m} P(A_i).$$

  This property is also called finite additivity.
- *For any $A, B \in \mathcal{A}$ such that $A \subseteq B$, we have $P(A) \leq P(B)$.*

Proof Take $A_i = \emptyset$ for $i > m$, and by countable additivity we have finite additivity.

Let $C = B - A$ be the difference between B and A. By countable additivity of probability measure,

$$P(B) = P(A \cup C) = P(A) + P(C).$$

By Definition 1.7.2, $P(C) \geq 0$. $\square$

Definition 1.7.3 Let Ω be a finite set. Let $\mathcal{A} = 2^{\Omega}$, the power set of Ω. A probability measure P on $(\Omega, \mathcal{A})$ is called *uniform* if

$$P(\{\omega\}) = \frac{1}{|\Omega|}, \quad \forall \omega \in \Omega.$$

We note that if P is a uniform probability measure on $(\Omega, \mathcal{A})$, then for any $A \in \mathcal{A}$, $P(A) = \frac{|A|}{|\Omega|}$.

Example 1.7.6 Let $\Omega = \{1, 2, 3, 4, 5, 6\}$ and $\mathcal{A} = 2^{\Omega}$. The uniform probability measure on $(\Omega, \mathcal{A})$ is given by P such that

$$P(\{i\}) = \frac{1}{6}, \quad \text{for } i \in \Omega.$$

Let $A = \{1, 2, 3\}$, $B = \{2, 4\}$, then

$$P(A) = \frac{3}{6} = \frac{1}{2}, \quad P(B) = \frac{2}{6} = \frac{1}{3}.$$

Take any $A, B \in \mathcal{A}$ such that $P(B) > 0$. We would like to compute the probability of A occurring given the knowledge that B has occurred. We do not need to consider $A \cap B^c$ since B has already occurred. Instead, we look at $A \cap B$, which occurs when both A and B occur. This leads to the definition of the *conditional probability of A given B*:

$$P(A|B) := \frac{P(A \cap B)}{P(B)}, \quad \text{where } P(B) > 0. \tag{1.27}$$

Example 1.7.7 Continuing Example 1.7.6,

$$A \cap B = \{2\}, \quad P(A \cap B) = \frac{1}{6}.$$

By Eq. 1.27,

$$P(A|B) = \frac{P(A \cap B)}{P(B)} = \frac{1/6}{1/3} = \frac{1}{2}.$$

Definition 1.7.4 Two events A and B are said to be *independent* if $P(A \cap B) = P(A)P(B)$. Otherwise, we say that they are *dependent*.

By Eq. 1.27, when $P(B) > 0$, the condition $P(A \cap B) = P(A)P(B)$ is equivalent to

$$P(A|B) = \frac{P(A \cap B)}{P(B)} = \frac{P(A)P(B)}{P(B)} = P(A). \tag{1.28}$$

That is, the probability of A occurring given the knowledge that B has occurred is the same as the probability of A occurring without the knowledge that B has occurred.

Example 1.7.8 Continuing Example 1.7.7,

$$P(A \cap B) = \frac{1}{6}, \quad P(A)P(B) = \frac{1}{2} \times \frac{1}{3} = \frac{1}{6}.$$

By Definition 1.7.4, A and B are independent. We also note that

$$P(A|B) = P(A) = \frac{1}{2}.$$

Next, we state a very useful theorem.

Theorem 1.7.1 (Bayes' Theorem)   *If $P(A) > 0$ and $P(B) > 0$, then*

$$P(B)P(A|B) = P(A)P(B|A).$$

Proof By Eq. 1.27, we have

$$P(B)P(A|B) = P(A \cap B), \quad P(A)P(B|A) = P(A \cap B).$$

$\square$

Definition 1.7.5 A set of events $\{ E_1, E_2, \ldots \mid E_i \in \mathcal{A} \}$ is called a *partition of* Ω if:

- They are pairwise disjoint.
- $P(E_i) > 0$ for all i.
- $\cup_i E_i = \Omega$.

If the set of events is finite, it is called a *finite partition of* Ω; otherwise, it is called a *countable partition of* Ω.

Example 1.7.9 Let $\Omega = \{1, 2, 3, 4, 5, 6\}$, $\mathcal{A} = 2^\Omega$, and P be the uniform probability measure on $(\Omega, \mathcal{A})$ (see Example 1.7.6). Let

$$E_1 = \{1, 2, 3\}, \quad E_2 = \{4, 5\}, \quad E_3 = \{6\}.$$

Then, $\{ E_1, E_2, E_3 \}$ is a finite partition of Ω. We can also calculate that

$$P(E_1) = \frac{1}{2}, \quad P(E_2) = \frac{1}{3}, \quad P(E_3) = \frac{1}{6}.$$

Lemma 1.7.2 *Let $\{ E_1, E_2, \ldots \mid E_i \in \mathcal{A} \}$ be a finite or countable partition of Ω. Then, for any $A \in \mathcal{A}$, we have*

$$P(A) = \sum_i P(A|E_i)P(E_i).$$

Proof First, we note that

$$A = A \bigcap \Omega = A \bigcap \left(\bigcup_i E_i \right) = \bigcup_i \left(A \bigcap E_i \right).$$

Since E_i are pairwise disjoint, $E_i \cap A$ are also pairwise disjoint. We have

$$P(A) = P\left(\bigcup_i \left(A \bigcap E_i \right) \right) = \sum_i P\left(A \bigcap E_i \right) = \sum_i P(A|E_i)P(E_i).$$

$\square$

Example 1.7.10 Continuing Example 1.7.9, let $A = \{2, 4\}$, then

$$P(A) = \frac{1}{3}, \quad A \cap E_1 = \{2\}, \quad A \cap E_2 = \{4\}, \quad A \cap E_3 = \emptyset.$$

By Eq. 1.27,

$$P(A|E_1) = \frac{1/6}{1/2} = \frac{1}{3}, \quad P(A|E_2) = \frac{1/6}{1/3} = \frac{1}{2}, \quad P(A|E_3) = 0.$$

Furthermore,

$$\sum_{i=1}^{3} P(A|E_i)P(E_i) = \frac{1}{3} \times \frac{1}{2} + \frac{1}{2} \times \frac{1}{3} = \frac{1}{3} = P(A).$$

Now we can state a generalized version of Bayes' Theorem (Theorem 1.7.1).

Theorem 1.7.2 *Let $\{E_1, E_2, \ldots \mid E_i \in \mathcal{A}\}$ be a finite or countable partition of Ω. For any $A \in \mathcal{A}$ with $P(A) > 0$ and any $m \geq 1$, we have*

$$P(E_m|A) = \frac{P(A|E_m)P(E_m)}{\sum_i P(A|E_i)P(E_i)}.$$

Proof By Bayes' Theorem (Theorem 1.7.1),

$$P(E_m|A) = \frac{P(A|E_m)P(E_m)}{P(A)}.$$

The result then follows from Lemma 1.7.2.   □

1.7.3 Random Variables

Let $(\Omega, \mathcal{A}, P)$ be a probability space. A random variable X represents an unknown quantity that varies with the outcome of a random experiment. Before the random experiment, we know all the possible values X can take, but we do not know which one it will take until we see the outcome of the experiment.

Definition 1.7.6 A *random variable X* is a function $X : \Omega \to \mathbb{R}$, such that

$$X^{-1}(B) = \{\omega : X(\omega) \in B\} \in \mathcal{A}, \quad \forall B \in \mathcal{R},$$

where $\mathcal{R}$ is the Borel σ-algebra (see Definition 1.7.1).

Example 1.7.11

- Fix $A \in \mathcal{A}$, and the *indicator function*, denoted 1_A, for A is defined as follows:

$$1_A : \mathcal{A} \to \mathbb{R}, \quad 1_A(\omega) = \begin{cases} 1 & \omega \in A \\ 0 & \omega \notin A. \end{cases}$$

1_A is a random variable.

- Consider the probability space from Example 1.7.5, then any function $X : \Omega \to \mathbb{R}$ is a random variable. In such a case, X is called a *discrete random variable*.
- Let us consider the probability space discussed in Example 1.7.4. Define $X : \Omega \to \mathbb{R}$ such that $X(H) = 0, \; X(T) = 1$. For any $B \in \mathcal{R}$, $X^{-1}(B)$ is always a subset of Ω, which is contained in $\mathcal{A}$. And X is a discrete random variable.

Let X be a random variable, and define P^X as follows:

$$P^X : \mathcal{R} \to [0, 1]$$

$$B \mapsto P(X^{-1}(B)). \tag{1.29}$$

It is easy to see that $P^X(\mathbb{R}) = 1$ and $P^X(\emptyset) = 0$. Take any $B_i \in \mathcal{B}$ that are pairwise disjoint. Then $X^{-1}(B_i)$ are also pairwise disjoint since X is a function. The countable additivity of P^X follows from the countable additivity of P. Thus, P^X is a probability measure on $(\mathbb{R}, \mathcal{R})$. We say that P^X is *induced* by X, and it is called the *distribution* of X. The *cumulative distribution function (CDF)* of X, denoted F, is defined as

$$F : \mathbb{R} \to [0, 1]$$

$$x \mapsto P^X((-\infty, x]) = P(X^{-1}((-\infty, x])). \tag{1.30}$$

For simplicity, we will write $P(X \in B)$ instead of $P(X^{-1}(B))$ in Eq. 1.29 and $P(X \leq x)$ instead of $P(X^{-1}((-\infty, x]))$ in Eq. 1.30.

On the other hand, the next lemma says if we start from a function F with certain properties, there always exists a random variable that has F as its CDF. The proof can be found in, e.g., [Dur19, page 9].

Lemma 1.7.3 *If a function F satisfies the following conditions, then it is the distribution function of some random variable.*

- *F is nondecreasing.*
- *$\lim_{x \to \infty} F(x) = 1, \; \lim_{x \to -\infty} F(x) = 0$.*
- *F is right continuous, i.e., $\lim_{y \downarrow x} F(y) = F(x)$.*

When X is a discrete random variable (see Example 1.7.11), the distribution of X is completely determined by the following numbers:

$$P(X = j) = \sum_{\omega : X(\omega) = j} P(\{\omega\}).$$

Let $T := X(\Omega)$ be the image of Ω in $\mathbb{R}$. The *probability mass function (PMF)* of X is defined to be the function

$$p_X : T \to [0, 1]$$
$$x \mapsto P(X = x).$$

We have the following relation between the PMF of X and the CDF of X:

$$F(a) = \sum_{x \leq a,\ x \in T} p_X(x).$$

Example 1.7.12 Let us consider the probability space defined in Example 1.7.4. We have discussed in Example 1.7.11 that

$$X : \Omega \to \mathbb{R}, \quad X(H) = 0, \quad X(T) = 1$$

is a discrete random variable. The image of X in $\mathbb{R}$ is $T = \{0, 1\}$. And the PMF of X is given by

$$p_X(0) = P(X = 0) = P(\{H\}) = \frac{1}{2}, \quad p_X(1) = P(X = 1) = P(\{T\}) = \frac{1}{2}.$$

When the distribution function $F(x) = P(X \leq x)$ has the form

$$F(x) = \int_{-\infty}^{x} f(y)dy,$$

we say that X has *probability density function (PDF)* f and X is called a *continuous random variable.*

Example 1.7.13 Define $f(x) = 1$ for $x \in (0, 1)$ and 0 otherwise.

$$F(x) = \int_{-\infty}^{x} f(y)dy$$

is given by

$$F(x) = \begin{cases} 0 & x \leq 0 \\ x & 0 \leq x \leq 1 \\ 1 & x > 1. \end{cases}$$

It is easy to show that F satisfies the conditions in Lemma 1.7.3. If X is a random variable that has F as its CDF, then we say that X induces a *uniform distribution* on $(0, 1)$.

Example 1.7.14 A random variable Z that induces a *standard normal distribution* has the probability density function

$$f(z) = \frac{1}{\sqrt{2\pi}} \exp\left(-\frac{z^2}{2}\right)$$

and the cumulative distribution function

$$\Phi(z) = \frac{1}{\sqrt{2\pi}} \int_{-\infty}^{z} \exp\left(-\frac{y^2}{2}\right) dy.$$

The standard normal distribution will be very useful in later parts of the book, and we use $\Phi(z)$ instead of $F(z)$ to denote its CDF. Moreover, we say that Z is a *standard normal random variable*. Figure 1.1 shows that $f(z)$ is a bell-shaped curve that is symmetric about 0. The symmetry is also apparent from the formula for $f(z)$.

Next, we would like to define expectations and variances for random variables. The exact formulas for discrete and continuous random variables are different, but the information carried by those notions is the same. In particular, the *expectation/mean* of a random variable X is the expected average value of X. And the *variance* of X is the average squared distance from the mean. By squaring the distances, the small deviations from the mean are reduced, and the big ones are enlarged. Thus the variance measures how the values of X vary from the mean or how "spread out" the values of X are.

When X is a discrete random variable $X : \Omega \to \mathbb{R}$ with PMF p_X and $T = X(\Omega)$ (the image of Ω in $\mathbb{R}$), its *expectation/mean* is defined as

$$E[X] = \sum_{x \in T} x p_X(x), \tag{1.31}$$

Fig. 1.1 Probability density function of the standard normal random variable

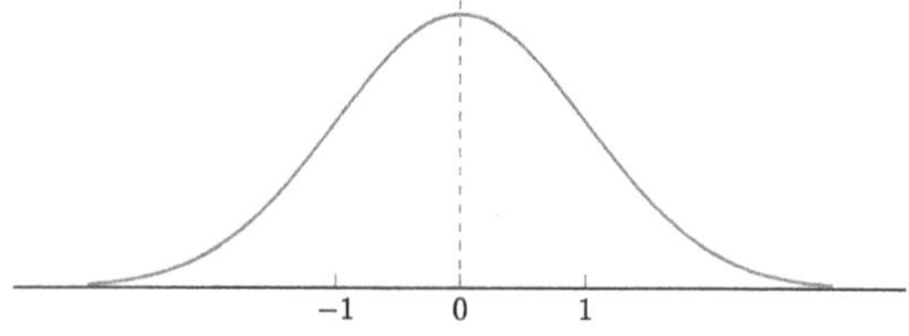

provided the sum exists.[6]

Example 1.7.15 Let us consider the discrete random variable discussed in Examples 1.7.4 and 1.7.12. By Eq. 1.31,

$$E[X] = 0 \times p_X(0) + 1 \times p_X(1) = 0 \times \frac{1}{2} + 1 \times \frac{1}{2} = \frac{1}{2}.$$

When X is a continuous random variable with PDF f, its *expectation/mean* is defined as

$$E[X] = \int_{-\infty}^{\infty} x f(x) dx, \tag{1.32}$$

provided the integral exists.

Example 1.7.16 Let X be a random variable that induces a uniform distribution on $(0, 1)$ (see Example 1.7.13), by Eq. 1.32,

$$E[X] = \int_{-\infty}^{\infty} x f(x) dx = \int_0^1 x dx = \frac{x^2}{2} \Big|_0^1 = \frac{1}{2}.$$

Example 1.7.17 Let Z be a random variable that induces the standard normal distribution (see Example 1.7.14), by Eq. 1.32,

$$E[Z] = \int_{-\infty}^{\infty} z f(z) dz = \frac{1}{\sqrt{2\pi}} \int_{-\infty}^{\infty} z \exp\left(-\frac{z^2}{2}\right) dz = 0.$$

As shown in Fig. 1.1, $f(x)$ is symmetric about 0, so it is not surprising that the expected average value of Z is 0.

Let g be a function $g : \mathbb{R} \to \mathbb{R}$. Then $g(X)$ is also a random variable.[7] It can be proven that if X is a discrete random variable with PMF p_X, then

$$E[g(X)] = \sum_x g(x) p_X(x).$$

If X is a continuous random variable with PDF f, then

$$E[g(X)] = \int_{-\infty}^{\infty} g(x) f(x) dx.$$

---

[6] For example, if Ω is finite or if Ω is countable and the series converges absolutely, the sum exists.

[7] To be more precise, g should be a *measurable function* for $g(X)$ to be a random variable. For the definition of measurable functions, we refer the readers to [Yeh14, page 72].

The proof can be found in, e.g., [Ros20, page 113].

Example 1.7.18 Define

$$g : \mathbb{R} \to \mathbb{R}$$
$$x \mapsto x^2.$$

Then the expectation of g is given by

$$E\left[X^2\right] = \sum_x x^2 p_X(x)$$

when X is a discrete random variable with PMF p_X. And

$$E\left[X^2\right] = \int_{-\infty}^{\infty} x^2 f(x) dx$$

when X is a continuous random variable with PDF $f(x)$.

Furthermore, given two random variables X, Y such that $E[|X|] < \infty$ and $E[|Y|] < \infty$, for any $a, b \in \mathbb{R}$,

$$E[X + Y] = E[X] + E[Y], \quad E[aX + b] = aE[X] + b, \quad E[b] = b. \tag{1.33}$$

The proof can be found in, e.g., [Dur19, page 24]

Example 1.7.19 Let X be a random variable, and let $\mu := E[X]$. By Eq. 1.33, we have

$$E\left[(X - \mu)^2\right] = E\left[X^2\right] + \mu^2 - 2E[X\mu] = E\left[X^2\right] + \mu^2 - 2\mu E[X]$$
$$= E\left[X^2\right] + \mu^2 - 2\mu^2$$
$$= E\left[X^2\right] - \mu^2. \tag{1.34}$$

Equation 1.34 provides the formula for computing the *variance* of a random variable X. More specifically, let X be a random variable with mean $E[X] = \mu$. If $E\left[X^2\right] < \infty$, then the *variance* of X is given by

$$\text{Var}(X) = E\left[(X - \mu)^2\right] = E\left[X^2\right] - \mu^2. \tag{1.35}$$

Example 1.7.20 Let us consider the discrete random variable discussed in Examples 1.7.4 and 1.7.12. By Eq. 1.35 and Examples 1.7.15 and 1.7.18,

$$\mathrm{Var}(X) = E[X^2] - \frac{1}{2^2} = \sum_x x^2 p_X(x) - \frac{1}{4} = 0 \times p_X(0) + 1 \times p_X(1) - \frac{1}{4} = \frac{1}{2} - \frac{1}{4} = \frac{1}{4}.$$

Example 1.7.21 Let X be a continuous random variable that induces the uniform distribution on $(0, 1)$ (see Example 1.7.13), by Eq. 1.35 and Examples 1.7.16 and 1.7.18,

$$\mathrm{Var}(X) = \int_{-\infty}^{\infty} x^2 f(x)dx - E[X]^2 = \int_0^1 x^2 dx - \frac{1}{2^2} = \left.\frac{x^3}{3}\right|_0^1 - \frac{1}{4} = \frac{1}{12}.$$

Example 1.7.22 Let Z be a random variable that induces the standard normal distribution (see Examples 1.7.14), by Eq. 1.35, Examples 1.7.17 and 1.7.18,

$$\mathrm{Var}(Z) = E\left[Z^2\right] - E[Z]^2 = \int_{-\infty}^{\infty} z^2 f(z)dz - 0 = \frac{1}{\sqrt{2\pi}} \int_{-\infty}^{\infty} z^2 \exp\left(-\frac{z^2}{2}\right) dz = 1.$$

We write $Z \sim \mathcal{N}(0, 1)$ to indicate that Z induces the standard normal distribution with mean 0 and variance 1.

Given two random variables X, Y such that $E[|X|] < \infty$ and $E[|Y|] < \infty$, take any $a, b \in \mathbb{R}$, then it follows from Eq. 1.33 that

$$\mathrm{Var}(aX + b) = E\left[(aX + b - E[aX + b])^2\right] = E\left[(aX + b - aE[X] - b)^2\right]$$

$$= a^2 E\left[(X - E[X])^2\right] = a^2 \mathrm{Var}(X). \tag{1.36}$$

In particular, we have

$$\mathrm{Var}(b) = 0, \quad \mathrm{Var}(X + b) = \mathrm{Var}(X), \quad \mathrm{Var}(aX) = a^2 \mathrm{Var}(X).$$

Example 1.7.23 Let $Z \sim \mathcal{N}(0, 1)$ be a standard normal random variable. Take any $\sigma, \mu \in \mathbb{R}$ with $\sigma^2 > 0$. Define $Y = \sigma Z + \mu$. Then by Eqs. 1.33 and 1.36,

$$E[Y] = \mu, \quad \mathrm{Var}(Y) = \sigma^2.$$

It can be shown (see, e.g., [Dur19, page 28]) that Y has PDF

$$f(y) = \frac{1}{\sigma\sqrt{2\pi}} \exp\left(-\frac{(y - \mu)^2}{2\sigma^2}\right). \tag{1.37}$$

We say that Y induces a *normal distribution with mean μ and variance σ^2*, written $Y \sim \mathcal{N}(\mu, \sigma^2)$. Y is also called *normal/a normal random variable*. We note that the mean and variance fully define a normal distribution.

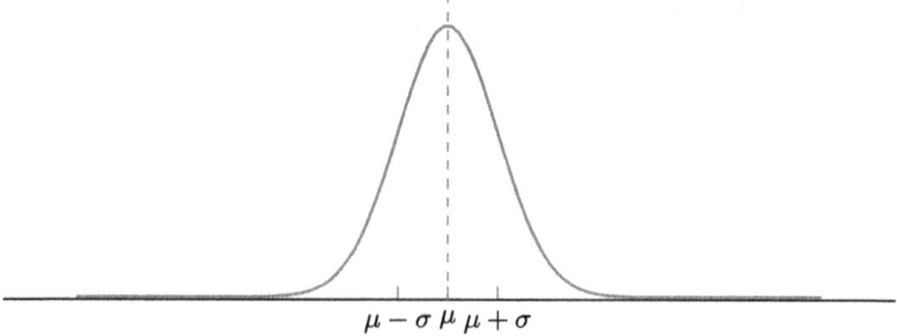

Fig. 1.2 Probability density function of a normal random variable

$f(y)$ is a bell-shaped curve symmetric about μ and obtains its maximum value of

$$\frac{1}{\sigma\sqrt{2\pi}} \approx \frac{0.399}{\sigma}$$

at $y = \mu$ (see Fig. 1.2).

Remark 1.7.1 On the other hand, if we let $Y \sim \mathcal{N}(\mu, \sigma)$ be a normal random variable, then

$$Z := \frac{Y - \mu}{\sigma}$$

is a standard normal random variable (for proof, see [Dur19, exercise 1.2.5]).

Next, let us look at the relations between two random variables. First, similar to Definition 1.7.4, we give the definition of independent random variables.

Definition 1.7.7 Given two random variables $X : \Omega \to \mathbb{R}, Y : \Omega \to \mathbb{R}$, they are said to be *independent* if for any $A, B \in \mathcal{R}$,

$$P(X \in A, Y \in B) = P(X \in A)P(Y \in B).$$

If two random variables $X : \Omega \to \mathbb{R}, Y : \Omega \to \mathbb{R}$ are independent, it can be proven that

$$\mathrm{E}[XY] = \mathrm{E}[X]\,\mathrm{E}[Y] \quad \text{if} \quad \mathrm{E}[|X|] < \infty \text{ and } \mathrm{E}[|Y|] < \infty. \tag{1.38}$$

The proof can be found in, e.g., [Dur19, page 41].

To further analyze the relation between two random variables X and Y, we define the *covariance* of X and Y to be

$$\mathrm{Cov}(X, Y) = \mathrm{E}[(X - \mu_X)(Y - \mu_Y)], \tag{1.39}$$

where μ_X and μ_Y denote expectations for X and Y, respectively. It is easy to see that

$$\text{Cov}(X, Y) = \text{Cov}(Y, X), \quad \text{Cov}(X, X) = \text{Var}(X).$$

In case $\text{E}[|X|] < \infty$ and $\text{E}[|Y|] < \infty$, by Eq. 1.33, we have

$$\begin{aligned}
\text{Cov}(X, Y) &= \text{E}[XY - \mu_X Y - \mu_Y X + \mu_x \mu_Y] \\
&= \text{E}[XY] - \mu_X \mu_Y - \mu_Y \mu_X + \mu_X \mu_Y = \text{E}[XY] - \text{E}[X]\text{E}[Y]. \quad (1.40)
\end{aligned}$$

Definition 1.7.8 Let X and Y be two random variables. If $\text{Cov}(X, Y) = 0$, we say that X and Y are *uncorrelated*. Otherwise, we say that X and Y are correlated.

Remark 1.7.2 By Eq. 1.38, if X and Y are two independent random variables such that $\text{E}[|X|] < \infty$ and $\text{E}[|Y|] < \infty$, then $\text{Cov}(X, Y) = 0$, and they are uncorrelated.

Let Z be another random variable such that $\text{E}[|Z|] < \infty$, by Eq. 1.38,

$$\text{E}[(X+Z)Y] - \text{E}[X+Z]\text{E}[Y] = \text{E}[XY] + \text{E}[ZY] - \text{E}[X]\text{E}[Y] - \text{E}[Z]\text{E}[Y].$$

Thus,

$$\text{Cov}(X + Z, Y) = \text{Cov}(X, Y) + \text{Cov}(X, Z).$$

It can be easily generalized to show that

$$\text{Cov}\left(\sum_{i=1}^{n} X_i, Y\right) = \sum_{i=1}^{n} \text{Cov}(X_i, Y),$$

where $X_1, X_2, \ldots, X_n$ are n random variables. Furthermore, by the symmetry of covariance ($\text{Cov}(X, Y) = \text{Cov}(Y, X)$), we have

$$\text{Cov}\left(\sum_{i=1}^{n} X_i, \sum_{j=1}^{m} Y_j\right) = \sum_{i=1}^{n} \sum_{j=1}^{m} \text{Cov}(X_i, Y_j),$$

where $Y_1, Y_2, \ldots, Y_m$ are m random variables. Set $m = n$, $Y_j = X_i$, and we have

$$\text{Var}\left(\sum_{i=1}^{n} X_i\right) = \sum_{i=1}^{n} \text{Var}(X_i) + \sum_{i=1}^{n} \sum_{j=1, j \neq i}^{n} \text{Cov}(X_i, X_j).$$

If we further assume X_i are independent with $\text{E}[|X_i|] < \infty$, by Remark 1.7.2,

$$\operatorname{Var}\left(\sum_{i=1}^{n} X_i\right) = \sum_{i=1}^{n} \operatorname{Var}(X_i). \tag{1.41}$$

Recall that we have defined Borel σ-algebra for $\mathbb{R}^n$ (Definition 1.7.1). Correspondingly, we can define a random vector similar to Definition 1.7.6.

Definition 1.7.9 A *random vector* X is a function $X : \Omega \to \mathbb{R}^d$, such that

$$X^{-1}(B) = \{\, \omega : X(\omega) \in B \,\} \in \mathcal{A}, \forall B \in \mathcal{R}^d.$$

Note that a random variable is a random vector for the case $d = 1$.

Definition 1.7.10 A random vector $X = (X_1, X_2, \ldots, X_n)$ induces a *Gaussian (or multivariate normal) distribution* if every linear combination

$$\sum_{j=1}^{n} a_j X_j, \quad a_j \in \mathbb{R}$$

is a normal random variable. The *mean vector* of X is

$$\boldsymbol{\mu} = (\mu_{X_1}, \mu_{X_2}, \ldots, \mu_{X_n}),$$

where μ_{X_i} is the mean of X_i. The *covariance matrix* of X is given by the matrix Q with entries $Q_{ij} = \operatorname{Cov}(X_i, X_j)$. When $\det Q \neq 0$, the *probability density function* for X is

$$f(x) = \frac{1}{(2\pi)^{\frac{n}{2}} \sqrt{\det Q}} \exp\left(-\frac{1}{2}(x - \mu)^\top Q^{-1}(x - \mu)\right).$$

We write $X \sim \mathcal{N}(\boldsymbol{\mu}, Q)$, and we say that X is *Gaussian*/a *Gaussian random vector*.

Example 1.7.24 If $X_1, \ldots, X_n$ are pairwise independent random variables and each $X_i \sim \mathcal{N}(\mu_i, \sigma_i^2)$ is normal, then $X = (X_1, X_2, \ldots, X_n)$ induces a Gaussian distribution with mean $\boldsymbol{\mu} = (\mu_1, \ldots, \mu_n)$ and covariance matrix Q a diagonal matrix with $Q_{ii} = \sigma_i^2$ (see [JP04, page 127] for a proof).

When we look at Gaussian random vectors, we have the following nice property for the components of the random vector. The proof can be found in, e.g., [JP04, page 128].

Theorem 1.7.3 *Let* $X = (X_1, X_2, \ldots, X_n)$ *be a Gaussian random vector. Then the components* X_i *are independent if and only if the covariance matrix* Q *of* X *is diagonal.*

A direct corollary is as follows.

Corollary 1.7.1 *Let* $X = (X_1, X_2, \ldots, X_n)$ *be a Gaussian random vector. Two components* X_i *and* X_j *are independent if and only if they are uncorrelated.*

Proof Let X_i and X_j be any two components of X.

$\Longrightarrow$ If X_i and X_j are independent, by Theorem 1.7.3, $\mathrm{Cov}(X_i, X_j) = 0$.

$\Longleftarrow$ If X_i, X_j are uncorrelated, the random vector $Y := (X_i, X_j)$ is Gaussian with a diagonal covariance matrix (see Example 1.7.24). Again by Theorem 1.7.3, X_i and X_j are independent. $\qquad\qquad\qquad\qquad\qquad\qquad\qquad\qquad\qquad\qquad\qquad\quad\square$

Corollary 1.7.2 *Two normal random variables* X *and* Y *are independent if and only if they are uncorrelated.*

Definition 1.7.11 Let X and Y be two random variables with finite variances. The *correlation coefficient* of X and Y is given by

$$\rho = \frac{\mathrm{Cov}(X, Y)}{\sqrt{\mathrm{Var}(X)\mathrm{Var}(Y)}}. \qquad\qquad (1.42)$$

It can be shown by the Cauchy–Schwarz inequality that $-1 \leq \rho \leq 1$ (see [JP04, p. 91]).

In general, the correlation coefficient is normally used to answer the question if large values of X tend to be paired with large or small values of Y. For example, if when X is large (or small), Y is also large (or small), then the signs of $X_i - \overline{X}$ and $Y_i - \overline{Y}$ will tend to be the same. Or if when X is large (or small), Y is small (or large), then the signs of $X_i - \overline{X}$ and $Y_i - \overline{Y}$ will tend to be different. In both cases, the absolute value of ρ will be big. On the other hand, in the special case when X and Y are uncorrelated (Definition 1.7.8), their correlation coefficient is $\rho = 0$. In particular, if X and Y are independent, then $\rho = 0$ (see Remark 1.7.2).

As another example, suppose X has finite expectation and variance. For $a, b \in \mathbb{R}$ and $a \neq 0$, if $Y = aX + b$, then by Eqs. 1.33 and 1.36,

$$
\begin{aligned}
\rho &= \frac{\mathrm{Cov}(X, Y)}{\sqrt{\mathrm{Var}(X)\mathrm{Var}(Y)}} = \frac{\mathrm{E}[XY] - \mathrm{E}[X]\mathrm{E}[Y]}{\sqrt{\mathrm{Var}(Y)\mathrm{Var}(X)}} \\[2mm]
&= \frac{\mathrm{E}[X(aX + b)] - \mathrm{E}[X]\mathrm{E}[aX + b]}{\sqrt{\mathrm{Var}(aX + b)\mathrm{Var}(X)}} \\[2mm]
&= \frac{\mathrm{E}\left[aX^2 + bX\right] - a\mathrm{E}[X]^2 - b\mathrm{E}[X]}{\sqrt{a^2\mathrm{Var}(X)^2}} = \frac{a\mathrm{E}\left[X^2\right] + b\mathrm{E}[X] - a\mathrm{E}[X]^2 - b\mathrm{E}[X]}{|a|\mathrm{Var}(X)} \\[2mm]
&= \frac{a\mathrm{Var}(X)}{|a|\mathrm{Var}(X)} = \frac{a}{|a|} = \begin{cases} 1 & a > 0 \\ -1 & a < 0. \end{cases}
\end{aligned}
$$

1.8 Statistics

In this section, we will first discuss a few important distributions (Sect. 1.8.1). Then we will introduce statistical methods for estimating the mean and variance of a normal distribution (Sect. 1.8.2) which utilize properties of those important distributions. Those methods will provide more insights into our analysis of device leakages in Sect. 4.2.3. Finally, we touch on some basics of hypothesis testing (Sect. 1.8.3) which justifies leakage assessment methods that will be introduced in Sect. 4.2.3.

We suggest the readers come back to this part later when they reach Chap. 4.

1.8.1 Important Distributions

Let Z denote a random variable that induces a standard normal distribution. We have discussed in Example 1.7.14 that Z has the probability density function

$$f(z) = \frac{1}{\sqrt{2\pi}} \exp\left(-\frac{z^2}{2}\right)$$

and the cumulative distribution function

$$\Phi(z) = \frac{1}{\sqrt{2\pi}} \int_{-\infty}^{z} \exp\left(-\frac{y^2}{2}\right) dy.$$

Furthermore, Z has expectation $E[Z] = 0$ (see Example 1.7.17) and variance $\mathrm{Var}(Z) = 1$ (see Example 1.7.22), and we write $Z \sim \mathcal{N}(0, 1)$.

Given any $\alpha \in (0, 1)$, we define z_α such that

$$P(Z > z_\alpha) = 1 - \Phi(z_\alpha) = \alpha, \quad \text{i.e.,} \quad \Phi(z_\alpha) = 1 - \alpha. \tag{1.43}$$

Those z_α values are useful for many applications, and there are tables listing the values of $\Phi(z)$ for small values of z (e.g., [Ros20, Table A1]). Given α, the approximated value of z_α can be found by examining such a table. In Table 1.4, we list a few values of z_α with corresponding α, which will be used later in the book.

By definition, $\Phi(z)$ is the integral of $f(z)$. As shown in Fig. 1.3, α corresponds to the area under $f(z)$ for $z > z_\alpha$. Furthermore, since $f(z)$ is symmetric about 0, we have

Table 1.4 Values of z_α (see Eq. 1.43) with corresponding α

α	0.1	0.05	0.01	0.005	0.001
$1 - \alpha$	0.900	0.950	0.990	0.995	0.999
z_α	1.282	1.645	2.326	2.576	3.090

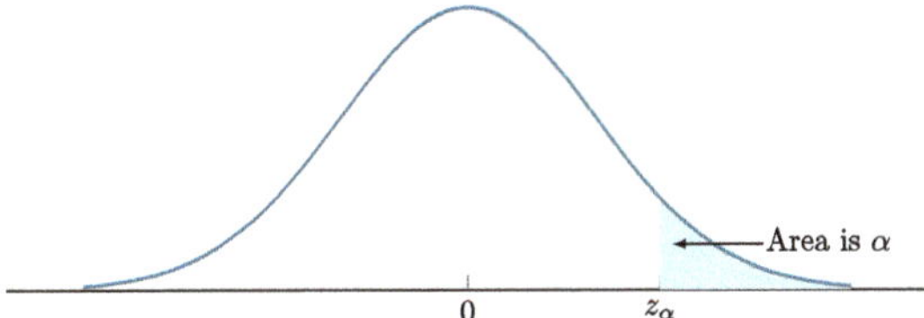

Fig. 1.3 Probability density function $f(z)$ for $Z \sim \mathcal{N}(0, 1)$. $P(Z > z_\alpha) = \alpha$, α corresponds to the area under $f(z)$ for $z > z_\alpha$

$$\Phi(-z_\alpha) = P(Z < -z_\alpha) = P(Z > z_\alpha) = \alpha,$$

and

$$P(-z_{\alpha/2} < Z < z_{\alpha/2}) = P(Z > -z_{\alpha/2}) - P(Z > z_{\alpha/2})$$
$$= 1 - P(Z \le -z_{\alpha/2}) - \frac{\alpha}{2} = 1 - \alpha. \tag{1.44}$$

Next, we look at χ^2-distributions.

Definition 1.8.1 Let $Z_1, Z_2, \ldots, Z_n$ be independent standard normal random variables. Define

$$X = \sum_{i=1}^{n} Z_i^2.$$

The distribution induced by X is called χ^2-*distribution with n degrees of freedom*. We write $X \sim \chi_n^2$.

Remark 1.8.1 We note that if $X_1 \sim \chi_{n_1}^2$, $X_2 \sim \chi_{n_2}^2$ are independent, then $X_1 + X_2 \sim \chi_{n_1+n_2}^2$.

For example, when $n = 8$, the probability density function for $X \sim \chi_8^2$ is shown in Fig. 1.4. Similarly to z_α (Eq. 1.43), for any $\alpha \in (0, 1)$, we define $\chi_{\alpha,n}^2$ to be the number such that

$$P(X \ge \chi_{\alpha,n}^2) = \alpha. \tag{1.45}$$

As shown in Fig. 1.4, α corresponds to the area under the PDF of X for $X \ge \chi_{\alpha,n}^2$. We also have

$$P(\chi_{1-\alpha,n}^2 < X < \chi_{\alpha,n}^2) = P(X \ge \chi_{1-\alpha,n}^2) - P(X \ge \chi_{\alpha,n}^2) = 1 - 2\alpha.$$

Finally, we provide some details on t-distributions.

Definition 1.8.2 Let $X \sim \chi_n^2$, $Z \sim \mathcal{N}(0, 1)$ be two independent random variables. Define a random variable

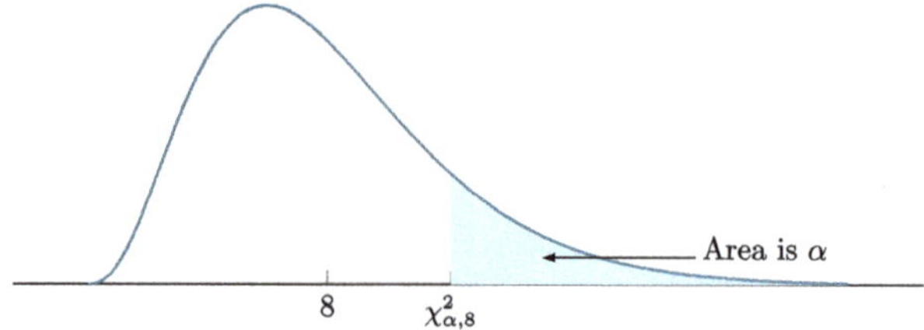

Fig. 1.4 Probability density function for $X \sim \chi^2_8$. $P(X \geq \chi^2_{\alpha,8}) = \alpha$

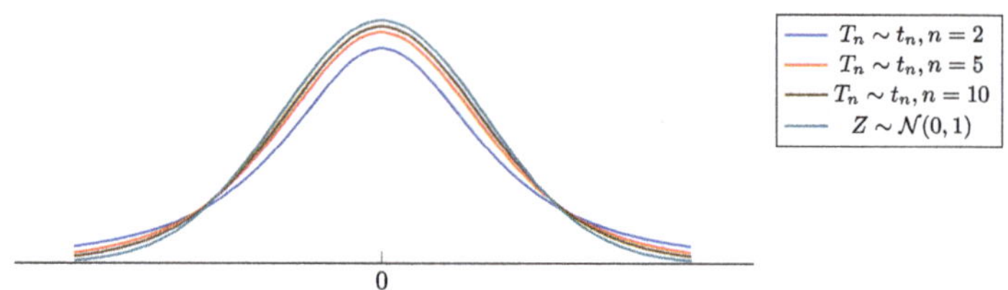

Fig. 1.5 Probability density functions for $T_n \sim t_n$ ($n = 2, 5, 10$) and for the standard normal random variable Z

$$T_n := \frac{Z}{\sqrt{X/n}}.$$

The distribution induced by T_n is called a *t-distribution with n degrees of freedom*. We write $T_n \sim t_n$.

It can be shown (see [Ros20, page 189]) that the PDF of T_n is symmetric about 0. And when n becomes larger, the PDF for T_n becomes more and more like that for a standard normal random variable. Furthermore,

$$E[T_n] = 0 \text{ for } n > 1, \quad \text{Var}(T_n) = \frac{n}{n - 2} \text{ for } n > 2.$$

For example, in Fig. 1.5 we can see the PDF of T_n for $n = 2, 5, 10$ and for $Z \sim \mathcal{N}(0, 1)$.

The same as for z_α (Eq. 1.43) and $\chi_{\alpha,n}$ (Eq. 1.45), given $\alpha \in (0, 1)$, we define $t_{\alpha,n}$ such that

$$P(T_n \geq t_{\alpha,n}) = \alpha. \tag{1.46}$$

By symmetry of the PDF for T_n, we have

$$P(T_n \leq -t_{\alpha,n}) = P(T_n \geq t_{\alpha,n}) = \alpha,$$

and

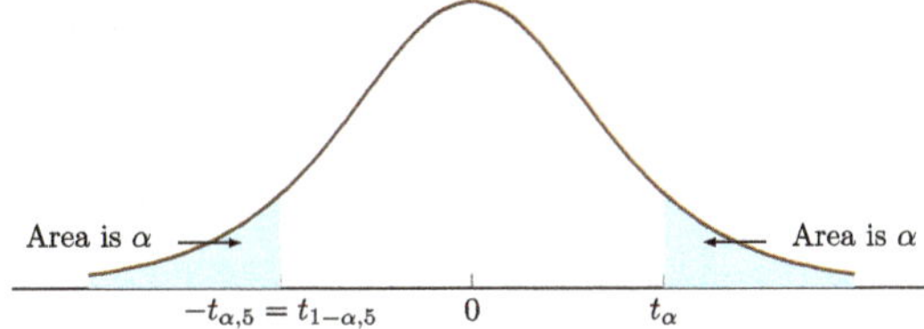

Fig. 1.6 Probability density function for T_5, $P(T_5 \geq t_{\alpha,5}) = \alpha$

$$P(T_n \geq -t_{\alpha,n}) = 1 - P(T_n < -t_{\alpha,n}) = 1 - \alpha \implies t_{1-\alpha,n} = -t_{\alpha,n}.$$

An illustration is shown in Fig. 1.6. We have

$$P(-t_{\alpha,n} \leq T \leq t_{\alpha,n}) = P(T_n \geq -t_{\alpha,n}) - P(T_n > t_{\alpha,n}) = 1 - 2\alpha,$$

which gives

$$P(-t_{\alpha/2,n} \leq T \leq t_{\alpha/2,n}) = 1 - \alpha, \tag{1.47}$$

or

$$P(|T| > t_{\alpha/2,n}) = \alpha. \tag{1.48}$$

The table of values of $t_{\alpha,n}$ can be found in standard books for statistics, see, e.g., [Ros20, Table A3].

Remark 1.8.2 For large values of n ($n \geq 30$), $t_{\alpha,n}$ can be approximated by z_α (see Table 1.4).

1.8.2 *Estimating Mean and Difference of Means of Normal Distributions*

In Sect. 1.7 we have discussed that a random experiment is an experiment whose output cannot be predicted with certainty in advance. However, if the experiment is repeated many times, we can see "regularity" in the average output. For a given random experiment, the *sample space*, denoted by Ω, is the set of all possible outcomes.

In this subsection, we are interested in a random variable $X : \Omega \to \mathbb{R}$ (see Definition 1.7.6) that induces a normal distribution. In particular, we assume $X \sim \mathcal{N}(\mu_x, \sigma_x^2)$ has mean μ_x and variance σ_x^2.

We will first discuss how to estimate μ_x. To do so, we repeat the random experiment n times and record the outcomes. Then the possible outcomes $\{X_1, X_2, \ldots, X_n\}$ are n independent identically distributed random variables. We refer to this set as a *sample*. An actual outcome for X_i, denoted x_i, is called a *realization* of X_i.

The *sample mean (empirical mean)*, denoted $\overline{X}$, is given by

$$\overline{X} := \frac{1}{n} \sum_{i=1}^{n} X_i. \tag{1.49}$$

Remark 1.8.3 It can be shown that the sum of independent normal random variables induces a normal distribution with mean (respectively, variance) given by the sum of the means (respectively, variances) of each random variable (see, e.g., [JP04, page 120] and also Eqs. 1.33 and 1.41).

Since $X_i \sim \mathcal{N}(\mu_x, \sigma_x^2)$ each are independent, together with Eqs. 1.33 and 1.41, we have

$$\mathrm{E}\left[\overline{X}\right] = \frac{1}{n} \sum_{i=1}^{n} \mathrm{E}\left[X_i\right] = \mu_x, \quad \mathrm{Var}(\overline{X}) = \frac{1}{n^2} \sum_{i=1}^{n} \mathrm{Var}(X_i) = \frac{\sigma_x^2}{n}.$$

By Remark 1.8.3,

$$\overline{X} \sim \mathcal{N}\left(\mu_x, \frac{\sigma_x^2}{n}\right), \tag{1.50}$$

i.e., the sample mean is a normal random variable with mean μ_x and variance σ_x^2/n. It follows from Remark 1.7.1 that

$$\frac{\overline{X} - \mu_x}{\sigma_x/\sqrt{n}} \sim \mathcal{N}(0, 1). \tag{1.51}$$

Similarly, for $i = 1, 2, \ldots, n$,

$$\frac{X_i - \mu_x}{\sigma_x} \sim \mathcal{N}(0, 1). \tag{1.52}$$

The *sample variance (empirical variance)*, denoted S_x^2, is given by

$$S_x^2 := \frac{1}{n-1} \sum_{i=1}^{n} (X_i - \overline{X})^2. \tag{1.53}$$

We note that

$$\sum_{i=1}^{n} (X_i - \mu_x)^2 = \sum_{i=1}^{n} ((X_i - \overline{X}) + (\overline{X} - \mu_x))^2$$

$$= n(\overline{X} - \mu_x)^2 + \sum_{i=1}^{n} (X_i - \overline{X})^2 + 2 \sum_{i=1}^{n} (X_i - \overline{X})(\overline{X} - \mu_x)$$

$$= n(\overline{X} - \mu_x)^2 + \sum_{i=1}^{n}(X_i - \overline{X})^2,$$

where

$$\sum_{i=1}^{n}(X_i - \overline{X}) = -n\overline{X} + \sum_{i=1}^{n} X_i = 0.$$

Dividing by σ_x^2, we get

$$\sum_{i=1}^{n}\left(\frac{X_i - \mu_x}{\sigma_x}\right)^2 = \left(\frac{\sqrt{n}(\overline{X} - \mu_x)}{\sigma_x}\right)^2 + \frac{\sum_{i=1}^{n}(X_i - \overline{X})^2}{\sigma_x^2}. \tag{1.54}$$

Since X_i are independent normal random variables, by Eq. 1.52 and Definition 1.8.1, the left-hand side of Eq. 1.54 induces a χ^2-distribution with n degrees of freedom. By Eq. 1.51 and Definition 1.8.1, the first term of the right-hand side of Eq. 1.54 induces a χ^2-distribution with 1 degree of freedom. By Remark 1.8.1, it is tempting to conclude that the two terms on the right-hand side of Eq. 1.54 are independent and the second term induces a χ^2-distribution with $n - 1$ degrees of freedom.

Such a result has indeed been proven. In particular, the proof of the following theorem was demonstrated in [Ros20, page 216][8]

Theorem 1.8.1 *The sample mean $\overline{X}$ and sample variance S_x^2 are independent random variables. Furthermore,*

$$\frac{(n-1)S_x^2}{\sigma_x^2} \sim \chi_{n-1}^2. \tag{1.55}$$

The above discussions give us the following useful result.

Lemma 1.8.1

$$\sqrt{n}\frac{\overline{X} - \mu_x}{S_x} \sim t_{n-1}.$$

Proof Since (see Eqs. 1.51 and 1.55)

$$\frac{\overline{X} - \mu_x}{\sigma_x/\sqrt{n}} \sim \mathcal{N}(0, 1), \qquad \frac{(n-1)S_x^2}{\sigma_x^2} \sim \chi_{n-1}^2,$$

by Definition 1.8.2,

---

[8] The results are only valid for a normal random variable X.

$$\frac{\sqrt{n}(\overline{X} - \mu_x)/\sigma_x}{\sqrt{S_x^2/\sigma_x^2}} = \sqrt{n}\frac{\overline{X} - \mu_x}{S_x} \sim t_{n-1}.$$

$\square$

Let Θ denote the subset of $\mathbb{R}$ that contains all possible values of μ_x. A *point estimator* of μ_x is a *function* with domain $\mathbb{R}^n$ and codomain Θ that is used to estimate the value of μ_x. We use a point in Θ for the estimation, hence the name point estimator.

Remark 1.8.4 For example, we can use the sample mean as a point estimator for μ_x. Similarly, we can use the sample variance as a point estimator for σ_x (see Example 4.2.1).

Example 1.8.1 (Sample correlation coefficient) Suppose U and W are two random variables. Let $\{(U_1, W_1), (U_2, W_2), \ldots, (U_n, W_n)\}$ be a sample for this pair of random variable (U, W). We further denote the sample mean and sample variance for $\{U_1, U_2, \ldots, U_n\}$ by $\overline{U}$ and S_u^2. Similarly, the sample mean and sample variance for $\{W_1, W_2, \ldots, W_n\}$ are denoted by $\overline{W}$ and S_w^2. Then, following Definition 1.7.11, we can define the *sample correlation coefficient*, denoted by r, as follows (see Eqs. 1.39 and 1.35):

$$r = \frac{\overline{UW} - \overline{U}\,\overline{W}}{\sqrt{S_u^2 S_w^2}} = \frac{\frac{1}{n}\sum_{i=1}^{n}(U_i - \overline{U})(W_i - \overline{W})}{\sqrt{\left(\frac{1}{n}\sum_{i=1}^{n}(U_i - \overline{U})^2\right)\left(\frac{1}{n}\sum_{i=1}^{n}(W_i - \overline{W})^2\right)}}$$

$$= \frac{\sum_{i=1}^{n}(U_i - \overline{U})(W_i - \overline{W})}{\sqrt{\sum_{i=1}^{n}(U_i - \overline{U})^2}\sqrt{\sum_{i=1}^{n}(W_i - \overline{W})^2}}. \tag{1.56}$$

Then, the sample correlation coefficient can be used as a point estimator for the correlation coefficient between U and W. We note that since the correlation coefficient analyzes the relations between U and W, we collect samples in pairs (U_i, W_i).

However, we do not expect μ_x to be exactly equal to the sample mean. Thus, we would like to specify an interval for which we have a certain degree of confidence that our parameter lies. We refer to such an estimator as an *interval estimator*.

For the rest of this part, let $\alpha \in (0, 1)$ be a real number. We recall the definitions of z_α and t_α from Eqs. 1.43 and 1.46, respectively.

Interval estimator for μ_x with known variance We first consider σ_x^2 to be known. By Eqs. 1.44 and 1.51,

$$P\left(-z_{\alpha/2} < \frac{\overline{X} - \mu_x}{\sigma_x/\sqrt{n}} < z_{\alpha/2}\right) = 1 - \alpha,$$

which gives

$$P\left(\overline{X} - z_{\alpha/2}\frac{\sigma_x}{\sqrt{n}} < \mu_x < \overline{X} + z_{\alpha/2}\frac{\sigma_x}{\sqrt{n}}\right) = 1 - \alpha.$$

Thus, the probability that μ_x lies between $\overline{X} - z_{\alpha/2}\frac{\sigma_x}{\sqrt{n}}$ and $\overline{X} + z_{\alpha/2}\frac{\sigma_x}{\sqrt{n}}$ is $1 - \alpha$. We say that

$$\left(\bar{x} - z_{\alpha/2}\frac{\sigma_x}{\sqrt{n}}, \quad \bar{x} + z_{\alpha/2}\frac{\sigma_x}{\sqrt{n}}\right) \tag{1.57}$$

is a $100(1 - \alpha)$ *percent confidence interval* for μ_x, where $\bar{x}$ is a realization of $\overline{X}$.

We define the *precision* of our estimate, denoted by c, to be

$$c := z_{\alpha/2}\frac{\sigma_x}{\sqrt{n}},$$

which is the length of half of the confidence interval. It measures how "close" is our estimate to μ_x. Consequently, to have an estimate with precision c and $100(1 - \alpha)$ confidence, the number of data in the sample should be at least (see Example 4.2.2)

$$n = \frac{\sigma_x^2}{c^2}z_{\alpha/2}^2. \tag{1.58}$$

Interval estimator for μ_x with unknown variance In case the variance is unknown, by Lemma 1.8.1 and Eq. 1.47, we have

$$P\left(-t_{\alpha/2,n-1} \leq \sqrt{n}\frac{\overline{X} - \mu_x}{S_x} \leq t_{\alpha/2,n-1}\right) = 1 - \alpha,$$

which gives

$$P\left(\overline{X} - t_{\alpha/2,n-1}\frac{S_x}{\sqrt{n}} \leq \mu_x \leq \overline{X} + t_{\alpha/2,n-1}\frac{S_x}{\sqrt{n}}\right) = 1 - \alpha.$$

Thus a $100(1-\alpha)$ *percent confidence interval* for μ_x is given by (see Example 4.2.2)

$$\left(\bar{x} - t_{\alpha/2,n-1}\frac{S_x}{\sqrt{n}}, \quad \bar{x} + t_{\alpha/2,n-1}\frac{S_x}{\sqrt{n}}\right). \tag{1.59}$$

Similarly, we can define the precision

$$c := t_{\alpha/2,n-1}\frac{s}{\sqrt{n}}.$$

Then to have an estimate with precision c and $100(1 - \alpha)$ confidence, the number of data required in the sample is given by

$$n = \frac{s_x^2}{c^2} t_{\alpha/2, n-1}^2.$$

By Remark 1.8.2, when n is large (≥ 30), $t_{\alpha, n}$ is close to z_α, and n can be estimated by (see Example 4.2.2)

$$n \approx \frac{s_x^2}{c^2} z_{\alpha/2}^2. \tag{1.60}$$

For the rest of this part, let Y be a normal random variable with mean μ_y and variance σ_y^2 that is independent from X. Let $\{Y_1, Y_2, \ldots, Y_m\}$ be a sample for Y with sample mean $\overline{Y}$ and sample variance S_y. We are interested in estimating $\mu_x - \mu_y$.

We note that since $\overline{X}$ and $\overline{Y}$ are point estimators for μ_x and μ_y, respectively, $\overline{X} - \overline{Y}$ is a point estimator for $\mu_x - \mu_y$.

Interval estimator for $\mu_x - \mu_y$ with known variances Suppose we know the values of σ_x^2 and σ_y^2. By Eq. 1.50,

$$\overline{X} \sim \mathcal{N}\left(\mu_x, \frac{\sigma_x^2}{n}\right), \quad \overline{Y} \sim \mathcal{N}\left(\mu_y, \frac{\sigma_y^2}{m}\right).$$

By Remark 1.8.3,

$$\mathrm{E}\left[\overline{X} - \overline{Y}\right] = \mu_x - \mu_y, \quad \mathrm{Var}\left(\overline{X} - \overline{Y}\right) = \frac{\sigma_x^2}{n} + \frac{\sigma_y^2}{m},$$

and

$$\overline{X} - \overline{Y} \sim \mathcal{N}\left(\mu_x - \mu_y, \frac{\sigma_x^2}{n} + \frac{\sigma_y^2}{m}\right) \implies \frac{\overline{X} - \overline{Y} - (\mu_x - \mu_y)}{\sqrt{\frac{\sigma_x^2}{n} + \frac{\sigma_y^2}{m}}} \sim \mathcal{N}(0, 1). \tag{1.61}$$

By Eq. 1.44, we have

$$P\left(-z_{\alpha/2} < \frac{\overline{X} - \overline{Y} - (\mu_x - \mu_y)}{\sqrt{\frac{\sigma_x^2}{n} + \frac{\sigma_y^2}{m}}} < z_{\alpha/2}\right) = 1 - \alpha.$$

A $100(1 - \alpha)$ confidence interval estimate for $\mu_x - \mu_y$ is then given by (see Example 4.2.3)

$$\left(\overline{x} - \overline{y} - z_{\alpha/2}\sqrt{\frac{\sigma_x^2}{n} + \frac{\sigma_y^2}{m}}, \quad \overline{x} - \overline{y} + z_{\alpha/2}\sqrt{\frac{\sigma_x^2}{n} + \frac{\sigma_y^2}{m}}\right). \tag{1.62}$$

The precision c is

$$c := z_{\alpha/2}\sqrt{\frac{\sigma_x^2}{n} + \frac{\sigma_y^2}{m}}.$$

If $m = n$, to have an estimate with precision c and $100(1 - \alpha)$ confidence, the number of data required in the sample is at least (see Example 4.2.3)

$$n = \frac{z_{\alpha/2}^2(\sigma_x^2 + \sigma_y^2)}{c^2}. \tag{1.63}$$

Interval estimator for $\mu_x - \mu_y$ with unknown equal variance Suppose $\sigma_x = \sigma_y$ is unknown. Let $\sigma = \sigma_x = \sigma_y$. By Eq. 1.55,

$$\frac{(n-1)S_x^2}{\sigma^2} \sim \chi_{n-1}^2, \qquad \frac{(m-1)S_y^2}{\sigma^2} \sim \chi_{m-1}^2.$$

Since we assume the samples are independent, those two χ^2 random variables are independent. By Remark 1.8.1, we have

$$\frac{(n-1)S_x^2}{\sigma^2} + \frac{(m-1)S_y^2}{\sigma^2} \sim \chi_{m+n-2}^2. \tag{1.64}$$

Let

$$S_p^2 := \frac{(n-1)S_x^2 + (m-1)S_y^2}{n+m-2}. \tag{1.65}$$

By Theorem 1.8.1, $\overline{X}$, S_x^2, $\overline{Y}$, S_y^2 are independent. By Definition 1.8.2 and Eqs. 1.61 and 1.64,

$$\frac{\overline{X} - \overline{Y} - (\mu_x - \mu_y)}{\sqrt{\frac{\sigma^2}{n} + \frac{\sigma^2}{m}}\sqrt{S_p^2/\sigma^2}} = \frac{\overline{X} - \overline{Y} - (\mu_x - \mu_y)}{S_p\sqrt{1/n + 1/m}} \sim t_{n+m-2}. \tag{1.66}$$

Then according to Eq. 1.47,

$$P\left(-t_{\alpha/2,n+m-2} \leq \frac{\overline{X} - \overline{Y} - (\mu_x - \mu_y)}{S_p\sqrt{1/n + 1/m}} \leq t_{\alpha/2,n+m-2}\right) = 1 - \alpha.$$

A $100(1 - \alpha)$ confidence interval estimate for $\mu_x - \mu_y$ is then given by

$$\left(\bar{x} - \bar{y} - t_{\alpha/2, n+m-2} s_p \sqrt{1/n + 1/m}, \quad \bar{x} - \bar{y} + t_{\alpha/2, n+m-2} s_p \sqrt{1/n + 1/m} \right).$$

If we assume $m = n$, to have an estimate with precision c and $100(1-\alpha)$ confidence, the number of data required in the sample is at least

$$n = \frac{2 t_{\alpha/2, 2n-2}^2 s_p^2}{c^2}.$$

For large n ($n \geq 30$), we can approximate n by (see Example 4.2.3)

$$n \approx \frac{2 z_{\alpha/2}^2 s_p^2}{c^2}. \tag{1.67}$$

1.8.3 Hypothesis Testing

By *statistical hypothesis*, we refer to a statement about the unknown parameters of a distribution (see Example 4.2.5). We call such a statement hypothesis because it is not known whether or not it is true. In this subsection, we will use samples from the distribution to draw certain conclusions regarding a given hypothesis about its unknown parameters. In particular, we will introduce a procedure for determining whether or not the values of a sample are consistent with the hypothesis. The decision will then be either to *accept* the hypothesis or to *reject* it. By accepting a hypothesis, we conclude that *the resulting data from the sample appear to be consistent with it.*

The same as in Sect. 1.8.2, we consider a normal random variable X with mean μ_x and variance σ_x^2. Furthermore, let $\{ X_1, X_2, \ldots X_n \}$ denote a sample from the distribution induced by X with sample mean $\bar{X}$ and sample variance S_x^2. We would like to test hypotheses about μ_x using data from this sample.

The hypothesis that we want to test is called the *null hypothesis*, denoted by H_0. For example,

$$H_0 : \mu_x = 1, \quad H_0 : \mu_x \geq 0.$$

We will test the null hypothesis against an *alternative hypothesis*, denoted by H_1. For example,

$$H_1 : \mu_x \neq 1, \quad H_1 : \mu_x > 1.$$

To test the hypothesis, we define a region C such that

if a sample $\{x_1, x_2, \ldots, x_n\} \notin C$, we accept the null hypothesis H_0.

And

if a sample $\{x_1, x_2, \ldots, x_n\} \in C$, we reject the null hypothesis H_0.

C is called the *critical region*. We also define the *level of significance of the test*, denoted by α, such that when H_0 is true, the probability of rejecting it is not bigger than α, namely

$$P\left(\{x_1, x_2, \ldots, x_n\} \in C \,|\, H_0 \text{ is true}\right) \leq \alpha.$$

Thus, the main procedure in our hypothesis testing is to find the critical region C given a level of significance α.

Two-sided hypothesis testing concerning μ_x Let $\mu_0 \in \mathbb{R}$ be a constant. We set the null hypothesis and the alternative hypothesis as follows:

$$H_0 : \mu_x = \mu_0, \quad H_1 : \mu_x \neq \mu_0. \tag{1.68}$$

Recall that the sample mean, $\overline{X}$, is a point estimator for μ_x (see Remark 1.8.4). Then it is reasonable to accept H_0 if $\overline{X}$ is not too far from μ_0. Given α, we choose the critical region to be

$$C = \left\{ (X_1, X_2, \ldots, X_n) \mid |\overline{X} - \mu_0| > c \right\}, \tag{1.69}$$

where c is a number such that if $\overline{X} = \mu_0$,

$$P(|\overline{X} - \mu_0| > c) = \alpha. \tag{1.70}$$

Then our main task is to find c that satisfies the above equation.

Suppose the variance σ_x^2 is known. If $\overline{X} = \mu_0$, then by Eq. 1.50,

$$\overline{X} \sim \mathcal{N}\left(\mu_0, \frac{\sigma^2}{n}\right).$$

Define

$$Z := \frac{\overline{X} - \mu_0}{\sigma/\sqrt{n}}, \tag{1.71}$$

by Remark 1.7.1, $Z \sim \mathcal{N}(0, 1)$. According to Eq. 1.70, we can choose c such that

$$P\left(|Z| > \frac{c\sqrt{n}}{\sigma}\right) = \alpha \implies 2P\left(Z > \frac{c\sqrt{n}}{\sigma}\right) = \alpha \implies P\left(Z > \frac{c\sqrt{n}}{\sigma}\right) = \frac{\alpha}{2}.$$

By the definition of z_α (Eq. 1.43),

$$\frac{c\sqrt{n}}{\sigma} = z_{\alpha/2} \implies c = \frac{z_{\alpha/2}\sigma}{\sqrt{n}}. \tag{1.72}$$

And the critical region for significance level α is given by

$$C = \left\{ (X_1, X_2, \ldots, X_n) \ \middle| \ |\overline{X} - \mu_0| > \frac{z_{\alpha/2}\sigma}{\sqrt{n}} \right\}. \tag{1.73}$$

Consequently, we reject the null hypothesis (Eq. 1.68) if

$$|\bar{x} - \mu_0| > z_{\alpha/2}\frac{\sigma}{\sqrt{n}}, \quad \text{i.e.,} \quad \frac{\sqrt{n}}{\sigma}|\bar{x} - \mu_0| > z_{\alpha/2}$$

and accept H_0 otherwise (see Example 4.2.6).

Remark 1.8.5 We can see that when $\mu_0 = 0$, the critical region corresponding to the level of significance α in Eq. 1.73 is the complement of the $100(1 - \alpha)$ percent confidence interval for μ_x (Eq. 1.57).

Suppose we do not know the variance σ_x^2. Recall that sample variance S_x^2 (Eq. 1.53) is a point estimator for σ_x^2. Similar to Eq. 1.71, we are interested in the following random variable:

$$T := \frac{\sqrt{n}(\overline{X} - \mu_0)}{S_x}. \tag{1.74}$$

We want to find c such that when $\mu_x = \mu_0$,

$$P(|T| > c) = \alpha.$$

Note that when $\mu = \mu_0$, T induces a t-distribution with $n - 1$ degrees of freedom. By Eq. 1.48, we choose

$$c = t_{\alpha/2,n-1}.$$

Hence to achieve the level of significance α, we reject H_0 (Eq. 1.68) if

$$\left| \frac{\sqrt{n}(\bar{x} - \mu_0)}{s_x} \right| > t_{\alpha/2,n-1}$$

and accept H_0 otherwise.

One-sided hypothesis testing concerning μ_x Now we consider the same null hypothesis with a different alternative hypothesis as follows:

$$H_0 : \mu = \mu_0, \quad H_1 : \mu > \mu_0. \tag{1.75}$$

We refer to such a test as *one-sided test.*

In this case, we will reject H_0 when $\overline{X}$ is much bigger than u_0 since when $\overline{X}$ is smaller, it is more likely for H_0 to be true than for H_1 to be true. In other words, the critical region is of the following form:

$$C = \left\{ (X_1, X_2, \ldots, X_n) \,\middle|\, \overline{X} - \mu_0 > c \right\}. \tag{1.76}$$

To find the value of c, we assume H_0 is true. Then by definition, c should be chosen such that

$$P(\overline{X} - \mu_0 > c) = \alpha.$$

In case the variance σ_x^2 is known, by the definition of Z (Eq. 1.71),

$$P\left(Z > \frac{c\sqrt{n}}{\sigma} \right) = \alpha.$$

By the definition of z_α (Eq. 1.43),

$$\frac{c\sqrt{n}}{\sigma} = z_\alpha \implies c = \frac{z_\alpha \sigma}{\sqrt{n}}. \tag{1.77}$$

The critical region for significance level α is then given by

$$C = \left\{ (X_1, X_2, \ldots, X_n) \,\middle|\, \overline{X} - \mu_0 > \frac{z_\alpha \sigma}{\sqrt{n}} \right\}.$$

Thus, we reject the null hypothesis (Eq. 1.75) if

$$\overline{x} - \mu_0 > \frac{z_\alpha \sigma}{\sqrt{n}}, \quad \text{i.e.,} \quad \frac{\sqrt{n}}{\sigma}(\overline{x} - \mu_0) > z_\alpha$$

and accept H_0 otherwise (see Example 4.2.7).

Suppose we know a good estimate of c for the critical region in Eq. 1.76. Let $\mu_0 = 0$. We have

$$C = \left\{ (X_1, X_2, \ldots, X_n) \,\middle|\, \overline{X} > c \right\}. \tag{1.78}$$

Then by Eq. 1.77, to test whether μ_x is different from 0 with significance level α, the number of data required is at least (see Example 4.2.7)

$$n = \frac{\sigma^2}{c^2} z_\alpha^2. \tag{1.79}$$

In case we do not know the variance σ_x^2. By the definition of T (Eq. 1.74), we have

$$P\left(T > \frac{c\sqrt{n}}{S_x}\right) = \alpha.$$

Then according to Eq. 1.46,

$$\frac{c\sqrt{n}}{S_x} = t_{\alpha,n-1} \implies c = \frac{t_{\alpha,n-1} S_x}{\sqrt{n}}.$$

Thus the significance level α test is to reject H_0 (Eq. 1.75) if

$$\frac{\sqrt{n}(\bar{x} - \mu_0)}{s_x} > t_{\alpha,n-1}$$

and accept H_0 otherwise. When n is large ($\geq$ 30), we reject H_0 if (see Example 4.2.7)

$$\frac{\sqrt{n}(\bar{x} - \mu_0)}{s_x} > z_\alpha. \tag{1.80}$$

Suppose we want to test if the mean μ_x is bigger than 0 with significance level α, and we have a good estimate for c. Set $\mu_0 = 0$. The number of data required is at least

$$n = \frac{s_x^2}{c^2} t_{\alpha,n-1}^2.$$

For large n ($n \geq 30$), we have (see Example 4.2.7)

$$n = \frac{s_x^2}{c^2} z_\alpha^2. \tag{1.81}$$

Two-sided hypothesis testing about μ_x and μ_y For the rest of this part, let Y denote a normal random variable independent from X with mean μ_y and variance σ_y^2. Furthermore, let $\{Y_1, Y_2, \ldots, Y_m\}$ denote a sample from the distribution induced by Y with sample mean $\bar{Y}$ and sample variance S_y^2.

We would like to test the following hypotheses:

$$H_0 : \mu_x = \mu_y, \quad H_1 : \mu_x \neq \mu_y. \tag{1.82}$$

Since $\bar{X}$ and $\bar{Y}$ are point estimators for μ_x and μ_y, respectively, $\bar{X} - \bar{Y}$ is a point estimator for $\mu_x - \mu_y$. Then it is reasonable to reject H_0 when $|\bar{X} - \bar{Y}|$ is far from zero. Given α, our critical region is of the form

$$C = \left\{ (X_1, X_2, \ldots, X_n, Y_1, Y_2 \ldots, Y_m) \mid |\overline{X} - \overline{Y}| > c \right\}, \tag{1.83}$$

where c is chosen such that

$$P(|\overline{X} - \overline{Y}| > c|H_0 \text{ is true}) = P(|\overline{X} - \overline{Y}| > c|\mu_x = \mu_y) = \alpha. \tag{1.84}$$

Our task is to decide how to choose the value of c.

In case the variances σ_x^2 and σ_y^2 are known, by Eq. 1.61, when $\mu_x = \mu_y$ (i.e., when H_0 is true), we have

$$\frac{\overline{X} - \overline{Y}}{\sqrt{\dfrac{\sigma_x^2}{n} + \dfrac{\sigma_y^2}{m}}} \sim \mathcal{N}(0, 1). \tag{1.85}$$

By Eq. 1.44,

$$P\left(-z_{\alpha/2} < \frac{\overline{X} - \overline{Y}}{\sqrt{\dfrac{\sigma_x^2}{n} + \dfrac{\sigma_y^2}{m}}} < z_{\alpha/2} \right) = 1 - \alpha \implies P\left(\frac{|\overline{X} - \overline{Y}|}{\sqrt{\dfrac{\sigma_x^2}{n} + \dfrac{\sigma_y^2}{m}}} > z_{\alpha/2} \right) = \alpha.$$

Thus, we let

$$c = z_{\alpha/2}\sqrt{\frac{\sigma_x^2}{n} + \frac{\sigma_y^2}{m}}. \tag{1.86}$$

To achieve significance level α, we reject H_0 if

$$|\overline{x} - \overline{y}| > z_{\alpha/2}\sqrt{\frac{\sigma_x^2}{n} + \frac{\sigma_y^2}{m}}$$

and accept H_0 otherwise (see Example 4.2.8).

Furthermore, suppose $m = n$, and we have a good estimate for c. To test if $\mu_x \neq \mu_y$ with significance level α, the number of data required is at least (see Example 4.2.8)

$$n = \frac{z_{\alpha/2}^2(\sigma_x^2 + \sigma_y^2)}{c^2}. \tag{1.87}$$

In case the variances are unknown but we know that $\sigma_x = \sigma_y$. Let $\sigma = \sigma_x = \sigma_y$. By Eq. 1.66, when $\mu_x = \mu_y$,

$$\frac{\overline{X} - \overline{Y}}{\sqrt{S_p^2(1/n + 1/m)}} \sim t_{n+m-2}.$$

According to Eq. 1.48,

$$P\left(\left|\frac{\overline{X} - \overline{Y}}{\sqrt{S_p^2(1/n + 1/m)}}\right| > t_{\alpha/2, n+m-2}\right) = \alpha.$$

Thus, we let

$$c = t_{\alpha/2, n+m-2}\sqrt{S_p^2(1/n + 1/m)}.$$

For a test with significance level α, we reject H_0 if

$$|\overline{x} - \overline{y}| > t_{\alpha/2, n+m-2}\sqrt{s_p^2(1/n + 1/m)}$$

and accept H_0 otherwise. Such a test is called the *student's t−test*.

For large n and m, we reject H_0 if (see Example 4.2.11)

$$|\overline{x} - \overline{y}| > z_{\alpha/2}\sqrt{s_p^2(1/n + 1/m)}, \quad \text{or equivalently,} \quad \frac{|\overline{x} - \overline{y}|}{\sqrt{s_p^2(1/n + 1/m)}} > z_{\alpha/2}.$$

$$(1.88)$$

Furthermore, when $n = m$, we have (see Eq. 1.65)

$$S_p^2(1/n + 1/m) = \frac{(n-1)S_x^2 + (n-1)S_y^2}{2n - 2} \times \frac{2}{n} = \frac{S_x^2 + S_y^2}{n},$$

and we reject H_0 if (see Examples 4.2.8 and 4.2.9)

$$\frac{|\overline{x} - \overline{y}|}{\sqrt{\frac{s_x^2 + s_y^2}{n}}} > z_{\alpha/2}. \tag{1.89}$$

In this case, suppose we have a good estimate for c, to have a student's t-test with significant level α, and the number of data we need for both samples is given by (see Examples 4.2.8 and 4.2.9)

$$n = z_{\alpha/2}^2 \frac{S_x^2 + S_y^2}{c^2}. \tag{1.90}$$

If we further assume that the unknown variances σ_x^2 and σ_y^2 are not equal, it can be shown that [Wel47]

$$\frac{\overline{X} - \overline{Y}}{\sqrt{\frac{S_x^2}{n} + \frac{S_y^2}{m}}} \sim t_\nu,$$

where

$$v \approx \frac{(S_x^2/n + S_y^2/m)^2}{(S_x^2/n)^2/(n-1) + (S_y^2/m)^2/(m-1)}.$$

And a test with significance level α rejects H_0 if

$$\frac{|\bar{x} - \bar{y}|}{\sqrt{\frac{s_x^2}{n} + \frac{s_y^2}{m}}} > t_{\alpha/2, v}.$$

Such a test is called *Welch's t-test*.

When n and m are big (≥ 30), we test if (see Example 4.2.13)

$$\frac{|\bar{x} - \bar{y}|}{\sqrt{\frac{s_x^2}{n} + \frac{s_y^2}{m}}} > z_{\alpha/2}. \tag{1.91}$$

Remark 1.8.6 Note that when $n = m$ is big (≥ 30), Welch's t-test and the student's t-test have the same formula (see Eq. 1.89 and 1.91).

Both student's t-test and Welch's t-test will be useful for leakage assessment in Sect. 4.2.3.

One-sided hypothesis testing about μ_x and μ_y For one-sided testing, we consider the following null and alternative hypotheses:

$$H_0 : \mu_x = \mu_y, \quad H_1 : \mu_x > \mu_y.$$

Similar to Eq. 1.76, our critical region is given by

$$C = \left\{ (X_1, X_2, \ldots, X_n, Y_1, Y_2, \ldots, Y_m) \mid \bar{X} - \bar{Y} > c \right\}, \tag{1.92}$$

where c is chosen such that

$$P\left(\bar{X} - \bar{Y} > c \mid \mu_x = \mu_y\right) = \alpha.$$

We will only discuss the case when σ_x^2 and σ_y^2 are known. For unknown variances, we refer the readers to [Wel47]. By Eqs. 1.85 and 1.43,

$$P\left(\frac{\bar{X} - \bar{Y}}{\sqrt{\frac{\sigma_x^2}{n} + \frac{\sigma_y^2}{m}}} > z_\alpha\right) = \alpha.$$

Thus, we choose

$$c = z_\alpha \sqrt{\frac{\sigma_x^2}{n} + \frac{\sigma_y^2}{m}}. \tag{1.93}$$

To achieve the level of significance α, we reject H_0 if

$$\overline{x} - \overline{y} > z_\alpha \sqrt{\frac{\sigma_x^2}{n} + \frac{\sigma_y^2}{m}}$$

and accept H_0 otherwise (see Example 4.2.10).

Furthermore, suppose $m = n$, we have a good estimate for c, and we know that $\mu_x \geq \mu_y$. To test if $\mu_x \neq \mu_y$, the number of data required is at least (see Example 4.2.10)

$$n = \frac{z_\alpha^2 (\sigma_x^2 + \sigma_y^2)}{c^2}. \tag{1.94}$$

1.9 Further Reading

For more detailed discussions on sets, functions, number theory, and abstract algebra, we refer the readers to [Her96, Chapters 1–6] and a series of lecture notes from Frédérique Oggier [Ogg].

[LX04] provides more in-depth studies for finite fields and coding theory.

For probability theory, we refer the readers to [Dur19] and [JP04] for a thorough analysis and [Ros20] for practical examples. [Ros20] also provides more insights on statistical methods presented in Sect. 1.8.

Chapter 2
Introduction to Cryptography

Before we dive into the modern cryptographic algorithms that are in use today (Chap. 3), we give an introduction to cryptography in general (Sect. 2.1) and discuss some classical ciphers that were designed a few centuries back (Sect. 2.2). In the end, we will discuss how cryptographic algorithms are actually used with different encryption modes (Sect. 2.3).

We start with a definition of cryptography.

Definition 2.0.1 *Cryptography* studies techniques that allow secure communication in the presence of adversarial behavior. These techniques are related to information security attributes such as *confidentiality, integrity, authentication*, and *non-repudiation*.

Below, we give more details on the information security attributes that can be achieved by using cryptography:

1. *Confidentiality* aims at preventing unauthorized disclosure of information. There are various technical, administrative, physical, and legal means to enforce confidentiality. In the context of cryptography, we are mostly interested in utilizing various encryption techniques to keep information private.
2. *Integrity* aims at preventing unauthorized alteration of data to keep them correct, authentic, and reliable. Similarly to confidentiality, while there are many means of ensuring data integrity, in cryptography we are looking at hash functions and message authentication codes.
3. *Authentication* aims at determining whether something or someone is who they claim they are. In communication, the entities should be able to identify each other. Similarly, the properties of the exchanged information, such as origin, content, and timestamp, should be authenticated. In cryptography, we are mostly interested in two aspects: entity authentication and data origin authentication. For these purposes, signatures, and identification, primitives are used.
4. *Non-repudiation* aims at assuring that the sender of the information is provided with proof of delivery, and the recipient is provided with proof of the sender's

© The Author(s), under exclusive license to Springer Nature Switzerland AG 2024
X. Hou, J. Breier, *Cryptography and Embedded Systems Security*,
https://doi.org/10.1007/978-3-031-62205-2_2

identity so that neither party can later deny the actions taken. Similarly to authentication, signatures, and identification, primitives are cryptographic means of supporting non-repudiation.

Note
CIA Triad is a widely utilized information security model, where the abbreviation stands for confidentiality, integrity, and availability. Therefore, a curious reader might be interested in knowing why we did not mention the availability. The answer is rather simple—there are no techniques within cryptography that could contribute in one way or another to ensure availability. Availability attribute ensures that information is consistently and readily accessible for authorized entities. One needs to look into other means of supporting this attribute.

2.1 Cryptographic Primitives

Cryptographic primitives are the tools that can be used to achieve the goals listed in Definition 2.0.1. The categorization of cryptographic primitives is depicted in Fig. 2.1. We have highlighted the ones that will be discussed in more detail in this book, especially regarding hardware attacks.

Let us briefly explain each primitive.

- Hash functions: Hash functions map data of arbitrary length to a binary array of some fixed length. We provide more details on hash functions in Sect. 2.1.1.
- Public key ciphers: Public key (or asymmetric) ciphers use a pair of related keys. This pair consists of a *private* key and a *public* key. These keys are generated by cryptographic algorithms that are based on mathematical problems called *one-*

Fig. 2.1 Categorization of cryptographic primitives. The ones highlighted in blue color will be discussed in this book

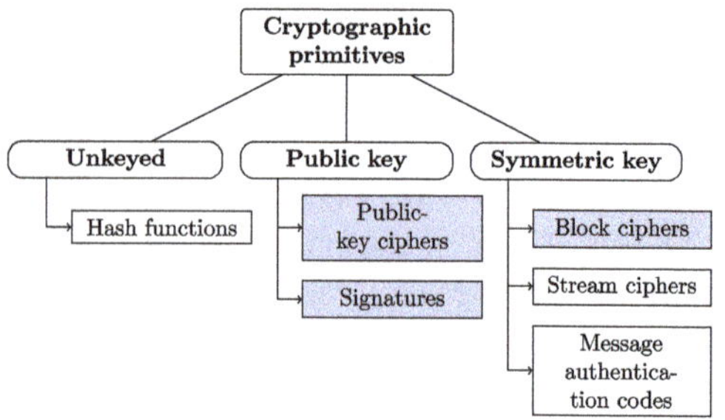

way functions. A one-way function is a function that is easy to compute on every input, but it is hard to compute its inverse.[1]

- Signatures: Digital signatures provide means for an entity to bind its identity to a message. This normally means that the sender uses their private key to sign the (hashed) message. Whoever has access to the public key can then verify the origin of the message.
- (Symmetric) block ciphers: Block ciphers are cryptographic algorithms operating on blocks of data of a fixed size (generally multiples of bytes for modern cipher designs). They use the same secret key for the encryption and decryption of data. Block ciphers are detailed in Sect. 2.1.2. Three modern block ciphers are discussed in Sect. 3.1.
- Stream ciphers: Stream ciphers are symmetric key ciphers that combine plaintext digits (usually bits) with the *keystream*, which is a stream of pseudorandom digits generated by the cipher. The combination is normally done by a bitwise XOR operation. The idea of stream ciphers comes from the one-time pad (Sect. 2.2.7).
- Message authentication codes (MACs): A message authentication code is a piece of information that is used to authenticate the origin of the message and to protect its integrity. MAC algorithms are commonly constructed from other cryptographic primitives, such as hash functions and block ciphers.

2.1.1 Hash Functions

A hash function is a computationally efficient function mapping data of arbitrary length to a binary array of some fixed length, called *hash values* or *message digests*.

The following are the properties that should be met in a properly designed cryptographic hash function:

(a) It is quick to compute a hash value for any given input.
(b) It is computationally infeasible to generate an input that yields a given hash value (a preimage).
(c) It is computationally infeasible to find a second input that maps to the same hash value when one input is already known (a second preimage).
(d) It is computationally infeasible to find any pair of different messages that produce the same hash value (a collision).

Cryptographic hash functions are mostly used for integrity and digital signatures. Message integrity use case of hash functions works as follows. The user creates a message digest of the original message at some point in time. At a later time (e.g., after a transmission), the digest is calculated again to check whether there have been any changes to the original message. In digital signatures, it is common to first

---

[1] It is worth noting that the existence of one-way functions is an open conjecture and depends on $P \neq NP$ inequality.

create a message digest that is afterward digitally signed, rather than signing the entire message which can be slow in case the message is large (see Sect. 3.4).

The current NIST standard for hash functions was released in 2015 and is called Secure Hash Algorithm 3 (SHA-3) [Dwo15]. It is based on Keccak permutation [BDPA13] which uses a previously developed *sponge* construction [BDPVA07].

2.1.2 Cryptosystems

We have mentioned three types of ciphers: public key ciphers, block ciphers, and stream ciphers. In this subsection, we will provide more discussions on ciphers, which are also called cryptosystems.

When we use ciphers, we normally assume insecure communication. A popular example setting is that Alice would like to send messages to Bob, but Eve is also listening to the communication. The goal of Alice is to make sure that even if Eve can intercept what was sent, she will not be able to find the original message. To do so, Alice will first *encrypt* the message, or the *plaintext*, and send the *ciphertext* to Bob, instead of the original message. Bob will then *decrypt* the ciphertext to get the plaintext. For this communication to work, there must be a *key* for encryption and decryption. It is clear that the decryption key should be secret from Eve, and a basic requirement is that the algorithm for *encryption/decryption* should be designed in a way that Eve cannot easily brute force the plaintext with the knowledge of the ciphertext.

Definition 2.1.1 A *cryptosystem* is a tuple $(\mathcal{P}, \mathcal{C}, \mathcal{K}, \mathcal{E}, \mathcal{D})$ with the following properties:

- $\mathcal{P}$ is a finite set of plaintexts, called *plaintext space*.
- $\mathcal{C}$ is a finite set of ciphertexts, called *ciphertext space*.
- $\mathcal{K}$ is a finite set of keys, called *key space*.
- $\mathcal{E} = \{\, E_k : k \in \mathcal{K}\,\}$, where $E_k : \mathcal{P} \to \mathcal{C}$ is an *encryption function*.
- $\mathcal{D} = \{\, D_k : k \in \mathcal{K}\,\}$, where $D_k : \mathcal{C} \to \mathcal{P}$ is a *decryption function*.
- For each $e \in \mathcal{K}$, there exists $d \in \mathcal{K}$ such that $D_d(E_e(p)) = p$ for all $p \in \mathcal{P}$.

If $e = d$, the cryptosystem is called a *symmetric (key) cryptosystem*. Otherwise, it is called a *public key/asymmetric cryptosystem*.

Take any $c_1 = E_e(p_1), c_2 = E_e(p_2)$ from the ciphertext space $\mathcal{C}$, where $e \in \mathcal{K}$. Let $d \in \mathcal{K}$ be the corresponding decryption key for e. If $c_1 = c_2$, then by definition,

$$p_1 = D_d(c_1) = D_d(c_2) = p_2.$$

Thus, E_e is an injective function (see Definition 1.1.2). We also note that if $\mathcal{P} = \mathcal{C}$, E_k is a permutation of $\mathcal{P}$ (see Definition 1.2.3).

There are mainly two types of symmetric ciphers: *block ciphers* and *stream ciphers*.

Definition 2.1.2 (Block Cipher) A *block cipher* is a symmetric key cryptosystem with $\mathcal{P} = \mathcal{C} = \mathcal{A}^n$ for some alphabet $\mathcal{A}$ and positive integer n. n is called the *block length*.

For classical ciphers that we will see in Sects. 2.2.1–2.2.5, $\mathcal{A} = \mathbb{Z}_{26}$. For modern cryptosystems that we will discuss in Sect. 3.1, $\mathcal{A} = \mathbb{F}_2 = \{0, 1\}$.

Now, if we have a long plaintext $\boldsymbol{p} = p_1 p_2 \ldots$, where each $p_i \in \mathcal{A}^n$ is one block of plaintext, and a key k, using a block cipher, we can obtain ciphertext string $\boldsymbol{c}$ as follows:[2]

$$\boldsymbol{c} = c_1 c_2 \cdots = e_k(p_1) e_k(p_2) \ldots.$$

But, for a *stream cipher*, $\mathcal{P} = \mathcal{C} = \mathcal{A}$ are single digits. Encryptions are computed on each digit of the plaintext. In particular, suppose we have a plaintext string $\boldsymbol{p} = p_1 p_2 \ldots$ (where $p_i \in \mathcal{A}$) and a key k. We first compute a key stream $\boldsymbol{z} = z_1 z_2 \ldots$ using the key k; then the ciphertext is obtained as follows:

$$\boldsymbol{c} = c_1 c_2 \cdots = e_{z_1}(p_1) e_{z_2}(p_2) \ldots$$

A stream cipher is said to be *synchronous* if the key stream only depends on the chosen key k but not on the encrypted plaintext. In this case, the sender and the receiver can both compute the keystream synchronously. In Sect. 2.2.7 we will see a classical synchronous stream cipher called *one-time pad*.

2.1.2.1 Converting Message to Plaintext

An important aspect to clarify is how the message that Alice intends to send is represented as plaintext.

For classical ciphers that we will discuss in Sect. 2.2, we will only consider messages consisting of English letters (A–Z), and we map each letter to an element in $\mathbb{Z}_{26}$. Table 2.1 lists the details of the mapping from letters to $\mathbb{Z}_{26}$. Thus the plaintext spaces are vector spaces over $\mathbb{Z}_{26}$.

Table 2.1 Converting English letters to elements in $\mathbb{Z}_{26}$

A	B	C	D	E	F	G	H	I	J	K	L	M	N	O	P	Q	R	S	T
0	1	2	3	4	5	6	7	8	9	10	11	12	13	14	15	16	17	18	19

U	V	W	X	Y	Z
20	21	22	23	24	25

---

[2] Such an encryption mode is called an ECB mode, and more encryption modes will be introduced in Sect. 2.3.

Table 2.2 Examples of methods for converting message symbols to bytes. The second column in each table is the binary representation of the byte value, and the third column is the corresponding hexadecimal representation

(a) ASCII		
A	01000001	41
B	01000010	42
a	01100001	61
b	01100010	62
$?$	00111111	3F

(b) UTF-8		
Á	11000001	C1
Ä	11000100	C4
Í	11001101	CD
×	11010111	D7
÷	11110111	F7

In modern computers, we store data in binary digits, which can be viewed as variables ranging over $\mathbb{F}_2$, or bits (see Definition 1.2.17). An 8-bit binary string is called a byte (see Definition 1.3.7). Computers often operate on a few bytes at a time. For example, a 64-bit processor operates on 8 bytes at a time. In computer architecture, a *word* is defined as the unit of data of (at most) a certain bit length that can be addressed and moved between storage and the processor. Therefore, for a 64-bit processor, the *word size* is 64 bits.

We have discussed that a byte can be represented as a decimal number between 0 and 255 or as a hexadecimal number between 00_{16} and FF_{16} (see Remark 1.3.3). When modern cryptographic algorithms are used, the messages are converted to plaintexts which are n-bit binary strings (i.e., vectors in $\mathbb{F}_2^n$), where n is a multiple of 8. For example, Table 2.2 lists the representation of some single symbols as bytes using ASCII and UTF-8 conversion methods. The second column gives the binary representation of the byte value, and the third column is the corresponding hexadecimal representation.

2.1.3   Security of Cryptosystems

When the security of a cryptosystem is analyzed, *Kerckhoffs' principle* is always followed.

Definition 2.1.3 (Kerckhoffs' Principle)   The security of a cryptosystem should depend only on the secrecy of the key.

In other words, everything is public knowledge except for the secret key.

To discuss the security of cryptosystems, we should also specify the attack assumptions. Normally, they consist of the attacker's knowledge and the attacker's goal. *Ciphertext-only attack* assumes the attacker has access to a collection of ciphertexts. *Known plaintext attack* assumes the attacker has a collection of plaintext and ciphertext pairs. And in *chosen plaintext attack*, the attacker has access to the encryption mechanism such that they can choose plaintexts and obtain the corresponding ciphertexts. The attacker's goal can be the recovery of the plaintext or the recovery of the key.

By Kerckhoffs' principle (Definition 2.1.3), we assume the attacker has knowledge of the cipher design and communication context, e.g., the sender is a student and might use words like "exam," "assignment," etc.

A ciphertext-only attack scenario is the weakest attacker model and also the most realistic one. For example, an intercepted encrypted network traffic falls into this category. As an example of a known plaintext attack scenario, one can think of the cryptanalysis of Enigma during World War II. There were situations when the German military broadcast the same message encrypted by different cryptosystems—for some recipients, it was encrypted by a so-called *dockyard cipher* (a manual cipher, relatively easy to cryptanalyze), and for some, it was encrypted by Enigma [Mah45]. If both messages were intercepted, the allies would possess both the plaintext and the ciphertext, thus making it a known plaintext attack on Enigma. When it comes to chosen plaintext attacks, one can imagine a scenario when an encryption device is captured, and the attacker can send queries to it and receive the ciphertexts. As the key would normally be stored in secure storage, the attacker needs to use the plaintext–ciphertext pairs to recover it. This is a common scenario for hardware attacks. While in the traditional cryptanalysis setting, a chosen plaintext attack is infeasible for modern ciphers, hardware attacks can recover the key relatively efficiently, depending on the attacker's assumptions and the attack type.

In this book, we say a cipher is *broken* if the secret key is recovered.[3] A cipher is said to be *perfectly secure* if, in a ciphertext-only attack setting, the attacker cannot obtain any information about the plaintext no matter how much computing power they has. A cipher is *secure in practice* if there is no known attack that can break it within a reasonable amount of time and with a reasonable amount of computing power. A cipher is said to be *computationally secure* if breaking it requires computing power that is not available in practice.

In Sect. 2.2.7, we will introduce a classical cipher that achieves perfect secrecy. However, we will see that the key management of this cipher makes it impractical for modern usage. Modern cryptosystems that are popular today are considered to be computationally secure. Most of the ciphers are designed in a way that the effort taken to break them grows exponentially with the number of bits of the secret key, which is called *key length*. Thus, key length is an important factor in the security of modern ciphers.

2.2 Classical Ciphers

In this section, we will discuss some classical ciphers and analyze their security. We focus on the case when messages consist of English letters. Those letters are

---

[3] In a more general sense, breaking a cipher means finding a weakness in the cipher algorithm that can be exploited with a complexity less than brute force [Sch00].

identified with elements in $\mathbb{Z}_{26}$ as shown in Table 2.1. For easy reading, we will not distinguish letters and elements in $\mathbb{Z}_{26}$. For example, when the message is A, we may say that the plaintext is A or the plaintext is 0, similarly for ciphertext.

2.2.1 Shift Cipher

Definition 2.2.1 (Shift Cipher) Let $\mathcal{P} = \mathcal{C} = \mathcal{K} = \mathbb{Z}_{26}$. For each $k \in \mathcal{K}$, define

$$E_k : \mathbb{Z}_{26} \to \mathbb{Z}_{26}, \quad p \mapsto p + k \bmod 26; \quad D_k : \mathbb{Z}_{26} \to \mathbb{Z}_{26}, \quad c \mapsto c - k \bmod 26.$$

The cryptosystem $(\mathcal{P}, \mathcal{C}, \mathcal{K}, \mathcal{E}, \mathcal{D})$, where $\mathcal{E} = \{ E_k : k \in \mathcal{K} \}$ and $\mathcal{D} = \{ D_k : k \in \mathcal{K} \}$, is called the *shift cipher*.

By Theorem 1.4.2, $\mathbb{Z}_{26}$ is a commutative ring with addition and multiplication modulo 26. We also discussed that subtracting k corresponds to adding the additive inverse of k (see Remark 1.4.2).

Example 2.2.1 Let $k = 2$, we have

$$- k = -2 \bmod 26 = 24 \bmod 26.$$

Suppose the message is A, then the corresponding plaintext is 0 (see Table 2.1). The ciphertext is given by

$$E_k(\mathsf{A}) = 0 + 2 \bmod 26 = 2 \bmod 26 = \mathsf{C}.$$

When we decrypt the ciphertext using the same key, we get our original message:

$$D_k(\mathsf{C}) = 2 - 2 \bmod 26 = 2 + 24 \bmod 26 = 0 \bmod 26 = \mathsf{A}.$$

We note that encrypting using a key k is the same as shifting the letters by k positions, hence the name "shift cipher."

Example 2.2.2 For example, when $k = 5$,

$$E_k(\mathsf{A}) = 0 + 5 \bmod 26 = \mathsf{F}, \quad E_k(\mathsf{Z}) = 25 + 5 \bmod 26 = 4 \bmod 26 = \mathsf{E}.$$

To encrypt a message, we can follow Table 2.3 and replace letters in the first row with those in the second row. Suppose the message is I STUDY IN BRATISLAVA. Then the corresponding ciphertext (omitting the white spaces) is NXYZIDNSGWFYNXQFAF.

When $k = 3$, the cipher is called the *Caesar Cipher*, which was used by Julius Caesar around 50 B.C. It is unknown how effective the Caesar cipher was at the time. But it is likely to have been reasonably secure since most of Caesar's enemies

Table 2.3 Shift cipher with $k = 5$. The second row represents the ciphertexts for the letters in the first row

A	B	C	D	E	F	G	H	I	J	K	L	M	N	O	P	Q	R	S	T
F	G	H	I	J	K	L	M	N	O	P	Q	R	S	T	U	V	W	X	Y

U	V	W	X	Y	Z
Z	A	B	C	D	E

would have been illiterate and they might have also assumed the messages were written in an unknown foreign language.

Now, suppose as an attacker, we know that the ciphertext is NXYZIDNSGWFYNXQ FAF. By Kerckhoffs' principle (Definition 2.1.3), we can assume that we also know the communication language is English, and how can we find the corresponding plaintext?

With a moment's thought, it is easy to see that we can simply try all the possible keys until we find a plaintext that makes sense. For example, let $k = 1$, then N should be decrypted to M, X to W, and so on. Eventually, we get MWXYHCMRFVEXMWPEZE, which does not make sense. So we continue, when $k = 2$, we get LVWXGBLQEUDWLVODYD. When $k = 3$, we have KUVWFAKPDTCVKUNCXC, and for $k = 4$, we get JTUVEYJOCSBUJTMBWB. Finally, letting $k = 5$, we get a proper sentence ISTUDYINBRATISLAVA. Since there are only 25 possible keys (the key is not equal to 0), with a known ciphertext, it is easy to find the original plaintext and the key.

Such a method of trying every possible key until the correct one is found is called an *exhaustive key search*. We have demonstrated that with an exhaustive key search, we can *break* the shift cipher, i.e., find the key.

2.2.2 Affine Cipher

Recall that $\mathbb{Z}_n^*$ is the set of elements $x \in \mathbb{Z}_n$ such that $\gcd(x, n) = 1$ (Definition 1.4.5).

Definition 2.2.2 (Affine Cipher) Let $\mathcal{P} = \mathcal{C} = \mathbb{Z}_{26}$ and $\mathcal{K} = \{ (a, b) \mid a \in \mathbb{Z}_{26}^*,\ b \in \mathbb{Z}_{26} \}$. For each key (a, b), define

$$E_{(a,b)} : \mathbb{Z}_{26} \to \mathbb{Z}_{26}, \quad p \mapsto ap + b \bmod 26; \quad D_{(a,b)} : \mathbb{Z}_{26} \to \mathbb{Z}_{26},$$

$$c \mapsto a^{-1}(c - b) \bmod 26.$$

The cryptosystem $(\mathcal{P}, \mathcal{C}, \mathcal{K}, \mathcal{E}, \mathcal{D})$, where $\mathcal{E} = \{ E_{(a,b)} : (a, b) \in \mathcal{K} \}$, $\mathcal{D} = \{ D_{(a,b)} : (a, b) \in \mathcal{K} \}$, is called the *affine cipher*.

Note that when $a = 1$, we have a shift cipher (Definition 2.2.1).

Next, we will verify that the affine cipher is well-defined. In particular, we will show the following:

- Decryption is always possible, i.e., given any $a \in \mathbb{Z}_{26}^*$ and $b, y \in \mathbb{Z}_{26}$, a solution for x such that

$$ax + b \equiv y \bmod 26$$

  always exists in $\mathbb{Z}_{26}$.
- Each encryption function E_k is injective, i.e., different plaintexts produce different ciphertexts, or equivalently, if the solution for $ax + b \equiv y \bmod 26$ exists, then it is unique.

Fix $a \in \mathbb{Z}_{26}^*$, $b, y \in \mathbb{Z}_{26}$. To solve the equation

$$ax + b \equiv y \bmod 26$$

is equivalent to solving the equation

$$ax \equiv y - b \bmod 26.$$

When y varies over $\mathbb{Z}_{26}$, $y - b$ also varies over $\mathbb{Z}_{26}$. Thus we can focus on solutions for

$$ax \equiv z \bmod 26, \tag{2.1}$$

where $z \in \mathbb{Z}_{26}$. Since $a \in \mathbb{Z}_{26}^*$, by Theorem 1.4.6, Eq. 2.1 has a unique solution. The existence of the solution proves that decryption is possible, and the uniqueness guarantees that encryption functions are injective.

Given a key (a, b), to find $a^{-1} \bmod 26$, we can apply the extended Euclidean algorithm (Algorithm 1.2).

Example 2.2.3 Suppose the key for affine cipher is $(3, 1)$, by the extended Euclidean algorithm, we can find $3^{-1} \bmod 26$:

$$26 = 3 \times 8 + 2, \, 3 = 2 + 1 \implies 1 = 3 - (26 - 3 \times 8) = 3 \times 9 - 26 \implies 3^{-1} \bmod 26 = 9.$$

To encrypt the word STROM,[4] we compute (see Table 2.1)

$$3 \times 18 + 1 = 55 \equiv 3 \bmod 26, \quad 3 \times 19 + 1 = 58 \equiv 6 \bmod 26,$$
$$3 \times 17 + 1 = 52 \equiv 0 \bmod 26, \quad 3 \times 14 + 1 = 43 \equiv 17 \bmod 26,$$
$$3 \times 12 + 1 = 37 \equiv 11 \bmod 26.$$

---

[4] Strom is a Slovak word which means tree.

So the ciphertext is DGARL. We can list the correspondence between plaintext and ciphertext as follows:

S	T	R	O	M
18	19	17	14	12
3	6	0	17	11
D	G	A	R	L

We know that $26 = 2 \times 13$. By Theorem 1.4.3,

$$\varphi(26) = 26 \times \left(1 - \frac{1}{2}\right)\left(1 - \frac{1}{13}\right) = 12.$$

So there are 12 possible values for $a \in \mathbb{Z}_{26}^*$. And there are 26 possible values for $b \in \mathbb{Z}_{26}$. Then the total number of possible keys (a, b) is $12 \times 26 = 312$. Similarly to shift cipher, knowing a ciphertext, we can try each of the 312 keys until we find a plaintext that makes sense. Thus we can break affine cipher by exhaustive key search.

2.2.3 Substitution Cipher

Recall that the symmetric group of degree n, denoted S_n, is the set of permutations of a set X with n elements (see Definition 1.2.4). We have discussed that a permutation is a bijective function, and its inverse exists with respect to the composition of functions (see Lemma 1.2.1). In particular, any permutation $\sigma \in S_{26}$ has an inverse σ^{-1}.

Definition 2.2.3 (Substitution Cipher) Let $\mathcal{P} = \mathcal{C} = \mathbb{Z}_{26}$, and $\mathcal{K} = S_{26}$. For any key $\sigma \in S_{26}$, define

$$E_\sigma : \mathbb{Z}_{26} \to \mathbb{Z}_{26}, \quad p \mapsto \sigma(p); \qquad D_\sigma : \mathbb{Z}_{26} \to \mathbb{Z}_{26}, \quad c \mapsto \sigma^{-1}(c).$$

The cryptosystem $(\mathcal{P}, \mathcal{C}, \mathcal{K}, \mathcal{E}, \mathcal{D})$, where $\mathcal{E} = \{ E_\sigma : \sigma \in \mathcal{K} \}$, $\mathcal{D} = \{ D_\sigma : \sigma \in \mathcal{K} \}$, is called the *substitution cipher*.

We note that an affine cipher (Definition 2.2.2) is also a substitution cipher.

Example 2.2.4 Define σ as in Table 2.4, then the corresponding table for decryption can be computed by flipping the two rows of the table (see Table 2.5). For example, to decrypt UIFJNJUWUJPOHWNF, using Table 2.5, we get THE IMITATION GAME.

Table 2.4 Definition of σ, a key for substitution cipher

A	B	C	D	E	F	G	H	I	J	K	L	M	N	O	P	Q	R	S	T
W	X	Y	Z	F	G	H	I	J	K	L	M	N	O	P	Q	R	S	T	U

U	V	W	X	Y	Z
V	A	B	C	D	E

Table 2.5 Definition of σ^{-1}, where $\sigma \in S_{26}$ is a key for substitution cipher shown in Table 2.4

A	B	C	D	E	F	G	H	I	J	K	L	M	N	O	P	Q	R	S	T
V	W	X	Y	Z	E	F	G	H	I	J	K	L	M	N	O	P	Q	R	S

U	V	W	X	Y	Z
T	U	A	B	C	D

We have discussed that $|S_n| = n!$ (see Example 1.2.9). So the size of key space for substitution cipher is $26! \approx 4 \times 10^{26}$. Modern computers run at a speed of a few GHz, which is $\sim 10^9$ instructions per second. There are $\sim 10^5$ seconds per day, so one computer can run $\sim 10^{14}$ instructions per day or $\sim 10^{16}$ instructions per year. If we would like to exhaust every key for substitution cipher, we will need $\sim 10^{10}$ years. Compared to the age of the universe, which is 13.8 billion, i.e., 1.38×10^{10} years, exhaustive key search is impossible with current computation power. However, we will show in Sect. 2.2.6 that other methods can be used to break substitution cipher.

2.2.4 Vigenère Cipher

For the substitution cipher, one alphabet is mapped to a unique alphabet. Hence such a cipher is also called a *monoalphabetic cipher*. Vigenère cipher, named after the French cryptographer Blaise Vigenère, is a *polyalphabetic cipher* where one alphabet can be encrypted to different alphabets depending on the key.

Let m be a positive integer, and let $\mathbb{Z}_{26}^m$ be the set of matrices with coefficients in $\mathbb{Z}_{26}$ of size $1 \times m$. In other words, $\mathbb{Z}_{26}^m$ is the set of $1 \times m$ row vectors with coefficients in $\mathbb{Z}_{26}$ (see Definition 1.3.1). As discussed in Eq. 1.4, for any $\mathbf{x} = (x_0, x_1, \ldots, x_{m-1})$, $\mathbf{y} = (y_0, y_1, \ldots, y_{m-1})$ in $\mathbb{Z}_{26}^m$, the addition $\mathbf{x} + \mathbf{y}$ is computed componentwise:

$$\mathbf{x} + \mathbf{y} = (x_0 + y_0, x_1 + y_1, \ldots, x_{m-1} + y_{m-1}),$$

where $x_i + y_i$ is computed with addition modulo 26. Recall that the additive inverse of an element a in $\mathbb{Z}_{26}$ is given by $-a$ (see Remark 1.4.2). $\mathbf{x} - \mathbf{y}$ is then computed componentwise using the additive inverses of y_is.

Definition 2.2.4 (Vigenère Cipher) Let m be a positive integer, and let $\mathcal{K} = \mathcal{P} = \mathcal{C} = \mathbb{Z}_{26}^m$. For each $k \in \mathcal{K}$, define

$$E_k : \mathbb{Z}_{26}^m \to \mathbb{Z}_{26}^m, \quad p \mapsto p + k; \qquad D_k : \mathbb{Z}_{26}^m \to \mathbb{Z}_{26}^m, \quad c \mapsto c - k.$$

The cryptosystem $(\mathcal{P}, \mathcal{C}, \mathcal{K}, \mathcal{E}, \mathcal{D})$, where $\mathcal{E} = \{ E_k : k \in \mathcal{K} \}$, $\mathcal{D} = \{ D_k : k \in \mathcal{K} \}$, is called the *Vigenère cipher*.

The key for a Vigenère cipher is also called a *keyword* since it can be written as a string of letters. By definition, a Vigenère cipher encrypts m alphabetic characters at a time.

Example 2.2.5 Let $m = 6$ and choose SECRET as the keyword. Thus the key is

$$k = \begin{pmatrix} 18 & 4 & 2 & 17 & 4 & 19 \end{pmatrix}.$$

To encrypt AN EXAMPLE, we write the plaintext in groups of six letters and add the keyword to each group letter by letter, modulo 26.

A	N	E	X	A	M		P	L	E
0	13	4	23	0	12		15	11	4
18	4	2	17	4	19		18	4	2
18	17	6	14	4	5		7	15	6
S	R	G	O	E	F		H	P	G

The ciphertext is given by SRGOEFHPG.

Example 2.2.6 Let the keyword be SKALA. So $m = 5$ and

$$k = \begin{pmatrix} 18 & 10 & 0 & 11 & 0 \end{pmatrix}.$$

To decrypt ZSLWCAZHPR, we write the ciphertext in groups of five letters and add the keyword to each group letter by letter modulo 26. We get the plaintext HILLCIPHER.

Z	S	L	W	C	A	Z	H	P	R
25	18	11	22	2	0	25	7	15	17
18	10	0	11	0	18	10	0	11	0
7	8	11	11	2	8	15	7	4	17
H	I	L	L	C	I	P	H	E	R

The size of the key space for Vigenère Cipher is given by 26^m. If $m = 6$, it is about $3.1 \times 10^8 \approx 2^{28.2}$, which is possible to search each key using a computer.

However, for larger m, it becomes much more difficult. If $m = 25$, $26^{25} \approx 2^{117}$, which is not feasible with current computation powers.

2.2.5   Hill Cipher

Definition 2.2.5 (Hill Cipher)   Let m be an integer such that $m \geq 2$. Let $\mathcal{P} = \mathcal{C} = \mathbb{Z}_{26}^m$ and

$$\mathcal{K} = \left\{ A \mid A \in \mathcal{M}_{m \times m}(\mathbb{Z}_{26}),\ \det(A) \in \mathbb{Z}_{26}^* \right\}.$$

For each $A \in \mathcal{K}$, define

$$E_A : \mathbb{Z}_{26}^m \to \mathbb{Z}_{26}^m, \quad p \mapsto pA; \qquad D_A : \mathbb{Z}_{26}^m \to \mathbb{Z}_{26}^m, \quad c \mapsto cA^{-1}.$$

The cryptosystem $(\mathcal{P}, \mathcal{C}, \mathcal{K}, \mathcal{E}, \mathcal{D})$, where $\mathcal{E} = \{ E_A : A \in \mathcal{K} \}$, $\mathcal{D} = \{ D_A : A \in \mathcal{K} \}$, is called the *Hill cipher*.

By Theorem 1.4.2, $\mathbb{Z}_{26}$ is a commutative ring. We have defined the *determinant* of a square matrix with coefficients from a commutative ring R in Sect. 1.3.1 (Eq. 1.6). We discussed that an $m \times m$ matrix A is invertible in $\mathcal{M}_{m \times m}(R)$ if and only if its determinant, $\det(A)$, is a unit (see Definition 1.2.10) in R. Furthermore, when A is invertible, its inverse can be calculated using the adjoint matrix of A (Theorem 1.3.2). By Lemma 1.4.3, a matrix $A \in \mathcal{M}_{n \times n}(\mathbb{Z}_{26})$ is invertible if and only if $\gcd(\det(A), 26) = 1$, i.e., $\det(A) \in \mathbb{Z}_{26}^*$. Therefore, in the definition of the Hill cipher, we require $\det(A) \in \mathbb{Z}_{26}^*$ so that the decryption can be computed.

Example 2.2.7   Let

$$A = \begin{pmatrix} 2 & 1 & 2 \\ 3 & 12 & 4 \\ 0 & 5 & 1 \end{pmatrix}$$

be a matrix in $\mathcal{M}_{3 \times 3}(\mathbb{Z}_{26})$. We denote by A_{ij} the matrix obtained from A by deleting the ith row and the jth column. Then

$$A_{00} = \begin{pmatrix} 12 & 4 \\ 5 & 1 \end{pmatrix}, \quad A_{01} = \begin{pmatrix} 3 & 4 \\ 0 & 1 \end{pmatrix}, \quad A_{02} = \begin{pmatrix} 3 & 12 \\ 0 & 5 \end{pmatrix}.$$

Following the discussions in Example 1.3.6, we have

$$\det(A_{00}) = 12 - 20 \bmod 26 = -8 \bmod 26,$$

$$\det(A_{01}) = 3 - 0 \bmod 26 = 3 \bmod 26,$$

$$\det(A_{02}) = 15 - 0 \bmod 26 = 15 \bmod 26.$$

Similarly, we can calculate

$$\det(A_{10}) = -9 \bmod 26, \ \det(A_{11}) = 2 \bmod 26, \ \det(A_{12}) = 10 \bmod 26,$$

$$\det(A_{20}) = -20 \bmod 26, \ \det(A_{21}) = 2 \bmod 26, \ \det(A_{22}) = 21 \bmod 26.$$

Let a_{ij} denote the entry of A at ith row and jth column, then by Eq. 1.6,

$$\det(A) = \sum_{j=0}^{2}(-1)^{j} a_{0j}\det(A_{0j}) \bmod 26$$

$$= (-1)^{0} \times 2 \times (-8) + (-1)^{1} \times 1 \times 3 + (-1)^{2} \times 2 \times 15 \bmod 26$$

$$= -16 - 3 + 30 \bmod 26 = 11.$$

By the Euclidean algorithm (Algorithm 1.1), we can find $\gcd(26, 11)$:

$$26 = 11 \times 2 + 4, \ 11 = 4 \times 2 + 3, \ 4 = 3 + 1, \ 3 = 1 \times 3 \Longrightarrow \gcd(11, 26) = 1.$$

Thus A is an invertible matrix in $\mathcal{M}_{3\times3}(\mathbb{Z}_{26})$.

By the extended Euclidean algorithm (Algorithm 1.2),

$$1 = 4 - 3 = 4 - (11 - 4 \times 2) = 4 \times 3 - 11 = (26 - 11 \times 2) \times 3 - 11$$

$$= 26 \times 3 - 11 \times 7 \Longrightarrow 11^{-1} \bmod 26 = 7.$$

By Theorem 1.3.2,

$$A^{-1} = -7 \begin{pmatrix} -8 & 9 & -20 \\ -3 & 2 & -2 \\ 15 & -10 & 21 \end{pmatrix} \bmod 26 = \begin{pmatrix} 56 & -63 & 140 \\ 21 & -14 & 14 \\ -105 & 70 & -147 \end{pmatrix}$$

$$\bmod 26 = \begin{pmatrix} 4 & 15 & 10 \\ 21 & 12 & 14 \\ 25 & 18 & 9 \end{pmatrix}.$$

Example 2.2.8 Let

$$A = \begin{pmatrix} 2 & 1 & 2 \\ 3 & 12 & 4 \\ 0 & 5 & 1 \end{pmatrix}$$

be a key for Hill cipher. Suppose the plaintext is CIPHER. By Table 2.1, this corresponds to $(2\ 8\ 15)$ and $(7\ 4\ 17)$. To encrypt, we calculate

$$(2\ 8\ 15) \begin{pmatrix} 2 & 1 & 2 \\ 3 & 12 & 4 \\ 0 & 5 & 1 \end{pmatrix} \bmod 26 = (2\ 17\ 25),$$

$$(7\ 4\ 17) \begin{pmatrix} 2 & 1 & 2 \\ 3 & 12 & 4 \\ 0 & 5 & 1 \end{pmatrix} \bmod 26 = (0\ 10\ 21).$$

And the ciphertext is CRZAKV.

Now suppose the ciphertext is DOSJBQ. By Table 2.1, this corresponds to $(3\ 14\ 18)$ and $(9\ 1\ 16)$. We have calculated in Example 2.2.7 that

$$A^{-1} = \begin{pmatrix} 4 & 15 & 10 \\ 21 & 12 & 14 \\ 25 & 18 & 9 \end{pmatrix}.$$

We can then compute the plaintext as follows:

$$(3\ 14\ 18) \begin{pmatrix} 4 & 15 & 10 \\ 21 & 12 & 14 \\ 25 & 18 & 9 \end{pmatrix} \bmod 26 = (756\ 537\ 388) \bmod 26 = (2\ 17\ 24),$$

$$(9\ 1\ 16) \begin{pmatrix} 4 & 15 & 10 \\ 21 & 12 & 14 \\ 25 & 18 & 9 \end{pmatrix} \bmod 26 = (457\ 435\ 248) \bmod 26 = (15\ 19\ 14).$$

And the plaintext is CRYPTO.

Remark 2.2.1 By Definition 2.1.2, shift cipher, affine cipher, and substitution cipher are block ciphers of block length 1. Vigenère cipher and Hill cipher are block ciphers of block length m.

2.2.6 Cryptanalysis of Classical Ciphers

In this subsection, we will discuss the cryptanalysis of the classical ciphers introduced in the previous subsections. Cryptanalysis comes from the Latin words kryptós (hidden) and analýein (to analyze). The goal of cryptanalysis is to decrypt the ciphertext without knowing the key. Successful cryptanalysis recovers the plaintext or even the key. We recall the different assumptions of attack described in Sect. 2.1.3.

Example 2.2.9 (Known Plaintext Attack—Hill Cipher) Let us consider a known plaintext attack on Hill cipher. Suppose we know $m = 2$, i.e., $A \in \mathcal{M}_{2\times2}(\mathbb{Z}_{26})$, and we have a string of plaintext ATTACK and its corresponding ciphertext FTMTIM. By Definition 2.2.5, we have

$$E_A\left((0\ 19)\right) = (5\ 19), \quad E_A\left((19\ 0)\right) = (12\ 19), \quad E_A\left((2\ 10)\right) = (8\ 12).$$

The first two plaintext–ciphertext pairs give us

$$\begin{pmatrix} 0 & 19 \\ 19 & 0 \end{pmatrix} A \bmod 26 = \begin{pmatrix} 5 & 19 \\ 12 & 19 \end{pmatrix}. \tag{2.2}$$

The inverse of a 2×2 matrix can be computed using Eq. 1.7, where the computations should be mod 26. We have

$$\begin{pmatrix} 0 & 19 \\ 19 & 0 \end{pmatrix}^{-1} = 3^{-1} \begin{pmatrix} 0 & 7 \\ 7 & 0 \end{pmatrix} \bmod 26 = \begin{pmatrix} 0 & 11 \\ 11 & 0 \end{pmatrix}.$$

Together with Eq. 2.2,

$$A = \begin{pmatrix} 0 & 11 \\ 11 & 0 \end{pmatrix} \begin{pmatrix} 5 & 19 \\ 12 & 19 \end{pmatrix} \bmod 26 = \begin{pmatrix} 2 & 1 \\ 3 & 1 \end{pmatrix}.$$

We can verify this key using the third plaintext–ciphertext pair

$$(2\ 10) \begin{pmatrix} 2 & 1 \\ 3 & 1 \end{pmatrix} \bmod 26 = (8\ 12).$$

We have seen that an exhaustive key search can be used to break affine cipher, where the attacker can find both the plaintext and the key. But this does not apply to substitution cipher or Vigenère cipher. Next, we will discuss other cryptanalysis methods that can be used to break those ciphers.

2.2.6.1 Frequency Analysis

By Kerckhoffs' principle (Definition 2.1.3), we assume we know the plaintext is an English text. We also know the cipher used for communication. We assume a ciphertext-only attacker model, and we will show how to recover both the plaintext and the key using *frequency analysis* for affine cipher and Vigenère cipher.

As the plaintext is an English text, we first analyze the probabilities for the appearance of each letter in a standard English text. For example, Table 2.6 lists the analysis results from [BP82]. In particular, we observe that E has the highest probability and the second most common letter is T. Similarly, [BP82] also shows

Table 2.6 Probabilities of each letter in a standard English text [BP82]

A	0.082	B	0.015	C	0.028	D	0.043	E	0.127	F	0.022
G	0.020	H	0.061	I	0.070	J	0.002	K	0.008	L	0.040
M	0.024	N	0.067	O	0.075	P	0.019	Q	0.001	R	0.060
S	0.063	T	0.091	U	0.028	V	0.010	W	0.023	X	0.001
Y	0.020	Z	0.001									

that the most common two consecutive letters are TH, HE, IN, ..., and the most common three consecutive letters are THE, ING, AND, ...

Given a ciphertext that is encrypted using a monoalphabetic cipher (i.e., one alphabet is mapped to a unique alphabet), we expect a permutation of the letters in the ciphertext to have similar frequencies as in Table 2.6.

Example 2.2.10 Suppose the cipher used is an affine cipher, and we have the following ciphertext:

$$\text{VCVIRSKPOFPNZOTHOVMLVYSATISKVNVLIVSZVR.}$$

We can calculate the frequencies of each letter that appear in the text:

V	S	I	O	R	K	P	N	Z	T	L	C	F	H	M	Y	A
8	4	3	3	2	2	2	2	2	2	2	1	1	1	1	1	1

The most frequent letter is V, and the second most frequent one is S. Thus, it makes sense to assume V is the ciphertext corresponding to E and S to T. Let the key be (a, b), and by Table 2.1 and Definition 2.2.2, we have the following equations:

$$4a + b = 21 \bmod 26, \quad 19a + b = 18 \bmod 26,$$

which gives

$$15a = 23 \bmod 26.$$

By the extended Euclidean algorithm,

$$26 = 15 \times 1 + 11, \quad 15 = 11 \times 1 + 4, \quad 11 = 4 \times 2 + 3, \quad 4 = 3 + 1,$$

and

$$1 = 4 - 3 = 4 - (11 - 4 \times 2) = -11 + 4 \times 3 = -11 + (15 - 11) \times 3$$
$$= 15 \times 3 - 11 \times 4 = 15 \times 3 - (26 - 15) \times 4 = 15 \times 7 - 26 \times 4.$$

Hence, we have $15^{-1} \bmod 26 = 7$ and

$$a = 23 \times 15^{-1} \bmod 26 = 23 \times 7 \bmod 26 = 5 \bmod 26.$$

Furthermore, we get

$$b = 21 - 4a \bmod 26 = 21 - 4 \times 5 \bmod 26 = 1.$$

To decrypt the message, we compute the decryption key by finding $a^{-1} \bmod 26 = 5^{-1} \bmod 26$:

$$26 = 5 \times 5 + 1 \implies 1 = 26 - 5 \times 5 \implies 5^{-1} \bmod 26 = -5 \bmod 26 = 21.$$

Applying the decryption key $(21, 1)$ to the ciphertext, we get the following plaintext:

EVERYTHING IS KNOWN EXCEPT FOR THE SECRET KEY.

We note that the same technique works for substitution cipher since it is also monoalphabetic. But a longer ciphertext might be needed since we do not have equations to solve for the key. Instead, we must guess the mapping between each distinct letter in the ciphertext to the 26 alphabets (see [Sti05] Section 1.2.2).

Remark 2.2.2 Suppose the length of the keyword m is determined for Vigenère cipher. We take every mth letter from the ciphertext and obtain m ciphertexts. Then each of them can be considered as the ciphertext of the shift cipher with a key given by the corresponding letter in the keyword.

Example 2.2.11 Suppose we have the following ciphertext generated with Vigenère cipher (Definition 2.2.4), and we know that the keyword length $m = 3$.

SJRRIBSWRKRAOFCDACORRGSYZTCKVYXGCCSDDLCCEKOAMBHGCEKEPRS
TJOSDWXFOGMBVCCTMXHGXKNKVRCMLDLCMMNRIPDIVDAGVPZOXFOWYWI.

Take every third letter, and we have the following three ciphertexts:

SRSKODOGZKXCDCOBCESOWOBCXXKCDMRDDVOOW,
JIWRFARSTVGSLEAHEPTSXGVTHKVMLMIIAPXWI,
RBRACCRYCYCDCKMGKRJDFMCMGNRLCNPVGZFY.

We note that each of them can be considered as the ciphertext of a shift cipher, where the keys correspond to each letter of the keyword for the Vigenère cipher (as mentioned in Remark 2.2.2). The frequencies of each letter in the first ciphertext are as follows:

O	D	C	S	K	X	R	B	W	G	Z	E	M	V
7	5	5	3	3	3	2	2	2	1	1	1	1	1

The most frequent letter is O, and we assume O (14) is the ciphertext corresponding to E (4). And this gives us the first letter of the keyword

$$14 - 4 \bmod 26 = 10 \bmod 26 = K.$$

The frequencies of each letter in the second ciphertext are as follows:

I	A	S	T	V	W	R	G	L	E	H	P	X	M	J	F	K
4	3	3	3	3	2	2	2	2	2	2	2	2	2	1	1	1

Similarly, we assume E (4) is encrypted as I (8). And the second letter of the keyword is

$$8 - 4 \bmod 26 = 4 \bmod 26 = E.$$

The frequencies of each letter in the third ciphertext are as follows:

C	R	Y	M	G	D	K	F	N	B	A	J	L	P	V	Z
7	5	5	3	3	3	2	2	2	1	1	1	1	1	1	1

And we have the third letter of the keyword

$$2 - 4 \bmod 26 = 24 \bmod 26 = Y.$$

Thus we have recovered the keyword KEY. Computing decryption with the keyword, we get the following plaintext:

```
IF THE DISTANCE BETWEEN TWO APPEARANCES OF THE SAME WORD
IS A MULTIPLE OF M, THE CORRESPONDING PARTS IN THE
CIPHERTEXT WILL BE THE SAME.
```

Next, we will discuss two methods to determine the length m of the keyword for a Vigenère cipher.

2.2.6.2   Kasiski Test: Vigenère Cipher

We observe that if the distance between two appearances of the same sequence of alphabets in the plaintext is a multiple of m, the corresponding parts in the ciphertext will be the same. Kasiski test looks for identical parts of ciphertext and records the distance between those parts. Then we know that m is a divisor for all the distance values.

Example 2.2.12 Suppose the plaintext is

```
THE MEETING WILL BE IN THE CAFE AND THE STARTING TIME IS TEN
```

and the keyword is KEY ($m = 3$). The encryption gives us

```
THE MEETING WILL BE IN THE CAFE AND THE STARTING TIME IS TEN
KEY KEYKEYK EYKE YK EY KEY KEYK EYK EYK EYKEYKEY KEYK EY KEY
DLC WICDMLQ AGVP ZO ML DLC MEDO ELN XFO WRKVRSRE DMKO MQ DIL.
```

The first two appearances of THE have distance 15, which is a multiple of 3, and hence the corresponding parts in the ciphertext are the same DLC. But the third appearance of THE has distance 7 from the second appearance, and the corresponding parts in the ciphertext are different.

On the other hand, if we have only the ciphertext, we can observe the two identical parts DLC with distance 15, and then we can conclude that very likely m is a divisor of 15, i.e., $m = 1, 3, 5, 15$. To decide the exact value of m, a longer ciphertext is needed, or frequency analysis (see Example 2.2.11) can be applied assuming different values of m until a meaningful plaintext is found.

2.2.6.3   Index of Coincidence: Vigenère Cipher

Definition 2.2.6 Let $x = x_1 x_2 \ldots x_n$ be a string of n alphabetic characters. The *index of coincidence* of x, denoted by $I_c(x)$, is the probability that two random elements of x are identical.

Example 2.2.13 Let x be a long random text. If we randomly choose a letter from x, we expect that the probability for each letter to be chosen is close to $1/26$. Then, if we randomly choose two letters from x, the probability for those two letters to be the same is close to $1/26^2$. The index of coincidence for x will be close to

$$I_c(x) \approx 26 \left(\frac{1}{26} \right)^2 = 0.038.$$

Example 2.2.14 Let x be a long English text. If we randomly choose a letter from x, we expect that the probabilities for each letter to be chosen are similar to the values listed in Table 2.6. If we randomly choose two letters from x, the probability for both letters to be A is then given by 0.082^2, the probability for both to be B is 0.015^2, etc. Thus, the index of coincidence for x can be approximated as

$$I_c(x) \approx \sum_{i=0}^{25} p_i^2 = 0.065.$$

Remark 2.2.3 If x is a ciphertext string obtained using any monoalphabetic cipher, we would expect $I_c(x)$ to be close to 0.065. The individual probabilities for different alphabets will be permuted, but the sum will be unchanged.

Let $f_0, f_1, \ldots, f_{25}$ denote the frequencies of letters A, B, $\ldots$, Z in x. If we randomly choose a letter, the probability of each letter appearing is then given by

$$\frac{\binom{f_i}{2}}{\binom{n}{2}}.$$

We have the following formula for $I_c(x)$:

$$I_c(x) = \frac{\sum_{i=0}^{25} \binom{f_i}{2}}{\binom{n}{2}} = \frac{\sum_{i=0}^{25} f_i(f_i - 1)}{n(n-1)}. \tag{2.3}$$

Example 2.2.15 Let x be the ciphertext from Example 2.2.11. The total number of letters is 110, and the frequencies of each letter are

C	R	O	D	S	K	G	M	V	X	I	W	A	B	F	Y	T	L	E	P	J	Z	H	N
12	9	7	7	7	6	6	6	5	5	4	4	4	3	3	3	3	3	3	3	2	2	2	2

By Eq. 2.3, the index of coincidence of x is

$$I_c(x) = \frac{1}{110 \times 109}(12 \times 11 + 9 \times 8 + \cdots + 2 \times 1) = 0.004454.$$

Given a ciphertext $c = c_1 c_2 \ldots c_n$ output from Vigenère cipher. To find the length of the keyword m, for each $m \geq 1$, we construct substrings of c by taking every mth letter.

$$c_1 = c_1 c_{m+1} \ldots$$

$$c_2 = c_2 c_{m+2} \ldots$$

$$\vdots$$

$$c_m = c_m c_{2m} \ldots$$

If m is the keyword length, we expect $I_c(c_i)$ to be close to 0.065 (see Remark 2.2.3). Otherwise, c_i will be more random and $I_c(c_i)$ will be closer to 0.038 (see Example 2.2.13)

Example 2.2.16 Suppose we have the same ciphertext as in Example 2.2.11, and we do not know the value of m.

Assume $m = 1$, we have calculated that

$$I_c(c) = 0.004454$$

in Example 2.2.15.

Assume $m = 2$, we have

$c_1 = $ SRISRROCAORSZCVXCSDCEOMHCKPSJSWFGBCTXGKKRMDCMRPIDGPOFWW

$c_2 = $ JRBWKAFDCRGYTKYGCDLCKABGEERTODXOMVCMHXNVCLLMNIDVAVZXOYI

and

$$I_c(c_1) = 0.05253, \quad I_c(c_2) = 0.03636.$$

Assume $m = 3$,

$c_1 = $ SRSKODOGZKXCDCOBCESOWOBCXXKCDMRDDVOOW

$c_2 = $ JIWRFARSTVGSLEAHEPTSXGVTHKVMLMIIAPXWI

$c_3 = $ RBRACCRYCYCDCKMGKRJDFMCMGNRLCNPVGZFY

and

$$I_c(c_1) = 0.07958, \quad I_c(c_2) = 0.04054, \quad I_c(c_3) = 0.06984.$$

Thus it is more likely that $m = 3$. The exact value can be verified by frequency analysis as shown in Example 2.2.11 to see if the recovered plaintext is meaningful.

2.2.7 One-Time Pad

In this subsection, we will discuss a type of synchronous stream cipher (see Sect. 2.1.2) called *one-time pad*, which was invented by Gilbert Vernam in 1917.

Definition 2.2.7 (One-Time Pad) Given a positive integer n, let $\mathcal{P} = \mathcal{C} = \mathcal{K} = \mathbb{F}_2^n$. For any $k \in \mathcal{K}$, define

$$E_k : \mathbb{F}_2^n \to \mathbb{F}_2^n, \quad p \mapsto p \oplus k \qquad D_k : \mathbb{F}_2^n \to \mathbb{F}_2^n, \quad c \mapsto c \oplus k.$$

The cryptosystem $(\mathcal{P}, \mathcal{C}, \mathcal{K}, \mathcal{E}, \mathcal{D})$, where $\mathcal{E} = \{\, E_k : k \in \mathcal{K} \,\}$, $\mathcal{D} = \{\, D_k : k \in \mathcal{K} \,\}$, is called the *one-time pad*.

Recall that vector addition in $\mathbb{F}_2^n$ is defined as bitwise XOR, denoted by $\oplus$ (see Definition 1.3.6).

For encryption, we require the key to be chosen randomly with uniform probability (see Definition 1.7.3) from $\mathcal{K}$. This requirement will be justified in Theorem 2.2.1. Furthermore, we note that if the attacker has knowledge of one pair of plaintext p and its corresponding ciphertext c, they can recover the key by computing $p \oplus c = p \oplus p \oplus k = k$. Thus each key can be used only once.

One distinct feature of the one-time pad from the previously introduced classical ciphers is that it achieves perfect secrecy (see Sect. 2.1.3). Before proving this, we will first formalize the notion of perfect secrecy.

Let $\mathcal{P}$, $\mathcal{C}$, and $\mathcal{K}$ denote the plaintext space, ciphertext space, and key space, respectively, for a given cryptosystem. The random experiment we are interested in is encryption using one key and one plaintext for communication. The sample space (see Sect. 1.7) is $\Omega = \mathcal{P} \times \mathcal{K}$.

Let $p := \{(p,k) \mid k \in \mathcal{K}\}$ denote the event that p is encrypted. Similarly, $k := \{(p,k) \mid p \in \mathcal{P}\}$ denotes the event that k is used for encryption. $c := \{(p,k) \mid E_k(p) = c\}$ denotes the event that c is the ciphertext. Note that p and k are independent.

By Kerckhoffs' principle, $P(p)$ and $P(k)$ are known to the attacker. Then the cryptosystem is perfectly secure if p and c are independent (Definition 1.7.4) for any p and c, or equivalently (Eq. 1.28)

$$P(p \cap c) = P(p)P(c), \text{ i.e., } P(p|c) = P(p).$$

Example 2.2.17 (An Example of Cipher that Is Not Perfectly Secure) Let

$$\mathcal{P} = \{0, 1\}, \quad \mathcal{K} = \{x, y\}, \quad \mathcal{C} = \{\alpha, \beta\}.$$

Define the encryption functions as follows:

$$E_x(0) = E_y(1) = \alpha, \quad E_x(1) = E_y(0) = \beta.$$

Suppose

$$P(0) = \frac{1}{3}, \quad P(1) = \frac{2}{3}, \quad P(x) = \frac{1}{5}, \quad P(y) = \frac{4}{5}.$$

Then

$$P(\alpha) = P(x \cap 0) + P(y \cap 1) = P(x)P(0) + P(y)P(1) = \frac{3}{5},$$

and

$$P(0|\alpha) = \frac{P(0)P(\alpha|0)}{P(\alpha)} = \frac{P(0)P(x)}{P(\alpha)} = \frac{1}{9}.$$

We have

$$P(1|\alpha) = 1 - P(0|\alpha) = \frac{8}{9},$$

and

$$P(\beta) = 1 - P(\alpha) = \frac{2}{5}.$$

Similarly, we get

$$P(0|\beta) = \frac{2}{3}, \quad P(1|\beta) = \frac{1}{3}.$$

Thus $P(p|c) \neq P(p)$ for all $p \in \mathcal{P}, c \in \mathcal{C}$, and the cipher is not perfectly secure.

In particular, if the attacker knows the ciphertext is α, they can conclude that it is more likely that the plaintext is 1 rather than 0, and if the ciphertext is β, they can conclude that it is more likely for the plaintext to be 0.

We recall uniform probability measures from Definition 1.7.3.

Theorem 2.2.1 *One-time pad is perfectly secure if and only if the probability measure on the key space is uniform.*

Proof Fix a positive integer n, and let $\mathcal{P} = \mathcal{C} = \mathcal{K} = \mathbb{F}_2^n$. For any $p \in \mathcal{P}$ and $c \in \mathcal{C}$, if c is the ciphertext corresponding to p, then we know the key used is $k_{p,c} := p \oplus c$. Thus

$$P(c|p) = P(k_{p,c}).$$

$\Longrightarrow$ Fix $c \in \mathcal{C}$, for any p, we have

$$P(k_{p,c}) = P(c|p) = \frac{P(p \cap c)}{P(p)} = \frac{P(p)P(c)}{P(p)} = P(c),$$

which shows that the probability of $k_{p,c}$ is not dependent on p and the probabilities of all $k_{p,c}$s are the same for this fixed c. When p takes all possible values in $\mathcal{P}$, we have all possible values of $k_{p,c} \in \mathcal{K}$. Thus we can conclude that $P(k)$ is the same for all $k \in \mathcal{K}$.

$\Longleftarrow$ Since $\{q \mid q \in \mathcal{P}\}$ is a finite partition of Ω, by Theorem 1.7.2, for any $c \in \mathcal{C}$ and any $p \in \mathcal{P}$,

$$P(p|c) = \frac{P(c|p)P(p)}{\sum_{q \in \mathcal{P}} P(c|q)P(q)} = \frac{P(k_{p,c})P(p)}{\sum_{q \in \mathcal{P}} P(k_{q,c})P(q)}.$$

Since the probability measure on the key space is uniform,

$$P(k) = \frac{1}{|\mathcal{K}|}, \quad \forall k \in \mathcal{K}.$$

Also, $\sum_{q \in \mathcal{P}} P(q) = 1$. We have

$$P(p|c) = \frac{P(k_{p,c})P(p)}{\sum_{q \in \mathcal{P}} P(k_{q,c})P(q)} = \frac{P(p)}{\sum_{q \in \mathcal{P}} P(q)} = P(p).$$

$\square$

We note that brute force of the key does not work for one-time pad—by brute force, the attacker can obtain any plaintext of the same length as the original plaintext.

However, key management is the bottle neck of one-time pad. With a plaintext of length n, we will also need a key of length n. Furthermore, as we have mentioned earlier, each key can only be used once. Thus it is necessary to share a key of the same length as the message each time before the communication. This makes it impractical to use one-time pad.

2.3   Encryption Modes

We have seen a few examples of classical block ciphers. For messages that are longer than the block length, the way we encrypted them (e.g., see Examples 2.2.8 and 2.2.5) can be described by Fig. 2.2. Similarly, the decryption method we have applied (e.g., see Examples 2.2.8 and 2.2.6) corresponds to Fig. 2.3.

In general, when we use a symmetric block cipher of block length n to encrypt a long message, we first divide this long message into blocks of plaintexts of length n. Then we apply certain *encryption mode* to encrypt the plaintext blocks. If the last block has a length of less than n, padding might be required. Different methods exist for padding, e.g., using a constant or using a random number.

The simplest encryption mode is the mode we have been using so far, which is called *electronic codebook (ECB) mode*. ECB mode is easy to use, but the main drawback is that the encryption of identical plaintext blocks produces identical ciphertext blocks. For an extreme case, if the plaintext is either all 0s or all 1s, it would be easy for the attacker to deduce the message given a collection of plaintext

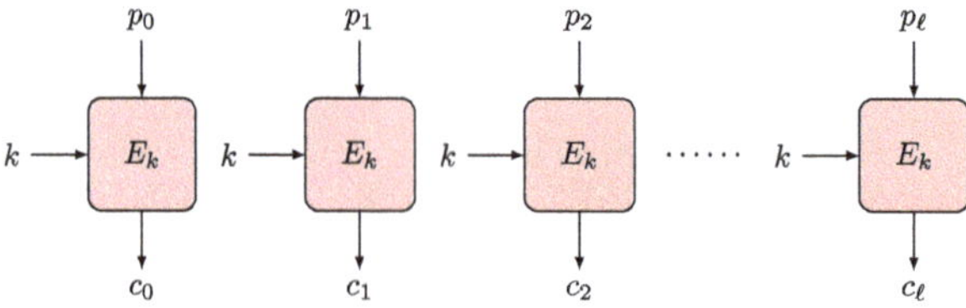

Fig. 2.2 ECB mode for encryption

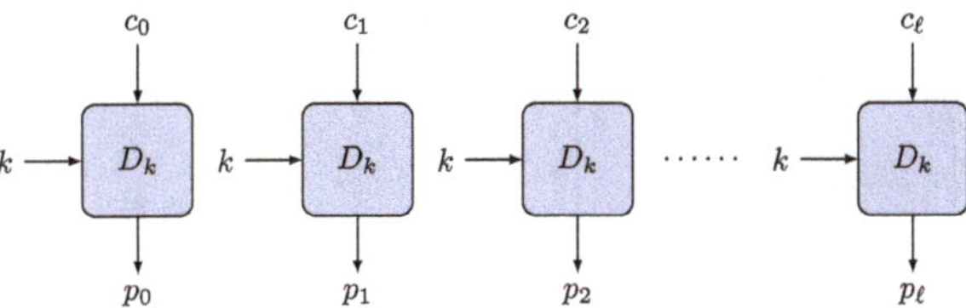

Fig. 2.3 ECB mode for decryption

(a) Original picture (b) ECB encrypted (c) CBC encrypted

Fig. 2.4 Original picture and encrypted picture with ECB and CBC modes

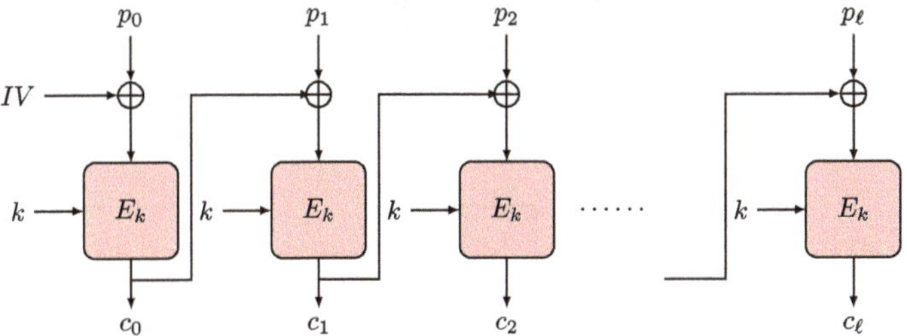

Fig. 2.5 CBC mode for encryption

and ciphertext pairs. Due to this property, it is also easy to recognize patterns of the plaintext in the ciphertext, which makes statistical attacks easier (e.g., frequency analysis of the affine cipher described in Example 2.2.10). For example, Fig. 2.4b gives an example for encryption using ECB mode. Compared to the original image in Fig. 2.4a, we can see a clear pattern of the plaintext from the ciphertext.

To avoid such problems, we can use the *cipherblock chaining (CBC) mode*. The encryption and decryption are shown in Figs. 2.5 and 2.6, respectively, where IV stands for *initialization vector*. An IV has the same length as the plaintext block and is public. We can see that with CBC, the same plaintext is encrypted differently with different IVs. Figure 2.4a encrypted with CBC mode is shown in Fig. 2.4c, where no clear pattern can be seen.

Furthermore, if a plaintext block is changed, the corresponding ciphertext block will also be changed, affecting all the subsequent ciphertext blocks. Hence CBC mode can also be useful for authentication.

However, with CBC mode, the receiver needs to wait for the previous ciphertext block to arrive to decrypt the next ciphertext block. In real-time applications, *output feedback (OFB) mode* can be used to make communication more efficient. As

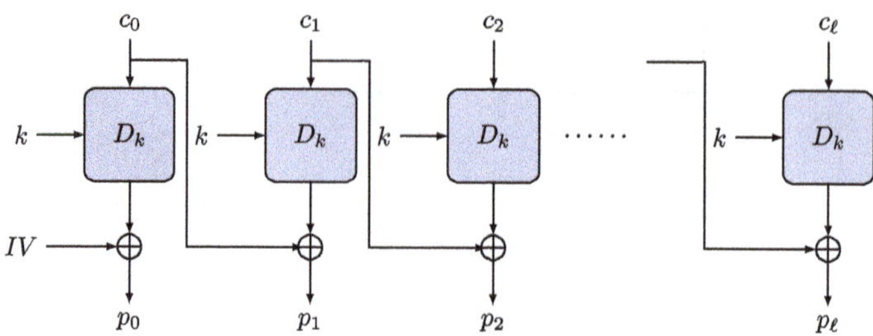

Fig. 2.6 CBC mode for decryption

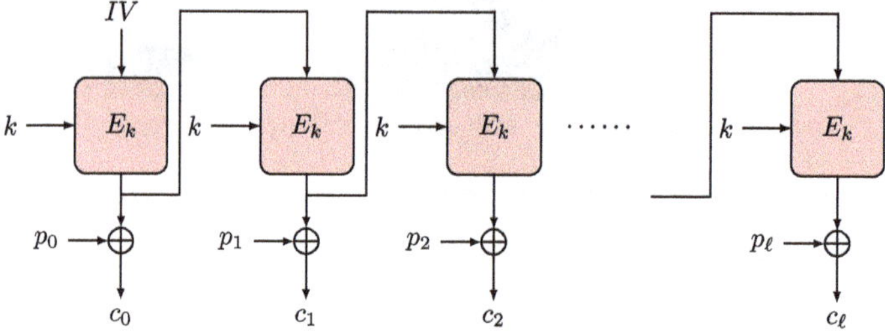

Fig. 2.7 OFB mode for encryption

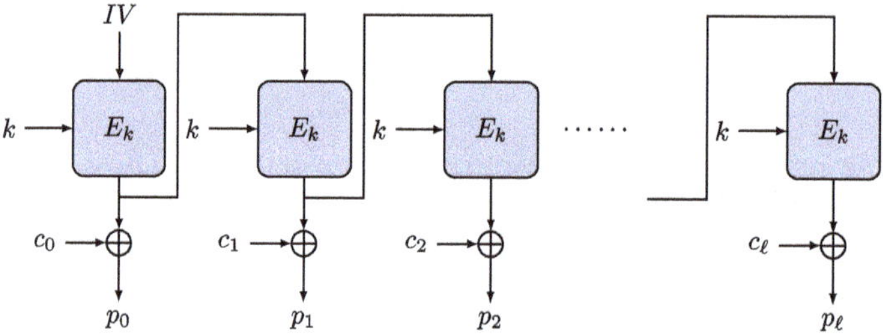

Fig. 2.8 OFB mode for decryption

shown in Figs. 2.7 and 2.8, the encryption function is not used for encrypting the plaintext blocks, rather it is used for generating a key sequence. Ciphertext blocks are computed by XORing the plaintext blocks and the key sequence. Such a design allows the receiver and the sender to generate the key sequence simultaneously before the ciphertext is sent.

In a way, OFB mode can be considered as a synchronous stream cipher (see Sect. 2.1.2). Another advantage of OFB mode is that padding is not needed. However, the encryption of a plaintext block does not depend on the previous blocks, which makes it easier for the attacker to modify the ciphertext blocks.

2.4 Further Reading

We refer the readers to [Sti05, Chapter 1] for more discussions on classical ciphers and to [MVOV18] for a detailed presentation on different cryptographic primitives. As for encryption modes and padding schemes, we refer the readers to [PP09, Chapter 5].

In Sect. 2.2.7 we introduced a classical stream cipher—one-time pad. The area of stream ciphers, albeit less discussed in the cryptography books than its block cipher counterpart, encompasses many modern algorithm designs. We do not go into detail in this book; interested readers will find more information in [KPP$^+$22].

The physical attacks we will present in Chaps. 4 and 5 are for symmetric block ciphers, one particular public key cipher (RSA), and RSA signatures. There is also plenty of research on physical attacks on other cryptographic primitives, e.g., hash functions [HH11, HLMS14, KMBM17], post-quantum public key algorithms [MWK$^+$22, PSKH18, XIU$^+$21, PPM17], or stream ciphers [BMV07, BT12, KDB$^+$22].

Chapter 3
Modern Cryptographic Algorithms and Their Implementations

We have defined cryptosystem/cipher in Definition 2.1.1. When the keys for encryption and decryption are the same, it is a *symmetric cipher*. Otherwise, it is a *public-key/asymmetric cipher*. In general, symmetric key ciphers are faster, but they require key exchange before communication.

In this chapter, we will detail the designs of three symmetric block ciphers—DES (Sect. 3.1.1), AES (Sect. 3.1.2), and PRESENT (Sect. 3.1.3) as well as one public key cipher—RSA (Sect. 3.3). We will also discuss how RSA can be used for digital signatures (Sect. 3.4). Moreover, we will present different techniques for implementing those algorithms (Sects. 3.2 and 3.5).

3.1 Symmetric Block Ciphers

For the construction of symmetric block ciphers, two important principles are followed by modern cryptographers—*confusion* and *diffusion*. Shannon first introduced them in his famous paper [Sha45].

Confusion obscures the relationship between the ciphertext and the key. To achieve this, each part of the ciphertext should depend on several parts of the key. For example, in Vigenère cipher, each letter of the plaintext and each letter of the key influence exactly one letter of the ciphertext. Consequently, we can use the Kasiski test (Sect. 2.2.6.2) or index of coincidence (Sect. 2.2.6.3) to attack the Vigenère cipher. Diffusion obscures the statistical relationship between the plaintext and the ciphertext. Each change in the plaintext is spread over the ciphertext, with the redundancies being dissipated. For example, monoalphabetic ciphers (Sect. 2.2.4) have very low diffusion—the distributions of letters in plaintext correspond directly to those in the ciphertext. That is also why frequency analysis (Sect. 2.2.6.1) can be applied to break those ciphers.

X. Hou, J. Breier, *Cryptography and Embedded Systems Security*,
https://doi.org/10.1007/978-3-031-62205-2_3

As mentioned in Sect. 2.1.2, for modern symmetric block ciphers, $\mathcal{P} = \mathcal{C} = \mathbb{F}_2^n$ for a positive integer n, which is called the *block length* of the cipher. Furthermore, the key space is also a vector space over $\mathbb{F}_2$ and its dimension is called the *key length* of the cipher. Each key $k \in \mathcal{K}$ is called a *master key*.

A symmetric block cipher design specifies a *round function* and a *key schedule*. Encryption of a plaintext block consists of a few *rounds* of round functions, possibly with minor differences. Each round function takes the cipher's current *state* as an input and outputs the next state. The key schedule takes the master key k and outputs the keys for each round, which are called *round keys*. In most cases, the key schedule is an invertible function. In particular, given one or more round keys, the master keys can be calculated.

By Kerckhoffs' principle, round functions and key schedule specifications are public, but the master key (hence also the round keys) are secret. In physical attacks that we will discuss in the later parts of the book, the attacker normally aims to recover some round key(s) and then use the inverse key schedule to find the master key.

To be more specific, suppose we have a symmetric block cipher with round function F and in total $\mathtt{Nr}$ number of rounds. Let K_i denote the round key for round i and S_i denote the cipher state at the end of round i. For a plaintext $p \in \mathbb{F}_2^n$, the corresponding ciphertext $c \in \mathbb{F}_2^n$ can be computed as follows:[1]

$$S_0 = p,$$

$$S_1 = F(S_0, K_1),$$

$$S_2 = F(S_1, K_2),$$

$$\vdots$$

$$S_{\mathtt{Nr}} = F(S_{\mathtt{Nr}-1}, K_{\mathtt{Nr}}),$$

$$c = S_{\mathtt{Nr}}.$$

To perform decryption, we require that for any given round key K_i, $F(\cdot, K_i)$ has an inverse, i.e.,

$$F^{-1}(F(x, K_i), K_i) = x, \quad \forall x \in \mathbb{F}_2^n.$$

In this case, given ciphertext c, plaintext p can be computed as follows:

$$S_{\mathtt{Nr}} = c,$$

$$S_{\mathtt{Nr}-1} = F^{-1}(S_{\mathtt{Nr}}, K_{\mathtt{Nr}}),$$

---

[1] The round function for the last round might be a bit different, as for the case of AES (see Sect. 3.1.2).

$$\vdots$$

$$S_1 = F^{-1}(S_2, K_2),$$

$$S_0 = F^{-1}(S_1, K_1),$$

$$p = S_0.$$

We recall for a vector space over $\mathbb{F}_2$, vector addition is given by bitwise XOR, denoted $\oplus$ (Definition 1.3.6). XOR with the round key is a common operation in round functions of symmetric block ciphers.

Another common function is a substitution function called *Sbox*, denoted SB,

$$\mathrm{SB} : \mathbb{F}_2^{\omega_1} \to \mathbb{F}_2^{\omega_2}.$$

Normally ω_1 or/and ω_2 is a divisor of the block length n and a few Sboxes are applied in one round function. When $\omega_1 = \omega_2$, SB is a permutation on $\mathbb{F}_2^{\omega_1}$ and we say that the Sbox is a ω_1-bit *Sbox*.

There are mainly two types of symmetric block ciphers—*Feistel cipher* and *Substitution–permutation network (SPN) cipher*.

For a Feistel cipher, the cipher state at the beginning/end of each round is divided into two halves of equal length. The cipher state at the end of round i is denoted as L_i and R_i, where L stands for left and R stands for right. The round function F is defined as

$$(L_i, R_i) = F(L_{i-1}, R_{i-1}, K_i), \text{ where } L_i = R_{i-1}, \ R_i = L_{i-1} \oplus f(R_{i-1}, K_i). \tag{3.1}$$

We note that f is a function that does not need to have an inverse since the function F defined as in Eq. 3.1 is always invertible:

$$L_{i-1} = R_i \oplus f(L_i, K_i), \quad R_{i-1} = L_i.$$

Furthermore, the ciphertext is normally given by $R_{\mathrm{Nr}}||L_{\mathrm{Nr}}$ (i.e., swapping the left and right side of the cipher state at the end of the last round). In this case, if we let R_i and L_i denote the right and left part of the cipher state at the end of round i in the decryption, then the decryption computation is the same as in Eq. 3.1 except that the round keys are in reverse order as that for encryption. An illustration of Feistel cipher can be seen in Fig. 3.1.

Let ω be a divisor of n, the block length, and let $\ell = n/\omega$. The design of an SPN cipher encryption is shown in Fig. 3.2, where SB is an ω-bit Sbox. In most cases, $\omega = 4, 8$.

Each round of an SPN cipher normally consists of bitwise XOR with the round key, application of ℓ parallel ω-bit Sboxes, and a permutation on $\mathbb{F}_2^n$. The encryption starts with XOR with a round key, also ends with XOR with a round key before outputting the ciphertext. Otherwise, the cipher states in the second (or the last)

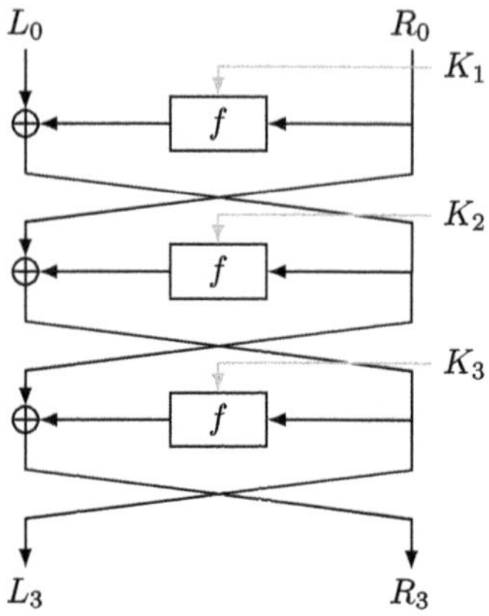

Fig. 3.1 An illustration of Feistel cipher encryption algorithm

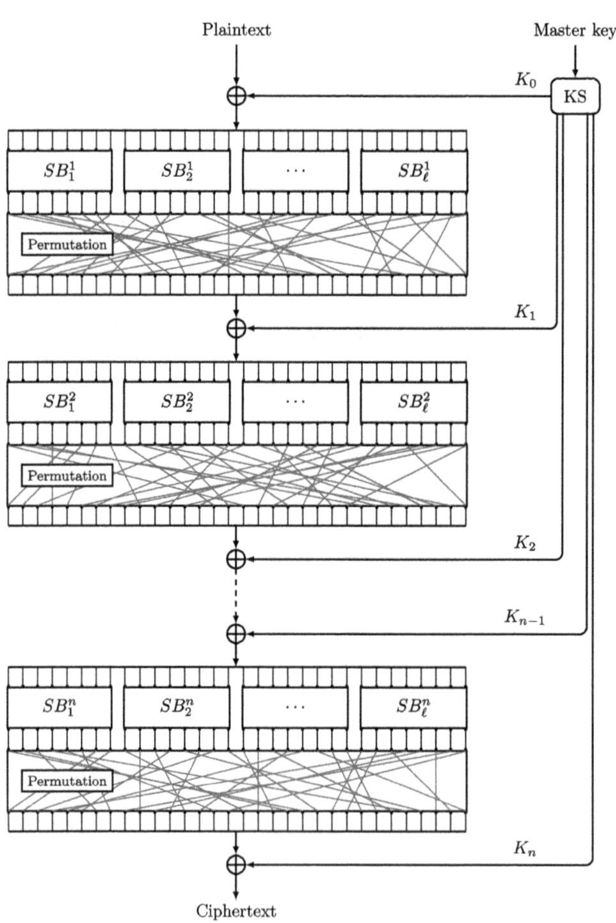

Fig. 3.2 An illustration of SPN cipher encryption algorithm

Fig. 3.3 An illustration of
DES encryption algorithm

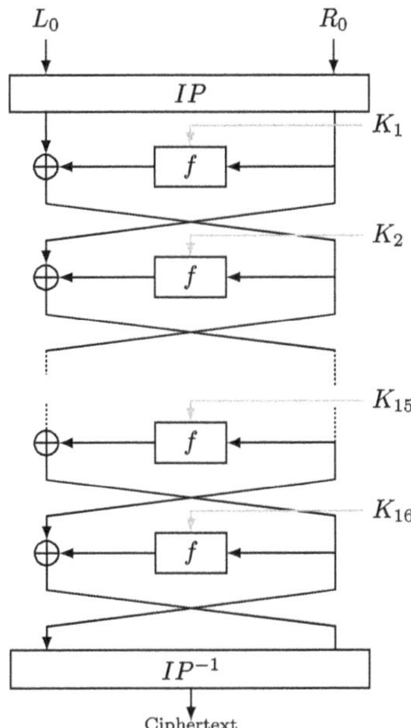

round are all known to the attacker. Those two operations are called *whitening*. For decryption, the inverse of Sbox and permutation are computed, and round keys are XOR-ed with the cipher state in reverse order compared to that for encryption.

3.1.1 DES

Let us first look at one Feistel cipher—Data Encryption Standard (DES). DES was developed at IBM by a team led by Horst Feistel and the design was based on Lucifer cipher [Sor84]. It was used as the NIST standard from 1977 to 2005. Furthermore, it has a significant influence on the development of cipher design.

The block length of DES is $n = 64$, i.e., $\mathcal{P} = \mathcal{C} = \mathbb{F}_2^{64}$. Hence $L_i, R_i \in \mathbb{F}_2^{32}$. The master key length is 56, i.e., $\mathcal{K} = \mathbb{F}_2^{56}$. The round key length is 48. The total number of rounds $\mathtt{Nr} = 16$. An illustration of DES encryption is shown in Fig. 3.3. Each DES round function follows the structure as described in Eq. 3.1.

Before the first round function, the encryption starts with an *initial permutation (IP)*. The inverse of IP, called the *final permutation* (IP^{-1}) is applied to the cipher state after the last round before outputting the ciphertext. Initial and final permutations are included for the ease of loading plaintext/ciphertext. Initial and final permutations are shown in Table 3.1. For example, in IP, the 1st bit of the output is from the 58th bit of the input. The 2nd bit of the output is from the 50th bit of the input.

Table 3.1 Initial permutation (IP) and final permutation (IP^{-1}) in DES algorithm

(a) IP							
58	50	42	34	26	18	10	2
60	52	44	36	28	20	12	4
62	54	46	38	30	22	14	6
64	56	48	40	32	24	16	8
57	49	41	33	25	17	9	1
59	51	43	35	27	19	11	3
61	53	45	37	29	21	13	5
63	55	47	39	31	23	15	7

(b) IP^{-1}							
40	8	48	16	56	24	64	32
39	7	47	15	55	23	63	31
38	6	46	14	54	22	62	30
37	5	45	13	53	21	61	29
36	4	44	12	52	20	60	28
35	3	43	11	51	19	59	27
34	2	42	10	50	18	58	26
33	1	41	9	49	17	57	25

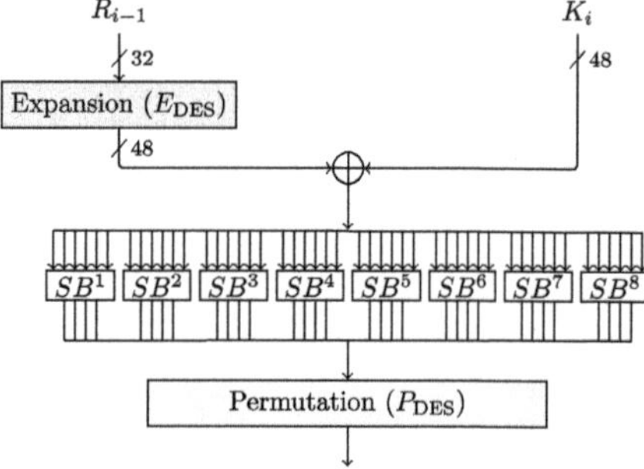

Fig. 3.4 Function f in DES round function

Note For DES specification, we consider the 1st bit of a value as the leftmost bit in its binary representation. For example, the 1st bit of $3 = 011_2$ is 0, the 2nd bit is 1 and the last bit is 1.

At the ith round, the function f in the round function of DES takes input $R_{i-1} \in \mathbb{F}_2^{32}$ and round key $K_i \in \mathbb{F}_2^{48}$, then outputs a 32-bit intermediate value as follows:

$$f(R_{i-1}, K_i) = P_{\mathrm{DES}}(\mathrm{Sboxes}(E_{\mathrm{DES}}(R_{i-1}) \oplus K_i)).$$

Firstly, R_{i-1} is passed to an *expansion function* $E_{\mathrm{DES}} : \mathbb{F}_2^{32} \to \mathbb{F}_2^{48}$. Then the output $E_{\mathrm{DES}}(R_{i-1})$ is XOR-ed with the round key K_i, producing a 48-bit intermediate value. This 48-bit value is divided into eight 6-bit subblocks. Eight distinct Sboxes, $\mathrm{SB}_{\mathrm{DES}}^{j} : \mathbb{F}_2^6 \to \mathbb{F}_2^4$ ($1 \le j \le 8$), are applied to each of the 6 bits. Finally, the resulting 32-bit intermediate value goes through a permutation function $P_{\mathrm{DES}} : \mathbb{F}_2^{32} \to \mathbb{F}_2^{32}$. An illustration of f is shown in Fig. 3.4.

Details of the expansion function E_{DES} are given in Table 3.2. 16 bits of the input are repeated and affect two bits of the output, which influence two Sboxes. Such a design makes the dependency of the output bits on the input bits spread faster and achieves higher diffusion.

Table 3.2 Expansion function $E_{DES} : \mathbb{F}_2^{32} \to \mathbb{F}_2^{48}$ in DES round function. The 1st bit of the output is given by the 32nd bit of the input. The 2nd bit of the output is given by the 1st bit of the input

32	1	2	3	4	5	4	5	6	7	8	9	8	9	10	11
12	13	12	13	14	15	16	17	16	17	18	19	20	21	20	21
22	23	24	25	24	25	26	27	28	29	28	29	30	31	32	1

Table 3.3 SB_{DES}^1 in DES found function

14	4	13	1	2	15	11	8	3	10	6	12	5	9	0	7
0	15	7	4	14	2	13	1	10	6	12	11	9	5	3	8
4	1	14	8	13	6	2	11	15	12	9	7	3	10	5	0
15	12	8	2	4	9	1	7	5	11	3	14	10	0	6	13

Table 3.4 Permutation function $P_{DES} : \mathbb{F}_2^{32} \to \mathbb{F}_2^{32}$ in DES round function. The 1st bit of the output is given by the 16th bit of the input. The 2nd bit of the output comes from the 7th bit of the input

16	7	20	21	29	12	28	17	1	15	23	26	5	18	31	10
2	8	24	14	32	27	3	9	19	13	30	6	22	11	4	25

The design of the first Sbox is shown in Table 3.3, and the rest of the Sboxes are detailed in Appendix C. To use those tables, take an input of one Sbox, say $b_1b_2b_3b_4b_5b_6$, the output corresponds to row b_1b_6 and column $b_2b_3b_4b_5$. We note that each row of each of the Sbox tables is a permutation of integers $0, 1, \ldots, 15$.

Example 3.1.1 Suppose the input of SB_{DES}^1 is

$$b_1b_2b_3b_4b_5b_6 = 100110.$$

According to Table 3.3, the row number is given by $b_1b_6 = 2$. The column number is given by $b_2b_3b_4b_5 = 0011 = 3$. Hence the output is $8 = 1000$. Similarly (see Table C.1 (b)),

$$SB_{DES}^3(100110) = 9 = 1001$$

The details of the permutation function P_{DES} are given in Table 3.4.

The key schedule of DES takes a 64-bit master key as input and outputs round keys of length 48. An illustration of the key schedule is in Fig. 3.5, where PC stands for *permuted choice*.

Each 8th bit of the master key is a parity-check bit of the previous 7 bits, i.e., the XORed value of those 7 bits. PC1 reduces 64-bit input to 56 bit by ignoring those parity-check bits and outputs a permutation of the remaining 56 bits. Then the output is divided into two 28-bit halves (see Table 3.5). Each half rotates left by one or two bits, depending on the round (see Table 3.6). Finally, PC2 selects 48 bits out of 56 bits, permutes them, and outputs the round key (see Table 3.7).

Fig. 3.5 DES key schedule

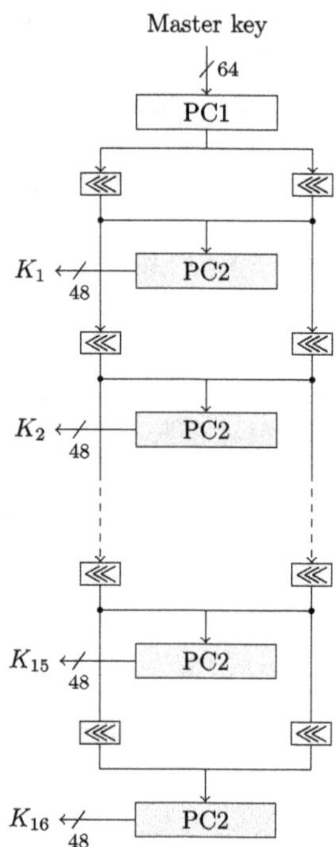

Table 3.5 Left and right part of the intermediate values in DES key schedule after PC1. The 1st bit of the left part comes from the 57th bit of the master key (input to PC1)

Left							Right						
57	49	41	33	25	17	9	63	55	47	39	31	23	15
1	58	50	42	34	26	18	7	62	54	46	38	30	22
10	2	59	51	43	35	27	14	6	61	53	45	37	29
19	11	3	60	52	44	36	21	13	5	28	20	12	4

Table 3.6 Number of key bits rotated per round in DES key schedule

Round	1	2	3	4	5	6	7	8	9	10	11	12	13	14	15	16
Rotation	1	1	2	2	2	2	2	2	1	2	2	2	2	2	2	1

Table 3.7 PC2 in DES key schedule

14	17	11	24	1	5	3	28	15	6	21	10	23	19	12	4
26	8	16	7	27	20	13	2	41	52	31	37	47	55	30	40
51	45	33	48	44	49	39	56	34	53	46	42	50	36	29	32

For some master keys, the key schedule outputs the same round keys for more than one round. Those master keys are called *weak keys*. Weak keys should not be used. It can be shown that there are in total four of them:

- `01010101 01010101`,

- `FEFEFEFE FEFEFEFE,`
- `E0E0E0E0 F1F1F1F1,`
- `1F1F1F1F 0E0E0E0E.`

Remark 3.1.1 From the design of the DES key schedule, we can see that with the knowledge of any round key, the attacker can recover 48 bits of the master key. The remaining 8 can be found by brute force. Alternatively, with the knowledge of another round key, the master key can be recovered.

3.1.2 AES

In 1997, NIST published a call for cryptographic algorithms as a replacement for DES. In October 2000, Rijndael was selected as the winner and certain versions of Rijndael are set as the Advanced Encryption Standard (AES). Rijndael was invented by Belgian cryptographers Joan Daemen and Vincent Rijmen and optimized for software efficiency on 8 and 32 bit processors.

For AES, block length $n = 128$, number of rounds $\mathtt{Nr} = 10, 12, 14$ with corresponding key lengths 128, 192, 256. The corresponding algorithms are hence named AES-128, AES-192, and AES-256 respectively. The original design of Rijndael also allows for other key lengths and block lengths. As shown in Table 3.8, where blue-colored values are specifications adopted by AES.

The encryption algorithm starts with an initial AddRoundKey operation. Then the round function for the first $\mathtt{Nr}-1$ rounds consists of four operations: SubBytes, ShiftRows, MixColumns, and AddRoundKey. Finally, the last round (round $\mathtt{Nr}$) consists of SubBytes, ShiftRows, and AddRoundKey. AddRoundKey is bitwise XOR with the round key and SubBytes is the application of 8-bit Sboxes. ShiftRows permutes the bytes and MixColumns is a function on 32-bit values (four bytes). Figure 3.6 illustrates the AES round function.

The inverse of SubBytes, ShiftRows, and MixColumns are denoted as InvSubBytes, InvShiftRows, and InvMixColumns respectively. The first round of AES decryption computes AddRoundKey, InvShiftRows, and InvSubBytes. Then the round function for the next $\mathtt{Nr}-1$ rounds consists of AddRoundKey,

Table 3.8 Specifications of Rijndael design, where blue-colored values are adopted by AES

Key length	Block length				
	128	160	192	224	256
128	10	11	12	13	14
160	11	11	12	13	14
192	12	12	12	13	14
224	13	13	13	13	14
256	14	14	14	14	14

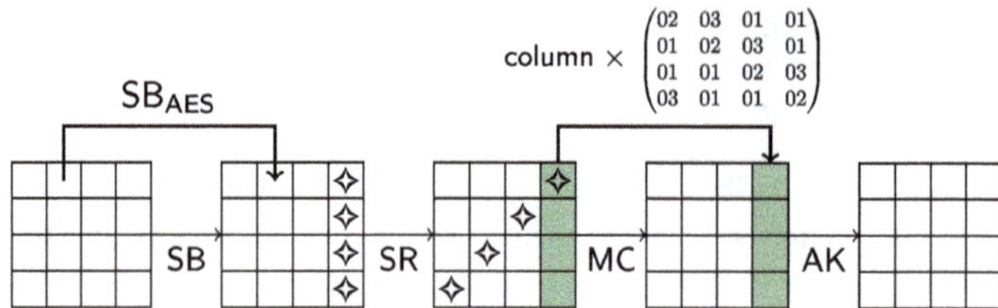

Fig. 3.6 AES round function for round i, $1 \leq i \leq \mathrm{Nr}-1$. SB, SR, MC, and AK stand for SubBytes, ShiftRows, MixColumns, and AddRoundKey respectively

InvMixColumns, InvShiftRows, and InvSubBytes. Finally, there is an additional AddRoundKey operation. The round keys for decryption are in reverse order as those for encryption.

To give more details on the AES round function, we represent the AES cipher state as a four-by-four matrix of bytes:

$$\begin{pmatrix} s_{00} & s_{01} & s_{02} & s_{03} \\ s_{10} & s_{11} & s_{12} & s_{13} \\ s_{20} & s_{21} & s_{22} & s_{23} \\ s_{30} & s_{31} & s_{32} & s_{33} \end{pmatrix}. \tag{3.2}$$

Recall that one byte is a vector in $\mathbb{F}_2^8$ and can be represented as a hexadecimal number between 00 and FF (see Definition 1.3.7 and Remark 1.3.3). As discussed in Sect. 1.5.1, a byte can also be identified as an element in $\mathbb{F}_2[x]/(f(x))$, where $f(x) = x^8 + x^4 + x^3 + x + 1 \in \mathbb{F}_2[x]$ is an irreducible polynomial over $\mathbb{F}_2$.

Remark 3.1.2 We refer to $\begin{pmatrix} s_{i0} & s_{i1} & s_{i2} & s_{i3} \end{pmatrix}$ as the $(i+1)$th row of the cipher state, and

$$\begin{pmatrix} s_{0j} \\ s_{1j} \\ s_{2j} \\ s_{3j} \end{pmatrix}$$

as the $(j+1)$th column of the cipher state.

The 8-bit Sbox in AES can be described using Table 3.9, for example, $\mathrm{SB_{AES}}(12) = C9$. Different from the eight Sboxes in DES, AES Sbox can also be defined algebraically. Let

Table 3.9 AES Sbox

	0	1	2	3	4	5	6	7	8	9	A	B	C	D	E	F
0	63	7C	77	7B	F2	6B	6F	C5	30	01	67	2B	FE	D7	AB	76
1	CA	82	C9	7D	FA	59	47	F0	AD	D4	A2	AF	9C	A4	72	C0
2	B7	FD	93	26	36	3F	F7	CC	34	A5	E5	F1	71	D8	31	15
3	04	C7	23	C3	18	96	05	9A	07	12	80	E2	EB	27	B2	75
4	09	83	2C	1A	1B	6E	5A	A0	52	3B	D6	B3	29	E3	2F	84
5	53	D1	00	ED	20	FC	B1	5B	6A	CB	BE	39	4A	4C	58	CF
6	D0	EF	AA	FB	43	4D	33	85	45	F9	02	7F	50	3C	9F	A8
7	51	A3	40	8F	92	9D	38	F5	BC	B6	DA	21	10	FF	F3	D2
8	CD	0C	13	EC	5F	97	44	17	C4	A7	7E	3D	64	5D	19	73
9	60	81	4F	DC	22	2A	90	88	46	EE	B8	14	DE	5E	0B	DB
A	E0	32	3A	0A	49	06	24	5C	C2	D3	AC	62	91	95	E4	79
B	E7	C8	37	6D	8D	D5	4E	A9	6C	56	F4	EA	65	7A	AE	08
C	BA	78	25	2E	1C	A6	B4	C6	E8	DD	74	1F	4B	BD	8B	8A
D	70	3E	B5	66	48	03	F6	0E	61	35	57	B9	86	C1	1D	9E
E	E1	F8	98	11	69	D9	8E	94	9B	1E	87	E9	CE	55	28	DF
F	8C	A1	89	0D	BF	E6	42	68	41	99	2D	0F	B0	54	BB	16

$$A = \begin{pmatrix} 1 & 1 & 1 & 1 & 1 & 0 & 0 & 0 \\ 0 & 1 & 1 & 1 & 1 & 1 & 0 & 0 \\ 0 & 0 & 1 & 1 & 1 & 1 & 1 & 0 \\ 0 & 0 & 0 & 1 & 1 & 1 & 1 & 1 \\ 1 & 0 & 0 & 0 & 1 & 1 & 1 & 1 \\ 1 & 1 & 0 & 0 & 0 & 1 & 1 & 1 \\ 1 & 1 & 1 & 0 & 0 & 0 & 1 & 1 \\ 1 & 1 & 1 & 1 & 0 & 0 & 0 & 1 \end{pmatrix}, \quad a = \begin{pmatrix} 0 \\ 1 \\ 1 \\ 0 \\ 0 \\ 0 \\ 1 \\ 1 \end{pmatrix},$$

then

$$\mathrm{SB}_{\mathrm{AES}}(z) = \begin{cases} A z^{-1} + a & z \neq 0 \\ a & z = 0 \end{cases} \tag{3.3}$$

where z^{-1} is the inverse of z as an element in $\mathbb{F}_2[x]/(f(x))$ (see Sect. 1.5.1).

Example 3.1.2 $\mathrm{SB}_{\mathrm{AES}}(00) = a = 01100011_2 = 63$.

Example 3.1.3 Suppose the input of AES Sbox is $03 = 00000011_2$, which corresponds to $x + 1 \in \mathbb{F}_2[x]/(f(x))$. We have shown in Example 1.5.21 that $03^{-1} = 11110110_2$. Then

$$
A \begin{pmatrix} 1 \\ 1 \\ 1 \\ 1 \\ 0 \\ 1 \\ 1 \\ 0 \end{pmatrix} + a = \begin{pmatrix} 1 1 1 1 1 0 0 0 \\ 0 1 1 1 1 1 0 0 \\ 0 0 1 1 1 1 1 0 \\ 0 0 0 1 1 1 1 1 \\ 1 0 0 0 1 1 1 1 \\ 1 1 0 0 0 1 1 1 \\ 1 1 1 0 0 0 1 1 \\ 1 1 1 1 0 0 0 1 \end{pmatrix} \begin{pmatrix} 1 \\ 1 \\ 1 \\ 1 \\ 0 \\ 1 \\ 1 \\ 0 \end{pmatrix} + \begin{pmatrix} 0 \\ 1 \\ 1 \\ 0 \\ 0 \\ 0 \\ 1 \\ 1 \end{pmatrix} = \begin{pmatrix} 0 \\ 0 \\ 0 \\ 1 \\ 1 \\ 0 \\ 0 \\ 0 \end{pmatrix} + \begin{pmatrix} 0 \\ 1 \\ 1 \\ 0 \\ 0 \\ 0 \\ 1 \\ 1 \end{pmatrix} = \begin{pmatrix} 0 \\ 1 \\ 1 \\ 1 \\ 1 \\ 0 \\ 1 \\ 1 \end{pmatrix}.
$$

So $\mathrm{SB}_{\mathrm{AES}}(03) = 01111011_2 = 7\mathrm{B}$, which agrees with Table 3.9.

For decryption, we need to compute the inverse of SubBytes, InvSubBytes. Let g denote the function

$$
g(z) = Az + a.
$$

Then by Eq. 3.3, InvSubBytes computes

$$
\mathrm{SB}_{\mathrm{AES}}^{-1}(z) = \begin{cases} (g^{-1}(z))^{-1} & g^{-1}(z) \neq 0 \\ 0 & g^{-1}(z) = 0 \end{cases},
$$

where $g^{-1}(z)$ is given by (see [DR02])

$$
g^{-1}(z) = \begin{pmatrix} 0 1 0 1 0 0 1 0 \\ 0 0 1 0 1 0 0 1 \\ 1 0 0 1 0 1 0 0 \\ 0 1 0 0 1 0 1 0 \\ 0 0 1 0 0 1 0 1 \\ 1 0 0 1 0 0 1 0 \\ 0 1 0 0 1 0 0 1 \\ 1 0 1 0 0 1 0 0 \end{pmatrix} z + \begin{pmatrix} 0 \\ 0 \\ 0 \\ 0 \\ 0 \\ 1 \\ 0 \\ 1 \end{pmatrix}.
$$

InvSubBytes can also be described using a table, as detailed in Table 3.10.

Example 3.1.4 Let $z = 63 = 01100011_2$. Then

Table 3.10 Inverse of AES Sbox

	0	1	2	3	4	5	6	7	8	9	A	B	C	D	E	F
0	52	09	6A	D5	30	36	A5	38	BF	40	A3	9E	81	F3	D7	FB
1	7C	E3	39	82	9B	2F	FF	87	34	8E	43	44	C4	DE	E9	CB
2	54	7B	94	32	A6	C2	23	3D	EE	4C	95	0B	42	FA	C3	4E
3	08	2E	A1	66	28	D9	24	B2	76	5B	A2	49	6D	8B	D1	25
4	72	F8	F6	64	86	68	98	16	D4	A4	5C	CC	5D	65	B6	92
5	6C	70	48	50	FD	ED	B9	DA	5E	15	46	57	A7	8D	9D	84
6	90	D8	AB	00	8C	BC	D3	0A	F7	E4	58	05	B8	B3	45	06
7	D0	2C	1E	8F	CA	3F	0F	02	C1	AF	BD	03	01	13	8A	6B
8	3A	91	11	41	4F	67	DC	EA	97	F2	CF	CE	F0	B4	E6	73
9	96	AC	74	22	E7	AD	35	85	E2	F9	37	E8	1C	75	DF	6E
A	47	F1	1A	71	1D	29	C5	89	6F	B7	62	0E	AA	18	BE	1B
B	FC	56	3E	4B	C6	D2	79	20	9A	DB	C0	FE	78	CD	5A	F4
C	1F	DD	A8	33	88	07	C7	31	B1	12	10	59	27	80	EC	5F
D	60	51	7F	A9	19	B5	4A	0D	2D	E5	7A	9F	93	C9	9C	EF
E	A0	E0	3B	4D	AE	2A	F5	B0	C8	BE	BB	3C	83	53	99	61
F	17	2B	04	7E	BA	77	D6	26	E1	69	14	63	55	21	0C	7D

$$g^{-1}(z) = \begin{pmatrix} 0\,1\,0\,1\,0\,0\,1\,0 \\ 0\,0\,1\,0\,1\,0\,0\,1 \\ 1\,0\,0\,1\,0\,1\,0\,0 \\ 0\,1\,0\,0\,1\,0\,1\,0 \\ 0\,0\,1\,0\,0\,1\,0\,1 \\ 1\,0\,0\,1\,0\,0\,1\,0 \\ 0\,1\,0\,0\,1\,0\,0\,1 \\ 1\,0\,1\,0\,0\,1\,0\,0 \end{pmatrix} \begin{pmatrix} 0 \\ 1 \\ 1 \\ 0 \\ 0 \\ 0 \\ 1 \\ 1 \end{pmatrix} + \begin{pmatrix} 0 \\ 0 \\ 0 \\ 0 \\ 0 \\ 1 \\ 0 \\ 1 \end{pmatrix} = \begin{pmatrix} 0 \\ 0 \\ 0 \\ 0 \\ 0 \\ 1 \\ 0 \\ 1 \end{pmatrix} + \begin{pmatrix} 0 \\ 0 \\ 0 \\ 0 \\ 0 \\ 1 \\ 0 \\ 1 \end{pmatrix} = \begin{pmatrix} 0 \\ 0 \\ 0 \\ 0 \\ 0 \\ 0 \\ 0 \\ 0 \end{pmatrix},$$

which is equal to 00. And we have $\mathrm{SB}_{\mathrm{AES}}^{-1}(63) = 00$.

Example 3.1.5 Let $z - 8\mathsf{C} = 10001100_2$. Then

$$g^{-1}(z) = \begin{pmatrix} 0\,1\,0\,1\,0\,0\,1\,0 \\ 0\,0\,1\,0\,1\,0\,0\,1 \\ 1\,0\,0\,1\,0\,1\,0\,0 \\ 0\,1\,0\,0\,1\,0\,1\,0 \\ 0\,0\,1\,0\,0\,1\,0\,1 \\ 1\,0\,0\,1\,0\,0\,1\,0 \\ 0\,1\,0\,0\,1\,0\,0\,1 \\ 1\,0\,1\,0\,0\,1\,0\,0 \end{pmatrix} \begin{pmatrix} 1 \\ 0 \\ 0 \\ 0 \\ 1 \\ 1 \\ 0 \\ 0 \end{pmatrix} + \begin{pmatrix} 0 \\ 0 \\ 0 \\ 0 \\ 0 \\ 1 \\ 0 \\ 1 \end{pmatrix} = \begin{pmatrix} 0 \\ 1 \\ 0 \\ 1 \\ 1 \\ 1 \\ 1 \\ 0 \end{pmatrix} + \begin{pmatrix} 0 \\ 0 \\ 0 \\ 0 \\ 0 \\ 1 \\ 0 \\ 1 \end{pmatrix} = \begin{pmatrix} 0 \\ 1 \\ 0 \\ 1 \\ 1 \\ 0 \\ 1 \\ 1 \end{pmatrix},$$

which corresponds to

$$x^6 + x^4 + x^3 + x + 1 \in \mathbb{F}_2[x]/(f(x)).$$

By the Euclidean algorithm

$$f(x) = (x^2 + 1)(x^6 + x^4 + x^3 + x + 1) + (x^5 + x^3 + x^2),$$
$$x^6 + x^4 + x^3 + x + 1 = x(x^5 + x^3 + x^2) + (x + 1),$$
$$x^5 + x^3 + x^2 = (x^4 + x^3 + x + 1)(x + 1) + 1.$$

Then by the extended Euclidean algorithm

$$
\begin{aligned}
1 &= (x^5 + x^3 + x^2) + (x^4 + x^3 + x + 1)(x + 1) \\
&= (x^5 + x^3 + x^2) + (x^4 + x^3 + x + 1)((x^6 + x^4 + x^3 + x + 1) \\
&\quad + x(x^5 + x^3 + x^2)) \\
&= (x^4 + x^3 + x + 1)(x^6 + x^4 + x^3 + x + 1) \\
&\quad + (x^5 + x^4 + x^2 + x + 1)(x^5 + x^3 + x^2) \\
&= (x^4 + x^3 + x + 1)(x^6 + x^4 + x^3 + x + 1) \\
&\quad + (x^5 + x^4 + x^2 + x + 1)(f(x) + (x^2 + 1)(x^6 + x^4 + x^3 + x + 1)) \\
&= (x^5 + x^4 + x^2 + x + 1)f(x) + (x^7 + x^6 + x^5 + x^4)(x^6 + x^4 + x^3 + x + 1).
\end{aligned}
$$

And we have

$$(x^6 + x^4 + x^3 + x + 1)^{-1} \bmod f(x) = x^7 + x^6 + x^5 + x^4 = 11110000_2 = \mathsf{F0},$$

which gives $\mathrm{SB}_{\mathrm{AES}}^{-1}(\mathsf{8C}) = \mathsf{F0}$

As the name suggests, the ShiftRows operation shifts the bytes in the rows of the cipher state. Recall the representation of the AES cipher state from Eq. 3.2. Then the ShiftRows operation can be described by the following transformation:

$$
\begin{pmatrix}
s_{00} & s_{01} & s_{02} & s_{03} \\
s_{10} & s_{11} & s_{12} & s_{13} \\
s_{20} & s_{21} & s_{22} & s_{23} \\
s_{30} & s_{31} & s_{32} & s_{33}
\end{pmatrix}
\rightarrow
\begin{pmatrix}
s_{00} & s_{01} & s_{02} & s_{03} \\
s_{11} & s_{12} & s_{13} & s_{10} \\
s_{22} & s_{23} & s_{20} & s_{21} \\
s_{33} & s_{30} & s_{31} & s_{32}
\end{pmatrix}.
$$

The first row does not change. The second row rotates left by one byte. The third row rotates left by two bytes. Finally, the last row rotates left by three bytes.

In another representation, let us denote the input of ShiftRows using cipher state representation in Eq. 3.2. Let the output of ShiftRows be a matrix B with entries b_{ij} $(0 \le i, j \le 3)$. Then

$$\begin{pmatrix} b_{0j} \\ b_{1j} \\ b_{2j} \\ b_{3j} \end{pmatrix} = \begin{pmatrix} s_{0j} \\ s_{1(j+1 \bmod 4)} \\ s_{2(j+2 \bmod 4)} \\ s_{3(j+3 \bmod 4)} \end{pmatrix}, \quad 0 \le j < 4. \tag{3.4}$$

For decryption, the inverse of ShiftRows, InvShiftRows, can be easily deduced.

The MixColumns function takes each of the four columns of the cipher state (Eq. 3.2)

$$\begin{pmatrix} s_{0j} \\ s_{1j} \\ s_{2j} \\ s_{3j} \end{pmatrix}, \quad j = 0, 1, 2, 3,$$

as input. The column is considered as a polynomial over $\mathbb{F}_2[x]/(f(x))$:

$$s_{3j}x^3 + s_{2j}x^2 + s_{1j}x + s_{0j}.$$

MixColumns multiplies $s_{3j}x^3 + s_{2j}x^2 + s_{1j}x + s_{0j}$ with another polynomial over $\mathbb{F}_2[x]/(f(x))$ given by

$$g(x) = 03x^3 + 01x^2 + 01x + 02.$$

The multiplication is computed modulo $x^4 + 1$. This design choice is based on specific diffusion and performance goals. We will not go into the details in this book, interested readers can refer to [DR02]. Let $d(x) = d_3x^3 + d_2x^2 + d_1x + d_0$ denote the product of $s_{3j}x^3 + s_{2j}x^2 + s_{1j}x + s_{0j}$ with $g(x)$ modulo $x^4 + 1$. We have

$$\begin{aligned}
d(x) &= (s_{3j}x^3 + s_{2j}x^2 + s_{1j}x + s_{0j})(03x^3 + 01x^2 + 01x + 02_{16}) \bmod (x^4 + 1) \\
&= 03s_{3j}x^6 + (01s_{3j} + 03s_{2j})x^5 + (01s_{3j} + 01s_{2j} + 03s_{1j})x^4 \\
&\quad + (02s_{3j} + 01s_{2j} + 01s_{1j} + 03s_{0j})x^3 + (02s_{2j} + 01s_{1j} + 01s_{0j})x^2 \\
&\quad + (02s_{1j} + 01s_{0j})x + 02s_{0j} \bmod (x^4 + 1) \\
&= (02s_{3j} + 01s_{2j} + 01s_{1j} + 03s_{0j})x^3 + (03s_{3j} + 02s_{2j} + 01s_{1j} + 01s_{0j})x^2 \\
&\quad + (01s_{3j} + 03s_{2j} + 02s_{1j} + 01s_{0j})x + 01s_{3j} \\
&\quad + 01s_{2j} + 03s_{1j} + 02s_{0j}.
\end{aligned} \tag{3.5}$$

Thus, MixColumns can be considered as multiplying the input column by a matrix:

$$\begin{pmatrix} d_0 \\ d_1 \\ d_2 \\ d_3 \end{pmatrix} = \begin{pmatrix} 02 & 03 & 01 & 01 \\ 01 & 02 & 03 & 01 \\ 01 & 01 & 02 & 03 \\ 03 & 01 & 01 & 02 \end{pmatrix} \begin{pmatrix} s_{0j} \\ s_{1j} \\ s_{2j} \\ s_{3j} \end{pmatrix}. \tag{3.6}$$

Example 3.1.6 Suppose

$$\begin{pmatrix} s_{0j} \\ s_{1j} \\ s_{2j} \\ s_{3j} \end{pmatrix} = \begin{pmatrix} D4 \\ BF \\ 5D \\ 30 \end{pmatrix}.$$

Then

$$\begin{pmatrix} 02 & 03 & 01 & 01 \\ 01 & 02 & 03 & 01 \\ 01 & 01 & 02 & 03 \\ 03 & 01 & 01 & 02 \end{pmatrix} \begin{pmatrix} D4 \\ BF \\ 5D \\ 30 \end{pmatrix} = \begin{pmatrix} 04 \\ 66 \\ 81 \\ E5 \end{pmatrix}.$$

For example, we have calculated in Examples 1.5.17 and 1.5.20 that

$$02 \times D4 = 10110011_2, \quad 03 \times BF = 11011010_2.$$

The first entry of the product is then given by

$$10110011 \oplus 11011010 \oplus 01011101 \oplus 00110000 = 00000100 = 04.$$

Remark 3.1.3 For any

$$a = \begin{pmatrix} a_0 \\ a_1 \\ a_2 \\ a_3 \end{pmatrix}, \quad b = \begin{pmatrix} b_0 \\ b_1 \\ b_2 \\ b_3 \end{pmatrix},$$

we have

$$\mathrm{MixColumns}(a + b) = \mathrm{MixColumns}(a) + \mathrm{MixColumns}(b),$$

where the addition is computed modulo $f(x)$. As discussed in Remark 1.5.2, this addition is equivalent to XOR. Consequently, we have

$$\mathrm{MixColumns}(a \oplus b) = \mathrm{MixColumns}(a) \oplus \mathrm{MixColumns}(b).$$

The inverse of MixColumns, InvMixColumns, is defined by multiplying each column of the cipher state by the inverse of $g(x)$ (Eq. 3.1.2) modulo $x^4 + 1$. We note that

$$x^4 + 1 = (x + 1)^4$$

as a polynomial over $\mathbb{F}_2[x]/(f(x))$. Since 1 is not a root of $g(x)$, $x + 1$ does not divide $g(x)$, which gives

$$\gcd(g(x), x^4 + 1) = 1.$$

We have shown that $\mathbb{F}_2[x]/(f(x))$ is a field in Sect. 1.5.1. $g(x)^{-1} \bmod x^4 + 1$ can be computed using the extended Euclidean algorithm, similarly to Example 1.5.10. We have

$$g(x)^{-1} \bmod x^4 + 1 = \mathtt{0B}x^3 + \mathtt{0D}x^2 + \mathtt{09}x + \mathtt{0E}.$$

It can be shown in the same way as in Eq. 3.5 that, multiplication by $g(x)^{-1} \bmod x^4 + 1$ is equivalent to multiplication by the following matrix

$$\begin{pmatrix} \mathtt{0E} & \mathtt{0B} & \mathtt{0D} & \mathtt{09} \\ \mathtt{09} & \mathtt{0E} & \mathtt{0B} & \mathtt{0D} \\ \mathtt{0D} & \mathtt{09} & \mathtt{0E} & \mathtt{0B} \\ \mathtt{0B} & \mathtt{0D} & \mathtt{09} & \mathtt{0E} \end{pmatrix}. \tag{3.7}$$

We will discuss the AES key schedule for key length 128, which corresponds to $\mathtt{Nr} = 10$. The algorithms for other key lengths are defined similarly (see [DR02] for more details). The key schedule algorithm is named KeyExpansion, shown in Algorithm 3.1. The master key k is written as a four-by-four array of bytes, denoted by $K[4][4]$ in the algorithm. KeyExpansion expands $K[4][4]$ to a 4×44 array of bytes, denoted by $W[4][44]$. Since $\mathtt{Nr} = 10$, in total we need 11 round keys. The ith round key is given by the columns $4i$ to $4(i + 1) - 1$ of W. Note that the 0th round key, i.e., the round key for whitening at the beginning of the encryption, is given by the first 4 columns of W, which are equal to the master key (lines 1–3). Round constants, denoted Rcon (line 6), is an array of ten bytes, computed as follows:

$$\mathrm{Rcon}[1] = x^0 = \mathtt{01}, \quad \text{and} \quad \mathrm{Rcon}[j] = x\mathrm{Rcon}[j - 1] = x^{j-1}, \text{ for } j > 1.$$

We have

$$\mathrm{Rcon} = \{\mathtt{01}, \mathtt{02}, \mathtt{04}, \mathtt{08}, \mathtt{10}, \mathtt{20}, \mathtt{40}, \mathtt{80}, \mathtt{1B}, \mathtt{36}\}.$$

Algorithm 3.1: KeyExpansion—AES-128 key schedule

Input: $K[4][4]$ `// master key written as a four-by-four array of bytes`
Output: $W[4][44]$
1 **for** $j = 0$, $j < 4$, $j++$ **do**
2   **for** $i = 0$, $i < 4$, $i++$ **do**
3       $W[i][j]=K[i][j]$

4 **for** $j = 4$, $j < 44$, $j++$ **do**
5   **if** $j \bmod 4 == 0$ **then**
6       $W[0][j] = W[0][j-4] \oplus \mathrm{SB}_{\mathrm{AES}}(W[1][j-1]) \oplus \mathrm{Rcon}[j/4]$
7   **for** $i = 1$, $i < 4$, $i++$ **do**
8          $W[i][j] = W[i][j-4] \oplus \mathrm{SB}_{\mathrm{AES}}(W[i+1 \bmod 4][j-1])$

9   **else**
10   **for** $i = 0$, $i < 4$, $i++$ **do**
11         $W[i][j] = W[i][j-4] \oplus W[i][j-1]$

12 **return** W

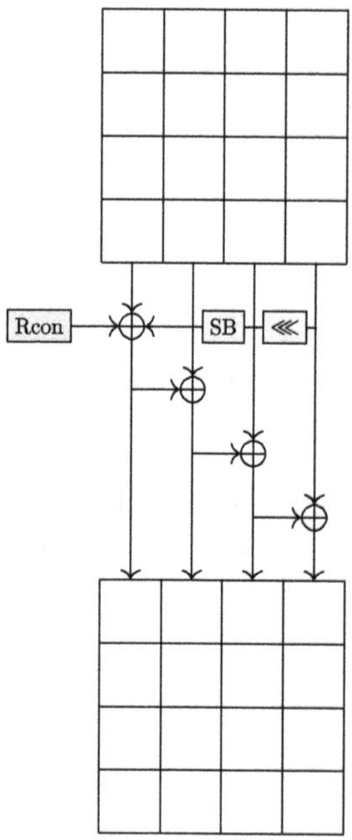

Fig. 3.7 Key schedule for AES-128

The key schedule is also depicted in Fig. 3.7, where the round keys are represented as four-by-four grids and each box corresponds to one byte. The rotation $\ll$ rotates the right-most column by one byte

$$\begin{pmatrix} y_0 \\ y_1 \\ y_2 \\ y_3 \end{pmatrix} \mapsto \begin{pmatrix} y_1 \\ y_2 \\ y_3 \\ y_0 \end{pmatrix}.$$

Remark 3.1.4 We note that with the knowledge of any round key for AES-128 encryption, the attacker can recover the master key using the inverse of the key schedule.

3.1.3 PRESENT

PRESENT was proposed in 2007 [BKL$^+$07] as a symmetric block cipher optimized for hardware implementation. It has block length $n = 64$, number of rounds Nr = 31, and a key length of either 80 or 128. The Sbox for PRESENT is a 4-bit Sbox. When the key length is 80, the algorithm is called PRESENT-80.

The round function of PRESENT consists of addRoundKey, sBoxLayer, and pLayer. After 31 rounds, addRoundKey is applied again before the ciphertext output (see Fig. 3.8).

Note As opposed to DES specification, for PRESENT specification, we consider the 0th bit of a value as the right-most bit in its binary representation. For example, the 0th bit of $3 = 011_2$ is 1, the 1st bit is 1 and the 2nd bit is 0.

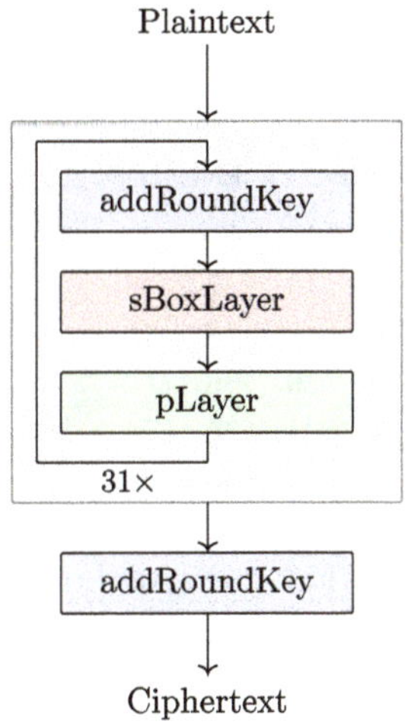

Fig. 3.8 An illustration of PRESENT encryption algorithm

addRoundKey takes the current 64-bit cipher state

$$b_{63}b_{62}\ldots b_0$$

and XOR it with the round key

$$K_i = \kappa^i_{63}\ldots\kappa^i_0, \quad (1 \le i \le 32)$$

bitwise

$$b_j = b_j \oplus \kappa^i_j, \quad 0 \le j \le 63.$$

sBoxLayer applies sixteen 4-bit Sboxes to each nibble of the current cipher state. The 4-bit Sbox is given by Table 3.11. For example, if the input is 0, the output is C. pLayer permutes the 64 bits of the cipher state using the following formula:

$$\text{pLayer}(j) = \left\lfloor \frac{j}{4} \right\rfloor + (j \bmod 4) \times 16,$$

where j denotes the bit position. For example, the 0th bit of the input stays as the 0th bit of the output, and the 1st bit of the input goes to the 16th bit of the output. It can also be described using Table 3.12.

Figure 3.9 shows two rounds of PRESENT.

Here we detail the key schedule for PRESENT-80. We refer the readers to [BKL$^+$07] for the key schedule for the 128-bit master key. Let us denote the variable storing the key by $k_{79}k_{78}\ldots k_0$. At round i, the round key is given by

$$K_i = \kappa^i_{63}\kappa^i_{62}\ldots\kappa^i_0 = k_{79}k_{78}\ldots k_{16}.$$

Table 3.11 PRESENT Sbox

0	1	2	3	4	5	6	7	8	9	A	B	C	D	E	F
C	5	6	B	9	0	A	D	3	E	F	8	4	7	1	2

Table 3.12 PRESENT pLayer

0	1	2	3	4	5	6	7	8	9	10	11	12	13	14	15
0	16	32	48	1	17	33	49	2	18	34	50	3	19	35	51
16	17	18	19	20	21	22	23	24	25	26	27	28	29	30	31
4	20	36	52	5	21	37	53	6	22	38	54	7	23	39	55
32	33	34	35	36	37	38	39	40	41	42	43	44	45	46	47
8	24	40	56	9	25	41	57	10	26	42	58	11	27	43	59
48	49	50	51	52	53	54	55	56	57	58	59	60	61	62	63
12	28	44	60	13	29	45	61	14	30	46	62	15	31	47	63

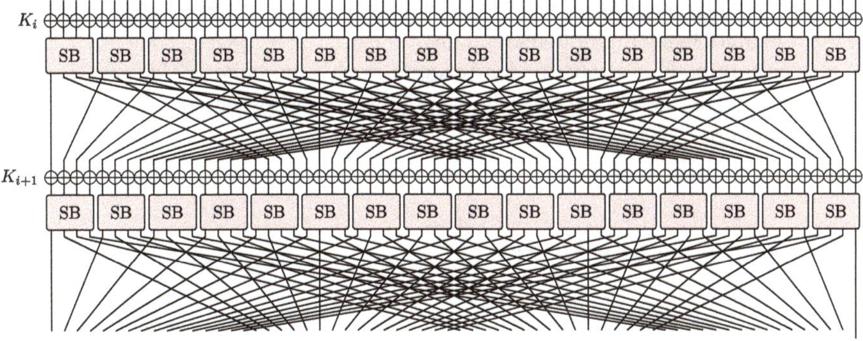

Fig. 3.9 Two rounds of PRESENT

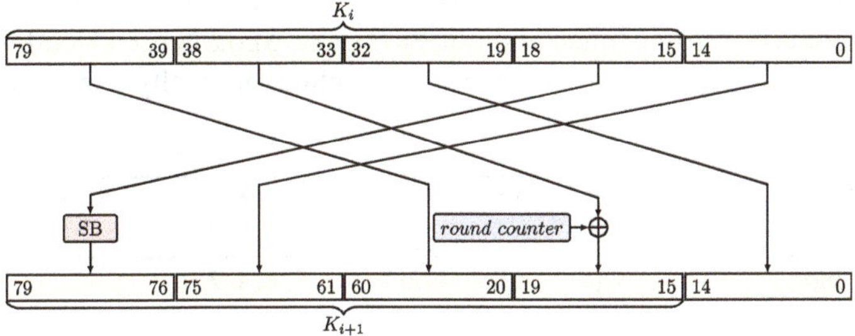

Fig. 3.10 PRESENT-80 key schedule

After extracting the round key, the variable $k_{79}k_{78}\ldots k_0$ is updated using the following steps:

1. Left rotate of 61 bits, $k_{79}k_{78}\ldots k_1k_0 \leftarrow k_{18}k_{17}\ldots k_{20}k_{19}$;
2. $k_{79}k_{78}k_{77}k_{76} = \mathrm{SB_{PRESENT}}(k_{79}k_{78}k_{77}k_{76})$;
3. $k_{19}k_{18}k_{17}k_{16}k_{15} = k_{19}k_{18}k_{17}k_{16}k_{15}\oplus \mathrm{round\_counter}$;

where $\mathrm{SB_{PRESENT}}$ stands for the PRESENT Sbox (Table 3.11) and round_counter $= 1, 2, \ldots, 31$. A graphical illustration is shown in Fig. 3.10.

Remark 3.1.5 With the knowledge of any round key for PRESENT-80, the attacker can recover 64 bits of the master key. The remaining 16 bits can be recovered by brute force. Alternatively, with the knowledge of another round key, the master key can also be revealed.

3.2 Implementations of Symmetric Block Ciphers

In Sect. 3.1, we saw that there are mainly three building blocks for a symmetric block cipher: bitwise XOR with round key, Sbox, and permutation. In this section, we will discuss how to implement each of them. While we mainly focus on the software implementations of PRESENT and AES, the main ideas apply in general to other ciphers with similar constructions.

It is easy in both software and hardware to implement bitwise XOR with a round key. In hardware, there is an XOR gate and almost every processor has a dedicated XOR instruction.

3.2.1   Implementing Sboxes

In software, a naïve way to implement Sbox is to use a lookup table. The table is stored as an array in random access memory or flash memory. The storage space required for an Sbox SB: $\mathbb{F}_2^{\omega_1} \rightarrow \mathbb{F}_2^{\omega_2}$ is $\omega_2 \times 2^{\omega_1}$. For example, PRESENT has a 4-bit Sbox (Table 3.11) and the storage required is $2^4 \times 4 = 64$ bits, or 8 bytes. A lookup table implementation of PRESENT Sbox in pseudocode is shown in Algorithm 3.2. As current computer architectures normally use word sizes of

---

Algorithm 3.2: A lookup table implementation of PRESENT Sbox in pseudocode

---

```
1  integer array [1..16] Sbox = {C, 5, 6, B, 9, 0, A, D, 3, E, F, 8, 4, 7, 1, 2}
2  s = Sbox[s] // table lookup
```

---

at least one byte (generally multiple bytes), it is not efficient to implement Sbox nibble-wise. To optimize the execution time, we can merge two PRESENT Sbox table lookups (Algorithm 3.3). However, even though we can utilize the space

---

Algorithm 3.3: A more efficient lookup table implementation of PRESENT Sbox in pseudocode

---

```
1  integer array [1..16] Sbox = {C, 5, 6, B, 9, 0, A, D, 3, E, F, 8, 4, 7, 1, 2}
2  integer big_s = Sbox[s & 0F] // lower nibble; & denotes bitwise AND (see
       Definition 1.3.6)
3  big_s = big_s ∨ (Sbox[(s≫4) & 0F] ≪4) // upper nibble; ∨ denotes bitwise OR
       (see Remark 1.3.2)
4  s = big_s // state update
```

---

more efficiently, the additional operations take extra computing time. To avoid the bit shifts and boolean operations, it is better to combine two 4×4 Sbox tables into one bigger 8×8 table (Algorithm 3.4):

$$SB(0)|SB(0) \ SB(0)|SB(1) \ \ldots \ SB(0)|SB(F)$$
$$SB(1)|SB(0) \ SB(1)|SB(1) \ \ldots \ SB(1)|SB(F)$$
$$\vdots \qquad\qquad \vdots \qquad\qquad \vdots$$
$$SB(F)|SB(0) \ SB(F)|SB(1) \ \ldots \ SB(F)|SB(F)$$

Algorithm 3.4: A lookup table implementation combining two PRESENT Sboxes in parallel in pseudocode

---

1 integer array [1..256] Sbox = {CC, C5, …, C1, C2, 5C, 55, …, 51, 52, … 2C, 25, …, 21, 22}
2 s = Sbox[s] // table lookup of two nibbles in parallel

3.2.2 Implementing Permutations

The efficiency of the implementation is highly dependent on the design of the permutation. For AES ShiftRows, the bytes are permuted, making it easier to implement. For PRESENT pLayer, the bit level permutations are "free" in hardware as we just need to reorder the wires, no new gates are required. However, in software, extracting each bit and putting it in the right position is time-consuming.

3.2.2.1 Implementing pLayer

In this part, we will discuss two methods for implementing PRESENT pLayer by combining it with sBoxLayer.

The first method is straightforward. We will construct sixteen 4×64 lookup tables, TB1, TB2, …, TB16. The input of TBi is given by the ith nibble of the cipher state at the input of sBoxLayer. The outputs are 64-bit values with mostly 0s except for 4 bits that are related to this ith input nibble through sBoxLayer and pLayer.

Let us consider TB1, whose input is the first nibble of the cipher state at the input of sBoxLayer. By Table 3.12, the Sbox output corresponding to this nibble should go to bits 0, 16, 32 and 48 of the output of pLayer. Thus, each entry of TB1 is a 64-bit value with bits in positions 0, 16, 32 and 48 given by the Sbox output, and the other bits are all 0.

Example 3.2.1 For example, if the input is A, the Sbox output should be $F = 1111_2$ and

$$TB1[A] = 0\ldots010\ldots010\ldots010\ldots1,$$

where the 0th, 16th, 32nd and 48th bits are 1. Similarly, PRESENT Sbox output for input B is 1000_2, and

$$TB1[B] = 0\ldots010\ldots0,$$

where the 48th bit is 1.

Example 3.2.2 TB2 takes the second nibble of the cipher state as input. The output bits should be positioned at 1, 17, 33 and 49. Thus

$$TB2[B] = 0\ldots010\ldots0,$$

where only the 49th bit is 1.

As for the memory consumption, a 4×64 table takes 64×2^4 bits and those sixteen tables take 16384 bits of memory. Compared to one Sbox table, which is 64 bits, this is much bigger, but these tables also implement pLayer of PRESENT. The speed can be further improved by merging two Sbox computations and constructing eight 8×64 lookup tables. The memory consumption will be the same. But the speed will be much faster.

The second method [GHNZ09, PV13] requires a deeper look at the pLayer design. The aim is to design four 8×8 tables that output the corresponding Sbox values and permutate the bits of each byte of the sBoxLayer input.

If we analyze Table 3.12 and Fig. 3.9, we can see that in round i:

- The 0th bits of bytes at positions $0, 1, 3, 5$ in pLayer output come from the 0th nibble of the input of pLayer, which corresponds to the 0th nibble of the cipher state at sBoxLayer input of round i;
- The 1st bits of bytes at positions $0, 1, 3, 5$ in pLayer output correspond to the 1st nibble of the cipher state at sBoxLayer input;
- The 2nd bits of bytes at positions $0, 1, 3, 5$ in pLayer output correspond to the 2nd nibble of the cipher state at sBoxLayer input;
- The 3rd bits of bytes at positions $0, 1, 3, 5$ in pLayer output correspond to the 3rd nibble of the cipher state at sBoxLayer input;
- $\ldots$
- The 7th bits of bytes $0, 1, 3, 5$ in pLayer output correspond to the 7th nibble of the cipher state at sBoxLayer input;

Similar observations hold for bytes at positions $2, 4, 6, 7$.

We can have the following four tables for the implementation of sBoxLayer and pLayer:

- Table one takes the 0th byte (bits $0-7$) of sBoxLayer input, the corresponding output will be the 0th and 1st bits for bytes at positions $0, 1, 3, 5$ (bits $0, 1, 16, 17, 32, 33, 48, 49$) in the output of pLayer;

- Table two takes the 1st byte (bits $8-15$) of sBoxLayer input, the corresponding output will be the 2nd and 3rd bits for bytes at positions $0, 1, 3, 5$ (bits $2, 3, 18, 19, 34, 35, 50, 51$) in the output of pLayer;
- Table three takes the 2nd byte (bits $16-23$) of sBoxLayer input, the corresponding output will be the 4th and 5th bits for bytes at positions $0, 1, 3, 5$ (bits $4, 5, 20, 21, 36, 37, 52, 53$) in the output of pLayer;
- Table four takes the 3rd byte (bits $24-31$) of sBoxLayer input, the corresponding output will be the 6th and 7th bits for bytes at positions $0, 1, 3, 5$ (bits $6, 7, 22, 23, 38, 39, 54, 55$) in the output of pLayer.

The same tables can also be used for the remaining four bytes of the cipher state:

- Table one takes the 4th byte (bits $32-39$) of sBoxLayer input, the corresponding output will be the 0th and 1st bits for bytes at positions $2, 4, 6, 7$ (bits $8, 9, 24, 25, 40, 41, 56, 57$) in the output of pLayer;
- Table two takes the 5th byte (bits $40-47$) of sBoxLayer input, the corresponding output will be the 2nd and 3rd bits for bytes at positions $2, 4, 6, 7$ (bits $10, 11, 26, 27, 42, 43, 58, 59$) in the output of pLayer;
- Table three takes the 6th byte (bits $48-55$) of sBoxLayer input, the corresponding output will be the 4th and 5th bits for bytes at positions $2, 4, 6, 7$ (bits $12, 13, 28, 29, 44, 45, 60, 61$) in the output of pLayer;
- Table four takes the 7th byte (bits $56-63$) of sBoxLayer input, the corresponding output will be the 6th and 7th bits for bytes at positions $2, 4, 6, 7$ (bits $14, 15, 30, 31, 46, 47, 62, 63$) in the output of pLayer.

Since the input for each table is one byte, we will be computing two Sboxes in parallel. In Algorithm 3.4 we have seen the algorithm for such a computation. To see how the four tables are computed, we will detail the first three entries of each table. The other entries are calculated with similar methods.

First, we note that to combine two Sboxes, the lookup table starts with

$$CC \quad C5 \quad C6 \quad \ldots$$

As mentioned above, one type of input intended for Table one is bits at positions $0-7$ of sBoxLayer input, those bits correspond to bits at positions $0-7$ at sBoxLayer output. The corresponding output of Table one are bits at positions $0, 1, 16, 17, 32, 33, 48, 49$ of pLayer output. According to pLayer (Table 3.12) design, we will need to permute bits at positions $0-7$ to $0, 4, 1, 5, 2, 6, 3, 7$ so that they will give us bits at positions $0, 1, 16, 17, 32, 33, 48, 49$ of pLayer output. For example, if the input of Table one is 00, the corresponding sBoxLayer output is CC $= 11001100$, where the 0th bit is 0. After permutation, we get $11110000 = $ F0. Similarly, we get that Table one starts with

$$F0 \quad B1 \quad B4 \quad \ldots \tag{3.8}$$

If we consider the other set of inputs intended for Table one, which are bits at positions $32-39$, they should be first permuted to $32, 36, 33, 37, 34, 38, 35, 39$ so that the output will be bits at positions $8, 9, 24, 25, 40, 41, 56, 57$. Then we arrive at the same values as in Eq. 3.8.

For Table two, the output will later be positioned at the 2nd and 3rd positions in the eight bytes of the pLayer output. A natural choice is to design it so that the output can be combined with the outputs of other tables with a binary operation, e.g., $\vee$. In particular, since the output of Table one starts with bits from positions $0, 1$ and $8, 9$, the output of Table two will put bits from positions $2, 3$ and $10, 11$ in the 2nd and 3rd positions. Thus, Table two permutes bits $8-15$ to $11, 15, 8, 12, 9, 13, 10, 14$, which then will give bits at $50, 51, 2, 3, 18, 19, 34, 35$ for pLayer output. Similarly, bits $40-47$ will be permuted to $43, 47, 40, 44, 41, 45, 42, 46$ and give bits at $58, 59, 10, 11, 26, 27, 42, 43$ for pLayer output. The first few entries of Table two are as follows:

$$3C \quad 6C \quad 2D \quad \ldots$$

Table three first permutes bits from $16-23$ (resp. $48-55$) to $18, 22, 19, 23, 16, 20, 17, 21$ (resp. $50, 54, 51, 55, 48, 52, 49$), which then give bits $36, 37, 52, 53, 4, 5, 20, 21$ (resp. $44, 45, 60, 61, 12, 13, 28, 29$) of pLayer output. The table starts with

$$0F \quad 1B \quad 4B \quad \ldots$$

Table four first permutes bits from $24-31$ (resp. $56-63$) to $25, 29, 26, 30, 27, 31, 24, 28$ (resp. $57, 61, 58, 62, 59, 63, 56, 60$), which then give bits $22, 23, 38, 39, 54, 55, 6, 7$ (resp. $30, 31, 46, 47, 62, 63, 14, 15$) of pLayer output. The table starts with

$$C3 \quad C6 \quad D2 \quad \ldots$$

A pseudocode for the implementation is detailed in Algorithm 3.5. We represent the ith byte of the cipher state at sBoxLayer input by b_i ($i = 0, 1, 2, \ldots, 7$). The algorithm demonstrates how the bits $0-7$ of the pLayer output can be computed. Other bits can be calculated similarly. In line 1, we pass the 0th byte of the cipher state at sBoxLayer input, b_0, to Table one. The table lookup result is stored in $a1$, which gives us bits $0, 1, 16, 17, 32, 33, 48, 49$ of pLayer output. In line 5, the leftmost two bits of $s1$ are given by the leftmost two bits of $a1$, which correspond to bits at positions 0 and 1 in pLayer output. Similarly, $s2$ (reps. $s3, s4$) stores bits at positions $2, 3$ (resp. $4, 5, 6, 7$) at of pLayer. Then those eight bits are combined together in line 9 to produce the 0th byte of the pLayer output.

3.2.2.2 AES T-tables

This part discusses an implementation method combining SubBytes, ShiftRows, and MixColumns for AES round function. Let SB denote the AES Sbox.

Algorithm 3.5: An implementation that combines sBoxLayer and pLayer for PRESENT

Input: $b_7, b_6, b_5, b_4, b_3, b_2, b_1, b_0$, Table_one, Table_two, Table_three, Table_four
// $b_7, b_6, b_5, b_4, b_3, b_2, b_1, b_0$ is the cipher state at the input of
   sBoxLayer, each bi represents one byte
// Table_one = {F0, B1, B4, ... }
// Table_two = {3C, 6C, 2D, ... }
// Table_three = {0F, 1B, 4B, ... }
// Table_four = {C3, C6, D2, ... }
Output: cipher state at the output of pLayer
// compute bytes at positions $0, 1, 3, 5$ in pLayer output----
1 $a1$ = Table_one[b_0]// look up bits 0, 1, 16, 17, 32, 33, 48, 49
2 $a2$ = Table_two[b_1]// look up bits 50, 51, 2, 3, 18, 19, 34, 35
3 $a3$ = Table_three[b_2]// look up bits 36, 37, 52, 53, 4, 5, 20, 21
4 $a4$ = Table_four[b_3]// look up bits 22, 23, 38, 39, 54, 55, 6, 7
// computing bits $0 - 7$ of pLayer output
5 $s1 = a1$ & C0// extract bits 0, 1. & denotes bitwise AND (see
   Definition 1.3.6)
6 $s2 = a2$ & 30// extract output bits 2, 3
7 $s3 = a3$ & 0C// extract output bits 4, 5
8 $s4 = a4$ & 03// extract output bits 6, 7
9 $b_0 = s1 \vee s2 \vee s3 \vee s4$// combine bits, $\vee$ denotes bitwise OR (see
   Remark 1.3.2)
// other bits of bytes at positions $0, 1, 3, 5$ in pLayer output
10 ...
// compute bytes at positions $2, 4, 6, 7$ in pLayer output---
11 $a1$ = Table_one[b_4]   // look up bits 8, 9, 24, 25, 40, 41, 56, 57
12 $a2$ = Table_two[b_5]   // look up bits 58, 59, 10, 11, 26, 27, 42, 43
13 $a3$ = Table_three[b_6]   // look up bits 44, 45, 60, 61, 12, 13, 28, 29
14 $a4$ = Table_four[b_7]   // look up bits 30, 31, 46, 47, 62, 63, 14, 15
15 ...

Recall that the cipher state of AES can be represented by a four-by-four matrix of bytes (Eq. 3.2). Let us denote the input of SubBytes by a matrix S. The outputs of SubBytes, ShiftRows, and MixColumns are represented by matrices A, B, and D respectively. By definition, $a_{ij} = \mathrm{SB}(s_{ij})$, $0 \le i, j < 4$. By Eqs. 3.4 and 3.5,

$$\begin{pmatrix} b_{0j} \\ b_{1j} \\ b_{2j} \\ b_{3j} \end{pmatrix} = \begin{pmatrix} a_{0j} \\ a_{1(j+1 \bmod 4)} \\ a_{2(j+2 \bmod 4)} \\ a_{3(j+3 \bmod 4)} \end{pmatrix}, \quad \begin{pmatrix} d_{0j} \\ d_{1j} \\ d_{2j} \\ d_{3j} \end{pmatrix} = \begin{pmatrix} 02 & 03 & 01 & 01 \\ 01 & 02 & 03 & 01 \\ 01 & 01 & 02 & 03 \\ 03 & 01 & 01 & 02 \end{pmatrix} \begin{pmatrix} b_{0j} \\ b_{1j} \\ b_{2j} \\ b_{3j} \end{pmatrix},$$

$$j = 0, 1, 2, 3.$$

We have

$$
\begin{pmatrix} d_{0j} \\ d_{1j} \\ d_{2j} \\ d_{3j} \end{pmatrix} = \begin{pmatrix} 02 & 03 & 01 & 01 \\ 01 & 02 & 03 & 01 \\ 01 & 01 & 02 & 03 \\ 03 & 01 & 01 & 02 \end{pmatrix} \begin{pmatrix} \mathrm{SB}(s_{0j}) \\ \mathrm{SB}(s_{1(j+1 \bmod 4)}) \\ \mathrm{SB}(s_{2(j+2 \bmod 4)}) \\ \mathrm{SB}(s_{3(j+3 \bmod 4)}) \end{pmatrix}
$$

$$
= \begin{pmatrix} 02 \\ 01 \\ 01 \\ 03 \end{pmatrix} \mathrm{SB}(s_{0j}) \oplus \begin{pmatrix} 03 \\ 02 \\ 01 \\ 01 \end{pmatrix} \mathrm{SB}(s_{1(j+1 \bmod 4)}) \oplus \begin{pmatrix} 01 \\ 03 \\ 02 \\ 01 \end{pmatrix} \mathrm{SB}(s_{2(j+2 \bmod 4)})
$$

$$
\oplus \begin{pmatrix} 01 \\ 01 \\ 03 \\ 02 \end{pmatrix} \mathrm{SB}(s_{3(j+3 \bmod 4)}),
$$

where $j = 0, 1, 2, 3$. For $a \in \mathbb{F}_2^8$, define

$$
T_0(a) := \begin{pmatrix} 02 \\ 01 \\ 01 \\ 03 \end{pmatrix} \mathrm{SB}(a), \quad T_1(a) := \begin{pmatrix} 03 \\ 02 \\ 01 \\ 01 \end{pmatrix} \mathrm{SB}(a), \quad T_2(a) := \begin{pmatrix} 01 \\ 03 \\ 02 \\ 01 \end{pmatrix} \mathrm{SB}(a),
$$

$$
T_3(a) := \begin{pmatrix} 01 \\ 01 \\ 03 \\ 02 \end{pmatrix} \mathrm{SB}(a).
$$

Then

$$
\begin{pmatrix} d_{0j} \\ d_{1j} \\ d_{2j} \\ d_{3j} \end{pmatrix} = T_0(s_{0j}) \oplus T_1(s_{1(j+1 \bmod 4)}) \oplus T_2(s_{2(j+2 \bmod 4)}) T_3(s_{3(j+3 \bmod 4)}),
$$

Thus the four tables T_0, T_1, T_2, T_3 of size 8×32 can be used to implement SubBytes, ShiftRows, and MixColumns. Those four tables are called *T-tables* for AES. We note that to store the T-tables we need processors with a word size of 32 or above. They cannot be used for the last round of AES as there is no Mixcolumns operation.

3.2.3 Bitsliced Implementations

Bitsliced implementation of symmetric block ciphers was first introduced by Eli Biham for implementing DES [Bih97]. The goal of a bitsliced implementation is to simulate a hardware implementation in software so that several plaintext blocks can be encrypted in parallel. The operations in symmetric block ciphers will be represented as a sequence of logical operations. Naturally, the implementations should be adjusted based on the specific underlying hardware—the word size of the architecture (see Sect. 2.1.2). We will see that with word size ω, we can encrypt ω blocks of plaintext in parallel.

3.2.3.1 Algebraic Normal Form

To introduce bitsliced implementation, we will need to discuss the algebraic normal form for a Boolean function. Let n be a positive integer in this part.

Definition 3.2.1 A *Boolean function* is a function $\varphi : \mathbb{F}_2^n \to \mathbb{F}_2$.

From the definition, we can see that a Boolean function has 2^n possible input values. For each input value, there are 2 possible output values. Thus, in total, we have 2^{2^n} possible Boolean functions defined for $\mathbb{F}_2^n \to \mathbb{F}_2$. In particular, a Boolean function can be specified by giving the output values for all inputs, such a table is called a *truth table*.

Example 3.2.3 The parity-check bit defined for 3 bits is a Boolean function

$$\varphi : \mathbb{F}_2^3 \to \mathbb{F}_2$$

$$x_2 x_1 x_0 \mapsto x_0 + x_1 + x_2.$$

Its truth table is given by:

x_2	0 0 0 0 1 1 1 1
x_1	0 0 1 1 0 0 1 1
x_0	0 1 0 1 0 1 0 1
$\varphi(x)$	0 1 1 0 1 0 0 1

Example 3.2.4 Now let use consider the Boolean function defined as follows:

$$\varphi_0 : \mathbb{F}_2^4 \to \mathbb{F}_2$$

$$x \mapsto \mathrm{SB}_{\mathrm{PRESENT}}(x)_0$$

where $\mathrm{SB}_{\mathrm{PRESENT}}(x)_0$ is the 0th bit of $\mathrm{SB}_{\mathrm{PRESENT}}(x)$, the PRESENT Sbox output corresponding to x. The truth table of φ_0 is given by the first five and the second

Table 3.13 The Boolean function φ_0 takes input x and outputs the 0th bit of $\mathrm{SB_{PRESENT}}(x)$. The second last row lists the output of φ_0 for different input values. The last row lists the coefficients (Eq. 3.10) for the algebraic normal form of φ_0

x	0	1	2	3	4	5	6	7	8	9	A	B	C	D	E	F
x_3	0	0	0	0	0	0	0	0	1	1	1	1	1	1	1	1
x_2	0	0	0	0	1	1	1	1	0	0	0	0	1	1	1	1
x_1	0	0	1	1	0	0	1	1	0	0	1	1	0	0	1	1
x_0	0	1	0	1	0	1	0	1	0	1	0	1	0	1	0	1
$\mathrm{SB_{PRESENT}}(x)$	C	5	6	B	9	0	A	D	3	E	F	8	4	7	1	2
$\varphi_0(x)$	0	1	0	1	1	0	0	1	1	0	1	0	0	1	1	0
λ_x	0	1	0	0	1	0	1	0	1	0	0	0	0	0	0	0

last (the row for $\varphi_0(x)$) rows in Table 3.13. For example, if the input is 0, the Sbox output is C = 1100. Then $\varphi_0(x) = 0$.

Definition 3.2.2 Fix $v = v_{n-1}v_{n-2}, \ldots, v_1 v_0 \in \mathbb{F}_2^n$, we define the *indicator function* for v, denoted 1_v, as follows:

$$1_v : \mathbb{F}_2^n \to \mathbb{F}_2$$

$$x \mapsto \prod_{i:v_i=1} x_i \prod_{i:v_i=0} (1 - x_i).$$

With this definition, for any $\varphi : \mathbb{F}_2^n \to \mathbb{F}_2$, we can express φ in the following polynomial expression:

$$\varphi(x) = \sum_{v \in \mathbb{F}_2^n} \varphi(v) 1_v(x).$$

After simplification, φ can be written as

$$\varphi(x) = \sum_{v \in \mathbb{F}_2^n} \left(\lambda_v \prod_{i=0}^{n-1} x_i^{v_i} \right),$$

which is called the *algebraic normal form* representation of the Boolean function φ.

Example 3.2.5 Continuing Example 3.2.3, we can find the algebraic normal form of φ as follows

$$\varphi(x) = \sum_{v \in \mathbb{F}_2^n} \varphi(v) 1_v(x) = 1_{001}(x) + 1_{010}(x) + 1_{100}(x) + 1_{111}(x)$$

$$= x_0(1 - x_1)(1 - x_2) + x_1(1 - x_0)(1 - x_2) + x_2(1 - x_0)(1 - x_1) + x_0 x_1 x_2$$

$$= x_0 + x_1 + x_2 - 2(x_0 x_1 + x_0 x_2 + x_1 x_2) + 4 x_0 x_1 x_2 = x_0 + x_1 + x_2.$$

It can be proven that the algebraic normal form of a Boolean function is unique.[2]

Theorem 3.2.1 *Every Boolean function* $\varphi : \mathbb{F}_2^n \to \mathbb{F}_2$ *has a unique algebraic normal form representation*

$$\varphi(\boldsymbol{x}) = \sum_{\boldsymbol{v} \in \mathbb{F}_2^n} \left(\lambda_{\boldsymbol{v}} \prod_{i=0}^{n-1} x_i^{v_i} \right). \tag{3.9}$$

The coefficients $\lambda_{\boldsymbol{v}} \in \mathbb{F}_2$ *are given by*

$$\lambda_{\boldsymbol{v}} = \sum_{\boldsymbol{w} \leq \boldsymbol{v}} \varphi(\boldsymbol{w}), \tag{3.10}$$

where $\boldsymbol{w} \leq \boldsymbol{v}$ *means that* $w_i \leq v_i$ *for all* $0 \leq i \leq n - 1$.

We note that there are 2^{2^n} Boolean functions defined for $\mathbb{F}_2^n \to \mathbb{F}_2$. Furthermore, there are 2^{2^n} choices for the coefficients $\lambda_{\boldsymbol{v}}$ ($\lambda_{\boldsymbol{v}} = 0, 1$ and there are 2^n distinct $\boldsymbol{v}$). Thus the number of distinct expressions on both sides of Eq. 3.9 coincides.

Example 3.2.6 Continuing Example 3.2.3. By Eq. 3.10,

$$\lambda_{110} = \varphi(000) + \varphi(100) + \varphi(010) + \varphi(110) = 0 + 1 + 1 + 0 = 0.$$

Similarly, we can calculate all the coefficients λ:

$$\lambda_{000} = 0, \ \lambda_{001} = 1, \ \lambda_{010} = 1, \ \lambda_{011} = 1 + 1 = 0,$$
$$\lambda_{100} = 1, \ \lambda_{101} = 0, \ \lambda_{110} = 0, \ \lambda_{111} = 0.$$

By Eq. 3.9,

$$\varphi(\boldsymbol{x}) = \sum_{\boldsymbol{v} \in \mathbb{F}_2^n} \left(\lambda_{\boldsymbol{v}} \prod_{i=0}^{n-1} x_i^{v_i} \right) = \lambda_{001} x_0 + \lambda_{010} x_1 + \lambda_{100} x_2 = x_0 + x_1 + x_2$$

which agrees with Example 3.2.5.

Example 3.2.7 Continuing Example 3.2.4, we can calculate $\lambda_{\boldsymbol{v}}$ using Eq. 3.10. Those values are given by the last row of Table 3.13. For example,

$$\lambda_{1100} = \varphi_0(0000) + \varphi_0(1000) + \varphi_0(0100) + \varphi_0(1100) = 0 + 1 + 1 + 0 = 0.$$

By Eq. 3.9,

$$\varphi_0(\boldsymbol{x}) = \sum_{\boldsymbol{v}\in\mathbb{F}_2^n}\left(\lambda_{\boldsymbol{v}}\prod_{i=0}^{n-1}x_i^{v_i}\right) = \lambda_{0001}x_0 + \lambda_{0100}x_2 + \lambda_{0110}x_1x_2 + \lambda_{1000}x_3$$

$$= x_0 + x_2 + x_1x_2 + x_3. \tag{3.11}$$

For example, if the input is $\boldsymbol{0} = 0000$, the PRESENT Sbox output is $\mathsf{C} = 1100$, then the output of φ_0 is 0 and

$$x_0 + x_2 + x_1x_2 + x_3 = 0 + 0 + 0 + 0 = 0.$$

If the input is $7 = 1110$, the PRESENT Sbox output is $\mathsf{D} = 1101$, then the output of φ_0 is 1 and

$$x_0 + x_2 + x_1x_2 + x_3 = 1 + 1 + 1 + 0 = 1.$$

Similarly, we can define $\varphi_i(\boldsymbol{x}) = \mathrm{SB}_{\mathrm{PRESENT}}(\boldsymbol{x})_i$ for $i = 1, 2, 3$, where $\mathrm{SB}_{\mathrm{PRESENT}}(\boldsymbol{x})_i$ is the ith bit of PRESENT Sbox output for $\boldsymbol{x}$. We can calculate the algebraic normal form for each of φ_i in a similar way (see Appendix D). They are given by:

$$\varphi_1(\boldsymbol{x}) = x_1 + x_3 + x_1x_3 + x_2x_3 + x_0x_1x_2 + x_0x_1x_3 + x_0x_2x_3, \tag{3.12}$$

$$\varphi_2(\boldsymbol{x}) = 1 + x_2 + x_3 + x_0x_1 + x_0x_3 + x_1x_3 + x_0x_1x_3 + x_0x_2x_3, \tag{3.13}$$

$$\varphi_3(\boldsymbol{x}) = 1 + x_0 + x_1 + x_3 + x_1x_2 + x_0x_1x_2 + x_0x_1x_3 + x_0x_2x_3. \tag{3.14}$$

3.2.3.2　Bitsliced Implementation of PRESENT

In this part, we will use PRESENT as a running example to show how the bitsliced implementation of a symmetric block cipher is designed.

First, we discuss how to transform the plaintext blocks into bitsliced format. As a simple example, let us consider block length 3 and a 4-bit architecture, which allows us to encrypt 4 blocks of plaintext simultaneously. We take 4 plaintext blocks, say

$$p_1 = 010,\ p_2 = 110,\ p_3 = 001,\ p_4 = 100.$$

The bitsliced format of p_js is given by a 3×4 array, denoted S, where each column is given by one block of plaintext:

$$S = \begin{pmatrix} 0 & 0 & 1 & 0 \\ 1 & 1 & 0 & 0 \\ 0 & 1 & 0 & 1 \end{pmatrix}.$$

In particular, if we let $S[x]$ denote the xth row of S, then $S[0]$ corresponds to the 0th bits of p_j. $S[1]$ corresponds to the 1st bits of p_j. And $S[2]$ corresponds to the 2nd bits of p_j.

Next, we will show how to encrypt 8 plaintext blocks in parallel with PRESENT assuming an 8-bit architecture. Let $p_1, p_2, \ldots p_8$ be 8 plaintext blocks, each of length 64. We convert them into bitsliced format as described above and store them in a 64×8 array S_0, where $S_0[y]$ contains the yth bits of each plaintext block. Furthermore, for each round key K_i, we construct a 64×8 array $\texttt{Keyi}$ whose columns are given by K_i, i.e.,

$$\texttt{Keyi}[y][z] = K_i[y] \quad \forall 0 \le z < 8. \tag{3.15}$$

The bitsliced implementation of the ith round of PRESENT is given in Algorithm 3.6. Line 1 implements addRoundKey. For example, when $i = 0$, the xth bit of each plaintext (row x of S_0) are XORed with the xth bit of K_1 (row x of $\texttt{Key0}$). To implement the Sbox in bitsliced format, we refer to the algebraic normal forms for each output bit of the Sbox as a function of the input bits (see Example 3.2.7). We recall that addition and multiplication in $\mathbb{F}_2$ can be implemented as logical XOR ($\oplus$) and logical AND (&) respectively (see Definition 1.2.17). There are in total 16 Sboxes and we consider each of them in one loop of line 2. x_0, x_1, x_2, x_3 defined in line 3 are arrays of size 8, each storing one bit of Sbox input from all eight encryption computations. Lines 4–7 compute eight Sboxes in parallel, each corresponding to the encryption of one plaintext block. The 0th bits of the Sbox outputs are given by the $4b$th bits of the cipher state at the end of sBoxLayer, where $0 \le b \le 15$. Line 4 computes the 0th bits of the Sbox outputs using Eq. 3.11. Similarly, lines 5, 6, 7 compute the 1st, 2nd and 3rd bit of Sbox outputs using Eqs. 3.12, 3.13 and 3.14 respectively. Finally, pLayer is implemented by line 8 onward using Table 3.12. We note that $S_i[0]$ (line 8) is an array of 8 bits and we are permuting the 0th bit of cipher state for 8 encryptions simultaneously. The same can be done for the remaining 63 bits.

It is easy to see that with 32-bit (resp. 64-bit) architecture, we can encrypt 32 (resp. 64) plaintext blocks in parallel. We note that bitsliced implementations are mostly used for bit-oriented ciphers (e.g., DES, PRESENT). For byte-oriented ciphers (e.g., AES), table-based implementations will likely give better performance.

3.3 RSA

In Sect. 2.1.2 we have mentioned that there are symmetric key and asymmetric cryptosystems. Up to now, we have only seen symmetric cryptosystems, both classical and modern designs. For symmetric key cipher, a prior communication of the master key (*key exchange*) is required before any ciphertext is transmitted. With only a symmetric key cipher, the key exchange may be difficult to achieve

---

Algorithm 3.6: Bitsliced implementation of round i of PRESENT, $1 \leq i \leq 31$

Input: S_{i-1}, Keyi// S_{i-1} is the output of round $i-1$. When $i=1$, S_0
    contains the plaintext blocks in bitsliced format.
    // Keyi is the ith round key K_i in bitsliced format given in Eq. 3.15.
Output: S_i: output of round i
    // addRoundKey---
1 $S_{i-1} = S_{i-1} \oplus \text{Keyi}$// bitwise XOR
    // sBoxLayer---
2 **for** $b = 0, b < 16, b++$ **do**
        // Bits of Sbox inputs
3 | $x_0 = S_{i-1}[4b]$, $x_1 = S_{i-1}[4b+1]$, $x_2 = S_{i-1}[4b+2]$, $x_3 = S_{i-1}[4b+3]$
        // 0th bit of Sbox output
4 | $\text{state}[4b] = x_0 \oplus x_2 \oplus (x_1 \& x_2) \oplus x_3$
        // 1st bit of Sbox output
5 | $\text{state}[4b+1] =$
   | $x_1 \oplus x_3 \oplus (x_1 \& x_3) \oplus (x_2 \& x_3) \oplus (x_0 \& x_1 \& x_2) \oplus (x_0 \& x_1 \& x_3) \oplus (x_0 \& x_2 \& x_3)$
        // 2nd bit of Sbox output
6 | $\text{state}[4b+2] =$
   | $1 \oplus x_2 \oplus x_3 \oplus (x_0 \& x_1) \oplus (x_0 \& x_3) \oplus (x_1 \& x_3) \oplus (x_0 \& x_1 \& x_3) \oplus (x_0 \& x_2 \& x_3)$
        // 3rd bit of Sbox output
7 | $\text{state}[4b+3] =$
   | $1 \oplus x_0 \oplus x_1 \oplus x_3 \oplus (x_1 \& x_2) \oplus (x_0 \& x_1 \& x_2) \oplus (x_0 \& x_1 \& x_3) \oplus (x_0 \& x_2 \& x_3)$
    // pLayer---
8 $S_i[0] = \text{state}[0]$
9 $S_i[16] = \text{state}[1]$
10 $S_i[32] = \text{state}[2]$
11 ...
12 **return** S_i

---

due to, e.g., far distance, and too many parties involved. In practice, this is where asymmetric key cryptosystem comes into use.

For example, Alice would like to communicate with Bob using AES. To exchange the master key, k, for AES, she will encrypt k by a public key cryptosystem using Bob's public key e. Let $c = E_e(k)$. The resulting ciphertext c will be sent to Bob, and Bob can decrypt it with his secret private key d, $k = D_d(c)$. Then Alice and Bob can communicate with key k using AES.

Clearly, we require that it is computationally infeasible to find the private key d given the public key e. In practice, this is guaranteed by some intractable problem.[3] However, the cipher might not be secure in the future. For example, if a quantum computer with enough bits is manufactured, it can break many public key cryptosystems [EJ96]. Furthermore, we note that a public key cipher is not perfectly secure (see Sect. 2.2.7) as the attacker can brute force the key.

In this section, we will be discussing one public key cryptosystem—RSA. It was published in 1977 and named after its inventors Ron Rivest, Adi Shamir, and

---

[3] A problem is intractable if there does not exist an efficient algorithm to solve it.

Leonard Adleman. RSA is the first public key cryptosystem, and still in use today. The security relies on the difficulty of finding the factorization of a composite positive integer.

Definition 3.3.1 (RSA) Let $n = pq$, where p, q are distinct prime numbers. Let $\mathcal{P} = \mathcal{C} = \mathbb{Z}_n$, $\mathcal{K} = \mathbb{Z}^*_{\varphi(n)} - \{1\}$. For any $e \in \mathcal{K}$, define encryption

$$E_e : \mathbb{Z}_n \to \mathbb{Z}_n, \quad m \mapsto m^e \bmod n,$$

and the corresponding decryption

$$D_d : \mathbb{Z}_n \to \mathbb{Z}_n, \quad c \mapsto c^d \bmod n,$$

where $d = e^{-1} \bmod \varphi(n)$.

The cryptosystem $(\mathcal{P}, \mathcal{C}, \mathcal{K}, \mathcal{E}, \mathcal{D})$, where $\mathcal{E} = \{E_e : e \in \mathcal{K}\}$, $\mathcal{D} = \{D_d : d \in \mathcal{K}\}$, is called *RSA*.

Recall by Theorem 1.4.3, $\varphi(n) = (p - 1)(q - 1)$ and by Definition 1.4.5, $\mathbb{Z}^*_{\varphi(n)}$ consists of elements in $\mathbb{Z}_{\varphi(n)}$ that are coprime to $\varphi(n)$, or equivalently, that have multiplicative inverses modulo $\varphi(n)$.

For encryption, the message sender needs to have knowledge of n and e. They are the public key for RSA. n is called *RSA modulus* and e is called the *encryption exponent*. The *private key* d for decryption is kept secret. In this case, only the private key owner can decrypt the message sent to him.

To generate the keys for RSA, we first generate randomly and independently two large primes p and q. Then we compute $n = pq$. Normally p and q are supposed to have equal lengths. For example, take p and q to be 512-bit primes, and n will be a 1024-bit modulus. Next $e \in \mathbb{Z}^*_{\varphi(n)}$ is chosen. Since $\varphi(n)$ is even, e is odd. Finally, we compute $d = e^{-1} \bmod \varphi(n)$.

Example 3.3.1 As a toy example, suppose Bob would like to generate his private and public keys for RSA. Bob randomly generates $p = 3$ and $q = 5$. Then he computes $n = 15$ and

$$\varphi(n) = (3 - 1) \times (5 - 1) = 2 \times 4 = 8.$$

From $\mathbb{Z}^*_8 = \{1, 3, 5, 7\}$, Bob chooses $e = 3$. By the extended Euclidean algorithm, he computes

$$8 = 3 \times 2 + 2, \ 3 = 2 \times 1 + 1 \implies 1 = 3 - 2 \times 1 = 3 - (8 - 3 \times 2) = -8 + 3 \times 3.$$

Hence his private key $d = 3^{-1} \bmod 8 = 3$.

Suppose Alice would like to send plaintext $m = 2$ to Bob, using Bob's public key $n = 15$ and $e = 3$. Alice computes

$$c = m^e \bmod n = 2^3 \bmod 15 = 8 \bmod 15.$$

After receiving the ciphertext c from Alice, Bob computes the plaintext using his private key

$$m = c^d \bmod n = 8^3 \bmod 15 = 512 \bmod 15 = 2 \bmod 15.$$

Example 3.3.2 Now we will look at a bit larger values for p and q. Let $p = 29$, $q = 41$, then $n = 1189$ and $\varphi(n) = 28 \times 40 = 1120$. It is easy to verify that $3 \nmid \varphi(n)$. Let us choose $e = 3$. By the extended Euclidean algorithm

$$1120 = 3 \times 373 + 1 \implies 1 = 1120 - 3 \times 373.$$

Hence

$$d = -373 \bmod 1120 = 747.$$

To send plaintext $m = 2$ to Bob. Alice computes

$$c = m^e \bmod n = 2^3 \bmod 1189 = 8 \bmod 1189.$$

To decrypt, Bob calculates

$$m = c^d \bmod n = 8^{747} \bmod 1189 = 2 \bmod 1189.$$

Since

$$747 = 512 + 128 + 64 + 32 + 8 + 2 + 1,$$

we compute

$$8^4 \bmod 1189 = 4096 \bmod 1189 = 529, \quad 8^8 \bmod 1189 = 529^2 \bmod 1189 = 426,$$
$$8^{16} \bmod 1189 = 426^2 \bmod 1189 = 748, \quad 8^{32} \bmod 1189 = 748^2 \bmod 1189 = 674,$$
$$8^{64} \bmod 1189 = 674^2 \bmod 1189 = 78, \quad 8^{128} \bmod 1189 = 78^2 \bmod 1189 = 139,$$
$$8^{256} \bmod 1189 = 139^2 \bmod 1189 = 297, \quad 8^{512} \bmod 1189 = 297^2 \bmod 1189 = 223.$$

And we have

$$8^{512+128} \bmod 1189 = 223 \times 139 \bmod 1189 = 83,$$
$$8^{64+32} \bmod 1189 = 78 \times 674 \bmod 1189 = 256$$
$$8^{8+2+1} \bmod 1189 = 426 \times 64 \times 8 \bmod 1189 = 525,$$
$$8^{747} \bmod 1189 = 83 \times 256 \times 525 \bmod 1189 = 2.$$

Next, we explain why the decryption works. By the choice of e and d,

$$ed \equiv 1 \bmod \varphi(n) \implies ed = \varphi(n)a + 1 \text{ for some } a \in \mathbb{Z}.$$

Then

$$c^d = (m^e)^d = m^{\varphi(n)a+1} = m^{(p-1)(q-1)a}m.$$

By Corollary 1.4.3,

$$c^d \equiv m \bmod p, \quad c^d \equiv m \bmod q.$$

Since p and q are distinct prime numbers and $n = pq$, by Chinese Remainder Theorem (see Theorem 1.4.7 and Example 1.4.19)

$$c^d \equiv m \bmod n.$$

We note that, if p or q is known to the attacker, they can factorize n and compute $\varphi(n)$. Then with e, d can be computed using the extended Euclidean algorithm (Algorithm 1.2). Thus all $p, q, \varphi(n)$ should be kept secret.

RSA can only be secure if computing d from n and e is intractable. Of course, if the attacker can factorize n with an efficient algorithm, then RSA is broken. However, there is no proof to conclude if factorizing n is intractable or not. Up to now, the best-known algorithm for integer factorization has been used to factorize RSA modulus of bit length 768 [KAF$^+$10]. In practice, the most commonly used RSA modulus n is 1024, 2048, or 4096 bit. Interestingly, it has been proved that if d is known, then n can be factorized with an efficient algorithm (see [Buc04, Page 172]). On the other hand, there is no proof that RSA is secure if factoring is computationally infeasible—there might be other ways to attack RSA [May03].

Normally e is chosen to be small to make the encryption efficient. However, e cannot be too small. It has been shown that only the $n/4$ least significant bits of d suffice to recover d in the case of a small e [BDF98]. Also, d cannot be too small, it was proven that if $d < n^{0.292}$, then RSA can be broken [BD00].

3.4 RSA Signatures

In this section, we discuss how RSA can be used for digital signatures.

As mentioned in Sect. 2.1, digital signatures provide a means for an entity to bind its identity to a message stored in electronic form. This normally means that the sender uses their private key to sign the (hashed) message. Whoever has access to the public key can then verify the origin of the message. For example, the message can be electronic contracts or electronic bank transactions.

In more detail, suppose Alice signs a message m with a private key d and generates signature s. The receiver Bob receives the message and the signature, he can then verify s with public key e and a *verification algorithm*. Given m and s, the verification algorithm returns true to indicate a valid signature and false otherwise.

To use RSA for digital signature, we again let p and q be two distinct primes. Let $n = pq$. We choose $e \in \mathbb{Z}^*_{\varphi(n)}$ and compute $d = e^{-1} \mod \varphi(n)$. Same as for RSA, the public key consists of e and n. And d is the private key. p, q and $\varphi(n)$ should be kept secret.

To sign a message m, Alice computes the signature

$$s = m^d \mod n.$$

Then Alice sends both m and s to Bob. To verify the signature, Bob computes

$$s^e \mod n.$$

If

$$s \equiv m \mod n,$$

then the verification algorithm outputs true, and false otherwise.

Up to now, the only method known to compute s from $m \mod n$ is using d, so if the verification algorithm outputs true, Bob can conclude that Alice is the owner of d.

Example 3.4.1 Alice chooses $p = 5$ and $q = 7$. Then $n = 35$ and $\varphi(n) = 24$. Suppose Alice chooses $e = 5$, which is coprime to 24. By the extended Euclidean algorithm

$$24 = 5 \times 4 + 4,\ 5 = 4 + 1 \Longrightarrow 1 = 5 - (24 - 5 \times 4) = 24 \times (-1) + 5 \times 5,$$

we have $d = e^{-1} \mod 24 = 5$. To sign message $m = 10$, Alice computes

$$s = m^d \mod n = 10^5 \mod 35 = 5.$$

Alice sends both the message $m = 10$ and signature $s = 5$ to Bob. Bob verifies the signature

$$s^e \mod n = 5^5 \mod 35 = 10 = m.$$

The most common attack for a digital signature is to create a valid signature for a message without knowing the secret key. Such an attack is called *forgery*. If the goal is to create a valid signature given a message that was not signed by Alice before, it is called *selective forgery*. If the goal is to create a valid signature for any message not signed by Alice before, then the attack is called *existential forgery*.

There are normally three attacker assumptions. *Key-only attack* assumes the attacker only has knowledge of e. *Known message attack* considers an attacker who has a list of messages previously signed by Alice. In a *chosen message attack*, the attacker can request Alice's signature on a list of messages.

Next, we discuss the security of RSA signatures with respect to forgery attacks.

First, we consider a known message existential forgery attack. Suppose the attacker, Eve, knows messages m_1, m_2 and their corresponding signatures s_1 and s_2. Eve computes

$$s = s_1 s_2 \bmod n, \quad m = m_1 m_2 \bmod n.$$

Since

$$s = m_1^d m_2^d \bmod n = (m_1 m_2)^d \bmod n = m^d \bmod n,$$

s is a valid signature for m.

A chosen message selective forgery attack works as follows. Eve chooses a message $m \in \mathbb{Z}_n$ and takes any message $m_1 \in \mathbb{Z}_n^*$ that is different from m. She computes

$$m_2 = m m_1^{-1} \bmod n.$$

Eve obtains valid signatures

$$s_1 = m_1^d \bmod n, \quad \text{and} \quad s_2 = m_2^d \bmod n$$

for m_1 and m_2. Then she computes

$$s = s_1 s_2 \bmod n.$$

Since

$$s = m_1^d m_2^d \bmod n = (m_1 m_2)^d \bmod n = m^d \bmod n,$$

s is a valid signature for m.

In view of those attacks, RSA signatures are commonly used together with a fast public hash function h (see Sect. 2.1.1). To sign a message m, Alice computes the signature

$$s = h(m)^d \bmod n.$$

Then she sends both m and s to Bob. Bob computes $s^e \bmod n$ and $h(m)$. If

$$s^e \bmod n = h(m),$$

then Bob concludes the signature is valid.

With a hash function, the two attacks discussed above will not work. Suppose Eve knows messages m_1, m_2 and their corresponding signatures s_1 and s_2. She can compute $h(m_1)$ and $h(m_2)$ as h is public. However, to repeat the known message existential forgery attack, she needs to find m such that $h(m) = h(m_1)h(m_2)$, which is computationally infeasible according to property (c) of hash functions listed in Sect. 2.1.1.

Suppose Eve chooses a message m, and computes $h(m)$. To repeat the chosen message selective forgery attack, she needs to find m_1 such that $h(m_1) = y$ for some $y \in \mathbb{Z}_n^*$. For the same reason as above, this is computationally infeasible.

3.5 Implementations of RSA Cipher and RSA Signatures

In this section, we discuss several methods for implementing RSA and RSA signature computations. Section 3.5.1 presents three methods for implementing modular exponentiation. As we will see, those methods will require the computations of other modular operations. Then in Sect. 3.5.2, we discuss how to efficiently implement modular multiplication.

3.5.1 Implementing Modular Exponentiation

To implement RSA or RSA signatures, we need to compute

$$a^d \bmod n$$

for some integer $a \in \mathbb{Z}_n$, where $n = pq$ is a product of two distinct primes and $d \in \mathbb{Z}_{\varphi(n)}^*$. We can compute $d - 1$ modular multiplications, but it will be inefficient for large d. In practice, the bit length of d ranges in thousands, thus making the calculation infeasible by this naïve method. We will discuss three methods to make modular exponentiation computations faster.

3.5.1.1 Square and Multiply Algorithm

In this part, let $n \geq 2$ be an integer and $d \in \mathbb{Z}_{\varphi(n)}$. We discuss how to calculate

$$a^d \bmod n$$

for $a \in \mathbb{Z}_n$.

By Theorem 1.1.1, we can write d in the following form

$$d = \sum_{i=0}^{\ell_d-1} d_i 2^i,$$

where $d_i = 0, 1$, for $0 \le i \le \ell_d - 1$, and

$$d = d_{\ell_d-1} \ldots d_2 d_1 d_0$$

is the binary representation of d. Then we have

$$a^d = a^{\sum_{i=0}^{\ell_d-1} d_i 2^i} = \prod_{i=0}^{\ell_d-1} (a^{2^i})^{d_i} = \prod_{0 \le i < \ell_d, d_i=1} a^{2^i}.$$

Thus, to compute $a^d \bmod n$, we can first compute a^{2^i} for $0 \le i < \ell_d$. Then a^d is the product of a^{2^i} for which $d_i = 1$. One can see that compared to the naïve calculation, requiring $d - 1$ multiplications, this method only needs $\approx \log_2 d$ multiplications.

This observation leads us to the *square and multiply algorithm* listed in Algorithm 3.7. Line 5 computes $a^{2^{i+1}}$ in loop i. We check each bit of d (line 3), if the ith bit of d is 1, then a^{2^i} is multiplied to the result (line 4). As this algorithm starts from the least significant bit of d, i.e., d_0, it is also called the *right-to-left square and multiply algorithm*. Accordingly, the *left-to-right square and multiply algorithm* is listed in Algorithm 3.8. We can see that compared to Algorithm 3.7, Algorithm 3.8 requires one less variable and hence less storage.

Algorithm 3.7: Right-to-left square and multiply algorithm for computing modular exponentiation

---

Input: n, a, d // $n \in \mathbb{Z}, n \ge 2$; $a \in \mathbb{Z}_n$; $d \in \mathbb{Z}_{\varphi(n)}$ has bit length ℓ_d
Output: $a^d \bmod n$
1 result $= 1, t = a$
2 **for** $i = 0, i < \ell_d, i++$ **do**
   // ith bit of d is 1
3   **if** $d_i = 1$ **then**
       // multiply by a^{2^i}
4   result $=$ result $* t \bmod n$
     // $t = a^{2^{i+1}}$
5    $t = t * t \bmod n$
6 **return** result

Example 3.5.1 Let $n = 15$, $d = 3 = 11_2$, $a = 2$. Computing

$$a^d \bmod n = 2^3 \bmod 15 = 8 \bmod 15 = 8$$

Algorithm 3.8: Left-to-right square and multiply algorithm for computing modular exponentiation

---

Input: n, a, d // $n \in \mathbb{Z}, n \geq 2$; $a \in \mathbb{Z}_n$; $d \in \mathbb{Z}_{\varphi(n)}$
Output: $a^d \bmod n$
1   $t = 1$
2   **for** $i = \ell_d, i \geq 0, i - - $ **do**
3      $t = t * t \bmod n$
      // ith bit of d is 1
4   **if** $d_i = 1$ **then**
5        $t = a * t \bmod n$

6   **return** t

---

using Algorithm 3.7, we get the values of the variables in each loop as follows:

i	d_i	t	result
0	1	4	2
1	1	1	8

The returned value is 8. Similarly, using Algorithm 3.8, the intermediate values are:

i	d_i	t
1	1	2
0	1	8

Where in the last loop, line 3 computes $t = 4$ and line 5 calculates $t = 8 \bmod 15 = 8$.

Example 3.5.2 Let $n = 23$, $d = 4 = 100_2$, $a = 5$. Computing

$$a^d \bmod n = 5^4 \bmod 23 = 625 \bmod 23 = 4$$

using Algorithm 3.7, we get the values of the variables in each loop as follows:

i	d_i	t	result
0	0	2	1
1	0	4	1
2	1	16	4

The final result is 4. Using Algorithm 3.8, in the first loop ($i = 2$), line 3 computes $t = 1 \bmod 23$ and line 5 calculates $t = 1 \times 5 \bmod 23 = 5 \bmod 23$. The intermediate values are:

i	d_i	t
2	1	5
1	0	2
0	0	4

The final output is 4.

3.5.1.2 Montgomery Powering Ladder

Same as in Sect. 3.5.1.1, in this part, let $n \geq 2$ be an integer and $d \in \mathbb{Z}_{\varphi(n)}$. We introduce another method, Montgomery powering ladder, to compute $a^d \bmod n$ for $a \in \mathbb{Z}_n$.

Montgomery powering ladder was first introduced for efficient computations of elliptic curve scalar multiplications [Mon87]. Then it was adopted for computing exponentiation in any abelian group [JY03]. We will present the details of the method used for modular exponentiation. In particular, the abelian group we consider here is $\mathbb{Z}_n$ with modular multiplication.

Recall that we have the following binary representation of d

$$d = \sum_{i=0}^{\ell_d - 1} d_i 2^i.$$

For $0 \leq j \leq \ell_d - 1$, define

$$L_j := \sum_{i-j}^{\ell_d - 1} d_i 2^{i-j}, \quad H_j := L_j + 1.$$

Then

$$2L_{j+1} = 2\sum_{i=j+1}^{\ell_d - 1} d_i 2^{i-(j+1)} = \sum_{i=j+1}^{\ell_d - 1} d_i 2^{i-j} = -d_j + \sum_{i=j}^{\ell_d - 1} d_i 2^{i-j} = -d_j + L_j.$$

We have

$$L_j = 2L_{j+1} + d_j = L_{j+1} + H_{j+1} + d_j - 1 = 2H_{j+1} + d_j - 2,$$

and

$$L_j = \begin{cases} 2L_{j+1} & \text{if } d_j = 0 \\ L_{j+1} + H_{j+1} & \text{if } d_j = 1 \end{cases}, \quad H_j = \begin{cases} L_{j+1} + H_{j+1} & \text{if } d_j = 0 \\ 2H_{j+1} & \text{if } d_j = 1 \end{cases}.$$

Then for any $a \in \mathbb{Z}_n$,

$$a^{L_j} = \begin{cases} (a^{L_{j+1}})^2 & \text{if } d_j = 0 \\ a^{L_{j+1}} a^{H_{j+1}} & \text{if } d_j = 1 \end{cases}, \qquad a^{H_j} = \begin{cases} a^{L_{j+1}} a^{H_{j+1}} & \text{if } d_j = 0 \\ (a^{H_{j+1}})^2 & \text{if } d_j = 1 \end{cases}. \qquad (3.16)$$

Since

$$L_0 = \sum_{i=0}^{\ell_d - 1} d_i 2^i = d,$$

to compute $a^d \bmod n$ is equivalent to computing $a^{L_0} \bmod n$. By Eq. 3.16,

$$a^{L_0} = \begin{cases} (a^{L_1})^2 & \text{if } d_0 = 0 \\ a^{L_1} a^{H_1} & \text{if } d_0 = 1 \end{cases}.$$

Similarly, a^{L_1} and a^{H_1} can be computed with a^{L_2} and a^{H_2}. Thus, we can start from the most significant bit of d, $d_{\ell_d - 1}$, compute $a^{L_{\ell_d - 1}}$ and $a^{H_{\ell_d - 1}}$, then calculate $a^{L_{\ell_d - 2}}$ and $a^{H_{\ell_d - 2}}$ with Eq. 3.16, and so on. Note that

$$L_{\ell_d - 1} = d_{\ell_d - 1}, \qquad H_{\ell_d - 1} = d_{\ell_d - 1} + 1$$

and

$$a^{L_{\ell_d - 1}} = \begin{cases} 1 & \text{if } d_{\ell_d - 1} = 0 \\ a & \text{if } d_{\ell_d - 1} = 1 \end{cases}, \qquad a^{H_{\ell_d - 1}} = \begin{cases} a & \text{if } d_{\ell_d - 1} = 0 \\ a^2 & \text{if } d_{\ell_d - 1} = 1 \end{cases}. \qquad (3.17)$$

Details of Montgomery powering ladder for implementing modular exponentiation are shown in Algorithm 3.9, where at the end of the jth iteration, R_0 and R_1 correspond to a^{L_j} and a^{H_j} respectively. When $j = \ell_d - 1$, lines 4–9 implement Eq. 3.17. For $j < \ell_d - 1$, lines 4–9 implement Eq. 3.16.

The computations of lines 5 and 6 (respectively lines 8 and 9) can be done in parallel by first storing the computation results in temporary variables and then assigning to R_1 and R_0 (respectively R_0 and R_1).

Example 3.5.3 Same as in Example 3.5.1, let $n = 15$, $d = 3 = 11_2$, $a = 2$. We have calculated that $a^d \bmod n = 8$. To compute it with Algorithm 3.9, the intermediate values are

$$j = 1, d_1 = 1, \ R_0 = R_0 R_1 \bmod n = 2,$$
$$R_1 = R_1^2 = 2^2 \bmod 15 = 4$$
$$\overline{j = 0, d_0 = 1, \ R_0 = R_0 R_1 \bmod n = 2 \times 4 \bmod 15 = 8}$$

and the final result is 8.

Algorithm 3.9: Montgomery powering ladder for computing modular exponentiation

---

   Input: n, a, d // $n \in \mathbb{Z}$, $n \geq 2$; $a \in \mathbb{Z}_n$, $d \in \mathbb{Z}_{\varphi(n)}$ has bit length ℓ_d

   Output: $a^d \bmod n$

1 $R_0 = 1$
2 $R_1 = a$
3   **for** $j = \ell_d - 1, \, j \geq 0, \, j - - $ **do**
4   **if** $d_j = 0$ **then**
5 $R_1 = R_0 R_1 \bmod n$ // $a^{H_j} = a^{L_{j+1} H_{j+1}}$ for $j < \ell_d - 1$
6 $R_0 = R_0^2 \bmod n$ // $a^{L_j} = \left(a^{L_{j+1}}\right)^2$ for $j < \ell_d - 1$
7   **else**
8 $R_0 = R_0 R_1 \bmod n$ // $a^{L_j} = a^{L_{j+1} H_{j+1}}$ for $j < \ell_d - 1$
9 $R_1 = R_1^2 \bmod n$ // $a^{H_j} = \left(a^{H_{j+1}}\right)^2$ for $j < \ell_d - 1$

10 **return** R_0

Example 3.5.4 Here we repeat the computation in Example 3.5.2. Let $n = 23$, $d = 4 = 100_2$, $a = 5$. We know that $a^d \bmod n = 4$. With Algorithm 3.9, the intermediate values are

$$
\begin{aligned}
j = 2, \; d_2 = 1, \; &R_0 = R_0 R_1 \bmod n = 5, \\
&R_1 = R_1^2 = 5^2 \bmod 23 = 25 \bmod 23 = 2 \\
\hline
j = 1, \; d_1 = 0, \; &R_1 = R_0 R_1 \bmod n = 5 \times 2 \bmod 23 = 10, \\
&R_0 = R_0^2 = 5^2 \bmod 23 = 2 \\
\hline
j = 0, \; d_0 = 0, \; &R_1 = R_0 R_1 \bmod n = 2 \times 10 \bmod 23 = 20, \\
&R_0 = R_0^2 = 2^2 \bmod 15 = 4
\end{aligned}
$$

and the final result is 4.

3.5.1.3   Chinese Remainder Theorem (CRT) Based RSA

In this part, we focus on the case when $n = pq$ is the RSA modulus (p, q are distinct odd primes) and $d \in \mathbb{Z}^*_{\varphi(n)}$ is the private key.

   By Chinese Remainder Theorem (see Theorem 1.4.7 and Example 1.4.19), finding the solution for

$$
x \equiv a^d \bmod n
$$

is equivalent to solving

$$
x \equiv a^d \bmod p, \quad x \equiv a^d \bmod q.
$$

By Corollary 1.4.3, we can compute

$$x_p := a^{d \bmod (p-1)} \bmod p, \quad x_q := a^{d \bmod (q-1)} \bmod q,$$

and solve for

$$x \equiv x_p \bmod p, \quad x \equiv x_q \bmod q. \tag{3.18}$$

An implementation that computes $a^d \bmod n$ by solving equation 3.18 is called *CRT-based RSA* implementation.

By Eqs. 1.19 and 1.20, we compute

$$M_q = q, \quad M_p = p, \quad y_q = M_q^{-1} \bmod p = q^{-1} \bmod p,$$

$$y_p = M_p^{-1} \bmod q = p^{-1} \bmod q,$$

and

$$x = x_p y_q q + x_q y_p p \bmod n \tag{3.19}$$

gives us the solution to Eq. 3.18.

Calculating x by Eq. 3.19 is called the *Gauss's algorithm* for CRT. While *Garner's algorithm* calculates

$$x = x_p + ((x_q - x_p)y_p \bmod q)p. \tag{3.20}$$

We will show that Eq. 3.20 indeed gives the solution to Eq. 3.18. First, it is straightforward to see $x \equiv x_p \bmod p$. Furthermore,

$$x \equiv x_p + (x_q - x_p) \equiv x_q \bmod q.$$

Since $x_p \in \mathbb{Z}_p$, $x_p < p$. Similarly, $(x_q - x_p)y_p \bmod q \leq q - 1$. And

$$x = x_p + ((x_q - x_p)y_p \bmod q)p < p + (q - 1)p = n.$$

Thus $x \in \mathbb{Z}_n$.

Example 3.5.5 Let us consider the toy example from Example 3.3.1. We have

$$p = 3, \quad q = 5, \quad n = 15, \quad \varphi(n) = 8, \quad e = 3, \quad d = 3.$$

Bob receives ciphertext $c = 8$ from Alice. Instead of computing the plaintext directly using

$$m = c^d \bmod n = 8^3 \bmod 15,$$

we compute

$$m_p = c^{d \bmod (p-1)} \bmod p = 8^{3 \bmod 2} \bmod 3 = 8 \bmod 3 = 2,$$

$$m_q = c^{d \bmod (q-1)} \bmod q = 8^{3 \bmod 4} \bmod 5 = 512 \bmod 5 = 2.$$

By the extended Euclidean algorithm,

$$5 = 3 \times 1 + 2, \quad 3 = 2 + 1 \implies 1 = 3 - (5 - 3) = 3 \times 2 - 5.$$

Thus

$$y_p = p^{-1} \bmod q = 3^{-1} \bmod 5 = 2 \bmod 5,$$

$$y_q = q^{-1} \bmod p = 5^{-1} \bmod 3 = -1 \bmod 3 = 2 \bmod 3.$$

By Gauss's algorithm,

$$m = m_p y_q q + m_q y_p p \bmod n = 2 \times 2 \times 5 + 2 \times 2 \times 3 = 32 \bmod 15 = 2.$$

By Garner's algorithm,

$$m = m_p + ((m_q - m_p)y_p \bmod q)p = 2 + 0 = 2.$$

Both algorithms give us the original plaintext from Alice.

Example 3.5.6 Here we look at Example 3.3.2. We have

$$p = 29, \quad q = 41, \quad n = 1189, \quad \varphi(n) = 1120, \quad e = 3, \quad d = 747,$$

and ciphertext $c = 8$. Then

$$m_p = c^{d \bmod (p-1)} \bmod p = 8^{747 \bmod 28} \bmod 29 = 8^{19} \bmod 29 = 2,$$

$$m_q = c^{d \bmod (q-1)} \bmod q = 8^{747 \bmod 40} \bmod 41 = 8^{27} \bmod 41 = 2.$$

By the extended Euclidean algorithm

$$41 = 29 + 12, \quad 29 = 12 \times 2 + 5, \quad 12 = 5 \times 2 + 2, \quad 5 = 2 \times 2 + 1,$$

and

$$1 = 5 - 2 \times (12 - 5 \times 2) = -2 \times 12 + (29 - 12 \times 2) \times 5$$
$$= 29 \times 5 - 12 \times (41 - 29) = -41 \times 12 + 29 \times 17.$$

We have

$$y_p = p^{-1} \bmod q = 29^{-1} \bmod 41 = 17 \bmod 41,$$

$$y_q = q^{-1} \bmod p = 41^{-1} \bmod 29 = -12 \bmod 29 = 17 \bmod 29.$$

By Gauss's algorithm,

$$m = m_p y_q q + m_q y_p p \bmod n = 2 \times 17 \times 41 + 2 \times 17 \times 29 \bmod 1189$$

$$= 2380 \bmod 1189 = 2.$$

By Garner's algorithm,

$$m = m_p + ((m_q - m_p) y_p \bmod q) p = 2 + 0 = 2.$$

Example 3.5.7 Same as in Example 3.5.6, we keep

$$p = 29, \quad q = 41, \quad n = 1189, \quad \varphi(n) = 1120, \quad e = 3, \quad d = 747.$$

Then we have

$$y_p = 17, \quad y_q = 17.$$

Let $c = 155$, then

$$m_p = c^{d \bmod (p-1)} \bmod p = 155^{747 \bmod 28} \bmod 29 = 10^{19} \bmod 29 = 21,$$

$$m_q = c^{d \bmod (q-1)} \bmod q = 155^{747 \bmod 40} \bmod 41 = 32^{27} \bmod 41 = 9.$$

To compute $10^{19} \bmod 29$, we note that

$$10^2 \bmod 29 = 100 \bmod 29 = 13,$$
$$10^4 \bmod 29 = 13^2 \bmod 29 = 24,$$
$$10^8 \bmod 29 = 24^2 \bmod 29 = 25,$$
$$10^{16} \bmod 29 = 25^2 \bmod 29 = 16.$$

Thus

$$10^{19} \bmod 29 = 10^{16} \times 10^2 \times 10 \bmod 29 = 16 \times 13 \times 10 \bmod 29 = 21.$$

Similarly,

$$32^2 \bmod 41 = 40$$
$$32^3 \bmod 41 = 32 \times 40 \bmod 41 = 9$$

$$32^9 \bmod 41 = 9^3 \bmod 41 = 32$$
$$32^{27} \bmod 41 = 32^3 \bmod 41 = 9.$$

By Gauss's algorithm,

$$m = m_p y_q q + m_q y_p p \bmod n = 21 \times 17 \times 41 + 9 \times 17 \times 29 \bmod 1189$$
$$= 19074 \bmod 1189 = 50.$$

By Garner's algorithm,

$$m = m_p + ((m_q - m_p) y_p \bmod q) p = 21 + ((9 - 21) \times 17 \bmod 41) \times 29$$
$$= 21 + 1 \times 29 = 50.$$

Example 3.5.8 Let us consider Example 3.4.1 for RSA signatures computation. We have

$$p = 5, \quad q = 7, \quad n = 35, \quad \varphi(n) = 24, \quad e = 5, \quad d = 5, \quad m = 10.$$

To sign message $m = 10$, Alice computes

$$s_p = m^{d \bmod (p-1)} \bmod p = 10^{5 \bmod 4} \bmod 5 = 0,$$
$$s_q = m^{d \bmod (q-1)} \bmod q = 10^{5 \bmod 6} \bmod 7 = 5.$$

By the extended Euclidean algorithm

$$7 = 5 + 2, \quad 5 = 2 \times 2 + 1 \Longrightarrow 1 = 5 - 2 \times (7 - 5) = 5 \times 3 - 2 \times 7$$

We have

$$y_p = p^{-1} \bmod q = 3 \bmod 7,$$
$$y_q = q^{-1} \bmod p = -2 \bmod 5 = 3.$$

By Gauss's algorithm,

$$s = s_p y_q q + s_q y_p p \bmod n = 5 \times 3 \times 5 \bmod 35 = 5.$$

By Garner's algorithm,

$$s = s_p + ((s_q - s_p) y_p \bmod q) p = 0 + (5 \times 3 \bmod 7) \times 5 = 1 \times 5 = 5.$$

Compared to Gauss's algorithm, Garner's algorithm does not require the final modulo n reduction.

CRT-based RSA implementation can improve the efficiency of the computation in many ways. Firstly, y_p and y_q can be precomputed, which saves time during communication. Secondly, the intermediate values during the computation are only half as big compared to those in the computation of $a^d \bmod n$ since they are in $\mathbb{Z}_p$ or $\mathbb{Z}_q$ rather than $\mathbb{Z}_n$. Moreover, $x_p = a^{d \bmod (p-1)} \bmod p$ and $x_q = a^{d \bmod (q-1)} \bmod q$ can be calculated by the square and multiply algorithm (Algorithms 3.7 and 3.8) or Montgomery powering ladder (Algorithm 3.9) to further improve the efficiency. In this case, $d \bmod (p-1)$ and $d \bmod (q-1)$ are much smaller than d, computing x_p or x_q requires fewer multiplications than computing $a^d \bmod p$ or $a^d \bmod q$.

3.5.2 *Implementing Modular Multiplication*

From the previous subsection, we see that to have more efficient modular exponentiation implementations, we need to compute modular addition, subtraction, inverse, and multiplications. For modular addition and subtraction, we can just compute the corresponding integer operations and then perform a single reduction modulo the modulus. For inverse modulo an integer, as has been mentioned a few times, we can utilize the extended Euclidean algorithm. Next, we will discuss two methods for implementing modular multiplication.

Throughout this subsection, let n be an integer of bit length ℓ_n, in particular

$$2^{\ell_n - 1} \le n < 2^{\ell_n}. \tag{3.21}$$

Let $a, b \in \mathbb{Z}_n$ be two integers. Then $0 \le a, b < n$. We would like to compute

$$R := ab \bmod n.$$

Let us assume the computer's word size (see Sect. 2.1.2) is ω. Define

$$\kappa := \left\lceil \frac{\ell_n}{\omega} \right\rceil, \quad \text{i.e.,} \quad (\kappa - 1)\omega < \ell_n \le \kappa\omega.$$

We can write

$$a = a_{\kappa-1} || a_{\kappa-2} || \dots || a_0, \quad b = b_{\kappa-1} || b_{\kappa-2} || \dots || b_0,$$

$$0 \le a_i, b_j < 2^\omega \text{ for } 0 \le i, j < \kappa.$$

where $||$ indicates concatenation. Note that some a_i or b_j might be 0 if the bit length of a or b is less than ℓ_n. Furthermore, we have

$$a = \sum_{i=0}^{\kappa-1} a_i (2^\omega)^i, \quad b = \sum_{j=0}^{\kappa-1} b_j (2^\omega)^j. \tag{3.22}$$

Then the product of a and b is given by

$$t = ab = t_{2\kappa-1}||t_{2\kappa-2}||\ldots||t_0,$$

where

$$t_x = \sum_{i,j,\ i+j=x} a_i b_j, \quad 0 \le x \le 2\kappa - 1.$$

Such a multiplication method can be described by Algorithm 3.10.

Algorithm 3.10: Standard multiplication

Input: a, b// $a, b \in \mathbb{Z}_n$, where $n \ge 2$ is an integer of bit length ℓ_n
Output: ab
1 for $i = 0, 1, 2\ldots, 2\kappa - 1, t_i = 0$// $\kappa = \lceil \ell_n/\omega \rceil$, where ω is the word size of the computer
// for each b_j
2 **for** $j = 0,\ j < \kappa,\ j + + $ **do**
3      $T_1 = 0$
    // for each a_i
4   **for** $i = 0,\ i < \kappa,\ i + + $ **do**
        // T_i has bit length at most ω
5          $T_1||T_0 = t_{i+j} + a_i b_j + T_1$
6          $t_{i+j} = T_0$
7      $t_{j+\kappa} = T_1$
8 **return** $t_{2\kappa-1}||t_{2\kappa-2}||\ldots||t_0$

One drawback of Algorithm 3.10 is that a variable with double word size is being processed in line 5. To see this, the maximum value of the right-hand side in line 5 is

$$2^\omega - 1 + (2^\omega - 1)(2^\omega - 1) + 2^\omega - 1 = 2^{2\omega} - 1.$$

Moreover, to compute $R = t \bmod n$, division by n will be required.

Example 3.5.9 As a simple example, let us consider word size $\omega = 2$ and let $n = 15$ be a 4-bit integer. Let $a = 13 = 1101_2$ and $b = 5 = 0101_2$. We have

$$a_0 = 01_2, \quad a_1 = 11_2, \quad b_0 = 01_2, \quad b_1 = 01_2, \quad \kappa = \left\lceil \frac{4}{2} \right\rceil = 2.$$

The product $t = t_3||t_2||t_1||t_0$ has bit length at most 8. Computations in lines 5–7 for each loop are as follows:

$$T_1||T_0 = t_0 + a_0b_0 + T_1 = 00 + 01 + 00 = 0001, \ t_0 = 01,$$
$$T_1||T_0 = t_1 + a_1b_0 + T_1 = 00 + 11 + 00 = 0011, \ t_1 = 11,$$
$$T_1||T_0 = t_1 + a_0b_1 + T_1 = 11 + 01 + 00 = 0100, \ t_1 = 00,$$
$$T_1||T_0 = t_2 + a_1b_1 + T_1 = 00 + 11 + 01 = 0100, \ t_2 = 00, \ t_3 = 01$$

The values for each variable in Algorithm 3.10 are listed below

j	i	$a_i b_j$	T_1	T_0	t_3	t_2	t_1	t_0
0	0	01	00	01	00	00	00	01
0	1	11	00	11	00	00	11	01
1	0	01	01	00	00	00	00	01
1	1	11	01	00	01	00	00	01

As expected, we get

$$t = 01000001 = 65 = 13 \times 5.$$

Furthermore, if we would like to continue the computation and find ab mod 15, we will divide 65 by 15 and calculate the remainder, which is 5.

3.5.2.1 Blakely's Method

First proposed in 1983 [Bla83], Blakely's method for computing modular multiplication interleaves the multiplication steps with the reduction steps. The product ab is computed as follows

$$t = ab = \left(\sum_{i=0}^{\kappa-1} a_i (2^\omega)^i \right) b = \sum_{i=0}^{\kappa-1} (2^\omega)^i a_i b,$$

where a_is are given in Eq. 3.22. Algorithm 3.11 lists the steps for computing

$$R = t \bmod n = ab \bmod n$$

with Blakely's method.

Note that in line 3,

$$R \leq 2^\omega (n-1) + (2^\omega - 1)(n-1) = (2^{\omega+1} - 1)n - (2^{\omega+1} - 1) < (2^{\omega+1} - 1)n.$$

Algorithm 3.11: Blakely's method for computing modular multiplication

---

Input: n, a, b // $n \in \mathbb{Z}, n \geq 2$ has bit length ℓ_n; $a, b \in \mathbb{Z}_n$
Output: $ab \bmod n$
1 $R = 0$
   // $\kappa = \lceil \ell_n / \omega \rceil$, where ω is the word size of the computer
2 **for** $i = \kappa - 1, i \geq 0, i - -$ **do**
3      $\quad R = 2^\omega R + a_i b$
4      $\quad R = R \bmod n$
5 **return** R

Thus, line 4 can be replaced by comparing R with n for $2^{\omega+1} - 2$ times and subtract n from R in case $R \geq n$:

---

1 **for** $j = 0, 1, 2 \ldots, 2^{\omega+1} - 2$ **do**
2      $\quad$ **if** $R \geq n$ **then**
3          $\quad\quad R = R - n$
4      $\quad$ **else** break

In this way, we can avoid dividing by n to compute the remainder. In particular, when $\omega = 1$, $2^{\omega+1} - 2 = 2$. And we have Algorithm 3.12, which is the original proposal from Blakely [Bla83, Koç94].

Algorithm 3.12: Blakely's method for computing modular multiplication by taking $\omega = 1$

Input: n, a, b // $n \in \mathbb{Z}, n \geq 2$ has bit length ℓ_n; $a, b \in \mathbb{Z}_n$
Output: $ab \bmod n$
1 $R = 0$
2 **for** $i = \ell_n - 1, i \geq 0, i - -$ **do**
3      $\quad R = 2R + a_i b$
4      $\quad$ **if** $R \geq n$ **then** $R = R - n$
5      $\quad$ **if** $R \geq n$ **then** $R = R - n$
6 **return** R

Example 3.5.10 Same as in Example 3.5.9, let the word size $\omega = 2$, and

$$a = 13 = 1101_2, \quad b = 5, \quad n = 15, \quad \ell_n = 4, \quad \kappa = 2.$$

We have

$$a_0 = 01_2 = 1, \quad a_1 = 11_2 = 3.$$

Let us calculate $ab \bmod n$ using Algorithm 3.11. For $i = 1$,

$$R = 0 + 3 \times 5 \bmod 15 = 0 \bmod 15.$$

And for $i = 0$,

$$R = 0 + 1 \times 5 \bmod 15 = 5 \bmod 15.$$

We have the final result $13 \times 5 \bmod 15 = 5$.

Example 3.5.11 Let

$$a = 55 = 110111_2, \quad b = 46, \quad n = 69, \quad \omega = 2.$$

n is a 7-bit integer. Then

$$a_0 = 11 = 3, \quad a_1 = 01 = 1, \quad a_2 = 11 = 3, \quad a_3 = 0, \quad \kappa = \left\lceil \frac{7}{2} \right\rceil = 4.$$

Computing $ab \bmod n$ with Algorithm 3.11 gives us the following intermediate values:

$$
\begin{aligned}
i = 3 \quad & \text{line 3, } R = 0, \\
& \text{line 4, } R = 0, \\
i = 2 \quad & \text{line 3, } R = 3 \times 46 = 138, \\
& \text{line 4, } R = 138 \bmod 69 = 0, \\
i = 1 \quad & \text{line 3, } R = 1 \times 46 = 46, \\
& \text{line 4, } R = 46 \bmod 69 = 46, \\
i = 0 \quad & \text{line 3, } R = 2^2 \times 46 + 3 \times 46 = 322, \\
& \text{line 4, } R = 322 \bmod 69 = 46.
\end{aligned}
$$

We have $ab \bmod n = 46$.

Now we can expand the modular multiplication computations in the square and multiply algorithm with Blakely's method. The details are listed in Algorithm 3.13 for right-to-left square and multiply algorithm, and in Algorithm 3.14 for left-to-right square and multiply algorithm.

Since ℓ_n is the bit length of n, the bit lengths of the variables "result," "t," and "a" are at most ℓ_n. We can write

$$\text{result} = \sum_{j=0}^{\kappa-1} h_j (2^\omega)^j, \quad t = \sum_{j=0}^{\kappa-1} t_j (2^\omega)^j, \quad a = \sum_{j=0}^{\kappa-1} a_j (2^\omega)^j.$$

Algorithm 3.13: Right-to-left square and multiply algorithm with Blakely's method for modular multiplication

---

Input: n, a, d // $n \in \mathbb{Z}, n \geq 2$ has bit length ℓ_n; $a \in \mathbb{Z}_n$; $d \in \mathbb{Z}_{\varphi(n)}$ has bit length ℓ_d

Output: $a^d \bmod n$

1 result $= 1$
2 $t = a$
3 **for** $i = 0, i < \ell_d, i + +$ **do**
    // ith bit of d is 1
4   **if** $d_i = 1$ **then**
        // lines 5-9 implement result $=$ result$*t$ mod n
5          $R = 0$
        // $\kappa = \lceil \ell_n/\omega \rceil$, where ω is the word size of the computer
6   **for** $j = \kappa - 1, j \geq 0, j - -$ **do**
7              $R = 2^{\omega} R + h_j t$
8              $R = R \bmod n$
9   result $= R$
    // lines 10-14 implement $t = t * t$ mod n
10      $R = 0$
11   **for** $j = \kappa - 1, j \geq 0, j - -$ **do**
12          $R = 2^{\omega} R + t_j t$
13          $R = R \bmod n$
14      $t = R$
15 **return** result

Then, in Algorithm 3.13, lines 5–9 implement result $=$ result $* t$ mod n (line 4 of Algorithm 3.7) and lines 10–14 implement $t = t * t$ mod n (line 5 of Algorithm 3.7).

Similarly, in Algorithm 3.14, lines 3–7 implement $t = t * t$ mod n (line 3 of Algorithm 3.8) and lines 9–13 implement $t = a * t$ mod n (line 5 of Algorithm 3.8).

Example 3.5.12 Let us repeat the computation in Example 3.5.2 with Blakley's method. We will calculate

$$a^d \bmod n = 5^4 \bmod 23 = 625 \bmod 23 = 4.$$

Suppose the computer word size $\omega = 2$. $n = 23 = 10111_2$ has $\ell_n = 5$ bits, then $\kappa = \lceil 5/2 \rceil = 3$. Lines 1 and 2 in Algorithm 3.13 give

$$\text{result} = 1, \quad h_0 = 01, \quad h_1 = 00, \quad h_2 = 00, \quad t = 5 = 0101_2,$$

$$t_0 = 01, \quad t_1 = 01, \quad t_2 = 00.$$

The intermediate values during the computation are:

Algorithm 3.14: Left-to-right square and multiply algorithm with Blakely's method for modular multiplication

Input: n, a, d // $n \in \mathbb{Z}, n \geq 2$ has bitlength ℓ_n; $a \in \mathbb{Z}_n$; $d \in \mathbb{Z}_{\varphi(n)}$ has bit length ℓ_d

Output: $a^d \bmod n$

1 $t = 1$
2 **for** $i = \ell_d - 1, i \geq 0, i - - $ **do**
      // lines 3-7 implement $t = t * t \bmod n$
3      $R = 0$
      // $\kappa = \lceil \ell_n / \omega \rceil$, where ω is the word size of the computer
4   **for** $j = \kappa - 1, j \geq 0, j - -$ **do**
5          $R = 2^\omega R + t_j t$
6          $R = R \bmod n$
7      $t = R$
      // ith bit of d is 1
8   **if** $d_i = 1$ **then**
         // lines 9-13 implement $t = a * t \bmod n$
9          $R = 0$
10   **for** $j = \kappa - 1, j \geq 0, j - -$ **do**
11             $R = 2^\omega R + a_j t$
12             $R = R \bmod n$
13         $t = R$

14 **return** t

$i = 0 \ d_0 = 0$

    loop line 11 $\quad j = 2 \quad\quad R = 0$

                $j = 1 \quad\quad R = 2^\omega R + t_1 t \bmod n = 5 \bmod 23$

                $j = 0 \quad\quad R = 2^\omega R + t_0 t \bmod n = 2^2 \times 5 + 1 \times 5 \bmod 23$

                                $= 2 \bmod 23$

    line 14 $\quad\quad\quad t = 2 \quad\quad t_0 = 10, \ t_1 = 00, \ t_2 = 00$

$i = 1 \ d_1 = 0$

    loop line 11 $\quad j = 2 \quad\quad R = 0$

                $j = 1 \quad\quad R = 0$

                $j = 0 \quad\quad R = t_0 t \bmod n = 2 \times 2 \bmod 23 = 4 \bmod 23$

    line 14 $\quad\quad\quad t = 4 \quad\quad t_0 = 00, \ t_1 = 01, \ t_2 = 00$

$i = 2 \ d_2 = 1$

    loop line 6 $\quad j = 2 \quad\quad R = 0$

                $j = 1 \quad\quad R = 0$

                $j = 0 \quad\quad R = h_0 t \bmod n = 4 \bmod 23$

    line 9 $\quad\quad\quad\quad\text{result} = 4$

And the output is 4. Similarly, with Algorithm 3.14, line 1 gives

$$t = 1, \quad t_0 = 01, \quad t_1 = 00, \quad t_2 = 00.$$

We also have

$$a = 5, \quad a_0 = 01, \quad a_1 = 01, \quad a_2 = 00.$$

The intermediate values are

$i = 2 \; d_2 = 1$
      loop line 4 $j = 2 \; R = 0$
                    $j = 1 \; R = 0$
                    $j = 0 \; R = t_0 t \bmod n = 1 \bmod 23$
      line 7 $t = 1 \; t_0 = 01, \; t_1 = 00, \; t_2 = 00$
      loop line 10 $j = 2 \; R = 0$
                    $j = 1 \; R = 2^{\omega} R + a_1 t = 1 \bmod 23$
                    $j = 0 \; R = 2^{\omega} R + a_0 t = 2^2 + 1 = 5 \bmod 23$
      line 13 $t = 5 \; t_0 = 01, \; t_1 = 01, \; t_2 = 00$

$i = 1 \; d_1 = 0$
      loop line 4 $j = 2 \; R = 0$
                    $j = 1 \; R = t_1 t \bmod n = 5 \bmod 23$
                    $j = 0 \; R = 2^{\omega} R + t_0 t \bmod n = 2^2 \times 5 + 5 \bmod 23$
                                $= 25 \bmod 23 = 2 \bmod 23$
      line 7 $t = 2 \; t_0 = 10, \; t_1 = 00, \; t_2 = 00$

$i = 0 \; d_0 = 0$
      loop line 4 $j = 2 \; R = 0$
                    $j = 1 \; R = 0$
                    $j = 0 \; R = t_0 t \bmod n = 2 \times 2 \bmod 23 = 4 \bmod 23$
      line 7 $t = 4$

The output is also 4.

Similarly, we can adopt Blakely's method in Montgomery powering ladder (Algorithm 3.9) and we get Algorithm 3.15 for computing modular exponentiation. Since ℓ_n is the bit length of n, the bit lengths of the variables R_0 and R_1 are at most ℓ_n. We can write

$$R_0 = \sum_{i=0}^{\kappa-1} R_{0i} (2^{\omega})^i, \quad R_1 = \sum_{i=0}^{\kappa-1} R_{1i} (2^{\omega})^i.$$

Then lines 5–9 implement $R_1 = R_0 R_1 \bmod n$ (line 5 of Algorithm 3.9). Lines 10–14 implement $R_0 = R_0^2 \bmod n$ (line 6 of Algorithm 3.9). Lines 16–20 implement $R_0 = R_0 R_1 \bmod n$ (line 8 of Algorithm 3.9). Lines 21–25 implement $R_1 = R_1^2 \bmod n$ (line 9 of Algorithm 3.9).

Algorithm 3.15: Montgomery powering ladder with Blakely's method for computing modular multiplication

---

Input: n, a, d // $n \in \mathbb{Z}$, $n \geq 2$; $a \in \mathbb{Z}_n$; $d \in \mathbb{Z}_{\varphi(n)}$ has bit length ℓ_d

Output: $a^d \bmod n$

1   $R_0 = 1$
2   $R_1 = a$
3   **for** $j = \ell_d - 1$, $j \geq 0$, $j --$ **do**
4   **if** $d_j = 0$ **then**
      // lines 5-9 implement $R_1 = R_0 R_1 \bmod n$
5       $R = 0$
6   **for** $i = \kappa - 1$, $i \geq 0$, $i --$ **do**
        // $\kappa = \lceil \ell_n / \omega \rceil$, where ω is the word size of the computer
7         $R = 2^\omega R + R_{0i} R_1$
8         $R = R \bmod n$
9       $R_1 = R$
      // lines 10-14 implement $R_0 = R_0^2 \bmod n$
10      $R = 0$
11   **for** $i = \kappa - 1$, $i \geq 0$, $i --$ **do**
12       $R = 2^\omega R + R_{0i} R_0$
13       $R = R \bmod n$
14      $R_0 = R$
15   **else**
      // lines 16-20 implement $R_0 = R_0 R_1 \bmod n$
16      $R = 0$
17   **for** $i = \kappa - 1$, $i \geq 0$, $i --$ **do**
18       $R = 2^\omega R + R_{0i} R_1$
19       $R = R \bmod n$
20      $R_0 = R$
      // lines 21-25 implement $R_1 = R_1^2 \bmod n$
21      $R = 0$
22   **for** $i = \kappa - 1$, $i \geq 0$, $i --$ **do**
23       $R = 2^\omega R + R_{1i} R_1$
24       $R = R \bmod n$
25      $R_1 = R$
26   **return** R_0

Example 3.5.13 Here we repeat the computation in Example 3.5.4 with Algorithm 3.15. Let

$$n = 23, \quad d = 4 = 100_2, \quad a = 5.$$

We have calculated that $a^d \bmod n = 4$. Same as in Example 3.5.12, we assume $\omega = 2$. Then we have $\ell_n = 5$ and $\kappa = 3$. With Algorithm 3.15, lines 1 and 2 give

$$R_0 = 1, \quad R_{00} = 01, \quad R_{01} = 00, \quad R_{02} = 00, \quad R_1 = 5, \quad R_{10} = 01,$$

$$R_{11} = 01, \quad R_{12} = 00.$$

The intermediate values are

$j = 2\ d_2 = 1$
    loop line 17 $i = 2$ $R = 0$
                $i = 1$ $R = 0$
                $i = 0$ $R = R_{00} R_1 \bmod n = 5 \bmod 23$
    line 20 $R_0 = 5$ $R_{00} = 01,\ R_{01} = 01,\ R_{02} = 00$
    loop line 22 $i = 2$ $R = 0$
                $i = 1$ $R = 2^{\omega} R + R_{11} R_1 \bmod n = 5 \bmod 23$
                $i = 0$ $R = 2^{\omega} R + R_{10} R_1 \bmod n = 2^2 \times 5 + 5 \bmod 23 = 2$
    line 25 $R_1 = 2$ $R_{10} = 10,\ R_{11} = 00,\ R_{12} = 00$

$j = 1\ d_1 = 0$
    loop line 6 $i = 2$ $R = 0$
                $i = 1$ $R = R_{01} R_1 \bmod n = 2 \bmod 23$
                $i = 0$ $R = 2^{\omega} R + R_{00} R_1 \bmod n = 2^2 \times 2$
                        $+2 \bmod 23 = 10$
    line 9 $R_1 = 10$ $R_{10} = 10,\ R_{11} = 10,\ R_{12} = 00$
    loop line 11 $i = 2$ $R = 0$
                $i = 1$ $R = 2^{\omega} R + R_{01} R_0 \bmod n = 5 \bmod 23$
                $i = 0$ $R = 2^{\omega} R + R_{00} R_0 \bmod n = 2^2 \times 5$
                        $+5 \bmod 23 = 2$
    line 14 $R_0 = 2$ $R_{00} = 10,\ R_{01} = 00,\ R_{02} = 00$

$j = 0\ d_0 = 0$
    loop line 6 $i = 2$ $R = 0$
                $i = 1$ $R = 0$
                $i = 0$ $R = R_{00} R_1 \bmod n = 2 \times 10 \bmod 23 = 20$
    line 9 $R_1 = 20$
    loop line 11 $i = 2$ $R = 0$
                $i = 1$ $R = 0$
                $i = 0$ $R = R_{00} R_0 \bmod n = 2 \times 2 \bmod 23 = 4$
    line 14 $R_0 = 4$

Hence the output is 4.

3.5.2.2   Montgomery's Method

In this part, we discuss another method for computing modular multiplication, attributed to Peter Montgomery [Mon85].

Suppose n is odd and let $r = 2^{\ell_n}$. In particular, $\gcd(n, r) = 1$. By Bézout's identity (Theorem 1.1.3), there exist integers r^{-1} and $\hat{n}$ such that

$$rr^{-1} - n\hat{n} = 1. \tag{3.23}$$

We have discussed that such a pair of integers r^{-1} and $\hat{n}$ can be found with the extended Euclidean algorithm.

Remark 3.5.1 We note that for any positive integer t,

$$rr^{-1} - n\hat{n} + trn - trn = 1 \implies r(r^{-1} + tn) - n(\hat{n} + tr) = 1. \tag{3.24}$$

Then $r^{-1} + tn, \hat{n} + tr$ can replace r^{-1} and $\hat{n}$ in Eq. 3.23.

For the rest of this part, we further require that $\hat{n}$ is positive.

Example 3.5.14 Let $n = 15$. Then $\ell_n = 4$ and $r = 2^4 = 16$. By the extended Euclidean algorithm

$$16 = 15 + 1 \implies 1 = 16 - 15,$$

we have $r^{-1} = 1$, and $\hat{n} = 1$.

Example 3.5.15 Let $n = 23$. Then $\ell_n = 5$ and $r = 2^5 = 32$. By the extended Euclidean algorithm

$$32 = 23 + 9, \quad 23 = 9 \times 2 + 5, \quad 9 = 5 + 4, \quad 5 = 4 + 1,$$

and

$$1 = 5 - (9 - 5) = -9 + (23 - 9 \times 2) \times 2 = 23 \times 2 - 5 \times (32 - 23) = 23 \times 7 - 32 \times 5.$$

Hence $r^{-1} = -5$ and $\hat{n} = -7$. To make $\hat{n}$ positive, we can take $t = 1$ as in Eq. 3.24, we have

$$r^{-1} = -5 + n = -5 + 23 = 18, \quad \hat{n} = -7 + r = -7 + 32 = 25.$$

We can check that

$$18r - 25n = 18 \times 32 - 25 \times 23 = 1.$$

Example 3.5.16 Let $n = 57$. Then $\ell_n = 6$ and $r = 2^6 = 64$. By the extended Euclidean algorithm

$$64 = 57 + 7, \quad 57 = 7 \times 8 + 1 \implies 1 = 57 - (64 - 57) \times 8 = -64 \times 8 + 57 \times 9,$$

and we have $r^{-1} = -8$, and $\hat{n} = -9$. To get a positive $\hat{n}$, we choose (see Remark 3.5.1)

$$r^{-1} = -8 + n = -8 + 57 = 49, \quad \hat{n} + r = -9 + 64 = 55.$$

We can check that

$$49r - 55n = 49 \times 64 - 55 \times 57 = 1.$$

Example 3.5.17 Let $n = 1189$. Then $\ell_n = 11$ and $r = 2^{11} = 2048$. By the extended Euclidean algorithm

$$2048 = 1189 + 859, \; 1189 = 859 + 330, \; 859 = 330 \times 2 + 199, \; 330 = 199 + 131,$$
$$199 = 131 + 68, \qquad 131 = 68 + 63, \qquad 68 = 63 + 5, \qquad\qquad 63 = 5 \times 12 + 3,$$
$$5 = 3 + 2, \qquad\qquad 3 = 2 + 1,$$

and

$$1 = 3 - 2 = (63 - 5 \times 12) \times 2 - 5 = 63 \times 2 - (68 - 63) \times 25$$
$$= (131 - 68) \times 27 - 68 \times 25$$
$$= 131 \times 27 - (199 - 131) \times 52 = (330 - 199) \times 79 - 199 \times 52$$
$$= 330 \times 79 - (859 - 330 \times 2) \times 131$$
$$= (1189 - 859) \times 341 - 859 \times 131 = 1189 \times 341 - (2048 - 1189) \times 472$$
$$= 2048 \times (-472) - 1189 \times (-813).$$

We have $r^{-1} = -472$ and $\hat{n} = -813$. To have a positive $\hat{n}$, we take

$$r^{-1} = -472 + 1189 = 717, \quad \hat{n} = 2048 - 813 = 1235.$$

Before computing $R = ab \bmod n$, we first introduce Algorithm 3.16, denoted MonPro, which calculates $abr^{-1} \bmod n$ given a, b, n, and $\hat{n}$.

By Eq. 3.23,

$$1 + n\hat{n} \equiv 0 \bmod r.$$

Then in line 3 of Algorithm 3.16,

$$t + mn = t + t\hat{n}n = t(1 + \hat{n}n)$$

is divisible by r, and the output u is an integer. By our choice of $r = 2^{\ell_n}$ and Eq. 3.21

$$t = ab < rn.$$

From line 2 we know $m < r$. Hence in line 3,

$$u < \frac{rn + rn}{r} = 2n,$$

which shows that lines 4–5 calculate $u \bmod n$. Furthermore,

$$u \equiv \frac{ab + mn}{r} \equiv abr^{-1} + mnr^{-1} \equiv abr^{-1} \bmod n.$$

Thus, Algorithm 3.16 indeed outputs $abr^{-1} \bmod n$.

Algorithm 3.16: `MonPro`, Montgomery product algorithm

Input: $n, r, \hat{n}, a, b$ `// n is an odd integer of bit length` ℓ_n`;` $r = 2^{\ell_n}$`;` $\hat{n}$ `is a positive integer satisfying equation 3.23;` $a, b \in \mathbb{Z}_n$

Output: $abr^{-1} \bmod n$

1 $t = ab$

2 $m = t\hat{n} \bmod r$

3 $u = \dfrac{t + mn}{r}$

4 **if** $u \geq n$ **then**

5 $\quad\lfloor\ u = u - n$

6 **return** u

Let $x = x_{\ell_x-1} x_{\ell_x-2} \ldots x_1 x_0$ be a positive integer of bit length ℓ_x. By definition (see Theorem 1.1.1), we know that

$$x = \sum_{i=0}^{\ell_x-1} x_i 2^i.$$

If $\ell_x \geq \ell_n$, for any $i \geq \ell_n$, $x_i 2^i$ is a multiple of $r = 2^{\ell_n}$. Thus,

$$x \bmod r = \sum_{i=0}^{\min\{\ell_x-1,\ell_n-1\}} x_i 2^i.$$

In other words, to compute $x \bmod r$, we just keep the least significant ℓ_n bits of x. Note that the integer $r - 1$ has binary representation given by a binary string with ℓ_n 1s. We have

$$x \bmod r = x \ \& \ (r - 1).$$

We know that $a, b \geq 0$. Since we also choose $\hat{n} > 0$, line 2 can be replaced by

$$m = t\hat{n} \ \& \ (r - 1).$$

In case x is a multiple of r. We have

$$2^{\ell_n} \ \bigg| \ \sum_{i=0}^{\ell_x - 1} x_i 2^i.$$

It is easy to show that $x_i = 0$ for $0 \le i < \ell_n$. And

$$\frac{x}{r} = \sum_{i=\ell_n}^{\ell_x - 1} x_i 2^i. \tag{3.25}$$

For any positive integer $s \le \ell_x$, we define *right shift x by s bits* to be the integer $x_{\ell_x - 1} x_{\ell_x - 2} \ldots x_s.$[4] We write

$$x \gg s := x_{\ell_x - 1} x_{\ell_x - 2} \ldots x_s. \tag{3.26}$$

Compared with Eq. 3.25, division by r is equivalent to right shift by ℓ_n. We have shown that $t + mn$ in line 3 is a multiple of r. Then line 3 can be replaced by

$$u = (t + mn) \gg \ell_n.$$

In summary, Algorithm 3.16 can be rewritten as Algorithm 3.17. The discussions above demonstrate the main advantage of using `MonPro` over a standard modular multiplication method—the operations modulo n is replaced by modulo r, which can be simplified to an `AND` operation. Furthermore, to compute division by r, we can simply do a right shift.

Example 3.5.18 Let $n = 15$, Then

$$\ell_n = 4, \quad r = 2^4 = 16, \quad r - 1 = 15.$$

We have

$$53 \bmod r = 5, \quad 53 \ \& \ 15 = 110101 \ \& \ 1111 = 101 = 5.$$

Furthermore,

$$\frac{240}{r} = \frac{240}{16} = 15, \quad 240 \gg 4 = 11110000 \gg 4 = 1111 = 15.$$

---

[4] Note that when $s = \ell_x$, we have $x \gg s = 0$.

Algorithm 3.17: `MonPro`, Montgomery product algorithm

Input: $n, r, \hat{n}, a, b$ `// ` n ` is an odd integer of bit length ` ℓ_n`; ` $r = 2^{\ell_n}$`; ` $\hat{n}$ ` is a`
`    positive integer satisfying equation 3.23; ` $a, b \in \mathbb{Z}_n$

Output: $abr^{-1} \bmod n$

1 $t = ab$

2 $m = t\hat{n}$ `&` $(r - 1)$ `// for a non-negative integer,   mod ` r ` is equivalent to`
`    computing AND with ` $r - 1$`. This line implements line 2 of`
`    Algorithm 3.16.`

3 $u = (t + mn) \gg \ell_n$ `// for a non-negative integer, shift right by ` ℓ_n ` bits`
`    is equivalent to division by ` $r = 2^{\ell_n}$`. This line implements line 3 of`
`    Algorithm 3.16.`

4 **if** $u \geq n$ **then**

5 $\quad\lfloor\ u = u - n$

6 **return** u

Example 3.5.19 Let $n = 23$. Then $\ell_n = 5$ and $r = 2^5 = 32$. In Example 3.5.15 we have discussed that $r^{-1} = 18$ and $\hat{n} = 25$. We will compute a few modular multiplications which will be useful for Example 3.5.27.

Let $a = 22, b = 22$. Following Algorithm 3.16, we have

$$t = ab = 22 \times 22 = 484,$$

$$m = t\hat{n} \bmod r = 484 \times 25 \bmod 32 = 4,$$

$$u = \frac{t + mn}{r} = \frac{484 + 4 \times 23}{32} = 18,$$

and the output is 18. Indeed, $abr^{-1} \bmod n = 22 \times 22 \times 18 \bmod 23 = 18$.

Let $a = 18, b = 18$. We have

$$t = ab = 18 \times 18 = 324,$$

$$m = t\hat{n} \bmod r = 324 \times 25 \bmod 32 = 4,$$

$$u = \frac{t + mn}{r} = \frac{324 + 4 \times 23}{32} = 13,$$

and the output is 13. We can verity that $abr^{-1} \bmod n = 18 \times 18 \times 18 \bmod 23 = 13$.

Let $a = 9, b = 13$. We have

$$t = ab = 9 \times 13 = 117,$$

$$m = t\hat{n} \bmod r = 117 \times 25 \bmod 32 = 13,$$

$$u = \frac{t + mn}{r} = \frac{117 + 13 \times 23}{32} = 13,$$

and the output is 13. We can verity that $abr^{-1} \bmod n = 9 \times 13 \times 18 \bmod 23 = 13$.

Let $a = 13, b = 13$. We have

$$t = ab = 169,$$

$$m = t\hat{n} \bmod r = 169 \times 25 \bmod 32 = 1,$$

$$u = \frac{t + mn}{r} = \frac{169 + 1 \times 23}{32} = 6,$$

and the output is 6. We can verity that $abr^{-1} \bmod n = 13 \times 13 \times 18 \bmod 23 = 6$.

Let $a = 13, b = 1$. We have

$$t = ab = 13,$$

$$m = t\hat{n} \bmod r = 13 \times 25 \bmod 32 = 5,$$

$$u = \frac{t + mn}{r} = \frac{13 + 5 \times 23}{32} = 4,$$

and the output is 4. We can verity that $abr^{-1} \bmod n = 13 \times 18 \bmod 23 = 4$.

Let $a = 9, b = 9$. We have

$$t = ab = 81,$$

$$m = t\hat{n} \bmod r = 81 \times 25 \bmod 32 = 9,$$

$$u = \frac{t + mn}{r} = \frac{81 + 9 \times 23}{32} = 9,$$

and the output is 9. We can verity that $abr^{-1} \bmod n = 9 \times 9 \times 18 \bmod 23 = 9$.

Let $a = 9, b = 22$. We have

$$t = ab = 198,$$

$$m = t\hat{n} \bmod r = 198 \times 25 \bmod 32 = 22,$$

$$u = \frac{t + mn}{r} = \frac{198 + 22 \times 23}{32} = 22,$$

and the output is 22. We can verity that $abr^{-1} \bmod n = 9 \times 22 \times 18 \bmod 23 = 22$.

Example 3.5.20 Let $n = 15, a = 3, \ b = 5$. We have discussed in Example 3.5.14 that $r = 2^4 = 16, r^{-1} = 1$ and $\hat{n} = 1$. Following Algorithm 3.16, we have

$$t = ab = 3 \times 5 = 15,$$

$$m = t\hat{n} \bmod r = 15 \times 1 \bmod 16 = 15,$$

$$u = \frac{t + mn}{r} = \frac{15 + 15 \times 15}{16} = 15,$$

and the output is 0. Indeed, $abr^{-1} \bmod n = 15 \bmod 15 = 0$.

Example 3.5.21 Let $n = 57$, $a = 3$, $b = 5$. We have discussed in Example 3.5.16 that $r = 64$, $r^{-1} = 49$ and $\hat{n} = 55$. Following Algorithm 3.16, we have

$$t = ab = 3 \times 5 = 15$$

$$m = t\hat{n} \bmod r = 15 \times 55 \bmod 64 = 57 \bmod 64,$$

$$u = \frac{t + mn}{r} = \frac{15 + 57 \times 57}{64} = 51,$$

and the output is 51. We can check that

$$abr^{-1} \bmod n = 3 \times 5 \times 49 \bmod 57 = 735 \bmod 57 = 51.$$

Example 3.5.22 Let $n = 57$, $a = 21$, $b = 5$. We know from Example 3.5.16 that

$$r = 64, \quad r^{-1} = 49, \quad \hat{n} = 55.$$

Following Algorithm 3.16, we have

$$t = ab = 21 \times 5 = 105$$

$$m = t\hat{n} \bmod r = 105 \times 55 \bmod 64 = 15$$

$$u = \frac{t + mn}{r} = \frac{105 + 15 \times 57}{64} = 15,$$

and the output is 15. We can check that

$$abr^{-1} \bmod n = 21 \times 5 \times 49 \bmod 57 = 5145 \bmod 57 = 15.$$

For any $a \in \mathbb{Z}_n$, we define the $n-$*residue of a with respect to r* as

$$a_r := ar \bmod n.$$

Example 3.5.23 Let $n = 15$, and $a = 3$. Then $r = 16$ and

$$a_r = ar \bmod n = 3 \times 16 \bmod 15 = (3 \bmod 15) \times (16 \bmod 15) = 3 \times 1 \bmod 15 = 3.$$

Example 3.5.24 Let $n = 57$, and $a = 3$, then $r = 64$ and

$$a_r = ar \bmod n = 3 \times 64 \bmod 57 = (3 \bmod 57)(64 \bmod 57) = 3 \times 7 \bmod 57 = 21.$$

To compute $R = ab \bmod n$, we note that

$$R = ab \bmod n = a_r br^{-1} \bmod n = \texttt{MonPro}(a_r, b).$$

We refer to such a computation as Montogomery's method for modular multiplication. Details are given in Algorithm 3.18.

Algorithm 3.18: Montgomery's method for computing modular multiplication

---

Input: n, r, a, b // n an odd integer of bit length ℓ_n; $r = 2^{\ell_n}$; $a, b \in \mathbb{Z}_n$
Output: $ab \bmod n$
1 Compute a positive $\hat{n}$ with the extended Euclidean algorithm (Algorithm 1.2)
2 $a_r = ar \bmod n$
3 $u = \texttt{MonPro}(n, r, \hat{n}, a_r, b)$ // Algorithm 3.16 or 3.17
4 **return** u

Example 3.5.25 Let $n = 15$, $a = 3$, and $b = 5$. We have discussed that $a_r = 3$ (see Example 3.5.23) and $\texttt{MonPro}(3, 5) = 0$ (see Example 3.5.20). Then by Algorithm 3.18, $ab \bmod n = 0$. Indeed, $ab \bmod n = 3 \times 5 \bmod 15 = 0$.

Example 3.5.26 Let $n = 57$, $a = 3$, and $b = 5$. We know that $a_r = 21$ (see Example 3.5.20) and $\texttt{MonPro}(21, 5) = 15$ (see Example 3.5.22). Then by Algorithm 3.18, $ab \bmod n = 15$. We can check that $ab \bmod n = 3 \times 5 \bmod 57 = 15$.

Utilizing $\texttt{MonPro}$ for computing modular multiplication as in Algorithm 3.18 is not optimal as it requires computing $ar \bmod n$ for each multiplication. Even though $\hat{n}$ can be precomputed by the extended Euclidean algorithm, it is time-consuming. $\texttt{MonPro}$ will be more useful when multiple multiplications are computed. We will discuss a more efficient way of using $\texttt{MonPro}$.

By Corollary 1.4.2, the set

$$Z_r^n := \{a_r = ar \bmod n \mid a \subset \mathbb{Z}_n\}$$

contains the same elements modulo n as in $\mathbb{Z}_n$. We define addition $+_{\texttt{Mon}}$ and multiplication $\times_{\texttt{Mon}}$ operation on Z_r^n as follows:

$$a_r +_{\texttt{Mon}} b_r := (a + b)_r, \quad a_r \times_{\texttt{Mon}} b_r := (ab)_r \bmod n.$$

Then we have the following lemma.

Lemma 3.5.1 $(Z_r^n, +_{\texttt{Mon}}, \times_{\texttt{Mon}})$ *is a commutative ring with additive identity* 0_r *and multiplicative identity* 1_r.

Proof Firstly, $(a + b)_r = (a + b)r \bmod n$ and $(ab)_r = abr \bmod n$ are both in Z_r^n. Thus Z_r^n is closed under $+_{\texttt{Mon}}$ and $\times_{\texttt{Mon}}$.

Associativity and commutativity of $+_{\texttt{Mon}}$ follows from that for addition in $\mathbb{Z}_n$. The identity element for $+_{\texttt{Mon}}$ is $0_r = 0 \bmod n$ since for any $a_r \in Z_r^n$,

$$a_r + 0_r = ar \bmod n + 0 \bmod n = a_r.$$

The inverse of a_r with respect to $+_{\text{Mon}}$ is $(-a)_r$, where $-a$ is the inverse of a in $\mathbb{Z}_n$ with respect to addition modulo n:

$$a_r + (-a)_r = ar \bmod n + (-a)r \bmod n = ar - ar \bmod n = 0 \bmod n = 0_r.$$

We have proved that $(Z_r^n, +_{\text{Mon}})$ is an abelian group.

Now, for any $a_r, b_r, c_r \in Z_r^n$.

$$(a_r \times_{\text{Mon}} b_r) \times_{\text{Mon}} c_r = (ab)_r \times_{\text{Mon}} c_r = (abc)_r = abcr \bmod n$$

$$a_r \times_{\text{Mon}} (b_r \times_{\text{Mon}} c_r) = a_r \times_{\text{Mon}} (bc)_r = (abc)_r = abcr \bmod n.$$

Hence

$$(a_r \times_{\text{Mon}} b_r) \times_{\text{Mon}} c_r = a_r \times_{\text{Mon}} (b_r \times_{\text{Mon}} c_r)$$

and $\times_{\text{Mon}}$ is associative. Moreover,

$$a_r \times_{\text{Mon}} (b_r +_{\text{Mon}} c_r) = a_r \times_{\text{Mon}} (b + c)_r = (a(b + c))_r$$

$$= (ab + ac)_r = (ab)_r +_{\text{Mon}} (bc)_r$$

$$= a_r \times_{\text{Mon}} b_r +_{\text{Mon}} a_r \times_{\text{Mon}} c_r,$$

so the distributive law holds for $\times_{\text{Mon}}$ and $+_{\text{Mon}}$. The identity element for $\times_{\text{Mon}}$ is $1_r = r \bmod n$ since

$$a_r \times_{\text{Mon}} 1_r = 1_r \times_{\text{Mon}} a_r = a_r.$$

Hence, $(Z_r^n, +_{\text{Mon}}, \times_{\text{Mon}})$ is a commutative ring (see Definition 1.2.8).

Remark 3.5.2 We note that

$$a_r \times_{\text{Mon}} b_r = (ab)_r = abr \bmod n = arbrr^{-1} \bmod n = a_r b_r r^{-1} \bmod n$$

$$= \texttt{MonPro}(a_r, b_r).$$

Thus, $\texttt{MonPro}(a_r, b_r)$ implements the multiplication in the ring $(Z_r^n, +_{\text{Mon}}, \times_{\text{Mon}})$.

Now we can apply the Montgomery product algorithm $\texttt{MonPro}$ (Algorithm 3.16 or 3.17) for computing multiplications in the right-to-left (Algorithm 3.7) and the left-to-right (Algorithm 3.8) square and multiply algorithms. The details are listed in Algorithms 3.19 and 3.20.

By Lemma 3.5.1 and Remark 3.5.2, lines 5 and 6 in Algorithm 3.19 compute

$$\text{result}_r = \text{result}_r \times_{\text{Mon}} t_r \quad \text{and} \quad t_r = t_r \times_{\text{Mon}} t_r$$

Algorithm 3.19: Montgomery right-to-left square and multiply algorithm

---

Input: $n, r, \hat{n}, a, d$ // n is an odd integer of bit length ℓ_n; $r = 2^{\ell_n}$; $\hat{n}$ is given by Eq. 3.23; $a \in \mathbb{Z}_n$; $d \in \mathbb{Z}_{\varphi(n)}$ has bit length ℓ_d

Output: $a^d \bmod n$

1 $\text{result}_r = r \bmod n$
2 $t_r = ar \bmod n$
3 **for** $i = 0, i < \ell_d, i++$ **do**
      // ith bit of d is 1
4   **if** $d_i = 1$ **then**
5      $\text{result}_r = \text{MonPro}(n, r, \hat{n}, \text{result}_r, t_r)$// $\text{result}_r = \text{result}_r \times_{\text{Mon}} t_r$
6    $t_r = \text{MonPro}(n, r, \hat{n}, t_r, t_r)$// $t_r = t_r \times_{\text{Mon}} t_r$
7 $t = \text{MonPro}(n, r, \hat{n}, \text{result}_r, 1,)$// $t = t_r \times_{\text{Mon}} 1 = \text{result}_r \times r^{-1} \bmod n$
8 **return** result

---

respectively. It follows from Algorithm 3.7 that lines 1–6 in Algorithm 3.19 calculate

$$t_r = (a_r)^d \bmod n = (a^d)_r \bmod n.$$

Then line 7 removes r from $(a^d)_r$ and outputs the final result.

Algorithm 3.20: Montgomery left-to-right square and multiply algorithm

---

Input: $n, r, \hat{n}, a, d$ // n is an odd integer of bit length ℓ_n; $r = 2^{\ell_n}$; $\hat{n}$ is given by Eq. 3.23; $a \in \mathbb{Z}_n$; $d \in \mathbb{Z}_{\varphi(n)}$ has bit length ℓ_d

Output: $a^d \bmod n$

1 $t_r = r \bmod n$
2 $a_r = ar \bmod n$
3 **for** $i = \ell_d - 1, i \geq 0, i--$ **do**
4    $t_r = \text{MonPro}(n, r, \hat{n}, t_r, t_r)$// $t_r = t_r \times_{\text{Mon}} t_r$
5   **if** $d_i = 1$ **then**
6      $t_r = \text{MonPro}(n, r, \hat{n}, t_r, a_r)$// $t_r = t_r \times_{\text{Mon}} a_r$
7 $t = \text{MonPro}(n, r, \hat{n}, t_r, 1)$// $t = t_r \times_{\text{Mon}} 1 = t_r r^{-1} \bmod n$
8 **return** t

Similarly, lines 4 and 6 in Algorithm 3.20 compute

$$t_r = t_r \times_{\text{Mon}} t_r \quad \text{and} \quad t_r = t_r \times_{\text{Mon}} a_r$$

respectively. It follows from Algorithm 3.8 that lines 1–6 in Algorithm 3.20 calculate

$$t_r = (a_r)^d \bmod n = (a^d)_r \bmod n.$$

Then line 7 removes r from $(a^d)_r \bmod n$ and outputs the final result.

Example 3.5.27 Let

$$n = 23, \quad d = 4 = 100_2, \quad a = 5.$$

In Example 3.5.2, we have computed

$$a^d \bmod n = 5^4 \bmod 23 = 625 \bmod 23 = 4$$

with square and multiply algorithm. In Example 3.5.12 we showed the steps when modular multiplications in the square and multiply algorithm are done with Blakely's method. Now we calculate the same modular exponentiation with the square and multiply algorithm and Montgomery's method for modular multiplication.

According to Example 3.5.15 that

$$r = 32, \quad r^{-1} = 18, \quad \hat{n} = 25.$$

For the detailed computations with `MonPro` below, we refer to Example 3.5.19.
Following Algorithm 3.19, lines 1 and 2 give

$$\mathrm{result}_r = 32 \bmod 23 = 9, \quad t_r = 5 \times 32 \bmod 23 = 22.$$

For $i = 0, d_0 = 0$, line 6 computes

$$t_r = \mathtt{MonPro}(23, 32, 25, 22, 22) = 18.$$

For $i = 1, d_1 = 0$, line 6 computes

$$t_r = \mathtt{MonPro}(23, 32, 25, 18, 18) = 13.$$

For $i = 2, d_2 = 1$, line 5 computes

$$t_r = \mathtt{MonPro}(23, 32, 25, 9, 13) = 13.$$

Then line 6 computes (note that this computation does not affect the final output)

$$t_r = \mathtt{MonPro}(23, 32, 25, 13, 13) = 6.$$

Finally line 7 computes

$$t = \mathtt{MonPro}(23, 32, 25, 13, 1) = 4.$$

Following Algorithm 3.20, lines 1 and 2 give

$$t_r = 32 \bmod 23 = 9, \quad a_r = 5 \times 32 \bmod 23 = 22.$$

For $i = 2, d_2 = 1$, line 4 computes

$$t_r = \mathtt{MonPro}(23, 32, 25, 9, 9) = 9.$$

Then line 6 computes

$$t_r = \mathtt{MonPro}(23, 32, 25, 9, 22) = 22.$$

For $i = 1, d_1 = 0$, line 4 computes

$$t_r = \mathtt{MonPro}(23, 32, 25, 22, 22) = 18.$$

For $i = 0, d_1 = 0$, line 4 computes

$$t_r = \mathtt{MonPro}(23, 32, 25, 18, 18) = 13.$$

Finally, line 7 computes the output

$$t_r = \mathtt{MonPro}(23, 32, 25, 13, 1) = 4.$$

We can also apply the Montgomery product algorithm (Algorithm 3.16 or 3.17) to Montgomery powering ladder (Algorithm 3.9) for computing modular exponentiation. We have Algorithm 3.21.

Algorithm 3.21: Montgomery powering ladder with Montgomery's method for modular multiplication

---

Input: $n, r, \hat{n}, a, d$ // n is an odd integer of bit length ℓ_n; $r = 2^{\ell_n}$; $\hat{n}$ is given by Eq. 3.23; $a \in \mathbb{Z}_n$; $d \in \mathbb{Z}_{\varphi(n)}$ has bit length ℓ_d

Output: $a^d \bmod n$

1 $R_0 = r \bmod n$
2 $R_1 = ar \bmod n$
3 **for** $j = \ell_d - 1, j \geq 0, j -- $ **do**
4   **if** $d_j = 0$ **then**
5           $R_1 = \mathtt{MonPro}(n, r, \hat{n}, R_0, R_1)$ // $R_1 = R_0 \times_{\mathrm{Mon}} R_1$
6           $R_0 = \mathtt{MonPro}(n, r, \hat{n}, R_0, R_0)$ // $R_0 = R_0 \times_{\mathrm{Mon}} R_0$
7   **else**
8           $R_0 = \mathtt{MonPro}(n, r, \hat{n}, R_0, R_1)$ // $R_0 = R_0 \times_{\mathrm{Mon}} R_1$
9           $R_1 = \mathtt{MonPro}(n, r, \hat{n}, R_1, R_1)$ // $R_1 = R_1 \times_{\mathrm{Mon}} R_1$

10 $R_0 = \mathtt{MonPro}(n, r, \hat{n}, R_0, 1)$ // $R_0 = R_0 \times_{\mathrm{Mon}} 1 = R_0 \times r^{-1} \bmod n$
11 **return** R_0

By Lemma 3.5.1 and Remark 3.5.2, lines 5 and 6 in Algorithm 3.21 compute

$$R_1 = R_0 \times_{\text{Mon}} R_1, \quad R_0 = R_0 \times_{\text{Mon}} R_0$$

respectively. Similarly, lines 8 and 9 in Algorithm 3.21 compute

$$R_0 = R_0 \times_{\text{Mon}} R_1, \quad R_1 = R_1 \times_{\text{Mon}} R_1.$$

It follows from Algorithm 3.9 that lines 1–9 in Algorithm 3.21 calculate

$$(a_r)^d \bmod n = (a^d)_r \bmod n.$$

Then line 10 removes r from $(a^d)_r \bmod n$ and outputs the final result.

Example 3.5.28 We repeat the computation in Example 3.5.27 with Algorithm 3.21. We have

$$n = 23, \quad a = 5, \quad d = 4 = 100_2, \quad r = 32, \quad r^{-1} = 18, \quad \hat{n} = 25.$$

For the detailed computations with `MonPro` below, we refer to Example 3.5.19.
   Lines 1 and 2 in Algorithm 3.21 give

$$R_0 = r \bmod n = 32 \bmod 23 = 9, \quad R_1 = ar \bmod n = 5 \times 32 \bmod 23 = 22.$$

For $j = 2, d_2 = 1$, line 8 computes

$$R_0 = \text{MonPro}(23, 32, 25, 9, 22) = 22.$$

Since $d_1 = d_0 = 0$, for the rest of the computations, only R_0 is relevant for the result. For $j = 1$, line 6 calculates

$$R_0 = \text{MonPro}(23, 32, 25, 22, 22) = 18.$$

For $j = 0$, line 6 calculates

$$R_0 = \text{MonPro}(23, 32, 25, 18, 18) = 13.$$

Finally, line 10 gives the output

$$R_0 = \text{MonPro}(23, 32, 25, 13, 1) = 4.$$

3.6 Further Reading

Figures We note that figures in this chapter are adjusted versions of drawings from [Jea16]. Jean [Jea16] includes plenty of source files for various cryptographic-related illustrations.

Implementation of symmetric block ciphers For more discussions on implementations of symmetric block ciphers, we refer the readers to [Osw]. For a detailed analysis of algebraic normal form and Boolean functions, we refer the readers to [O'D14].

Bitsliced implementation of DES can be found in, e.g., [MPC00, Kwa00]. For AES, [KS09] discusses a bitsliced implementation for $64-$architecture and [SS16] presents the design for 32-bit architecture. More efficient bitsliced implementations of PRESENT can be found in [BGLP13] for 64-bit architecture and in [RAL17] for 32-bit architecture.

A related novel way of implementing symmetric block ciphers called *Fixslicing* was introduced in 2020 [ANP20, AP20] to achieve efficient software constant-time implementations. The main idea is to have an alternative representation of several rounds of the cipher by fixing the bits within a certain register to never move.

RSA security Currently, a few hundred qubits (a quantum counterpart to the classical bit) are possible for a quantum computer [Cho22]. To break RSA, thousands of qubits are required [GE21]. Nevertheless, post-quantum public key cryptosystems are being proposed (see, e.g., [HPS98, BS08]) to protect communications after a quantum computer is built.

Implementations of RSA For more discussions on different methods for implementing RSA, we refer the readers to [Koç94]. Koç [Koç94] also discusses how Garner's algorithm (Eq. 3.20) can be designed for solving simultaneous linear congruences in general. For a more efficient way to implement the extended Euclidean algorithm, see [Sti05, Algorithm 5.3].

Digital Signatures There are other digital signatures based on different public key cryptosystems. For more discussions, we refer the readers to [Buc04, Chapter 12].

Secret key In Sect. 2.2.6 we have seen that exhaustive key search can be used to break shift cipher and affine cipher. The lesson is that the key space should be big enough so the attacker cannot brute force the secret key. This size is determined by the current computation power. For example, the 56-bit secret key of DES was successfully broken in 1998 [Fou98]. The U.S. National Institute for Standards and Technology (NIST) issues recommendations for key sizes for government institutions in the USA. According to those, 80-bit keys were "retired" in 2010 [BBB$^+$07], and lesser than 112-bit keys were considered insufficient from 2015 onward [BD16]. National Security Agency (NSA) currently requires AES-256 for *Top Secret* classification since 2015 due to the emergence of quantum computing [Age15].

Chapter 4
Side-Channel Analysis Attacks
and Countermeasures

Side-channel analysis attacks target cryptographic implementations passively. The attacks exploit the possibility of the attacker observing the physical characteristics of a device that is running a cryptographic algorithm. The attacker obtains the so-called side-channel information, e.g., power consumption, electromagnetic emanation, execution time, etc., and then utilizes such information to recover the secret key.

In this chapter, we will focus on power analysis attacks that exploit power consumption information. The attack methodologies can be used in a similar manner when electromagnetic emanation (EM) is analyzed.

Although side-channel analysis attacks can refer to a wide range of attacks, including timing analysis [Koc96], cache attacks [GMWM16], etc., in this book, we use the terminology *side-channel analysis* attacks only in the narrower meaning which refers to power analysis attacks. In short, we also write side-channel analysis as *SCA*.

Device under test The device that the attacker takes measurements of is called the *device under test (DUT)*. For example, it can be a microcontroller, running a software implementation, an FPGA, or an ASIC, realizing a hardware implementation.

For power analysis attacks we study in this chapter, we assume the attacker has certain knowledge of the implementation, for example, how to interface with the encryption routine, whether the computation is executed serially or in parallel, whether the implementation is round-based or bit-sliced, or whether some types of countermeasures are present. Generally, this type of information can also be obtained by reverse engineering, visual inspection of the side-channel measurements, or sometimes just with a simple trial-and-error technique.

Attacker goal The ultimate goal of the attacker is to recover the master key of a symmetric block cipher or the private key of a public key cipher.

Attacker's assumptions Based on the assumption of whether the attacker can obtain a similar device to the target device, we distinguish two types of SCA attacks:

X. Hou, J. Breier, *Cryptography and Embedded Systems Security*,
https://doi.org/10.1007/978-3-031-62205-2_4

- **Non-profiled SCA.** If the attacker does not have access to a similar device, just the target device or just the measurements coming from the target device, we talk about a *non-profiled SCA*. In a general scenario, this attack utilizes a set of measurements where a fixed secret key is used to encrypt multiple (random) plaintexts.
- **Profiled SCA.** If we assume the attacker has access to a clone device of the target device, then they can carry out a *profiled SCA*. This attack operates in two phases. In the *profiling phase*, the attacker acquires side-channel measurements for known plaintext/ciphertext and known key pairs. This set of data is used to characterize or model the device. Then in the *attack phase*, the attacker acquires a few measurements from the target device, which is usually identical to the clone device, with known plaintext/ciphertext and an unknown key. These measurements from the target device are then tested against the characterized model from the clone device.

> **Source Code**
> The source code and measurement data for this chapter can be found in the following link:
>   https://github.com/XIAOLUHOU/SCA-measurements-and-analysis----Experimental-results-for-textbook

4.1 Experimental Setting

Power analysis measures the *power consumption* of the DUT in the form of a voltage change. The most convenient device to capture the voltage change over time is a digital sampling oscilloscope—a device that takes samples of the measured voltage signal over time. We refer to each sample point as a *time sample*. More information on measurement setups is provided in Sect. 6.1.

To be able to target the correct time slot, in our experiments, a trigger signal is raised to high (5V) during the computation that we want to capture and lowered afterward. One measurement consists of the voltage values for each *time sample* in this duration. It can be stored in an array of length equal to the total number of time samples in the measured time interval. It can also be drawn in a graph where the x-axis corresponds to time samples and the y-axis records the voltage values.[1] Thus, we refer to the result of one measurement as a *(power) trace*.

---

[1] Note that, in the case of ChipWhisperer, which will be used for our experiments and analysis, the y-axis does not show the actual voltage value but a 10-bit value proportional to the current going through the shunt resistor.

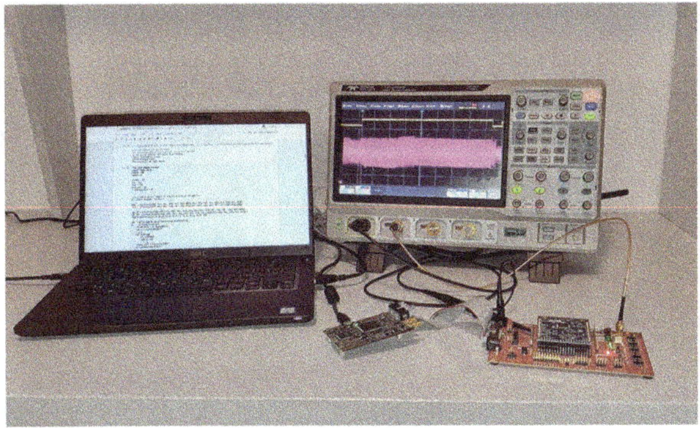

Fig. 4.1 Side-channel measurement setup used for the experiments: a laptop, the ChipWhisperer-Lite measurement board (black), and the CW308 UFO board (red) with the mounted ARM Cortex-M4 target board (blue). Note that the benchtop oscilloscope in the back was only used for the initial analysis—all the measurements were done by the ChipWhisperer

Device under test and oscilloscope For the experiments in this chapter, we used a ready-to-use measurement platform NewAE ChipWhisperer-Lite. The program code was running on a 32-bit ARM Cortex-M4 microcontroller (STM32F3) with a clock speed of $\approx$ 7.4 MHz. ADC was set to capture the samples at $4\times$ that speed, i.e., $\approx$ 29.6 MHz with a 10-bit resolution. However, for plotting purposes, we normally reduced the number of time samples. The measurement setup is depicted in Fig. 4.1. The ChipWhisperer-Lite board is in the middle of the picture in black color, handling the communication with the DUT and the acquisition. The red PCB on the right is the CW 308 UFO board—a breakout board with the DUT—and ARM Cortex-M4 (blue board) mounted on top. The controlling and data processing were done from a laptop, from the Jupyter environment available for the ChipWhisperer platform. In the back, there is a Teledyne T3DSO3504 benchtop oscilloscope that was used mainly for convenience purposes—to precisely locate the time intervals in the initial analysis stage.

Figure 4.2 shows one power trace for the first five rounds of PRESENT encryption. In order to see the trace more clearly, we have added a sequence nop instructions before and after the five rounds of cipher computation. This trace has in total 18,500 time samples. Certain patterns can be seen from the trace, and we can deduce the corresponding operations in each time interval. For example, from time sample 0–1434 and from time sample 17,514–18,500, we have nop instructions. We can also see the five repeated patterns in the figure and deduce the duration of each round, as indicated in the figure by red dotted lines. In terms of time samples, one round takes on average 3216 time samples. In this particular case, we reduced the number of samples by a factor of 3 (simply by taking every third sample) so that the patterns would still be visible to the reader. That means, with the ADC speed of $\approx$ 29.6 MHz, one round takes $(3216 \times 3)/29.6 \approx 325.9\,\mu s$. It is important to note

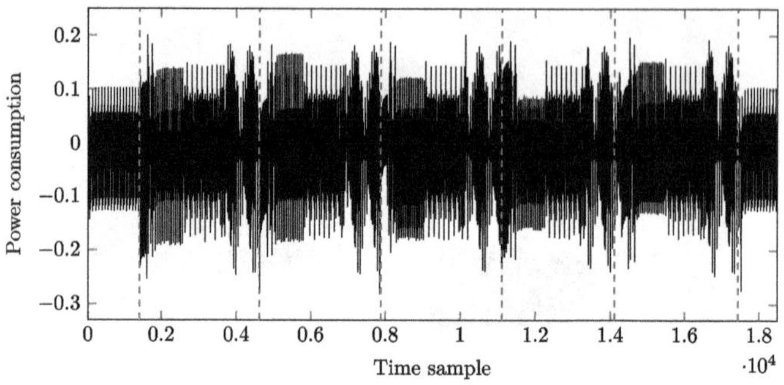

Fig. 4.2 Power trace of the first five rounds of PRESENT encryption. A sequence of nop instructions was executed before and after the cipher computation to clearly distinguish the operations

that for presentation purposes, we used an unoptimized software implementation of PRESENT.

Datasets Four datasets will be analyzed in more detail in the later parts of this chapter. All the datasets capture one round of software implementation of PRESENT. The description of each of them is given below:

- *Fixed dataset A*: This dataset contains 5000 traces with a fixed round key FEDCBA0123456789 and a fixed plaintext ABCDEF1234567890.
- *Fixed dataset B*: This dataset contains 5000 traces with a fixed round key FEDCBA0123456789 and a fixed plaintext 84216BA484216BA4.
- *Random plaintext dataset*: This dataset contains 5000 traces with a fixed round key

$$FEDCBA0123456789 \tag{4.1}$$

and a random plaintext for each trace.
- *Random dataset*: This dataset contains 10,000 traces with a random round key and a random plaintext for each trace.

In each case, the execution of the cipher is surrounded by nop instructions so that the round operation patterns can be clearly distinguished from the provided plots. While the raw traces are all 5000 time samples long, for plotting and analysis purposes, we shorten them to 3600 time samples as the later parts correspond to nop instructions and do not contain any useful information. We also note that for these datasets, we reduced the number of collected time samples by a factor of 3.

4.1.1 Attack Methods

There are two main classical power analysis attack methods, *simple power analysis (SPA)* and *differential power analysis (DPA)*. SPA assumes the attacker has access to only one or a few measurements corresponding to some fixed inputs. In DPA, we assume the attacker can take measurements for a potentially unlimited number of different inputs. We will present several DPA attacks on symmetric block ciphers (Sects. 4.3.1 and 4.3.2) and DPA (Sect. 4.4.1) and SPA (Sect. 4.4.2) attacks on RSA.

We will also discuss a newly proposed *side-channel assisted differential plaintext attack (SCADPA)* on SPN ciphers (Sect. 4.3.3). Similar to SPA, the attack does not require statistical analysis of the traces; only visual inspection is enough. The amount of traces needed is in between that for SPA and DPA, mostly dependent on the measurement equipment.

4.2 Side-Channel Leakages

In the later parts of the chapter, we will see that by analyzing the power consumption, we can deduce the secret key. Consequently, we also refer to the power consumption as the *leakage* of the device. We consider the leakage consists of two parts: *signal* and *noise*. Signal refers to the part of the leakage containing useful information for our attack; the rest is noise. For example, suppose we would like to recover the hamming weight of an intermediate value. In that case, the part of the leakage correlated to the hamming weight of that intermediate value is our signal.

Before we see how leakage can be defined and modeled, we show that it is dependent on the operations being executed and the data being processed.

We first take the *Fixed dataset A* described in Sect. 4.1. The average of those 5000 traces is shown in Fig. 4.3. As mentioned in Sect. 4.1, each trace in this dataset corresponds to one round of PRESENT computation surrounded by nop operations. By visual inspection, we can deduce that the beginning (time samples 0–209) and the ending (time samples 3381–3600) parts that consist of relatively uniform patterns correspond to nop instructions. Other than that, we can see three distinct patterns between them. Since one round of PRESENT consists of addRoundKey, sBoxLayer, and pLayer (see Fig. 3.8), we can roughly identify each of these three operations in the trace—they correspond to the blue (time samples 210–382), pink (time samples 383–567), and green (time samples 568–3380) parts of the trace, respectively. In this case, one round computation corresponds to 3170 time samples, which is fewer than that in Fig. 4.2. Such a difference can be caused by round counter and loop operations, register updates of round keys, etc., which are additionally computed in the five-round PRESENT implementation. These observations demonstrate that the leakage is dependent on the operations being executed in the DUT.

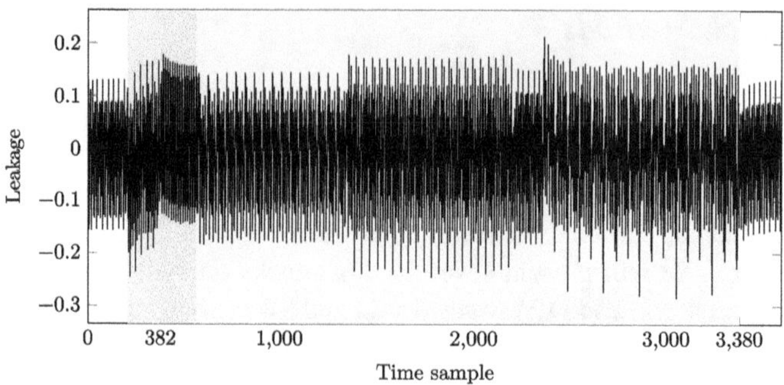

Fig. 4.3 The averaged trace for 5000 traces from the *Fixed dataset A* (see Sect. 4.1). The blue, pink, and green parts of the trace correspond to addRoundKey, sBoxLayer, and pLayer, respectively

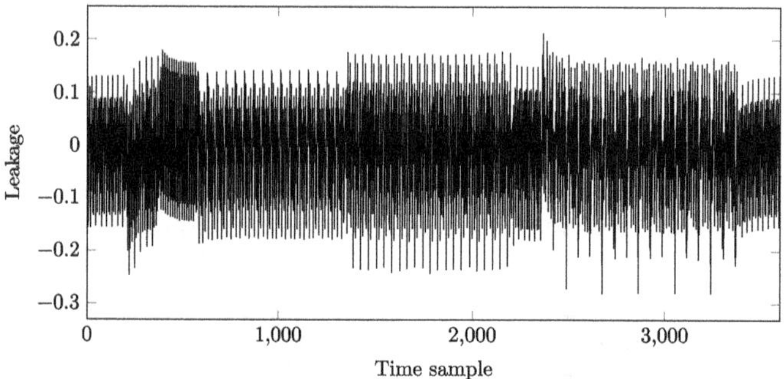

Fig. 4.4 The averaged trace for 1000 plaintexts with the 0th bit equal to 0. The computation corresponds to one round of PRESENT with a fixed round key

For another experiment, with the experimental setup described in Sect. 4.1, we have conducted measurements for one round of PRESENT with a fixed round key. A total of 1000 traces were collected, each for a random plaintext with the 0th bit equal to 0. The averaged trace is shown in Fig. 4.4. With the same key, we collected traces for 1000 plaintexts with the 0th bit equal to 1. And the averaged trace is shown in Fig. 4.5. We can see that those two averaged traces are very similar. Unsurprisingly, they also look similar to the trace in Fig. 4.3. Thus, the time interval for each operation in the first round of PRESENT corresponds to that in Fig. 4.3 as well.

We can gain more information when we take the difference between traces in Figs. 4.4 and 4.5. The difference trace is shown in Fig. 4.6. There are a few peaks in this difference trace, and apart from those peaks, most of the points are close to zero. Those peaks indicate that the 0th bit of the plaintext is related to the computations at the corresponding time samples. Compared with Fig. 4.3, we can see that the first and second peaks correspond to addRoundKey and pLayer operations. In particular,

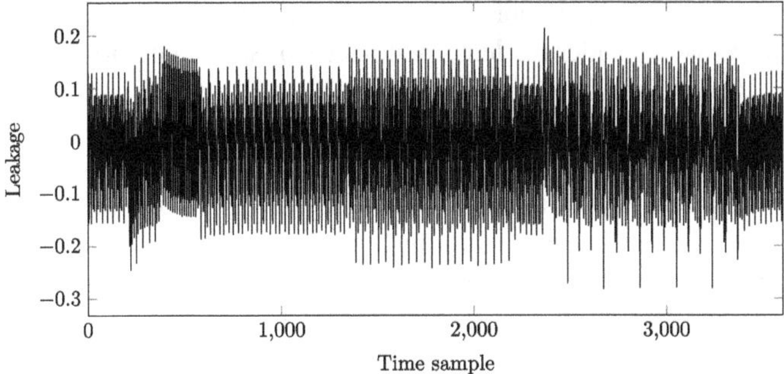

Fig. 4.5 The averaged trace for 1000 plaintexts with the 0th bit equal to 1. The computation corresponds to one round of PRESENT with a fixed round key

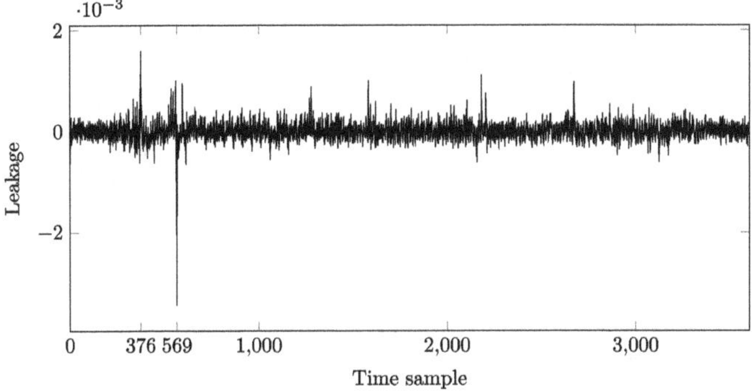

Fig. 4.6 The difference between traces from Figs. 4.4 and 4.5

these observations show that the leakage is dependent on the data being processed in the DUT.

> **Note**
> In the SCA attacks we will see in this book, we will only be interested in operation or/and data-dependent leakages.
>
> SPA typically exploits the relationship between the executed operations and the leakage (power consumption). DPA and SCADPA focus on the relationship between the processed data and the leakage (power consumption).

To analyze the leakage better, we model the leakage, signal, and noise at a given point in time as random variables. In particular, for a fixed time sample t, let L_t, X_t, and N_t denote the random variables corresponding to the leakage, signal, and noise, respectively. As we consider the leakage consists of signal and noise, we can write

$$L_t = X_t + N_t. \tag{4.2}$$

Since X_t contains the part of the leakage that is useful to us and the rest is noise, we make the "independent noise assumption" (see, e.g., [Pro13]) and assume N_t and X_t are independent random variables. When X_t is a constant, according to Eqs. 1.36 and 1.33, we have

$$\text{Var}(L_t) = \text{Var}(N_t), \quad X_t = \text{E}\,[L_t] - \text{E}\,[N_t]. \tag{4.3}$$

But how do we decide when X_t is constant? That depends on the information we would like to obtain from the traces. Let us consider one round of PRESENT computation. Suppose we are interested in the 0th Sbox output of the sBoxLayer (the right-most Sbox in Fig. 3.9), denoted by v. If we want information revealing the exact value of v, then for any given time sample t, the signal X_t is considered to be constant across the following dataset: measurements for computations of one round of PRESENT with a fixed master key and plaintexts with a fixed 0th nibble. The identical 0th nibble in plaintexts and fixed key guarantees the same 0th Sbox output. We can also use random master keys that result in the first round key having the same 0th nibble. If we want information revealing the Hamming weight of v, then measurements with master keys and plaintexts that result in a fixed wt (v) would correspond to constant X_ts.

4.2.1 Distribution of the Leakage

Since we are only interested in either data or operation-related leakages, for a given point in time t, if we fix the operation and the data, we get a constant signal, i.e., X_t is a constant. In the following, we will show that, in this case, the experimental results (histograms) demonstrate that it is reasonable to consider the distribution induced by L_t to be a normal distribution (see Example 1.7.23).

We note that since the noise comes from many sources, e.g., environment, other components in the DUT, setup, etc., it can be considered as a combination of various independent random variables. Thus, according to the *central limit theorem* [Dud14],[2] it is reasonable to assume the distribution induced by the noise is normal.

Let us take the *Fixed dataset A* described in Sect. 4.1. Figure 4.7 shows a small part from five randomly selected traces. We can see that they are very similar, with

---

[2] Roughly speaking, the central limit theorem says that if we combine different independent random variables, the resulting distribution tends to be normal.

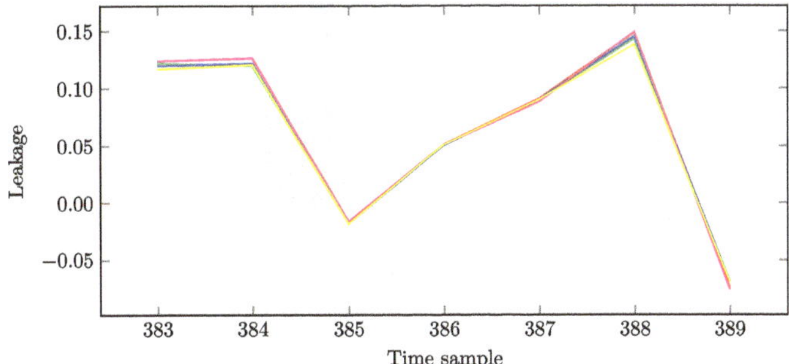

Fig. 4.7 Part of five random traces from the *Fixed dataset A* (see Sect. 4.1)

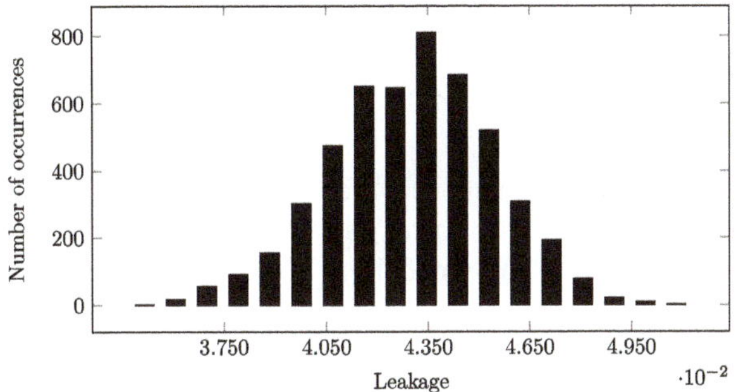

Fig. 4.8 Histogram of leakages at time sample $t = 3520$ across 5000 traces from the *Fixed dataset A*

minor differences. As the signal is the same, the minor differences are caused by the noise. We will further characterize the noise using histograms.

Recall that the averaged trace of those 5000 traces in *Fixed dataset A* is shown in Fig. 4.3. Take $t = 3520$. As we have mentioned in the discussion regarding Fig. 4.3, this time sample corresponds to nop operations. If we plot the histogram of leakages $l_{,3520}$ across those 5000 traces, we get Fig. 4.8. Most leakages are around 0.0435, and very few are below 0.037 or above 0.049.

Now we take another time sample $t = 2368$, which gives the highest peak in Fig. 4.3 and corresponds to pLayer computation. The histogram of leakages L_{2368} across those 5000 traces is shown in Fig. 4.9. Most leakages are around 0.213, and very few are below 0.207 or above 0.219. Compared to Fig. 4.8, we have much higher leakage values. This is because $t = 3520$ corresponds to nop operations, and for $t = 2368$, we have PRESENT round computations.

For both cases, the shapes of the histograms are similar to the PDF of a normal distribution (see Fig. 1.2). If we take a different time sample, the histogram will be similar, with differences in the values on the x-axis. In other words, the distribution induced by the leakages can be approximated by normal distributions. As mentioned, all traces correspond to the same operation and data in the DUT for

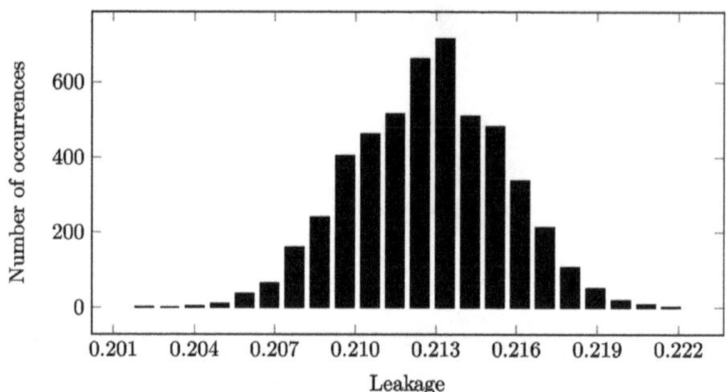

Fig. 4.9 Histogram of leakages at time sample $t = 2368$ across 5000 traces from the *Fixed dataset A*

a fixed time sample, resulting in a constant X_t. Thus, the variants in the leakage are caused by the noise.

Leakage Models One important concept for a power analysis attack is the leakage model, namely a model that estimates how the leakage is related to the data being processed. A good leakage model can make the attack more efficient (see Sect. 4.3.2).

Three commonly used leakage models are *identity leakage model, Hamming distance leakage model*, and *Hamming weight leakage model*. Assume a value v is being processed in the DUT, and right before it, another value u was used by the DUT. Then, according to the identity leakage model, the leakage is correlated (see Definition 1.7.8) to v. The Hamming distance leakage model assumes that the leakage is correlated to dis (v, u), the Hamming distance between v and u (see Eq. 1.24). Following the Hamming weight leakage model, the leakage will then be correlated to wt (v), the Hamming weight of v (see Definition 1.6.10). We refer the readers to Sect. 6.1.1 for more explanations of why there are side-channel leakages when the data in the DUT is changed.

In particular, let noise $\sim \mathcal{N}(0, \sigma^2)$ be a normal random variable with mean 0 and variance σ^2. For the identity leakage model, the modeled leakage is given by

$$\mathcal{L}(v) = v + \text{noise}.$$

For the Hamming distance leakage model, we have

$$\mathcal{L}(v) = \text{dis}(v, u) + \text{noise}.$$

Similarly, for the Hamming weight leakage model,

$$\mathcal{L}(v) = \text{wt}(v) + \text{noise}. \tag{4.4}$$

Even though the actual leakage may not be exactly equal to the modeled leakage $\mathcal{L}(v)$, those leakage models can be used to approximate the behavior of the actual leakages or for statistical analysis (see Sect. 4.3.1). For example, our previous experiments have demonstrated that the identity leakage model is realistic since when the data is fixed, the distribution of leakages is close to a normal distribution. It can be shown that the other two leakage models are also realistic (see [MOP08, Section 4.3]).

In this book, we will focus on two leakage models: the identity leakage model and the Hamming weight leakage model.

4.2.2 Estimating Leakage Distributions

In this subsection, we look at the analysis of leakages from a statistical point of view and provide concrete examples of methods for analyzing unknown distribution parameters discussed in Sect. 1.8.2. We consider the DUT computing PRESENT encryption with a fixed plaintext and a fixed key. We have shown that in this case, we can assume L_t induces a normal distribution for a given time sample t. Let μ_t and σ_t^2 denote the mean and variance of this normal distribution.

For the running example, we focus on the *Fixed dataset A* as described in Sect. 4.1. Let $t = 2368$, which gives the highest peak in Fig. 4.3 and corresponds to the computation of pLayer. We have seen the histogram of the leakages at this time sample in Fig. 4.9. With the terminologies from Sect. 1.8.2, a sample for L_{2368} is given by all leakage values at $t = 2368$ from those 5000 traces in *Fixed dataset A*. We can use our sample to estimate the mean (μ_t) and variance (σ_t^2) of the distribution induced by L_{2368} (assuming it is normal) using point estimators given by sample mean and sample variance (see Remark 1.8.4). Let $M = 5000$ denote our sample size.

Example 4.2.1 (Example of approximating mean and variance with sample mean and sample variance) By Eq. 1.49, the sample mean is given by the average of leakages at time sample 2368 across those 5000 traces. We have

$$\overline{l_{2368}} \approx 0.2132.$$

Following Eq. 1.53, we have also computed the sample variance

$$s_{2368}^2 \approx 8.5196 \times 10^{-6}. \tag{4.5}$$

Then the sample mean 0.2132 is an estimate for μ_{2368}, and the sample variance 8.5196×10^{-6} is an estimate for σ_{2368}^2.

We can also estimate the mean with an interval estimator.

Example 4.2.2 (Example of interval estimator for the mean) Since we do not know the variance of L_{2368}, by Eq. 1.59, a $100(1-\alpha)$ percent confidence interval for μ_{2368} is given by

$$\left(\overline{l_{2368}} - t_{\alpha/2,M-1} \frac{s_{2368}}{\sqrt{M}}, \quad \overline{l_{2368}} + t_{\alpha/2,M-1} \frac{s_{2368}}{\sqrt{M}} \right).$$

Take $\alpha = 0.01$. Then according to Remark 1.8.2 and Table 1.4, we get

$$t_{0.005,4999} \approx z_{0.005} = 2.576.$$

By Eq. 4.5,

$$s_{2368} = \sqrt{8.5196 \times 10^{-6}} \approx 2.9188 \times 10^{-3}.$$

And a 99% confidence interval for μ_{2368} is

$$\left(0.2132 - 2.576 \times \frac{2.9188 \times 10^{-3}}{\sqrt{5000}}, \quad 0.2132 + 2.576 \times \frac{2.9188 \times 10^{-3}}{\sqrt{5000}} \right)$$

$$\approx (0.2131, \, 0.2133). \tag{4.6}$$

Assume we know the variance of L_{2368} is actually given by $\sigma_{2368}^2 = 8.5196 \times 10^{-6}$. Suppose we want to find an estimate for μ_t with precision $c = 0.001$ and 99% of confidence. By Eq. 1.58, the number of traces we need to collect is given by

$$\frac{\sigma_{2368}^2}{c^2} z_{\alpha/2}^2 = \frac{8.5196 \times 10^{-6}}{0.001^2} \times 2.576^2 \approx 57, \tag{4.7}$$

where $1 - \alpha = 0.99$ gives $\alpha = 0.01$, and as mentioned above, $z_{0.005} = 2.576$. Thus, we should collect at least 57 traces to get a 99% percent confidence interval for μ_{2368}.

Since the number of traces to be collected is more than 30, according to Eq. 1.60, if we do not know the variance of L_t, we can use the sample variance s_{2368}^2 to compute the number of traces required. In this case, we will get the same result as in Eq. 4.7 since we have assumed the variance to be equal to this sample variance.

Now we take the *Fixed dataset B* described in Sect. 4.1. We again look at the time sample $t = 2368$. Let L'_{2368} denote the random variable corresponding to the leakage at time sample 2368 for one round encryption of the plaintext 84216BA484216BA4 with round key FEDCBA0123456789. Let μ'_{2368} and σ'^2 denote the mean and variance of L'_{2368}, respectively. Then the *Fixed dataset B* provides a sample for L'_{2368}. Similarly to Example 4.2.1, we can compute the sample mean and sample variance for L'_{2368} with this sample, and we have

$$\overline{l'_{2368}} \approx 0.2133, \quad s'^2_{2368} \approx 8.6198 \times 10^{-6}. \tag{4.8}$$

Example 4.2.3 (Example of interval estimator for the mean) Let us assume L_{2368} and L'_{2368} are independent. We further assume that we know that the actual variances for L_{2368} and L'_{2368} are equal to the sample variances we have computed. Suppose we want to find an estimation for $\mu_{2368} - \mu'_{2368}$. By Eq. 1.62, a 99% confidence interval estimate for $\mu_{2368} - \mu'_{2368}$ is given by

$$\left(\overline{l_{2368}} - \overline{l'_{2368}} - z_{0.005}\sqrt{\frac{\sigma^2_{2368}}{M} + \frac{\sigma'^2_{2368}}{M}}, \quad \overline{l_{2368}} - \overline{l'_{2368}} + z_{0.005}\sqrt{\frac{\sigma^2_{2368}}{M} + \frac{\sigma'^2_{2368}}{M}} \right)$$

$$= \left(0.2132 - 0.2133 - 2.576\sqrt{(8.5196 \times 10^{-6} + 8.6198 \times 10^{-6})/5000}, \right.$$

$$\left. 0.2132 - 0.2133 + 2.576\sqrt{(8.5196 \times 10^{-6} + 8.6198 \times 10^{-6})/5000} \right)$$

$$= \left(-2.5082 \times 10^{-4}, \quad 5.0820 \times 10^{-5} \right).$$

On the other hand, by Eq. 1.63, to achieve an estimation with precision, say $c = 0.001$, and $100(1 - \alpha)$ confidence, the number of data required to collect is given by

$$\frac{z^2_{\alpha/2}(\sigma^2_{2368} + \sigma'^2_{2368})}{c^2}.$$

Take $\alpha = 0.01$, then $z_{0.005} = 2.576$, and we have

$$\frac{z^2_{0.005}(\sigma^2_{2368} + \sigma'^2_{2368})}{c^2} = \frac{2.576^2 \times (8.5196 \times 10^{-6} + 8.6198 \times 10^{-6})}{0.001^2} \approx 114.$$

If we assume we do not know the variances, but we know that $\sigma_{2368} = \sigma'_{2368}$, by Eq. 1.67, the number of traces to collect is given by

$$\frac{2z^2_{\alpha/2}s^2_p}{c^2} = \frac{2 \times 2.576^2 \times 8.5697 \times 10^{-6}}{0.001^2} \approx 114,$$

where

$$s^2_p = \frac{(M-1)s^2_x + (M-1)s^2_y}{M + M - 2} = \frac{s^2_x + s^2_y}{2} = \frac{8.5196 \times 10^{-6} + 8.6198 \times 10^{-6}}{2}$$

$$= 8.5697 \times 10^{-6}.$$

Remark 4.2.1 We note that the sample variances of L_{2368} and L'_{2368} are very close. This is expected as it has been shown in Eq. 4.3 that the variances σ^2_{2368} and σ'^2_{2368} are both equal to the variance of the noise at time sample 2368.

For now, we have seen how to analyze the leakage at one particular time sample by approximating its distribution with a normal distribution. Similarly, we can also approximate the distribution of leakages across different time samples. In this case, we consider a random vector (see Definition 1.7.9) instead of a random variable. Thus, we would approximate the distributions induced by the leakages as multivariate normal distributions (Gaussian distributions). It can be seen from Definition 1.7.10 that to find a good Gaussian distribution for approximating the noise/leakage, we just need to approximate the mean vector and covariance matrix. We will see in Sect. 4.3.2.3 that the profiling phase of the template attack is exactly to calculate estimations for the mean vector and the covariance matrix. In reality, leakages at different time samples are correlated (see Definition 1.7.8). However, the effort to calculate the covariance matrix grows quadratically with the number of considered time samples. Thus, in practice, only a small part of the traces would be profiled with a non-diagonal covariance matrix (see Example 1.7.24).

4.2.3 Leakage Assessment

In the rest of this chapter, we will see various attacks on cryptographic implementations. As a developer, one might want to evaluate the implementation and conclude if it is vulnerable to SCA. On the other hand, different new attacks are being developed, and it is impractical to verify the security of our implementation against all of them. Leakage assessment aims to solve this problem by analyzing the power trace and answering whether any data-dependent information can be detected in the traces of the DUT.

> **Note**
> We note that the leakage assessment methods do not provide any conclusions in cases where data-dependent leakage is not detected. Therefore, the absence of data-dependent leakage indicated by a particular method does not prove that the implementation is leakage-free.

In this part, we discuss a method for leakage assessment based on student's t-test and Welch's t-test (see Sect. 1.8.3). The methodology is also referred to as *test vector leakage assessment (TVLA)*.

Consider a DUT running PRESENT encryption, and fix a time sample t. We also fix an intermediate value $\boldsymbol{v}$, for example, the input plaintext or one Sbox output.

We take the signal as the part of the leakage related to v. Let L_t and L_t' denote the leakages at time sample t corresponding to two encryptions with different fixed values of v.

Example 4.2.4 (Example of L_t and L_t') If we take v to be the plaintext, following the convention, we require the key to be the same, then L_t and L_t' would correspond to encryptions of two different fixed plaintexts with the same key. For example, we can take *Fixed dataset A* and *Fixed dataset B* as samples of L_t and L_t' for the first round of the encryption.

If we take v to be the 0th Sbox output in the first round of PRESENT, then L_t and L_t' would correspond to encryptions that result in two different 0th Sbox outputs. For example, let us take *Random dataset*; we have 634 traces corresponding to $v = 0$ and 651 traces corresponding to $v = F$. Those two sets of traces provide us with samples of L_t and L_t' for the first round of the encryption.

As discussed before, when the signal is fixed, we assume a normal distribution can approximate the distribution induced by L_t. We can write

$$L_t = X_t + N_t, \quad L_t' = X_t' + N_t',$$

with

$$L_t \sim \mathcal{N}(\mu_t, \sigma_t^2), \quad L_t' \sim \mathcal{N}(\mu_t', \sigma_t'^2).$$

Since (see Eq. 4.2)

$$L_t = X_t + N_t$$

and the signal X_t is a constant, the variance of L_t is given by the variance of N_t, and the mean of L_t is given by the sum of the constant X_t and the mean of N_t, as shown in Eq. 4.3. In other words,

$$\mu_t = X_t + \mathrm{E}\,[N_t], \quad \sigma_t^2 = \mathrm{Var}(N_t). \tag{4.9}$$

Similarly, we have

$$\mu_t' = X_t' + \mathrm{E}\big[N_t'\big], \quad \sigma_t'^2 = \mathrm{Var}\big(N_t'\big).$$

As the noise is independent of the signal, we have $N_t = N_t'$. Consequently,

$$\mu_t - X_t = \mu_t' - X_t', \quad \sigma_t^2 = \sigma_t'^2. \tag{4.10}$$

Before going into details about the TVLA methodology, we recall hypothesis testing techniques from Sect. 1.8.3. We can use those techniques to test hypotheses about μ_t and μ_t'.

Example 4.2.5 (Example of a hypothesis) If we are interested in whether $\mu_t = 0$, we can set a hypothesis that $\mu_t = 0$.

Example 4.2.6 (Example of two-sided hypothesis testing concerning μ_x) Let v be the plaintext, and we use *Fixed dataset A* (see Sect. 4.1) as a sample for L_t (i.e., L_t denotes the leakage at time sample t for one round encryption of the plaintext ABCDEF1234567890 with round key FEDCBA0123456789). Fix $t = 2368$, which gives the highest peak in Fig. 4.3 and corresponds to the computation of pLayer. Recall that in Example 4.2.1, we have calculated a sample mean of $\overline{l_{2368}} \approx 0.2132$ for L_{2368}. We would like to know if $\mu_{2368} = 0$. Following Eq. 1.68, we have null and alternative hypotheses given by

$$H_0 : \mu_{2368} = 0, \quad H_1 : \mu_{2368} \neq 0.$$

Suppose we know the variance is equal to the sample variance we have computed in Example 4.2.1, namely we assume

$$\sigma_{2368}^2 = 8.5196 \times 10^{-6}, \text{ which gives } \sigma_{2368} \approx 2.9188 \times 10^{-3}.$$

There are in total 5000 traces in *Fixed dataset A*. For a test with significance level $\alpha = 0.01$, the critical region is given by Eq. 1.69, with (see Eq. 1.72)

$$c = \frac{z_{\alpha/2}\sigma}{\sqrt{5000}} = 2.576 \times \frac{2.9188 \times 10^{-3}}{\sqrt{5000}} \approx 1.06 \times 10^{-4},$$

where $z_{0.005=2.576}$ (see Table 1.4). Since the sample mean

$$\overline{l_{2368}} \approx 0.2132 > c,$$

we reject the null hypothesis and conclude that $\mu_{2368} \neq 0$. The probability that our decision is wrong is given by $\alpha = 0.01$.

Example 4.2.7 (Example of one-sided hypothesis testing concerning μ_x) With the same notation as in Example 4.2.6, suppose we know that the mean of L_{2368}, μ_{2368}, is at least 0; we would like to know if it is bigger than 0. We set $\mu_0 = 0$ in Eq. 1.75 and get the following null hypothesis and the alternative hypothesis:

$$H_0 : \mu_{2368} = 0, \quad H_1 : \mu_{2368} > 0.$$

First, let us assume we know the variance is equal to the sample variance we computed in Example 4.2.1. There are in total 5000 traces in *Fixed dataset A*, and for a test with significance level $\alpha = 0.01$, the critical region is given by Eq. 1.76, with (see Eq. 1.77)

$$c = z_{0.01} \frac{\sigma_{2368}}{\sqrt{5000}} = \frac{2.326 \times 2.9188 \times 10^{-3}}{\sqrt{5000}} \approx 9.601 \times 10^{-5},$$

where $z_{0.01} = 2.326$ (see Table 1.4). Since our sample mean

$$\overline{l_{2368}} \approx 0.2132 > c,$$

we reject the null hypothesis and conclude that $\mu_{2368} > 0$. The probability that our decision is wrong is given by $\alpha = 0.01$.

Furthermore, we also would like to check how many traces are required for a test with significance level $\alpha = 0.01$. For this, we need to choose a value of c. Considering the value of the sample mean and sample variance, let us choose $c = 0.001$ in Eq. 1.78. According to Eq. 1.79, the number of traces to collect is then

$$\frac{\sigma_{2368}^2}{c^2} z_\alpha^2 = \frac{8.5196 \times 10^{-6}}{0.001^2} \times 2.326^2 \approx 46. \tag{4.11}$$

Now, suppose we do not know the variance σ_{2368}^2. Since the number of traces is big, according to Eq. 1.80, we compute

$$\sqrt{5000} \times \frac{\overline{l_{2368}}}{s_{2368}} = \sqrt{5000} \times \frac{0.2132}{2.9188 \times 10^{-3}} \approx 5165,$$

which is bigger than $z_{0.01} = 2.326$. Thus, we can reject the null hypothesis and conclude that $\mu_{2368} > 0$. The probability of a wrong decision is given by $\alpha = 0.01$. As for the number of traces needed, by Eq. 1.81, we will use the sample variance and reach the same result as in Eq. 4.11.

Example 4.2.8 (Example of two-sided hypothesis testing about μ_x and μ_y) The same as in Example 4.2.3, we take the leakages at $t = 2368$ from the *Fixed dataset B* as a sample for L'_{2368}. We have computed the sample mean and sample variance for this random variable, given in Eq. 4.8. We would like to know if the mean of L_{2368} (μ_{2368}) and the mean of L'_{2368} (μ'_{2368}) are the same. We set the following hypotheses (see Eq. 1.82):

$$H_0 : \mu'_{2368} = \mu_{2368}, \qquad H_1 : \mu'_{2368} \neq \mu_{2368}.$$

Assume we know the variances for both random variables are equal to the sample variances that we have computed (see Eqs. 4.5 and 1.82). There are in total 5000 traces in both *Fixed dataset A* and *Fixed dataset B*, and for a test with significance level $\alpha = 0.01$, the critical region is given by Eq. 1.83, with (see Eq. 1.86)

$$c = z_{0.005} \sqrt{\frac{\sigma_{2368}^2}{5000} + \frac{\sigma_{2368}'^2}{5000}} = 2.576 \times \sqrt{\frac{8.5196 \times 10^{-6} + 8.6198 \times 10^{-6}}{5000}} \approx 0.00015,$$

where $z_{0.005} = 2.576$ (see Table 1.4). Since our sample mean

$$\overline{l'_{2368}} - \overline{l_{2368}} \approx 0.0001 < c,$$

we accept the null hypothesis and conclude that $\mu'_{2368} = \mu_{2368}$. The probability that our decision is wrong is given by $\alpha = 0.01$.

Moreover, to check how many traces are needed for a test with significance level $\alpha = 0.01$, we choose $c = 0.001$ in Eq. 1.86. According to Eq. 1.87, the number of traces to collect is then

$$z_{\alpha/2}^2 \frac{\sigma_{2368}^2 + \sigma_{2368}'^2}{c^2} = 2.576^2 \times \frac{8.5196 \times 10^{-6} + 8.6198 \times 10^{-6}}{0.001^2} \approx 114. \qquad (4.12)$$

In case we do not know the variances, since the number of traces in both datasets is 5000, following the student's t-test, we compute (see Eq. 1.89)

$$\frac{|\overline{l'_{2368}} - \overline{l_{2368}}|}{\sqrt{\frac{s_{2368}^2 + s_{2368}'^2}{5000}}} = \frac{0.0001}{\sqrt{\frac{8.5196 \times 10^{-6} + 8.6198 \times 10^{-6}}{5000}}} \approx 1.7 < z_{0.005}.$$

We accept the null hypothesis and conclude that $\mu'_{2368} = \mu_{2368}$. The probability that our decision is wrong is given by $\alpha = 0.01$.

Set $c = 0.001$, then the number of traces needed for a student's t-test with significance level $\alpha = 0.01$ is given by (see Eq. 1.90)

$$z_{0.005}^2 \frac{s_{2368}^2 + s_{2368}'^2}{c^2} = 2.576^2 \times \frac{8.5196 \times 10^{-6} + 8.6198 \times 10^{-6}}{0.001^2} \approx 114.$$

Example 4.2.9 (Another example of two-sided hypothesis testing about μ_x and μ_y) Similar to Example 4.2.8, let us now look at a different time sample $t = 392$. We can compute the sample mean and sample variance of L_{392} with *Fixed dataset A*. They are given by

$$\overline{l_{392}} \approx -0.0525, \quad s_{392}^2 \approx 1.5141 \times 10^{-6}.$$

With *Fixed dataset B*, we get the sample mean and sample variance of L'_{392} as follows:

$$\overline{l'_{392}} \approx -0.0501, \quad s_{392}'^2 \approx 1.4801 \times 10^{-6}.$$

Similar to Example 4.2.8, we set the following hypotheses (see Eq. 1.82):

$$H_0 : \mu'_{392} = \mu_{392}, \quad H_1 : \mu'_{392} \neq \mu_{392}.$$

Let $\alpha = 0.01$. Then according to student's t-test with significance level α, we compute (see Eq. 1.89)

$$\frac{|\overline{l'_{392}} - \overline{l_{392}}|}{\sqrt{\frac{s^2_{392} + s'^2_{392}}{5000}}} = \frac{0.0024}{\sqrt{\frac{1.5141 \times 10^{-6} + 1.4801 \times 10^{-6}}{5000}}} \approx 98.1 > z_{0.005} \quad (z_{0.005} = 2.576).$$

We reject the null hypothesis and conclude that $\mu'_{392} \neq \mu_{392}$. The probability that our decision is wrong is given by $\alpha = 0.01$.

Set $c = 0.001$, then the number of traces needed for a student's t-test with significance level $\alpha = 0.01$ is given by (see Eq. 1.90)

$$z^2_{0.005} \frac{s^2_{392} + s'^2_{392}}{c^2} = 2.576^2 \times \frac{1.5141 \times 10^{-6} + 1.4801 \times 10^{-6}}{0.001^2} \approx 20.$$

Example 4.2.10 (Example of one-sided hypothesis testing about μ_x and μ_y) With the same notations as in Example 4.2.9, suppose we know that

$$\mu'_{392} \geq \mu_{392}.$$

We would like to know if $\mu'_{392} > \mu_{392}$. Then we have the following hypotheses:

$$H_0 : \mu'_{392} = \mu_{392}, \quad H_1 : \mu'_{392} > \mu_{392}.$$

Firstly, suppose we know the variances for both random variables are equal to the sample variances that we have computed. There are 5000 traces in both *Fixed dataset A* and *Fixed dataset B*. For a test with significance level $\alpha = 0.01$, the value of c in the critical region given by Eq. 1.92 is (see Eq. 1.93)

$$c = z_\alpha \sqrt{\frac{\sigma^2_{392} + \sigma'^2_{392}}{5000}} = 2.326 \times \sqrt{\frac{1.5141 \times 10^{-6} + 1.4801 \times 10^{-6}}{5000}} \approx 5.692 \times 10^{-5},$$

where $z_{0.01} = 2.326$. Since

$$\overline{l'_{392}} - \overline{l_{392}} = 0.0024 > c,$$

we reject the null hypothesis and conclude that $\mu'_{392} > \mu_{392}$. The probability of this choice being wrong is given by $\alpha = 0.01$.

Set $c = 0.001$. Then, the number of traces to collect for a hypothesis test with a level of significance $\alpha = 0.01$ ($z_\alpha = 2.326$) is given by (see Eq. 1.94)

$$\frac{z^2_\alpha (\sigma^2_{392} + \sigma'^2_{392})}{c^2} = \frac{2.326^2 \times (1.5141 \times 10^{-6} + 1.4801 \times 10^{-6})}{0.001^2} \approx 17.$$

Remark 4.2.2 According to Eq. 4.10,

$$X_t = X'_t, \quad \Longleftrightarrow \quad \mu_t = \mu'_t. \tag{4.13}$$

Then Example 4.2.8 concludes that when we take the signal to be part of the leakage related to the plaintext value, the signals at time sample 2368 for one round encryption of plaintexts ABCDEF1234567890 and 84216BA484216BA4 with the same round key FEDCBA0123456789 are very likely to be equal, according to our measurements *Fixed dataset A* and *Fixed dataset B*. The probability of the conclusions being wrong is 0.01. On the other hand, Example 4.2.9 concludes that the signals at time sample 392 are likely to be different (with a probability of 0.01 being wrong).

Furthermore, we see that to decide if the signals are different at a particular time sample with $c = 0.001^3$ and significance level 0.01 (i.e., probability of making wrong conclusions) we do not need that many traces.

Next, let us consider v being the 0th Sbox output in the first round of PRESENT. In this case, we can take L_t to be the leakages for a fixed value of v at time sample t and L'_t to be the leakages for another fixed value of v at t.

Example 4.2.11 When we consider v to be the 0th Sbox output in the first round of PRESENT, there are 16 different values of v that we can consider. Let L_t and L'_t denote the random variable for leakages corresponding to $v = 0$ and $v = F$ at time sample t. We would like to know if the signals at time sample $t = 392$ are the same for those two values of v.

Take the *Random dataset*. As mentioned in Example 4.2.4, we have 634 traces for $v = 0$ and 651 traces for $v = F$. We take those 634 (respectively, 651) traces as a sample for L_t (respectively, L'_t). The same as in Examples 4.2.8 and 4.2.9, we make the following hypotheses:

$$H_0 : \mu'_{392} = \mu_{392}, \quad H_1 : \mu'_{392} \neq \mu_{392}.$$

Firstly, we compute the sample means and sample variances for L_{392} and L'_{392}:

$$\overline{l_{392}} \approx -0.0425, \quad s^2_{392} \approx 2.2962 \times 10^{-6}, \quad \overline{l'_{392}} \approx -0.0539, \quad s'^2_{392} \approx 2.7378 \times 10^{-6}.$$

Let $\alpha = 0.01$. Then following student's t-test with significance level α, we compute (see Eq. 1.65)

---

[3] We note that this value of c can be considered as a precision.

$$s_p^2 = \frac{(634-1)s_{392}^2 + (651-1)s_{392}'^2}{634+651-2}$$

$$= \frac{633 \times 2.2962 \times 10^{-6} + 650 \times 2.7378 \times 10^{-6}}{1283} \approx 2.5199 \times 10^{-6}$$

and (see Eq. 1.88)

$$\frac{|\overline{l_{392}} - \overline{l'_{392}}|}{\sqrt{s_p^2(\frac{1}{634}+\frac{1}{651})}} \approx \frac{|-0.0425+0.0539|}{\sqrt{2.5199 \times 10^{-6} \times 3.1134 \times 10^{-3}}}$$

$$\approx 128.7 > z_{0.005} \quad (z_{0.005} = 2.576).$$

We reject the null hypothesis and conclude that $\mu'_{392} \neq \mu_{392}$. The probability that our decision is wrong is given by $\alpha = 0.01$.

Remark 4.2.3 Note that in Examples 4.2.8 and 4.2.9, the sample sizes (number of traces) are the same (both are 5000), but in Example 4.2.11, the sample sizes are different for L_t and L'_t. Thus, instead of using Eq. 1.89 as in Examples 4.2.8 and 4.2.9, we applied Eq. 1.88. But those two equations are the same when the sample sizes are equal.

We have seen before that the leakage L_t is dependent on the data being processed in the device. In fact, as mentioned at the beginning of Sect. 4.2, some SCA attacks (see Sects. 4.3.1, 4.3.3, and 4.4.2) exploit the dependency of the leakage on certain intermediate values. If the leakage is *not* exploitable, we would expect, at least, that the signals at time sample t should be the same when the only difference is the values of the data being processed. With our notations above, this means that we would like to test if $X_t = X'_t$, or equivalently $\mu_t = \mu'_t$ (see Remark 4.2.2), for two different fixed values of a certain intermediate value v.

Example 4.2.12 Continuing Remark 4.2.2 and the above discussion, we can conclude that the leakages at time sample 392 for our implementation of PRESENT on our DUT are very likely to be vulnerable to SCA attacks.

Another approach to analyzing whether the leakage is exploitable is to consider the signals for a fixed value of v and that for random values of v. Let L_t^r denote the random variable corresponding to the leakage at time sample t for encryptions corresponding to random values of v. Let X_t^r and N_t^r be the random variables for the corresponding signal and noise. We have

$$L_t^r = X_t^r + N_t^r.$$

With our assumptions and modeling, the signal X_t is a constant for a fixed value of v at time t. When the value of v is random, X_t^r is itself a random variable that varies depending on v. It is not easy to approximate the distribution induced by L_t^r

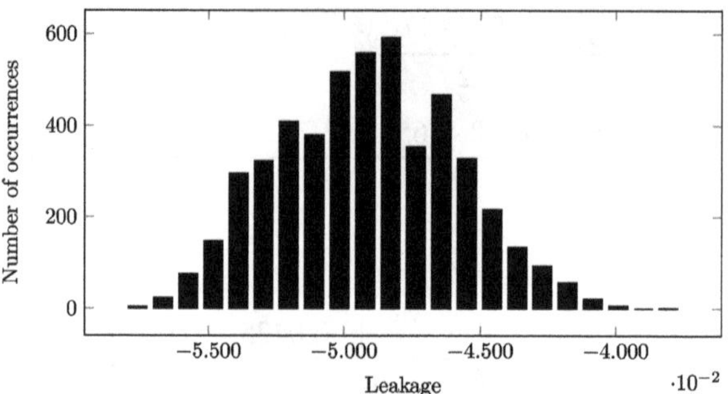

Fig. 4.10 Histogram of leakages at time sample $t = 392$ across 5000 traces from the *Random plaintext dataset*

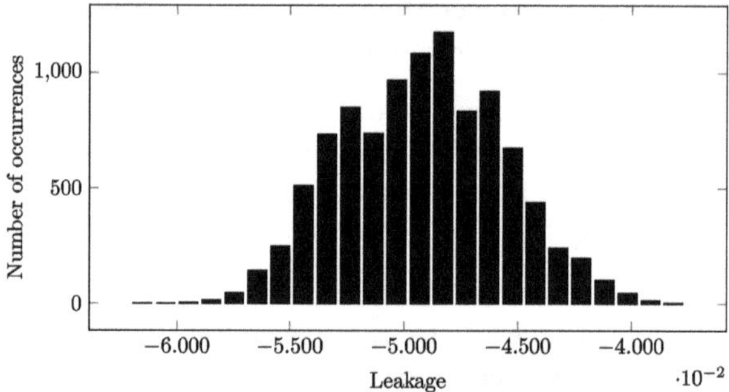

Fig. 4.11 Histogram of leakages at time sample $t = 392$ across 10,000 traces from the *Random dataset*

in this case. However, following the convention, we still use a normal distribution for the approximation.

To see that this makes sense, let us take the *Random plaintext dataset* and plot the histogram of leakages at $t = 392$ across 5000 traces from this dataset. We get Fig. 4.10. This corresponds to random values of v when v is taken to be the plaintext. As another example, the histogram for leakages at $t = 392$ across the 10,000 traces from the *Random dataset* is shown in Fig. 4.11. In this case, we can consider the leakage corresponds to random values of v when v is taken to be the 0th Sbox output. Those two figures demonstrate that it is reasonable to approximate the distribution induced by the leakage L_t^r with a normal distribution.

Suppose

$$L_t^r \sim \mathcal{N}(\mu_t^r, \sigma_t^{r2}).$$

Since the noise is independent of the signal, we have $N_t = N_t^r$. By Eqs. 1.33, 1.36, and 4.9,

$$\mu_t^r = \mathrm{E}\left[X_t^r\right] + \mathrm{E}\left[N_t^r\right], \quad \sigma_t^{r2} = \mathrm{Var}(X_t^r) + \mathrm{Var}(N_t^r) = \mathrm{Var}(X_t^r) + \sigma_t^2.$$

We have

$$\mu_t - X_t = \mu_t^r - \mathrm{E}\left[X_t^r\right], \quad \sigma_t^{r2} \neq \sigma_t^2. \tag{4.14}$$

Same as before, in case the leakage is not exploitable at time sample t, we expect the signal to be a constant at t, namely

$$X_t = X_t^r, \quad \text{and equivalently} \quad \mu_t^r = \mu_t. \tag{4.15}$$

Consequently, our hypotheses will be the same as in Examples 4.2.8, 4.2.9, and 4.2.11. The difference is that in this case, $\sigma_t^{r2} \neq \sigma_t^2$. For this reason, we apply Welch's t-test instead of the student's t-test.

Example 4.2.13 We consider the signal given by the plaintext value, i.e., $v =$ plaintext. Let $t = 392$. Then we can take *Fixed dataset A* as a sample for L_{392} and *Random plaintext dataset* as a sample for L_{392}^r We know that (see Example 4.2.11)

$$\overline{l_{392}} \approx -0.0525, \quad s_{392}^2 \approx 1.5141 \times 10^{-6}.$$

We can also compute

$$\overline{l_{392}^r} \approx -0.0488, \quad s_{392}^{r2} \approx 1.1700 \times 10^{-5}.$$

We would like to test if $\mu_t = \mu_t^r$. Thus, we set the following hypotheses:

$$H_0 : \mu_{392}^r = \mu_{392}, \quad H_1 : \mu_{392}^r \neq \mu_{392}.$$

Let $\alpha = 0.01$. Then following Welch's t-test with significance level α, we compute (see Eq. 1.91)

$$\frac{|\overline{l_{392}} - \overline{l_{392}^r}|}{\sqrt{\frac{s_{392}^2}{5000} + \frac{s_{392}^{r2}}{5000}}} = \frac{|-0.0525 + 0.0488|}{\sqrt{\frac{1.5141 \times 10^{-6}}{5000} + \frac{1.1700 \times 10^{-5}}{5000}}} \approx 72.0 > z_{0.005}.$$

We reject the null hypothesis and conclude that $\mu_{392}^r \neq \mu_{392}$. The probability that our decision is wrong is equal to $\alpha = 0.01$.

Example 4.2.14 Now we consider the signal to be given by the 0th Sbox output. For the fixed signal we choose $v = 0$. Take the *Random dataset*. Let L_t and L_t^r denote the random variables corresponding to leakages for $v = 0$ and random values of v at time sample t, respectively.

We know that there are 634 traces for $v = 0$. Fix $t = 392$. In Example 4.2.11 we have computed

$$\overline{l_{392}} \approx -0.0425, \quad s_{392}^2 \approx 2.2962 \times 10^{-6}.$$

For the random values of $\boldsymbol{v}$, we can take the whole dataset, which contains 10,000 traces, as a sample for L_t^r. We have

$$\overline{l_{392}^r} \approx -0.0487, \quad s_{392}^{r2} \approx 1.1624 \times 10^{-5}.$$

Let $\alpha = 0.01$. Then according to Welch's t-test with significance level α, we compute (see Eq. 1.91)

$$\frac{|\overline{l_{392}} - \overline{l_{392}^r}|}{\sqrt{\frac{s_{392}^2}{634} + \frac{s_{392}^{r2}}{5000}}} = \frac{|-0.0425 + 0.0487|}{\sqrt{\frac{2.2962 \times 10^{-6}}{634} + \frac{1.1624 \times 10^{-5}}{10,000}}} \approx 89.6 > z_{0.005}.$$

We reject the null hypothesis and conclude that $\mu_{392}^r \neq \mu_{392}$. The probability that our decision is wrong is $\alpha = 0.01$.

The rationale of the TVLA methodology is that if the leakage is not exploitable, the encryptions corresponding to two different intermediate values (or the encryption corresponding to one fixed intermediate value and that to a random intermediate value) should exhibit identical signals. Then according to Eq. 4.13 (or Eq. 4.15), the corresponding leakages will have the same means. With the help of the student's t-test (or Welch's t-test), we make hypotheses about means of leakages and test if they are equal.

Recall that for student's t-test and Welch's t-test (when the sample size is big), we need to choose a significance level α and compare computations using our samples with a threshold $z_{\alpha/2}$ (see Eqs. 1.88 and 1.91). For TVLA, following the convention, we set $z_{\alpha/2} = 4.5$. By Eq. 1.43, this threshold corresponds to

$$\frac{\alpha}{2} = 1 - \Phi(z_\alpha) = 1 - \Phi(4.5) = 1 - 0.9999966023268753 \approx 3.4 \times 10^{-6}.$$

The significance level is given by

$$\alpha \approx 6.8 \times 10^{-6}.$$

This means that there is a 6.8×10^{-4} percent chance that we would reject the null hypothesis (i.e., conclude that the means are different) in case it is true (i.e., the means are in fact the same).

The steps for TVLA are as follows:

TVLA Step 1 **Identify the cryptographic implementation for analysis**. In principle, TVLA can be used for analyzing leakages of implementations for any type of algorithm. In practice, they are mostly used for the analysis of symmetric block cipher implementations.

TVLA Step 2 **Choose the intermediate value v**. The choice of v determines how we measure our traces. TVLA tests if different values of v result in different signals.

TVLA Step 3 **Experimental setup and measure leakages**. As we can imagine, for the actual attacks, experimental setups are crucial factors for success. For leakage assessment, it would be better to carry out measurements with equipment that is expected to be used by attackers that we would like to protect against.

We will prepare two datasets, denoted by $\mathcal{T}_1$ and $\mathcal{T}_2$. To get the first dataset $\mathcal{T}_1$, we choose a fixed value for v. Then we randomly take M_1 inputs for the cryptographic implementation such that the value of v is equal to this fixed value. One trace is taken for each input.

For the second dataset $\mathcal{T}_2$, there are two options.

(a) *Fixed versus fixed*. Choose a different fixed value for v. Then randomly take M_2 inputs for the cryptographic implementation such that the value of v is equal to this fixed value. One trace is collected for each input.

(b) *Fixed versus random*. Randomly take M_2 inputs for the cryptographic implementation so that the value of v is random. One trace is collected for each input.

Let us represent those two sets of traces as follows:

$$\mathcal{T}_1 = \{\boldsymbol{\ell}_1^{(1)}, \boldsymbol{\ell}_2^{(1)}, \ldots, \boldsymbol{\ell}_{M_1}^{(1)}\}, \quad \mathcal{T}_2 = \{\boldsymbol{\ell}_1^{(2)}, \boldsymbol{\ell}_2^{(2)}, \ldots, \boldsymbol{\ell}_{M_2}^{(2)}\}.$$

Each trace $\boldsymbol{\ell}_j^{(i)} = \left(l_{j1}^{(i)}, l_{j2}^{(i)}, \ldots, l_{jq}^{(i)}\right)$ contains q time samples ($i = 1, 2$).

For our illustrations, we will consider two choices of v—the plaintext and the 0th Sbox output. When v is given by the plaintext, we take

$$\mathcal{T}_1 = \textit{Fixed dataset A}, \quad \mathcal{T}_2 = \textit{Fixed dataset B}$$

for the fixed versus fixed setting and

$$\mathcal{T}_1 = \textit{Fixed dataset A}, \quad \mathcal{T}_2 = \textit{Random plaintext dataset}$$

for the fixed versus random setting. For both cases, we will demonstrate the results for $M_1 = M_2 = 5000$ and $M_1 = M_2 = 50$.

When v is given by the 0th Sbox output, we take

$$\mathcal{T}_1 = \text{traces in } \textit{Random dataset} \text{ for } v = 0,$$

$$\mathcal{T}_2 = \text{traces in } \textit{Random dataset} \text{ for } v = \mathbf{F}$$

for the fixed versus fixed setting. As discussed in Example 4.2.4, $M_1 = 634$, $M_2 = 651$. For the fixed versus random setting, we choose

$$\mathcal{T}_1 = \text{traces in } \textit{Random dataset} \text{ for } v = 0, \quad \mathcal{T}_2 = \textit{Random dataset}$$

and $M_1 = 634$, $M_2 = 10{,}000$. For all our traces, $q = 3600$.

TVLA Step 4   **t-Test for one time sample.** Fix a time sample t. Let $L_t^{(1)}$ and $L_t^{(2)}$ denote the random variable corresponding to leakages at time sample t for computations resulting in datasets $\mathcal{T}_1$ and $\mathcal{T}_2$, respectively. Suppose

$$L_t^{(1)} \sim \mathcal{N}(\mu_t^{(1)}, \sigma_t^{(1)2}), \quad L_t^{(2)} \sim \mathcal{N}(\mu_t^{(2)}, \sigma_t^{(2)2}).$$

By definition (see Eqs. 1.49 and 1.53), we compute the sample mean and sample variance for $L_t^{(1)}$ (respectively, $L_t^{(2)}$), denoted by $\overline{l_t^{(1)}}$ and $s_t^{(1)2}$ (respectively, $\overline{l_t^{(2)}}$ and $s_t^{(2)2}$):

$$\overline{l_t^{(1)}} = \frac{1}{M_1} \sum_{j=1}^{M_1} l_{jt}^{(1)}, \quad \overline{l_t^{(2)}} = \frac{1}{M_2} \sum_{j=1}^{M_2} l_{jt}^{(2)},$$

and

$$s_t^{(1)2} = \frac{1}{M_1 - 1} \sum_{j=1}^{M_1} \left(l_{jt}^{(1)} - \overline{l_t^{(1)}} \right)^2, \quad s_t^{(2)2} = \frac{1}{M_2 - 1} \sum_{j=1}^{M_2} \left(l_{jt}^{(2)} - \overline{l_t^{(2)}} \right)^2.$$

Then we propose the following null and alternative hypotheses:

$$H_0 : \mu_t^{(1)} = \mu_t^{(2)}, \quad H_1 : \mu_t^{(1)} \neq \mu_t^{(2)}. \tag{4.16}$$

Depending on our setting, we choose between the student's t-test and Welch's t-test. As we have discussed above, for the fixed versus fixed setting, the noise for both cases is assumed to be the same and (see Eq. 4.10)

$$\sigma_t^{(1)2} = \sigma_t^{(2)2},$$

hence the usage of student's t-test. In the fixed versus random setting, the noises are different, and we have (see Eq. 4.14)

$$\sigma_t^{(1)2} \neq \sigma_t^{(2)2},$$

hence the application of Welch's t-test.

(a) *Student's t-test for the fixed versus fixed setting.* When the second dataset $\mathcal{T}_2$ is measured according to the fixed versus fixed setting, following the student's t-test, we compute (see Eq. 1.65)

$$s_p^2 = \frac{(M_1 - 1)s_t^{(1)2} + (M_2 - 1)s_t^{(2)2}}{M_1 + M_2 - 2}$$

and (see Eq. 1.88)

$$t - \text{value}_t := \frac{\overline{l_t^{(1)}} - \overline{l_t^{(2)}}}{\sqrt{s_p^2(1/M_1 + 1/M_2)}}. \tag{4.17}$$

(b) *Welch's t-test for the fixed versus random setting.* When the second dataset $\mathcal{T}_2$ is measured according to the fixed versus random setting, following Welch's t-test, we compute

$$t - \text{value}_t := \frac{\overline{l_t^{(1)}} - \overline{l_t^{(2)}}}{\sqrt{\frac{s_t^{(1)2}}{M_1} + \frac{s_t^{(2)2}}{M_2}}}. \tag{4.18}$$

Then we compare the t-value$_t$ with our threshold 4.5. In case

$$t - \text{value}_t > 4.5, \quad \text{or} \quad t - \text{value}_t < -4.5,$$

we reject the null hypothesis. Following the previous discussions, this means that the signals at time sample t are different for computations with two fixed values of v (or for a fixed value of v and random values of v). We conclude that there is a high chance that data-dependent leakage appears at time sample t.

TVLA Step 5 **Repeat TVLA Step 4 for all time samples** t.

We note that when the t-value is between -4.5 and 4.5 for all time samples $1, 2, \ldots, q$, we cannot conclude that the implementation is safe. As there might be other attacks that do not exploit the dependency of leakages on the chosen v.

Now, we show some results of the TVLA on our datasets. Let us first take v to be the plaintext. For the fixed versus fixed setting, we take *Fixed dataset A* and *Fixed dataset B* as samples for our analysis. The t-values with the student's t-test (Eq. 4.17) are shown in Fig. 4.12, where we have used the entire datasets and $M_1 = M_2 = 5000$. We can see that most of the time samples have t-values outside

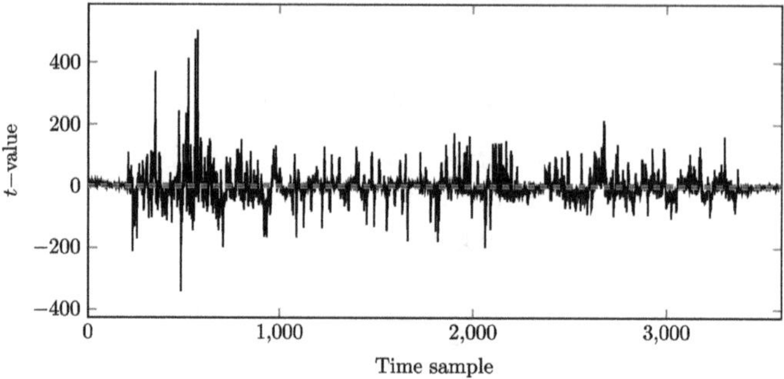

Fig. 4.12 t-Values (Eq. 4.17) for all time samples $1, 2, \ldots, 3600$ computed with *Fixed dataset A* and *Fixed dataset B*. The signal is given by the plaintext value, and the fixed versus fixed setting is chosen. Blue dashed lines correspond to the threshold 4.5 and -4.5

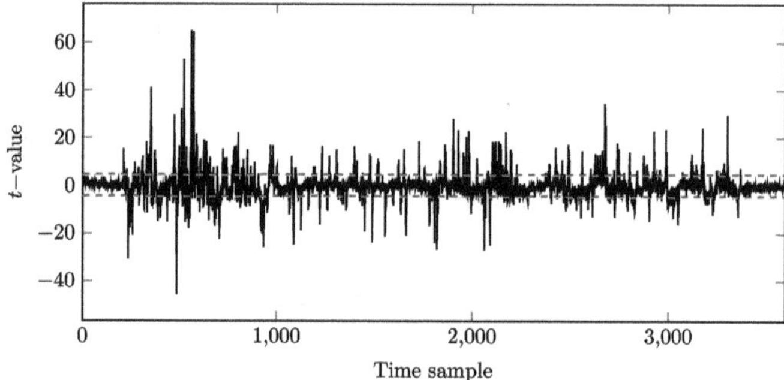

Fig. 4.13 t-Values (Eq. 4.17) for all time samples $1, 2, \ldots, 3600$ computed with 50 traces from *Fixed dataset A* and 50 traces from *Fixed dataset B*. The signal is given by the plaintext value, and the fixed versus fixed setting is chosen. Blue dashed lines correspond to the threshold 4.5 and -4.5

of the threshold. This is not surprising as the implementation does not have any countermeasures. In Sect. 4.3.1 we will see that using this implementation, with just a few traces, we can recover the first round key. If we reduce the number of traces for computing the t-values, we will get different results. For example, when we take 50 traces, i.e., $M_1 = M_2 = 50$, we have Fig. 4.13. Compared to Fig. 4.12, the absolute values of t-values are much smaller. This shows that when the sample size is bigger, it is more likely for us to capture information about the inputs from the leakages.

For the fixed versus random setting, t-values with Welch's t-test (Eq. 4.18) are computed with *Fixed dataset A* and *Random plaintext dataset*. The results are shown in Figs. 4.14 and 4.15 for $M_1 = M_2 = 5000$ and $M_1 = M_2 = 50$, respectively. Similarly, we also observe higher $|t|$-values with more traces. Furthermore, compared to Figs. 4.12 and 4.13, the $|t|$-values are much lower. This shows that it is more likely for us to distinguish between leakages corresponding

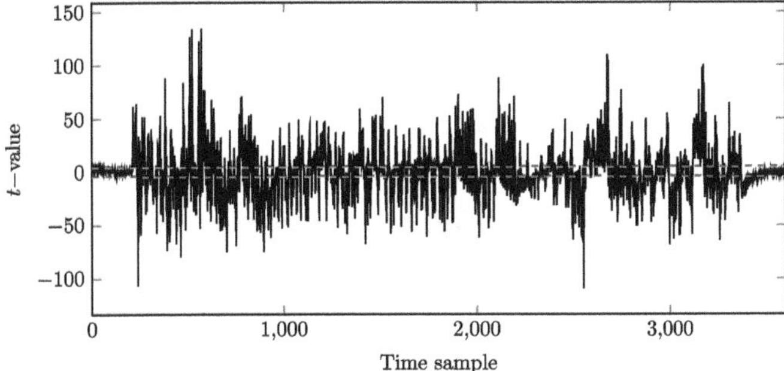

Fig. 4.14 t-Values (Eq. 4.18) for all time samples $1, 2, \ldots, 3600$ computed with *Fixed dataset A* and *Random plaintext dataset*. The signal is given by the plaintext value, and the fixed versus random setting is chosen. Blue dashed lines correspond to the threshold 4.5 and -4.5

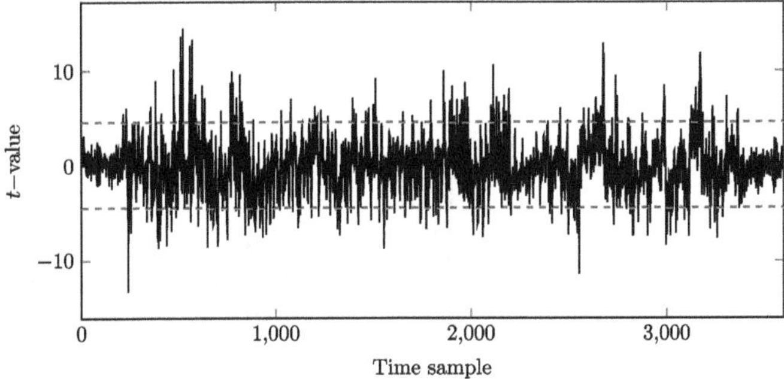

Fig. 4.15 t-Values (Eq. 4.18) for all time samples $1, 2, \ldots, 3600$ computed with 50 traces from *Fixed dataset A* and 50 traces from *Random plaintext dataset*. The signal is given by the plaintext value, and the fixed versus random setting is chosen. Blue dashed lines correspond to the threshold 4.5 and -4.5

to two fixed plaintexts rather than between leakages for a fixed plaintext and for random plaintexts.

Next, we take v to be the 0th Sbox output. We use *Random dataset* as samples for our random variables corresponding to leakages. For the fixed versus fixed setting, we take the $M_1 = 634$ traces for $v = \textbf{0}$ as $\mathcal{T}_1$ and $M_2 = 651$ traces for $v = F$ as $\mathcal{T}_2$. The t-values with the student's t-test (Eq. 4.17) are shown in Fig. 4.16. For the fixed versus random setting, we take the $M_1 = 634$ traces for $v = \textbf{0}$ as $\mathcal{T}_1$ and the whole dataset as $\mathcal{T}_2$ ($M_2 = 10,000$). Following Welch's t-test, the t-values (Eq. 4.18) are shown in Fig. 4.17. Again, we also show the results when fewer traces are used for the computations. The t-values can be found in Figs. 4.18 and 4.19.

In summary, we have the following observations:

- When more traces are used (i.e., when the sample size is bigger), it is more likely for us to capture information about the intermediate values from the leakages.

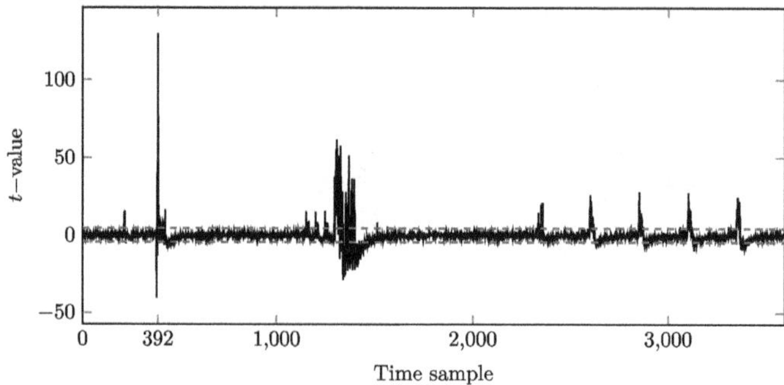

Fig. 4.16 t-Values (Eq. 4.17) for all time samples $1, 2, \ldots, 3600$ computed with traces from *Random dataset*. $\mathcal{T}_1$ contains $M_1 = 634$ traces and $\mathcal{T}_2$ contains $M_2 = 651$ traces. The signal is given by the 0th Sbox output, and the fixed versus fixed setting is chosen. Blue dashed lines correspond to the threshold 4.5 and -4.5

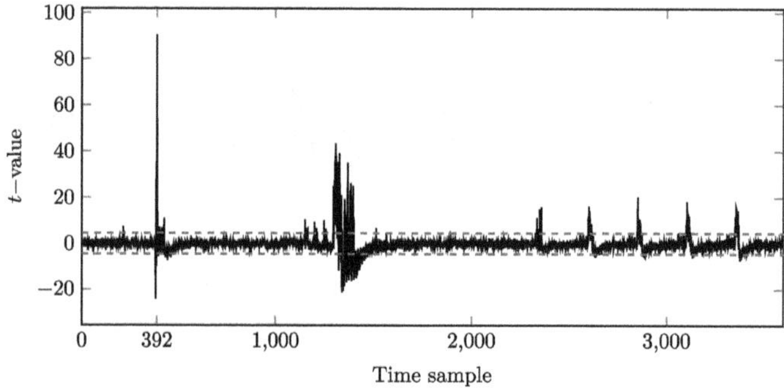

Fig. 4.17 t-Values (Eq. 4.18) for all time samples $1, 2, \ldots, 3600$ computed with traces from *Random dataset*. $\mathcal{T}_1$ contains $M_1 = 634$ traces and $\mathcal{T}_2$ contains $M_2 = 10,000$ traces. The signal is given by the 0th Sbox output, and the fixed versus random setting is chosen. Blue dashed lines correspond to the threshold 4.5 and -4.5

We will see in Sect. 4.3.2.4 that more traces indeed indicate higher chances for the attacks to be successful.

- When v is given by the 0th Sbox output, the highest $|t|$-value is obtained at 392 for all cases we have analyzed. We will see that this is the *point of interest (POI)* for our attack (Sect. 4.3.2).
- Compared to v being the plaintext, the $|t|$-values are in general smaller with much fewer time samples crossing the threshold when v is given by the 0th Sbox output. This is unsurprising as we would expect more computations to be correlated with the plaintext rather than a single Sbox output.

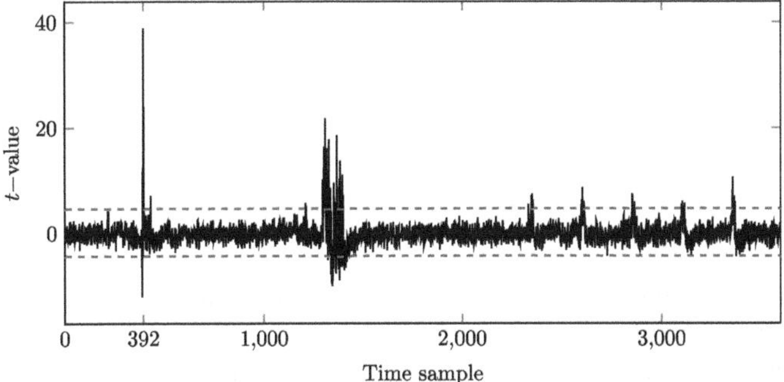

Fig. 4.18 t-Values (Eq. 4.17) for all time samples $1, 2, \ldots, 3600$ computed with traces from *Random dataset*. Both $\mathcal{T}_1$ and $\mathcal{T}_2$ contain 50 traces (i.e., $M_1 = M_2 = 50$). The signal is given by the 0th Sbox output, and the fixed versus fixed setting is chosen. Blue dashed lines correspond to the threshold 4.5 and -4.5

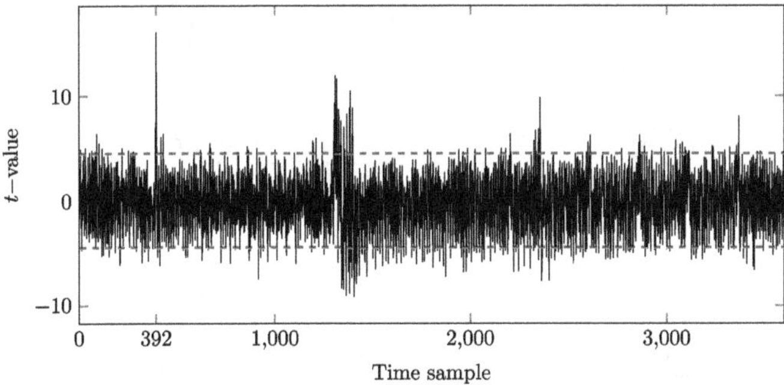

Fig. 4.19 t-Values (Eq. 4.17) for all time samples $1, 2, \ldots, 3600$ computed with traces from *Random dataset*. Both $\mathcal{T}_1$ and $\mathcal{T}_2$ contain 50 traces (i.e., $M_1 = M_2 = 50$). The signal is given by the 0th Sbox output, and the fixed versus random setting is chosen. Blue dashed lines correspond to the threshold 4.5 and -4.5

4.2.4 Signal-to-Noise Ratio

Signal-to-noise ratio (SNR) is commonly used in electrical engineering and signal processing, and the general definition is

$$
\text{SNR} = \frac{\text{Var(signal)}}{\text{Var(noise)}},
$$

where Var refers to the variance of a random variable (see Eq. 1.35).

In our case, for a fixed time sample t, X_t represents the signal, which is part of the leakage relevant to our attack. And the SNR at time t is given by

$$\mathrm{SNR}_t = \frac{\mathrm{Var}(X_t)}{\mathrm{Var}(N_t)}. \tag{4.19}$$

$\mathrm{Var}(X_t)$ measures how much the leakage varies at time sample t due to the signal. $\mathrm{Var}(N_t)$ measures how much the leakage varies due to the noise. Thus, SNR quantifies how much information is leaked at time sample t from the measurements. The higher the SNR, the lower the noise.

Example 4.2.15 Suppose we are interested in the Hamming weight of an 8-bit intermediate value at time sample t. In particular, the intermediate value we would like to analyze is from $\mathbb{F}_2^8$. We further assume that the leakage L_t is equal to the modeled leakage following the Hamming weight leakage model (Eq. 4.4). Thus $X_t = \mathrm{wt}(\boldsymbol{v})$ for $\boldsymbol{v} \in \mathbb{F}_2^8$. Then the variance of the signal is given by $\mathrm{Var}(\mathrm{wt}(\boldsymbol{v}))$ for $\boldsymbol{v} \in \mathbb{F}_2^8$. By definition (Eq. 1.31),

$$\mathrm{E}\,[\mathrm{wt}\,(\boldsymbol{v})] = \frac{1}{|\mathbb{F}_2^8|} \sum_{\boldsymbol{v} \in \mathbb{F}_2^8} \mathrm{wt}\,(\boldsymbol{v}) = \frac{1}{2^8} \sum_{i=1}^{8} i \binom{8}{i} = \frac{1}{2^8} \sum_{i=1}^{8} \frac{8!}{(i-1)!(8-i)!}$$

$$= \frac{8}{2^8} \sum_{i=1}^{8} \frac{7!}{(i-1)!(7-(i-1))!} = \frac{8}{2^8} \sum_{j=0}^{7} \binom{7}{j} = \frac{8 \times 2^7}{2^8} = 4.$$

And

$$\mathrm{E}\left[\mathrm{wt}\,(\boldsymbol{v})^2\right] = \frac{1}{|\mathbb{F}_2^8|} \sum_{\boldsymbol{v} \in \mathbb{F}_2^8} \mathrm{wt}\left(\boldsymbol{v}^2\right) = \frac{1}{2^8} \sum_{i=1}^{8} i^2 \binom{8}{i} = \frac{1}{2^8} \sum_{i=1}^{8} i \frac{8!}{(i-1)!(8-i)!}$$

$$= \frac{8}{2^8} \left(\sum_{i=1}^{8} (i-1) \frac{7!}{(i-1)!(8-i)!} + \sum_{i=1}^{8} \frac{7!}{(i-1)!(7-(i-1))!} \right)$$

$$= \frac{1}{2^5} \left(7 \sum_{i=2}^{8} \frac{6!}{(i-2)!(6-(i-2))!} + \sum_{j=0}^{7} \binom{7}{j} \right)$$

$$= \frac{1}{2^5} \left(2^7 + 7 \sum_{j=0}^{6} \binom{6}{j} \right)$$

$$= \frac{1}{2^5} (2^7 + 7 \times 2^6) = 2^2 + 7 \times 2 = 18.$$

By Eq. 1.35,

$$\mathrm{Var}(\mathrm{wt}\,(\boldsymbol{v})) = \mathrm{E}\left[\mathrm{wt}\,(\boldsymbol{v})^2\right] - \mathrm{E}\,[\mathrm{wt}\,(\boldsymbol{v})]^2 = 18 - 4^2 = 2.$$

Let σ_t^2 denote the variance of the noise N_t. We have

$$\text{SNR} = \frac{\text{Var}(X_t)}{\text{Var}(N_t)} = \frac{\text{Var}(\text{wt}(\boldsymbol{v}))}{\sigma_t^2} = \frac{2}{\sigma_t^2}.$$

Example 4.2.16 In this example, let L_t denote the random variable corresponding to the leakage of one round of PRESENT encryption at time t. We take the *Random dataset* (see Sect. 4.1) as a sample for L_t. Suppose we are interested in the exact value of the 0th Sbox output in the first round of PRESENT. Let us denote this intermediate value by $\boldsymbol{v}$.

Fix a time sample t. X_t is given by the part of the leakage related to the value of $\boldsymbol{v}$. To compute $\text{Var}(X_t)$, we first divide the traces in *Random dataset* into 16 sets according to the value of $\boldsymbol{v}$. Let us denote those 16 sets of traces by $A_1, A_2, \ldots, A_{16}$, where A_s contains traces corresponding to $\boldsymbol{v} = s - 1$.

As discussed in Sect. 4.2.1, for a fixed value of $\boldsymbol{v}$, X_t is a constant, and the leakage and the noise can be modeled by normal random variables. Let $L_{t,s}$ and $N_{t,s}$ denote the random variables corresponding to leakage and noise at time sample t for $\boldsymbol{v} = s - 1$. Let $X_{t,s}$ denote the constant leakage in this case.

Similar to Example 4.2.1, we can approximate the mean of $L_{t,s}$ using sample mean computed with set A_s. For example, take $t = 600$, and we have

$$\overline{l_{600,1}} \approx 0.08212, \quad \overline{l_{600,2}} \approx 0.08221, \quad \overline{l_{600,3}} \approx 0.08209, \quad \ldots$$

By Eq. 4.3, for any s,

$$X_{t,s} = \text{E}\left[L_{t,s}\right] - \text{E}\left[N_{t,s}\right]. \tag{4.20}$$

Variance of X_t is given by variance of $X_{t,s}$ values, and we have

$$\text{Var}(X_t) = \text{Var}\big(\text{E}\left[L_{t,s}\right] - \text{E}\left[N_{t,s}\right]\big).$$

Since any information related to $\boldsymbol{v}$ is contained in X_t, which is independent of N_t, $\text{E}\left[N_{t,s}\right]$ is a constant for all s. We have (see Eq. 1.36)

$$\text{Var}(X_t) = \text{Var}\big(\text{E}\left[L_{t,s}\right]\big),$$

which can be estimated with the sample variance of $\text{E}\left[L_{t,s}\right]$. For $t = 600$, we have

$$s_{X_{600}}^2 \approx 1.0088 \times 10^{-8}.$$

By Eq. 4.2,

$$\text{Var}(N_t) = \text{Var}(L_t - X_t).$$

On the other hand, since $\mathrm{E}\left[N_{t,s}\right]$ is a constant for different values of s, by Eqs. 1.36 and 4.20,

$$\mathrm{Var}\left(L_t - \mathrm{E}\left[L_{t,s}\right]\right)=\mathrm{Var}\left(L_t - X_{t,s} - \mathrm{E}\left[N_{t,s}\right]\right) = \mathrm{Var}\left(L_t - X_{t,s}\right) = \mathrm{Var}(L_t - X_t).$$

Thus $\mathrm{Var}(N_t)$ can be approximated by the sample variance of $L_t - \mathrm{E}\left[L_{t,s}\right]$. For $t = 600$, we have

$$s^2_{N_{600}} \approx 6.4184 \times 10^{-6}.$$

And the SNR at time sample 600 is given by

$$\mathrm{SNR}_{600} = \frac{\mathrm{Var}(X_{600})}{\mathrm{Var}(N_{600})} \approx \frac{s^2_{X_{600}}}{s^2_{N_{600}}} = \frac{1.0088 \times 10^{-8}}{6.4184 \times 10^{-6}} \approx 0.00157.$$

Example 4.2.17 For now, we have discussed the definition of SNR for one point in time. With the same method as in Example 4.2.16, we can compute the sample variance for $\mathrm{Var}(X_t)$ and $\mathrm{Var}(N_t)$, as well as SNR values for all time samples. They are shown in Figs. 4.20, 4.21, and 4.22, respectively.

We can see that the shape of variance of noise has similarities to one round of PRESENT computations (e.g., Fig. 4.3). This is reasonable since most of the leakage is not related to v.

Furthermore, the peaks for the variance of signal and SNR correspond to each other. The first two peaks are likely related to AddRoundKey and sBoxLayer. The peaks after 1000 are probably caused by the permutation of 4 bits of v (the 0th Sbox output). These observations can be confirmed by comparing them to Fig. 4.3. In particular, we can deduce that the peak at $t = 392$ is related to the 0th Sbox computation—as observed in Fig. 4.3, sBoxLayer starts from around time sample 382.

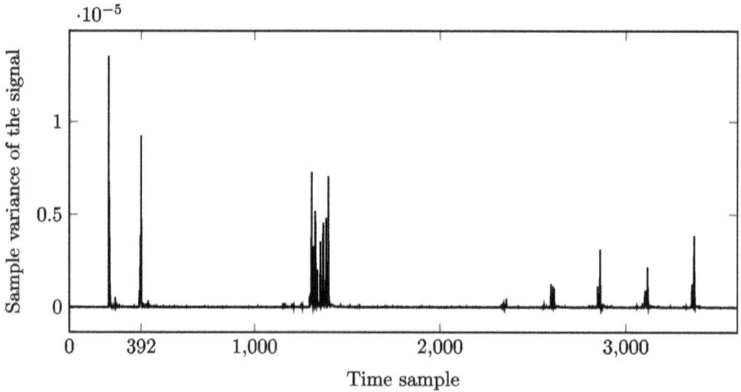

Fig. 4.20 Sample variance of the signal for each time sample, computed using *Random dataset*. The signal is given by the exact value of the 0th Sbox output

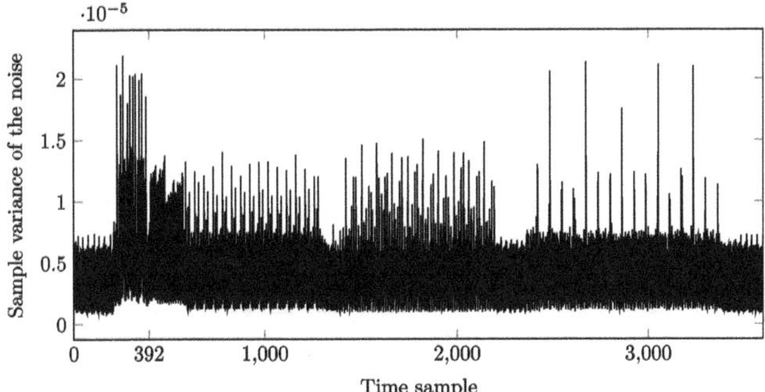

Fig. 4.21 Sample variance of the noise for each time sample, computed using *Random dataset*. The signal is given by the exact value of the 0th Sbox output

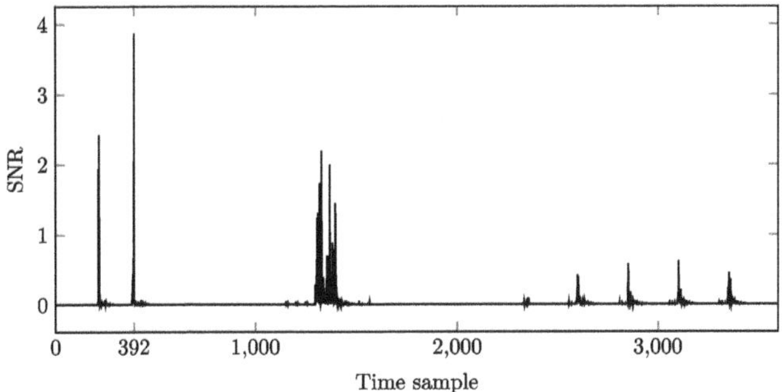

Fig. 4.22 SNR for each time sample, computed using *Random dataset*. The signal is given by the exact value of the 0th Sbox output

Example 4.2.18 We again look at the *Random dataset*. Instead of the exact values of v as in Example 4.2.16, we focus on the Hamming weight of the 0th Sbox output, i.e., $\mathrm{wt}\,(v)$. Then, in this case, for a fixed time sample t, we divide the traces into five sets according to the value of $\mathrm{wt}\,(v)$. Let us denote those five sets of traces by $A_1, A_2, \ldots, A_5$, where A_s contains traces corresponding to $\mathrm{wt}\,(v) = s - 1$. Following similar computations as in Example 4.2.16, for $t = 600$, we have

$$\overline{l_{600,1}} \approx 0.08212, \quad \overline{l_{600,2}} \approx 0.08206, \quad \overline{l_{600,3}} \approx 0.08214,$$

$$\overline{l_{600,4}} \approx 0.08211, \quad \overline{l_{600,5}} \approx 0.08206.$$

And

$$s^2_{X_{600}} \approx 1.1043 \times 10^{-9}, \quad s^2_{N_{600}} \approx 6.4271 \times 10^{-5}, \quad \mathrm{SNR}_{600} \approx 0.0001718.$$

The results for all time samples are shown in Figs. 4.23, 4.24, and 4.25.

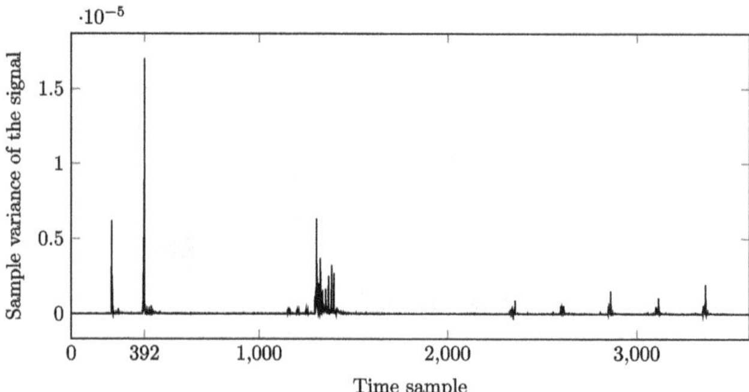

Fig. 4.23 Sample variance of the signal for each time sample, computed using *Random dataset*. The signal is given by the Hamming weight of the 0th Sbox output

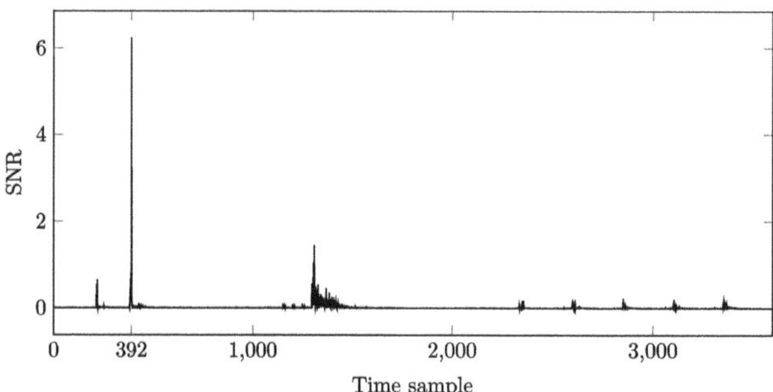

Fig. 4.24 SNR for each time sample, computed using *Random dataset*. The signal is given by the Hamming weight of the 0th Sbox output

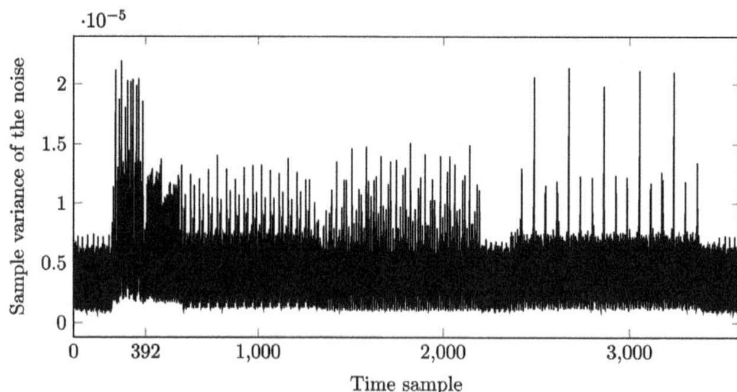

Fig. 4.25 Sample variance of the noise for each time sample, computed using *Random dataset*. The signal is given by the Hamming weight of the 0th Sbox output

The sample variance of the noise is very similar to Fig. 4.21 and also resembles the leakage of PRESENT computation since most of the leakage is not related to wt (v). The peaks in the variance of signal and SNR also correspond to each other. Compared to Fig. 4.22, the locations of the peaks are similar. It is worth noting that the highest peak in both Figs. 4.22 and 4.24 is at time sample 392. As mentioned in Example 4.2.17, this time sample corresponds to the computation of the 0th Sbox in sBoxLayer. We also note that Fig. 4.24 has a higher SNR value than Fig. 4.22 at this point. This suggests that the Hamming weight leakage model is closer to our DUT leakage than the identity leakage model.

Normally in DPA attacks, we would like to focus on time samples where the corresponding SNRs are high. We refer to those time samples as *points of interest (POIs)*.

Example 4.2.19 Continuing Example 4.2.17, the time sample with the highest SNR is given by $t = 392$. We can then take this point as our POI. Or, we can also take a few time samples that achieve the higher SNRs. For example, the top three SNRs are obtained at $t = 392, 218, 1328$.

Similarly, suppose we focus on the Hamming weight of the 0th Sbox output. Following the results from Example 4.2.18, in case we take just one POI, we have $t = 392$. And for three POIs, we have $t = 392, 1309, 1304$.

Those POIs will be further used for our attacks in Sect. 4.3.2.

Example 4.2.20 As another example, suppose instead of the exact value or Hamming weight of the 0th Sbox output v, we are interested in the 0th bit of v. With the same dataset *Random dataset*, we divide the traces into two sets A_1, A_2, corresponding to the 0th bit of v equal to 0 and 1, respectively. Following similar computations as in Example 4.2.16, for $t = 600$, we have

$$\overline{l_{600,1}} \approx 0.08206, \quad \overline{l_{600,2}} \approx 0.08216.$$

And

$$s^2_{X_{600}} \approx 2.6879 \times 10^{-9}, \quad s^2_{N_{600}} \approx 6.4256 \times 10^{-6}, \quad \mathrm{SNR}_{600} \approx 0.0004183.$$

The results for all time samples are shown in Figs. 4.26, 4.27, and 4.28.

We can see that Fig. 4.28 is similar to Figs. 4.21 and 4.25. Compared to Figs. 4.22 and 4.24, there are fewer peaks in Fig. 4.27. Furthermore, the highest peak is not around the sBoxLayer, but during pLayer computation. This is expected since now we only consider 1 bit instead of 4 bits of v.

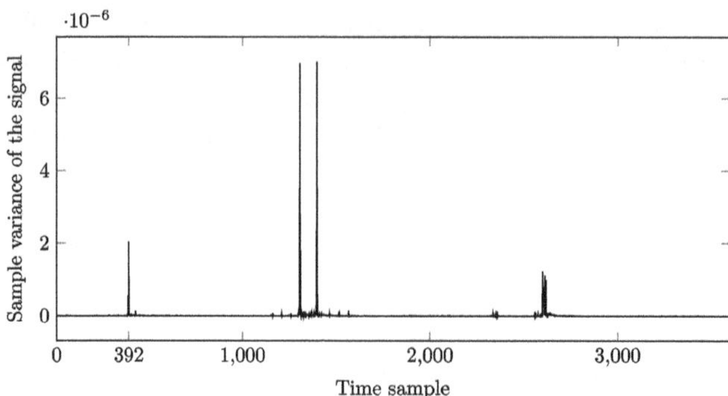

Fig. 4.26 Sample variance of the signal for each time sample, computed using *Random dataset*. The signal is given by the 0th bit of the 0th Sbox output

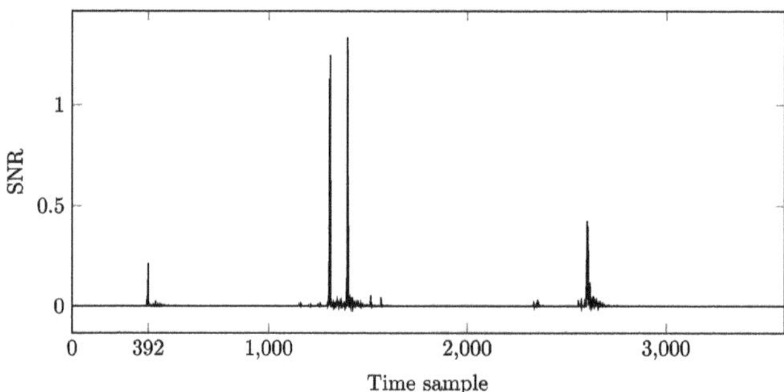

Fig. 4.27 SNR for each time sample, computed using *Random dataset*. The signal is given by the 0th bit of the 0th Sbox output

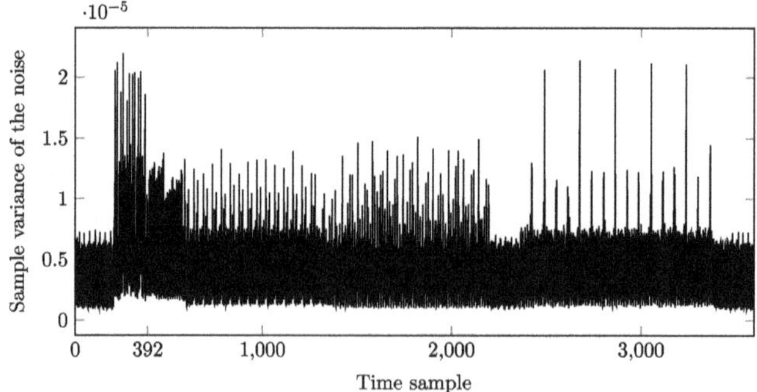

Fig. 4.28 Sample variance of the noise for each time sample, computed using *Random dataset*. The signal is given by the 0th bit of the 0th Sbox output

4.3 Side-Channel Analysis Attacks on Symmetric Block Ciphers

In this section, we will discuss two types of attacks on symmetric block cipher implementations: differential power analysis (DPA) in Sects. 4.3.1 and 4.3.2 and side-channel assisted differential plaintext attack (SCADPA) in Sect. 4.3.3.

4.3.1 Non-profiled Differential Power Analysis Attacks

As mentioned in Sect. 4.2, DPA exploits the relationship between leakages at specific time samples and the data being processed in the DUT. In this subsection, we will focus on the non-profiled setting, where we assume the attacker only has access to the target device (or measurements from the target device) and they aims to analyze the side-channel leakages to recover the master key of a symmetric block cipher.

Attacker assumption In more detail, we assume the attacker has the knowledge of the plaintext, and the goal is to recover the very first round key used at the beginning of a symmetric block cipher—for some ciphers, e.g., PRESENT, this is the first round key; for some ciphers, e.g., AES-128, this is the whitening key, which is equal to the master key. We note that after getting this round key, for some ciphers, e.g., AES-128 (see Remark 3.1.4), the master key can be found. For some ciphers, e.g., DES and PRESENT-80 (see Remarks 3.1.1 and 3.1.5), part of the master key can be found, and the remaining bits can be recovered by brute force. Otherwise, with the knowledge of this round key, the same attack method can be used to recover the next round key. In most cases, two round keys are enough to reveal the full master key using the reverse key schedule.

Similar attack strategies apply if we assume the attacker has the knowledge of the ciphertext and aims to recover the last round key. Furthermore, we also assume that the attacker has certain knowledge of the implementation for example, how to interface with the encryption routine, whether the implementation is round-based or bit-sliced, whether the computation is executed serially or in parallel, or whether some types of countermeasures are present.

4.3.1.1 Non-profiled DPA Attack Steps

A non-profiled DPA attack on symmetric block cipher implementations consists of the following steps:

DPA Step 1 **Identify the target cryptographic implementation**. DPA attacks can be applied to unprotected implementations of any symmetric block ciphers that have been proposed so far. As a running example, we will look at the computation of PRESENT.

DPA Step 2 **Experimental setup and measure leakages**. The efficiency and success of the attack are highly dependent on the measurement devices the attacker has access to. For our illustrations, we follow the experimental settings as described in Sect. 4.1.

Suppose we have taken measurements of the target implementation with M_p plaintexts. For $j = 1, \ldots, M_p$, let $\ell_j = (l_1^j, l_2^j, \ldots, l_q^j)$ denote the power trace corresponding to the jth plaintext, where q is the total number of time samples in one trace. For our attacks, we will use the *Random plaintext dataset* (see Sect. 4.1). In particular, we have $q = 3600$ and $M_p = 5000$.

DPA Step 3 **Choose the part of the key to recover**. The DPA attack is normally carried out in a divide-and-conquer manner. In particular, we focus on a small part (e.g., a nibble and a byte) of a round key in each attack, and each part of the round key can be recovered independently. With the inverse key schedule, one (e.g., for AES) or two (e.g., for PRESENT, DES) round keys will reveal the master key (see Remarks 3.1.1, 3.1.4, and 3.1.5). Let k denote the target part of the key, and let M_k denote the number of possible values of k. For our attacks, we will focus on the 0th nibble of the first round key for PRESENT and $M_k = 16$.

DPA Step 4 **Choose the target intermediate value**. To recover the part of the key chosen in the last step, we exploit relationships between leakages and a certain intermediate value being processed in the DUT. The goal is to gain information about this intermediate value, which reveals information about our chosen part of the key. Let v denote the target intermediate value. We require that there is a function φ, such that

$$v = \varphi(k, p),$$

where p denotes (part of) the plaintext. For our attack, to recover the 0th nibble of the first round key of PRESENT, we will target the 0th Sbox output of the first round. Then we have

$$v = \mathrm{SB}_{\mathrm{PRESENT}}(k \oplus p),$$

where k and p denote the 0th nibble of the first round key and that of the plaintext.

DPA Step 5 **Compute hypothetical target intermediate values**. By our choice of the target intermediate value, a small part of the key is related to it. Thus, when we make a guess of this part of the key, with knowledge of the plaintext we can obtain a hypothetical value for our target intermediate value. In particular, for each key hypothesis $\hat{k}_i$ of k and each (part of the) plaintext p_j, we can compute a hypothesis for v, denoted $\hat{v}_{ij}$, as follows:

$$\hat{v}_{ij} = \varphi(\hat{k}_i, p_j), \quad i = 1, 2, \ldots, M_k, \quad j = 1, 2, \ldots, M_p.$$

For our illustration, with each key hypothesis of the 0th nibble of the first round key and each plaintext, we have a hypothetical value for the 0th Sbox output:

$$\hat{v}_{ij} = \mathrm{SB}_{\mathrm{PRESENT}}(\hat{k}_i \oplus p_j), \quad i = 1, 2, \ldots, 16, \quad j = 1, 2, \ldots, 5000,$$

where p_j is the 0th nibble of the plaintext corresponding to the attack trace ℓ_j. Furthermore, we set

$$\hat{k}_i = i - 1, \quad i = 1, 2, \ldots, 16.$$

DPA Step 6 **Choose the leakage model**. For each hypothetical target intermediate value, we can compute the hypothetical signal depending on our leakage model

$$\mathcal{H}_{ij} := \mathcal{L}(\hat{v}_{ij}) - \text{noise}, \quad i = 1, 2, \ldots, M_k, \quad j = 1, 2, \ldots, M_p,$$

where we subtract the noise component from the leakage model. For example, if we choose the Hamming weight leakage model, according to Eq. 4.4, we have

$$\mathcal{H}_{ij} = \mathrm{wt}\left(\hat{v}_{ij}\right), \quad i = 1, 2, \ldots, M_k, \quad j = 1, 2, \ldots, M_p.$$

In our analysis, we will consider the identity leakage model and the Hamming weight leakage model. In Sect. 4.3.2.2 we will discuss another leakage model obtained by profiling the device.

DPA Step 7 **Statistical analysis**. In this step, we aim to use a statistical distinguisher to distinguish the correct key hypotheses from the rest. In this book, we will focus on correlation coefficient (see Definition 1.7.11). For other methodologies, we refer the readers to, e.g., [MOP08, Chapter 6].

For a fixed key hypothesis $\hat{k}_i$, we view the modeled signal as a random variable $\mathcal{H}_i$ that varies when the plaintext changes. If we fix a time sample t, we also consider the leakage at this time sample as a random variable L_t. Then a sample for this pair of random variable $(\mathcal{H}_i, L_t)$ is given by

$$\left\{ (\mathcal{H}_{ij}, l_t^j) \mid j = 1, 2, \ldots, M_p \right\}.$$

We would like to know how good the modeled signals are compared to the actual leakages for each key hypothesis. For the correct key hypothesis and the time samples corresponding to POIs, we expect the modeled signals to be "most" correlated to the actual leakages as compared to other key hypotheses and time samples. To measure how correlated are the leakages and modeled signals, we adopt the notion of correlation coefficient for further analysis. For each key hypothesis $\hat{k}_i$ ($i = 1, 2, \ldots, M_k$) and each time sample t ($t = 1, 2, \ldots, q$), we compute the sample correlation coefficient (see Example 1.8.1), denoted by $r_{i,t}$, of $\mathcal{H}_i$ and L_t:

$$r_{i,t} := \frac{\sum_{j=1}^{M_p} (\mathcal{H}_{ij} - \overline{\mathcal{H}_i})(l_t^j - \overline{l_t})}{\sqrt{\sum_{j=1}^{M_p} (\mathcal{H}_{ij} - \overline{\mathcal{H}_i})^2}\sqrt{\sum_{j=1}^{M_p} (l_t^j - \overline{l_t})^2}},$$

$$i = 1, 2, \ldots, M_k, \quad t = 1, 2, \ldots, q.$$

In our case,

$$r_{i,t} = \frac{\sum_{j=1}^{5000} (\mathcal{H}_{ij} - \overline{\mathcal{H}_i})(l_t^j - \overline{l_t})}{\sqrt{\sum_{j=1}^{5000} (\mathcal{H}_{ij} - \overline{\mathcal{H}_i})^2}\sqrt{\sum_{j=1}^{5000} (l_t^j - \overline{l_t})^2}},$$

$$i = 1, 2, \ldots, 16, \quad t = 1, 2, \ldots, 3600. \tag{4.21}$$

Since the target intermediate value v we have chosen will be processed in our DUT at certain points in time, we expect the leakages at those corresponding time samples to be correlated to v. Those time samples are our POIs. If a good leakage model (i.e., a model that is close to the actual leakage of the DUT) is chosen, we expect $\mathcal{H}_i$ and L_t to be correlated for the correct key hypothesis $\hat{k}_i$ and POIs t. Thus, the key hypothesis that achieves the highest *absolute value* of $r_{i,t}$ is expected to be the correct key. Furthermore, the time samples that achieve higher absolute values of $r_{i,t}$ will be our POIs in the attack.

In practice, if all r_{it}s are low, we will need more traces for the attack.

Note
According to Eq. 4.1, the correct value of the 0th nibble of the first round key is given by 9.

Example 4.3.1 As a simple example to illustrate how the sample correlation coefficient can be computed, suppose we obtained a sample

$$\{(1, 11), (0, 9), (1, 12), (1, 14), (0, 9)\}$$

for a pair of random variables (U, W). Then the sample mean for U is given by

$$\bar{u} = \frac{1 + 0 + 1 + 1 + 0}{5} = \frac{3}{5}.$$

And the sample mean for W is given by

$$\bar{w} = \frac{11 + 9 + 12 + 14 + 9}{5} = \frac{55}{5} = 11.$$

The sample correlation coefficient for U and W is given by

$$r = \frac{\sum_{i=1}^{5}(u_i - \bar{u})(w_i - \bar{w})}{\sqrt{\sum_{i=1}^{5}(u_i - \bar{u})^2}\sqrt{\sum_{i=1}^{5}(w_i - \bar{w})^2}}$$

$$= \frac{0.4 \times 0 + (-0.6) \times (-2) + 0.4 \times 1 + 0.4 \times 3 + (-0.6) \times (-2)}{\sqrt{0.4^2 \times 3 + 0.6^2 \times 2}\sqrt{2^2 + 1 + 3^2 + 2^2}} \approx 0.861.$$

4.3.1.2   Identity Leakage Model

Let us first consider the identity leakage model. Then in DPA Step 6, we have

$$\mathcal{H}_{ij} = \hat{v}_{ij}, \quad i = 1, 2, \ldots, 16, \quad j = 1, 2, \ldots, 5000.$$

Example 4.3.2 For the *Random plaintext dataset*, we have

$$p_1 = 9, \quad p_2 = \mathsf{C}.$$

As mentioned in DPA Step 5, $\hat{k}_1 = \mathsf{0}$, $\hat{k}_2 = \mathsf{1}$. Then according to Table 3.11,

$$\mathcal{H}_{11} = \hat{v}_{11} = \mathsf{SB_{PRESENT}}(\hat{k}_1 \oplus p_1) = \mathsf{SB_{PRESENT}}(\mathsf{0} \oplus 9) = \mathsf{SB_{PRESENT}}(9) = \mathsf{E} = 14,$$

$$\mathcal{H}_{12} = \hat{v}_{12} = \mathsf{SB_{PRESENT}}(\hat{k}_1 \oplus p_2) = \mathsf{SB_{PRESENT}}(\mathsf{0} \oplus \mathsf{C}) = \mathsf{SB_{PRESENT}}(\mathsf{C}) = 4 = 4,$$

$$\mathcal{H}_{21} = \hat{v}_{21} = \mathsf{SB_{PRESENT}}(\hat{k}_2 \oplus p_1) = \mathsf{SB_{PRESENT}}(\mathsf{1} \oplus 9) = \mathsf{SB_{PRESENT}}(8) = 3 = 3,$$

$$\mathcal{H}_{22} = \hat{v}_{22} = \mathsf{SB_{PRESENT}}(\hat{k}_2 \oplus p_2) = \mathsf{SB_{PRESENT}}(\mathsf{1} \oplus \mathsf{C}) = \mathsf{SB_{PRESENT}}(\mathsf{D}) = 7 = 7.$$

The sample correlation coefficients $r_{i,t}$ ($t = 1, 2, \ldots, 3600$) for $i = 1, 2, \ldots, 16$ are shown in Fig. 4.29. We can see that the blue plot has much bigger peaks than the rest, which correspond to $\hat{k}_{10} = 9$. This is the correct 0th nibble of the round key as given in Eq. 4.1. The plot of $r_{10,t}$ (corresponding to the correct key hypothesis 9) is shown in Fig. 4.30. We can also deduce that time samples that achieve those peaks in Fig. 4.30 correspond to the time when v (the 0th Sbox output) is being processed. The first cluster of peaks is most likely caused by sBoxLayer computation, and

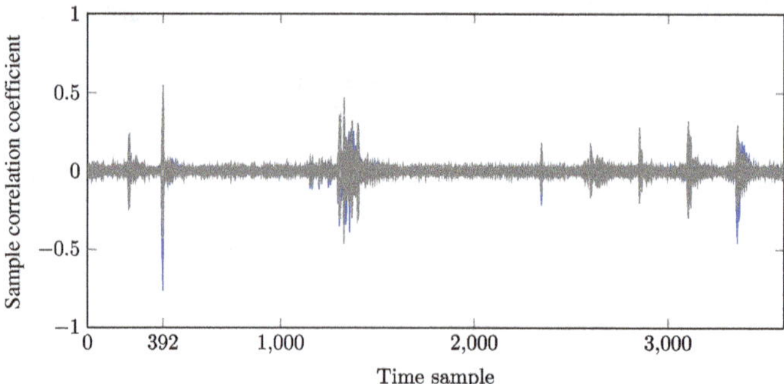

Fig. 4.29 Sample correlation coefficients $r_{i,t}$ ($i = 1, 2, \ldots, 16$) for all time samples $t = 1, 2, \ldots, 3600$. Computed following Eq. 4.21 with the identity leakage model and the *Random plaintext dataset*. The blue line corresponds to the correct key hypothesis $\hat{k}_{10} = 9$

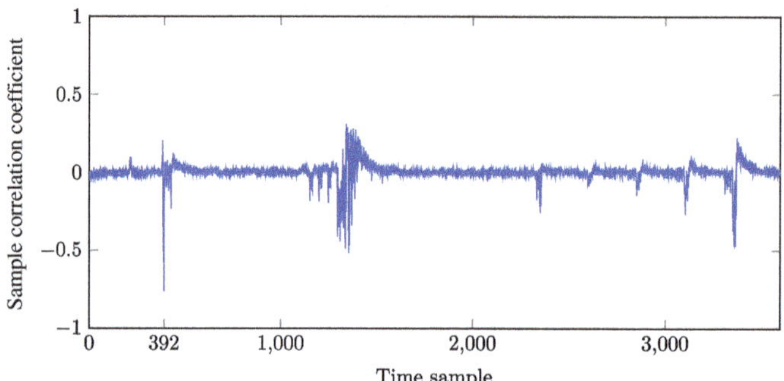

Fig. 4.30 Sample correlation coefficients $r_{10,t}$ (corresponds to the correct key hypothesis 9) for all time samples $t = 1, 2, \ldots, 3600$. Computed following Eq. 4.21 with the identity leakage model and the *Random plaintext dataset*

the other peak clusters are related to permutations of bits of v in pLayer. Those observations agree with the duration of each PRESENT round operation in Fig. 4.3. We also notice that the biggest peak in Fig. 4.30 is obtained at $t = 392$, which corresponds to the point with the highest SNR from Fig. 4.22 (Example 4.2.17).

For further illustration, the plots of $r_{i,t}$ ($t = 1, 2, \ldots, 3600$) for $i = 1, 5, 14$ (corresponding to key hypotheses 0, 4, D) are shown in Figs. 4.31, 4.32, and 4.33, respectively. Comparing those figures with Fig. 4.30, we can see some peaks appear at similar time samples in all figures. This is due to the fact that $\mathcal{H}_i$s are not independent random variables, and for those time samples t, $\mathcal{H}_i$s are also correlated with l_t for $i \neq 10$.

Remark 4.3.1 The correlation between $\mathcal{H}_i$s also influences the magnitude of the correlation coefficients for the wrong key hypotheses. If the correlation between $\mathcal{H}_i$s is higher, we would also see higher peaks in some wrong key hypotheses. For AES, the correlations between the first AddRoundKey outputs are higher than correlations

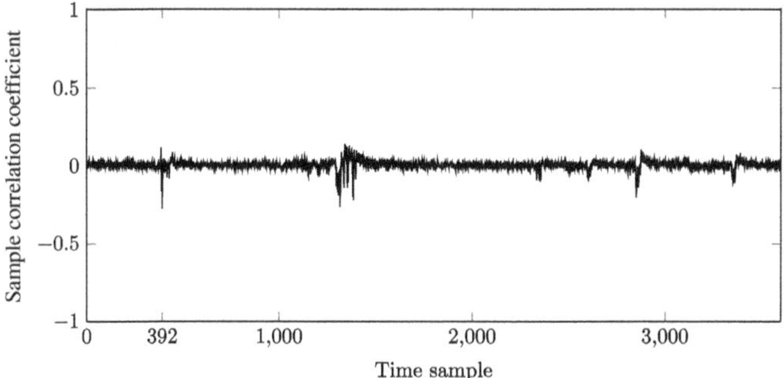

Fig. 4.31 Sample correlation coefficients $r_{1,t}$ (corresponds to a wrong key hypothesis 0) for all time samples $t = 1, 2, \ldots, 3600$. Computed following Eq. 4.21 with the identity leakage model and the *Random plaintext dataset*

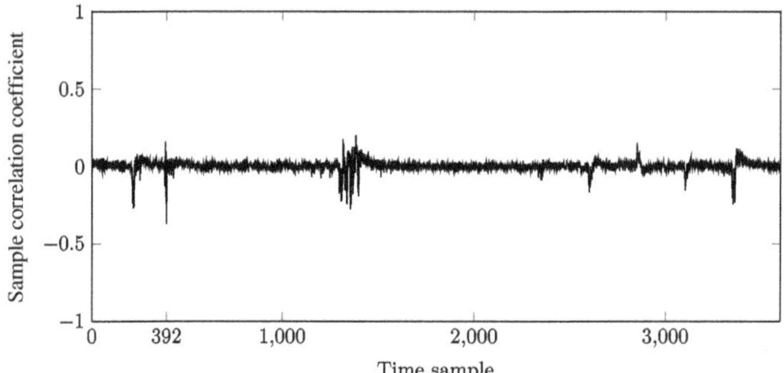

Fig. 4.32 Sample correlation coefficients $r_{5,t}$ (corresponds to a wrong key hypothesis 4) for all time samples $t = 1, 2, \ldots, 3600$. Computed following Eq. 4.21 with the identity leakage model and the *Random plaintext dataset*

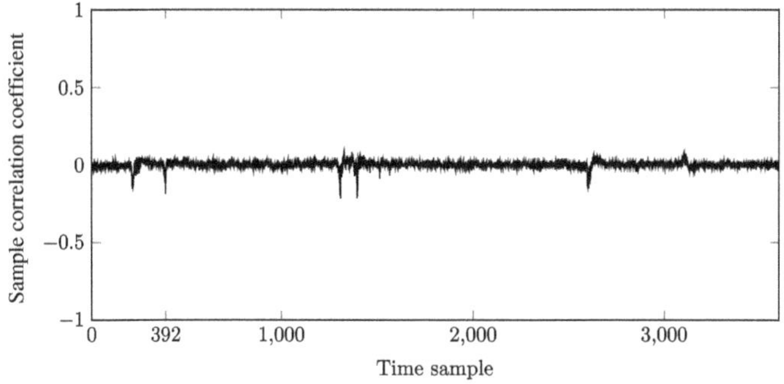

Fig. 4.33 Sample correlation coefficients $r_{14,t}$ (corresponds to a wrong key hypothesis D) for all time samples $t = 1, 2, \ldots, 3600$. Computed following Eq. 4.21 with the identity leakage model and the *Random plaintext dataset*

between the first SubBytes operation outputs, which is why in DPA Step 4 we chose the target intermediate value to an Sbox output.

4.3.1.3 Hamming Weight Leakage Model

In this part, let us consider the Hamming weight leakage model. In DPA Step 6, we have

$$\mathcal{H}_{ij} = \mathrm{wt}\left(\hat{\boldsymbol{v}}_{ij}\right), \quad i = 1, 2, \ldots, 16, \quad j = 1, 2, \ldots, 5000.$$

Example 4.3.3 Continuing Example 4.3.2, in this case, we have

$$\mathcal{H}_{11} = \mathrm{wt}\left(\hat{\boldsymbol{v}}_{11}\right) = \mathrm{wt}\,(\mathrm{E}) = 3$$
$$\mathcal{H}_{12} = \mathrm{wt}\left(\hat{\boldsymbol{v}}_{11}\right) = \mathrm{wt}\,(4) = 1$$
$$\mathcal{H}_{21} = \mathrm{wt}\left(\hat{\boldsymbol{v}}_{11}\right) = \mathrm{wt}\,(3) = 2$$
$$\mathcal{H}_{22} = \mathrm{wt}\left(\hat{\boldsymbol{v}}_{11}\right) = \mathrm{wt}\,(7) = 3.$$

The sample correlation coefficients $r_{i,t}$ ($t = 1, 2, \ldots, 3600$) for $i = 1, 2, \ldots, 16$ are shown in Fig. 4.34. The same as in Fig. 4.29, the blue plot has much bigger peaks than the rest, which corresponds to $\hat{k}_{10} = 9$. The plot of $r_{10,t}$ is shown in Fig. 4.35. The time samples that achieve peaks in this plot are similar to those in Fig. 4.30. Plots of $r_{i,t}$ for $i = 1, 5, 14$ are shown in Figs. 4.36, 4.37, and 4.38.

Remark 4.3.2 We note that the attacks we have seen recover one nibble of the first-round key. The other nibbles can be recovered independently with a similar method using the same traces.

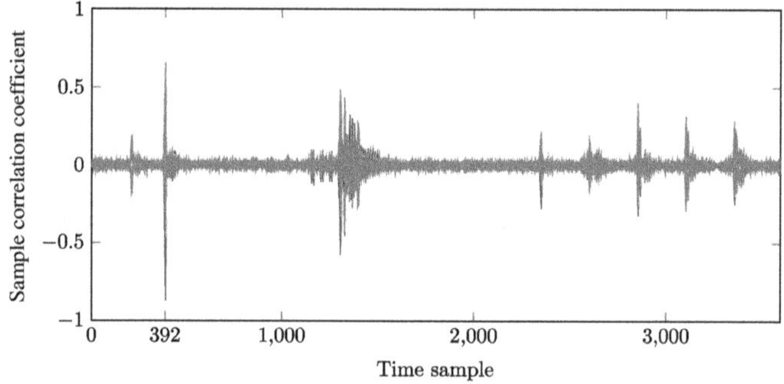

Fig. 4.34 Sample correlation coefficients $r_{i,t}$ ($i = 1, 2, \ldots, 16$) for all time samples $t = 1, 2, \ldots, 3600$. Computed following Eq. 4.21 with the Hamming leakage model and the *Random plaintext dataset*. The blue line corresponds to the correct key hypothesis $\hat{k}_{10} = 9$

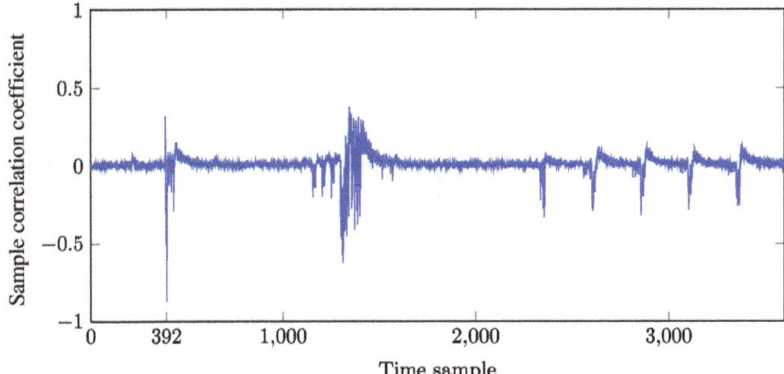

Fig. 4.35 Sample correlation coefficients $r_{10,t}$ (corresponds to the correct key hypothesis 9) for all time samples $t = 1, 2, \ldots, 3600$. Computed following Eq. 4.21 with the Hamming leakage model and the *Random plaintext dataset*

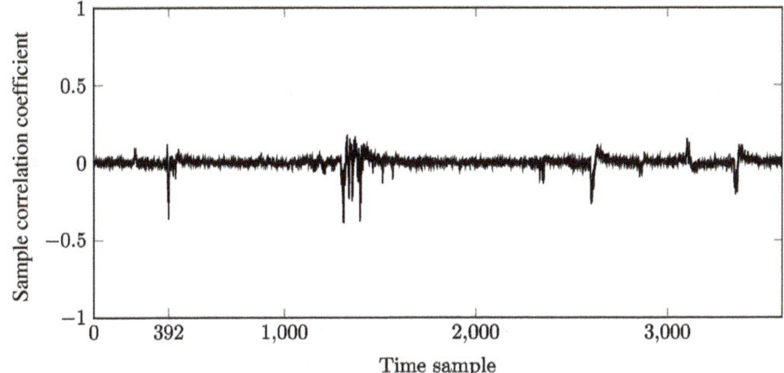

Fig. 4.36 Sample correlation coefficients $r_{1,t}$ (corresponds to a wrong key hypothesis 0) for all time samples $t = 1, 2, \ldots, 3600$. Computed following Eq. 4.21 with the Hamming leakage model and the *Random plaintext dataset*

4.3.2 *Profiled Differential Power Analysis*

In this subsection, we will consider a profiled setting. In particular, we assume the attacker has access to a clone device and can characterize the leakages of the clone device in the *profiling phase* before attacking the target device in the *attack phase*.

Attacker assumption We assume the attacker has the knowledge of the plaintext, and the goal is to recover the very first round key used in the encryption of a symmetric block cipher—for some ciphers, e.g., PRESENT, this is the first round key, and for some ciphers, e.g., AES, this is the whitening key, which is equal to the master key. Similar attack strategies apply if we assume the attacker has the knowledge of the ciphertext and aims to recover the last round key. We also assume the attacker has the knowledge of the detailed implementation so that the same program can be implemented by the attacker on the clone device. This is

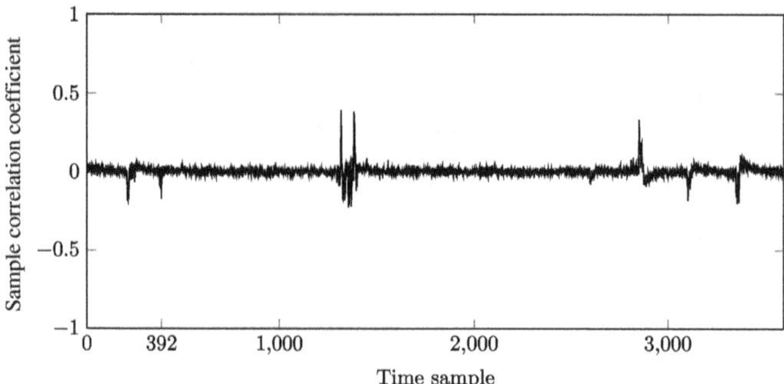

Fig. 4.37 Sample correlation coefficients $r_{5,t}$ (corresponds to a wrong key hypothesis 4) for all time samples $t = 1, 2, \ldots, 3600$. Computed following Eq. 4.21 with the Hamming leakage model and the *Random plaintext dataset*

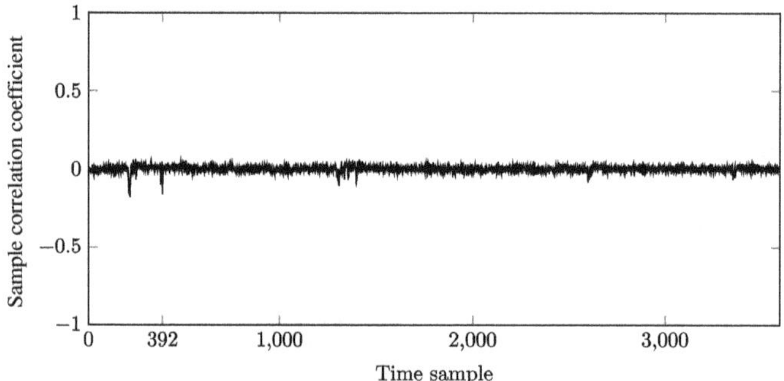

Fig. 4.38 Sample correlation coefficients $r_{14,t}$ (corresponds to a wrong key hypothesis D) for all time samples $t = 1, 2, \ldots, 3600$. Computed following Eq. 4.21 with the Hamming leakage model and the *Random plaintext dataset*

different from the non-profiled setting where only certain basic knowledge of the implementation is required.

For our illustrations, we suppose the *Random dataset* is obtained from a clone device, and the *Random plaintext dataset* is from the target device. Then before the attack, we can analyze the *Random dataset* to obtain more information about the leakage behavior of the DUT in the profiling phase.

The first major step in the profiling phase is to find the POIs, namely, time samples that will give us more information or with better signal. After identifying the POIs, in the attack phase, instead of computing the sample correlation coefficients for all time samples, we can just focus on the POIs.

4.3.2.1 Profiled DPA Attack Steps

The detailed steps for a profiled DPA attack are as follows:

P-DPA Step 1 **Identify the target cryptographic implementation**. This step is the same as DPA Step 1 in Sect. 4.3.1.1. As a running example, we will look at the computation of PRESENT.

P-DPA Step 2 **Measurement of profiling traces**. We first collect a set of traces for profiling using the clone device with random plaintexts and random keys. Those traces are called the *profiling traces*. Note that we assume the attacker has the knowledge of the plaintexts and the keys. Suppose there are in total M_{pf} profiling traces, and each trace contains q time samples. For our illustrations, we will use the *Random dataset* as profiling traces, then $M_{pf} = 10,000$, and $q = 3600$.

P-DPA Step 3 **Choose the part of the key to recover**. This step is the same as DPA Step 3 in Sect. 4.3.1.1. Let k denote the target part of the key, and let M_k denote the number of possible values of k. For our attacks, the same as in Sect. 4.3.1.1, we will focus on the 0th nibble of the first round key for PRESENT and $M_k = 16$.

P-DPA Step 4 **Choose the target intermediate value**. This step is the same as DPA Step 4 in Sect. 4.3.1.1. Let v denote the target intermediate value. We require that there is a function φ, such that

$$v = \varphi(k, p),$$

where p denotes (part of) the plaintext. For our attack, to recover the 0th nibble of the first round key of PRESENT, we will target the 0th Sbox output of the first round. Then we have

$$v = \mathrm{SB}_{\mathrm{PRESENT}}(k \oplus p),$$

where k and p denote the 0th nibble of the first round key and that of the plaintext.

P-DPA Step 5 **Decide on the target signal**. Before we do further analysis of the profiling traces, we need to choose what information related to the target intermediate value v we are looking for, for example, the Hamming weight of v or the 0th bit of v. In our illustrations, we will look at two types of target signals, one given by the exact value of v and the other one given by wt (v), the Hamming weight of v.

P-DPA Step 6 **Group the profiling traces**. We take our set of profiling traces and divide them into M_{signal} sets according to the target signal from P-DPA Step 5. Let us denote those sets by $A_1, A_2, \ldots, A_{M_{signal}}$.

For our illustrations, when the target signal is given by v, the exact value of the output of the 0th Sbox in PRESENT, we will divide our profiling traces *Random dataset* into 16 sets, $A_1, A_2, \ldots, A_{16}$, where A_s contains traces corresponding to $v = s - 1$. When the target signal is given by $\mathrm{wt}(v)$, the Hamming weight of v, we will divide the profiling traces into five sets, $A_1, A_2, \ldots, A_5$, where A_s contains traces corresponding to $\mathrm{wt}(v) = s - 1$.

P-DPA Step 7 **Modeling leakage, signal, and noise.** Let us fix a time sample t $(1 \leq t \leq q)$, and let L_t, X_t, and N_t denote the random variables corresponding to leakage, signal, and noise at t, respectively. When we fix the signal, as discussed in Sect. 4.2.1, the leakage L_t and the noise N_t can be modeled by normal random variables. When we focus on one particular target signal, i.e., when we only consider computations that result in the traces belonging to a particular set A_s, let $L_{t,s}$ and $N_{t,s}$ denote the random variables corresponding to leakage and noise at t, respectively. We further denote the constant signal as $X_{t,s}$. Then $L_{t,s}$ and $N_{t,s}$ can be modeled by normal random variables, and traces from A_s give us a sample to analyze $L_{t,s}$ and $N_{t,s}$.

For example, in our attack, if we only look at computations that result in $v = 1$, we denote the leakage and noise at a given time sample t as $L_{t,2}$ and $N_{t,2}$.

P-DPA Step 8 **Compute SNR.** The SNR values for each time sample $t = 1, 2, \ldots, q$ can be computed in a similar manner as in Example 4.2.16. In more detail, by Eq. 4.3, for any s,

$$X_{t,s} = \mathrm{E}\left[L_{t,s}\right] - \mathrm{E}\left[N_{t,s}\right]. \tag{4.22}$$

Hence

$$\mathrm{Var}(X_t) = \mathrm{Var}\left(X_{t,s}\right) = \mathrm{Var}\left(\mathrm{E}\left[L_{t,s}\right] - \mathrm{E}\left[N_{t,s}\right]\right).$$

Since any information related to the target signal is contained in X_t, which is independent of N_t, $\mathrm{E}\left[N_{t,s}\right]$ is a constant for all s. Consequently, we have (see Eq. 1.36)

$$\mathrm{Var}(X_t) = \mathrm{Var}\left(\mathrm{E}\left[L_{t,s}\right]\right).$$

Together with Eqs. 1.36, 4.22, and 4.2, we also have

$$\mathrm{Var}\left(L_t - \mathrm{E}\left[L_{t,s}\right]\right) = \mathrm{Var}\left(L_t - X_{t,s} - \mathrm{E}\left[N_{t,s}\right]\right)$$
$$= \mathrm{Var}\left(L_t - X_{t,s}\right) = \mathrm{Var}(L_t - X_t) = \mathrm{Var}(N_t).$$

Using our profiling traces, $\mathrm{Var}(X_t)$ and $\mathrm{Var}(N_t)$ can be approximated by the sample variances of $\mathrm{E}\left[L_{t,s}\right]$ and $L_t - \mathrm{E}\left[L_{t,s}\right]$, respectively. We can then approximate the SNR at time sample t using

$$\mathrm{SNR}_t = \frac{\mathrm{Var}(X_t)}{\mathrm{Var}(N_t)} \approx \frac{\text{sample variance of } \mathrm{E}\left[L_{t,s}\right]}{\text{sample variance of } L_t - \mathrm{E}\left[L_{t,s}\right]}.$$

The same computations can be done for all time samples t.

For our attacks, SNR values following the above steps have been computed in Example 4.2.17 when the target signal is the exact value of v and in Example 4.2.18 when the target signal wt (v).

P-DPA Step 9   **Identify the point of interest**. The point of interest is given by the time sample that achieves the highest SNR value. For our attacks, in Example 4.2.19, we have analyzed the *Random dataset* and identified one POI for the target signal given by the exact value of v: $t = 392$. The same POI is also for the case when the target signal is given by wt (v).

P-DPA Step 10   **Measurement of attack traces**. After getting our POI, we are ready to carry out the attack. This step is the same as in DPA Step 2 from Sect. 4.3.1.1. Suppose we have taken measurements of our target device with M_p plaintexts. For $j = 1, \ldots, M_p$, let $\boldsymbol{\ell}_j = (l_1^j, l_2^j, \ldots, l_q^j)$ denote the corresponding power trace, where q is the total number of time samples in one attack trace. Note that the measurements should be done in such a way that attack traces and profiling traces (see P-DPA Step 2) are aligned in the time domain so that the POI we have identified is the actual POI. In particular, one attack trace contains the same number of time samples as one profiling trace. We argue that this is achievable since we assume the attacker has the knowledge of the implementation and is in procession of a clone device. For our illustrations, we will use the *Random plaintext dataset* as our attack traces. We have $M_p = 5000$.

P-DPA Step 11   **Compute hypothetical target intermediate values**. This step is the same as DPA Step 5 from Sect. 4.3.1.1. For each key hypothesis $\hat{k}_i$ of k and each (part of the) plaintext p_j, we compute a hypothesis for v, which is given by

$$\hat{v}_{ij} = \varphi(\hat{k}_i, p_j), \quad i = 1, 2, \ldots, M_k, \quad j = 1, 2, \ldots, M_p.$$

For our attacks, with each key hypothesis of the 0th nibble of the first round key and each known plaintext, we have a hypothetical value for the 0th Sbox output:

$$\hat{v}_{ij} = \mathrm{SB}_{\mathrm{PRESENT}}(\hat{k}_i \oplus p_j), \quad i=1, 2, \ldots, 16, \quad j=1, 2, \ldots, 5000,$$

where p_j is the 0th nibble of the plaintext corresponding to the attack trace $\boldsymbol{\ell}_j$. Furthermore, we set

$$\hat{k}_i = i - 1, \quad i = 1, 2, \ldots, 16.$$

P-DPA Step 12 **Identify the leakage model and compute the hypothetical signals**. By our choice of the target signal from P-DPA Step 5, we have a corresponding leakage model. For example, if the target signal is the exact value of v, a natural choice of leakage model will be the identity leakage model.

For each hypothetical target intermediate value, we can compute the hypothetical signal depending on our leakage model

$$\mathcal{H}_{ij} := \mathcal{L}(\hat{v}_{ij}) - \text{noise}, \quad i = 1, 2, \ldots, M_k, \quad j = 1, 2, \ldots, M_p.$$

The main difference in this step as compared to DPA Step 6 in Sect. 4.3.1.1 is that our leakage model cannot be randomly chosen. We should choose a leakage model based on our target signal chosen in P-DPA Step 5. In our illustrations, we will consider the identity leakage model and the Hamming weight leakage model corresponding to the signal given by v and wt (v), respectively.

P-DPA Step 13 **Statistical analysis**. For a fixed key hypothesis $\hat{k}_i$, we view the modeled signal as a random variable $\mathcal{H}_i$ that varies when the plaintext changes. Take the time sample $t = \text{POI}$ as identified in P-DPA Step 9. We consider the leakage at this time sample as a random variable L_{POI}.

Then a sample for this pair of random variables $(\mathcal{H}_i, L_{\text{POI}})$ is given by

$$\left\{ (\mathcal{H}_{ij}, l^j_{\text{POI}}) \mid j = 1, 2, \ldots, \hat{M}_p \right\},$$

where l^j_{POI} is the POI-th entry of the attack trace ℓ_j obtained in P-DPA Step 10 ($j = 1, 2, \ldots, M_p$) and $2 \le \hat{M}_p \le M_p$.[4] With this sample, we can compute the sample correlation coefficient between $\mathcal{H}_i$ and L_{POI} for each key hypothesis $\hat{k}_i$ ($i = 1, 2, \ldots, M_k$):

$$r^{\hat{M}_p}_{i,\text{POI}} := \frac{\sum_{j=1}^{\hat{M}_p} (\mathcal{H}_{ij} - \overline{\mathcal{H}_i})(l^j_{\text{POI}} - \overline{l_{\text{POI}}})}{\sqrt{\sum_{j=1}^{\hat{M}_p} (\mathcal{H}_{ij} - \overline{\mathcal{H}_i})^2} \sqrt{\sum_{j=1}^{\hat{M}_p} (l^j_{\text{POI}} - \overline{l_{\text{POI}}})^2}}. \tag{4.23}$$

Figure 4.39 presents the values of $r^{\hat{M}_p}_{i,\text{POI}}$ ($i = 1, 2, \ldots, 16$) for POI $= 392$ computed with the identity leakage model. The x-axis indicates the number of

---

[4] When $\hat{M}_p = 1$, the denominator in Eq. 4.23 is equal to 0.

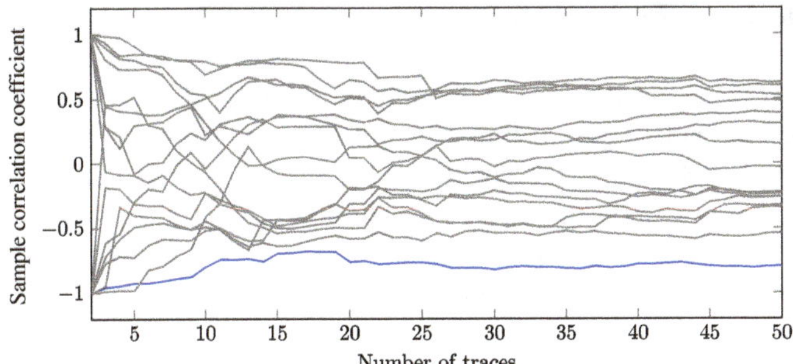

Fig. 4.39 Sample correlation coefficients $r_{i,\text{POI}}^{\hat{M}_p}$ ($i = 1, 2, \ldots, 16$) for POI $= 392$. Computed following Eq. 4.23 with the identity leakage model and the *Random plaintext dataset*. The blue line corresponds to the correct key hypothesis $\hat{k}_{10} = 9$

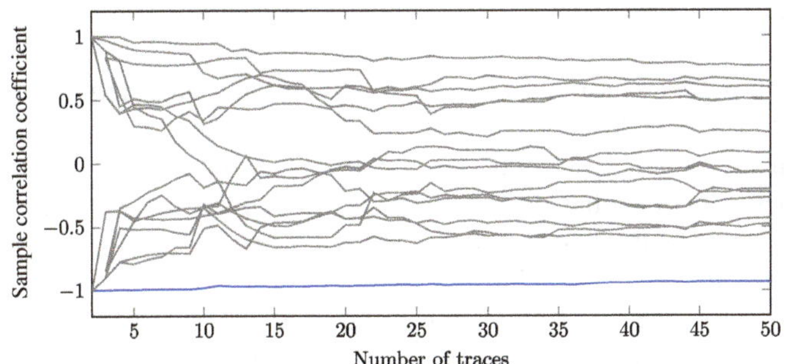

Fig. 4.40 Sample correlation coefficients $r_{i,\text{POI}}^{\hat{M}_p}$ ($i = 1, 2, \ldots, 16$) for POI $= 392$. Computed following Eq. 4.23 with the Hamming weight leakage model and the *Random plaintext dataset*. The blue line corresponds to the correct key hypothesis $\hat{k}_{10} = 9$

traces $\hat{M}_p$ used. The figure shows that with just roughly 20 traces, we can clearly distinguish the correct key hypothesis from the wrong ones.

Similarly, the results for the Hamming weight model are shown in Fig. 4.40. In this case, we need less than five traces to identify the correct key. This indicates that the Hamming weight leakage model is closer to our DUT leakage compared to the identity leakage model.

> **Note**
> A good leakage model is beneficial to our attack.

Remark 4.3.3 Except for computing SNR in P-DPA Step 8 to identify the POIs, other methods, e.g., *t*-test (Sect. 4.2.3) with a properly chosen intermediate value, can also be used for this purpose.

4.3.2.2 Stochastic Leakage Model

To fully utilize the cloned device in the profiled setting, we can further characterize the leakages instead of just identifying the POI. In this part, we will study a leakage model that assumes each bit of the target intermediate value (P-DPA Step 4 from Sect. 4.3.2.1) results in a different signal. In particular, suppose the target intermediate value $v = v_{m_v-1}v_{m_v-2}\dots v_1 v_0$ has bit length at most m_v,[5] the *stochastic leakage model* assumes that

$$\mathcal{L}(v) = \sum_{s=0}^{m_v-1} \alpha_s v_s + \text{noise}, \tag{4.24}$$

where noise $\sim \mathcal{N}(0, \sigma^2)$ denotes the noise with mean 0 and variance σ^2. α_s ($s = 0, 1, \dots, m_v-1$) are real numbers. We refer to α_s as the *coefficients* of the stochastic leakage model.

The attack with stochastic leakage model follows the same steps as described in Sect. 4.3.2.1. The only difference is in P-DPA Step 12, where we need extra effort to find our leakage model by profiling. We note that since the stochastic leakage model assumes each value of v has different signals, to identify the POI, we will choose the target signal to be the exact value of v in P-DPA Step 5. Then, using the leakages at the POI, we will find estimations for α_s values. Those estimated values together with Eq. 4.24 provide us with hypothetical signals in P-DPA Step 12.

To estimate α_s, we adopt the *ordinary least square method* from linear regression [DPRS11]. Let

$$\boldsymbol{\ell}_j^{pf} = \left(l_1^{j,pf}, l_2^{j,pf}, \dots, l_q^{j,pf} \right)$$

denote the jth profiling trace, where $j = 1, 2, \dots, M_{pf}$. The steps for computing estimations of coefficients α_s for the stochastic leakage model are as follows:

SLM Step a **Compute the vector of leakages**. We only focus on the leakage at the POI from each profiling trace. Let

$$\boldsymbol{\ell}_{pf} := \left(l_{\text{POI}}^{1,pf}, l_{\text{POI}}^{2,pf}, \dots, l_{\text{POI}}^{M_{pf},pf} \right)$$

be the vector of leakages at time sample $t = \text{POI}$ from all M_{pf} profiling traces.

For our illustrations, we aim to recover the same part of the key and take the same target intermediate value as in Sect. 4.3.2.1. We also use the *Random dataset* as our profiling traces, hence $M_{pf} = 10{,}000$. As

---

[5] When the bit length of v is less than m_v, some bits $v_{m_v-1}, \dots$ are zero.

discussed in P-DPA Step 9, our POI $= 392$, which corresponds to the target signal being the exact value of v.

SLM Step b **Construct matrix M_v for the target intermediate values.** For the jth profiling trace ℓ_j^{pf}, let

$$v_j^{pf} = v_{j(m_v-1)}^{pf} \cdots v_{j1}^{pf} v_{j0}^{pf}, \quad j = 1, 2, \ldots, M_{pf}$$

be the corresponding target intermediate value. Then the matrix M_v is given by

$$M_v := \begin{pmatrix} v_{10}^{pf} & v_{11}^{pf} & \cdots & v_{1(m_v-1)}^{pf} \\ v_{20}^{pf} & v_{21}^{pf} & \cdots & v_{2(m_v-1)}^{pf} \\ \vdots & \vdots & \ddots & \vdots \\ v_{M_{pf}0}^{pf} & v_{M_{pf}1}^{pf} & \cdots & v_{M_{pf}(m_v-1)}^{pf} \end{pmatrix}. \tag{4.25}$$

Since the stochastic leakage essentially assumes each value of v has a different leakage, we require that all possible values of v appear in M_v. Furthermore, in this case, we can guarantee that the matrix $M_v^\top M_v$ is invertible (see Appendix A.2). In particular, we should take enough random plaintexts so that all values of v appear. For our illustrations, v is the 0th Sbox output. Hence $m_v = 4$, and we need all 16 values of v to appear.

SLM Step c **Compute estimated values of coefficients α_s.** The estimated values $\hat{\alpha}_s$ for α_s are given by

$$\left(\hat{\alpha}_0 \, \hat{\alpha}_1 \, \ldots \, \hat{\alpha}_{m_v-1} \right)^\top = \left(M_v^\top M_v \right)^{-1} M_v^\top \ell_{pf}^\top. \tag{4.26}$$

For each actual leakage $l_t^{j,pf}$, define

$$\hat{l}_t^{j,pf} = \sum_{s=0}^{m_v-1} \hat{\alpha}_s v_{js}^{pf}.$$

And let

$$\hat{\ell}_{pf} := (\hat{l}_t^{1,pf}, \hat{l}_t^{2,pf}, \ldots, \hat{l}_t^{M_{pf},pf}).$$

Then by the ordinary least square method from linear regression, $\hat{\alpha}_s$ values computed with Eq. 4.26 minimize the Euclidean distance (Definition A.2.1) between $\hat{\ell}_{pf}$ and ℓ_{pf} (see, e.g., [Ros20, Section 9.8]).

Example 4.3.4 The first trace in *Random dataset* corresponds to the plaintext with the 0th nibble= 4 and the key with the 0th nibble= 7. Then in SLM Step b we have (see Table 3.11 for PRESENT Sbox)

$$v_1^{pf} = \mathrm{SB}_{\mathrm{PRESENT}}(4 \oplus 7) = \mathrm{SB}_{\mathrm{PRESENT}}(3) = \mathrm{B} = 1011_2.$$

And the first row of our matrix M_v is given by

$$\begin{pmatrix} 1 & 1 & 0 & 1 \end{pmatrix}.$$

With POI $= 392$ and the *Random dataset*, we get the following estimated values for the coefficients α_s:

$$\hat{\alpha}_0 \approx -0.02019, \quad \hat{\alpha}_1 \approx -0.02027, \quad \hat{\alpha}_2 \approx -0.01920, \quad \hat{\alpha}_3 \approx -0.02039.$$

According to the stochastic leakage model in Eq. 4.24, the estimated leakage of $v = v_3 v_2 v_1 v_0$ is given by

$$\mathcal{L}(v) = \hat{\alpha}_0 v_0 + \hat{\alpha}_1 v_1 + \hat{\alpha}_2 v_2 + \hat{\alpha}_3 v_3 + \text{noise}.$$

For example,

$$\mathcal{L}(\mathrm{E}) = \mathcal{L}(1110) = \hat{\alpha}_1 + \hat{\alpha}_2 + \hat{\alpha}_3 + \text{noise} = -0.05986 + \text{noise}.$$

And the estimated signal of $\mathrm{E} = 1110$ according to the stochastic leakage model is given by -0.052. Similarly, we can compute the estimated signals for all 16 possible values of the target intermediate value $0, 1, \ldots, \mathrm{F}$:

$$
\begin{array}{cccccccc}
0 & -0.02020 & -0.02027 & -0.04046 & -0.01920 & -0.03940 & -0.03947 \\
 & -0.05966 \\
-0.02039 & -0.04059 & -0.04066 & -0.06086 & -0.03959 & -0.05979 & -0.05986 \\
 & -0.08006.
\end{array}
\tag{4.27}
$$

We take the *Random plaintext dataset* as our attack traces.

Example 4.3.5 As mentioned in Example 4.3.2, $\hat{k}_1 = 0$, $\hat{k}_2 = 1$, and for the *Random plaintext dataset*,

$$p_1 = 9, \quad p_2 = \mathrm{C}.$$

Then following computations from Example 4.3.2 and the estimated signals given in Eq. 4.27, with the profiled stochastic leakage model, in P-DPA Step 12, we have

$$\mathcal{H}_{11} = \mathcal{L}(\mathrm{E}) - \text{noise} = -0.05986,$$

$$\mathcal{H}_{12} = \mathcal{L}(\mathrm{C}) - \text{noise} = -0.03959,$$

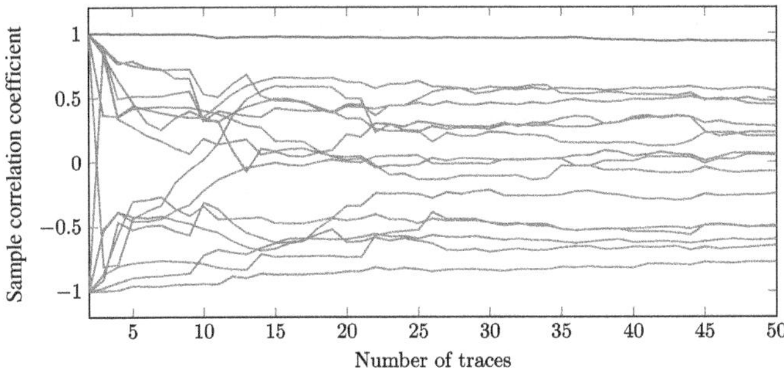

Fig. 4.41 Sample correlation coefficients $r_{i,\text{POI}}^{\hat{M}_p}$ ($i = 1, 2, \ldots, 16$) for POI $= 392$. Computed following Eq. 4.23 with the stochastic leakage model and the *Random plaintext dataset*. The blue line corresponds to the correct key hypothesis $\hat{k}_{10} = 9$

$$\mathcal{H}_{21} = \mathcal{L}(8) - \text{noise} = -0.02039,$$

$$\mathcal{H}_{22} = \mathcal{L}(D) - \text{noise} = -0.05979.$$

Following P-DPA Step 13 from Sect. 4.3.2.1, the attack results are shown in Fig. 4.41. Compared to Figs. 4.39 and 4.40, the attack results based on the stochastic leakage model are similar to that based on the Hamming weight leakage model, better than the results based on the identity leakage model. This shows that both the stochastic and the Hamming weight leakage models are better approximations of the DUT leakage than the identity leakage model.

4.3.2.3 Template-Based DPA

We have seen how to characterize the leakage assuming each bit of the target intermediate value leaks differently focusing on one POI. We can also characterize/profile the leakages of each possible value of the target intermediate value (P-DPA Step 4 from Sect. 4.3.2.1) at several POIs. The result of this profiling process is a set of *templates*. Then during the attack phase, instead of computing correlation coefficients, we use those templates to see which of them fits better to the measured power trace and deduce a probability for each key hypothesis.

As discussed in Sect. 4.2, for a computation with constant signal, the distribution of leakages at a single time sample can be modeled with a normal distribution. And leakages at a few time samples can be considered as a Gaussian random vector. The goal of profiling in template-based DPA is to estimate the mean and variance (respectively, mean vector and covariance matrix) of the normal random variable (respectively, Gaussian random vector). The resulting estimations are our templates.

The steps for template-based DPA are similar to those in Sect. 4.3.2.1, except for P-DPA Step 9, P-DPA Step 12, and P-DPA Step 13. P-DPA Step 9 will be replaced by two steps (Template Step a and Template Step b below), P-DPA Step

12 will be removed, and P-DPA Step 13 will be replaced by the following Template Step c:

Template Step a   **Identify point(s) of interest**. The same as in P-DPA Step 9, POIs are given by time samples that achieve the highest SNR values. The difference is that we can choose more than one POI. With more POIs, the effort for building the templates will increase, but the attack results will be better. Normally the attacker decides on the number of POIs based on experience.

Let q_{POI} denote the total number of chosen POIs, and let $t_1, t_2, \ldots, t_{q_{\mathrm{POI}}}$ denote the time samples that have been identified as POIs. For our illustrations, we will discuss the results of using just one POI and using three POIs. It follows from Example 4.2.19 that when the target signal is the exact value of v, the three POIs are given by

$$t_1 = 392, \quad t_2 = 218, \quad t_3 = 1328.$$

And when the target signal is wt (v), we have

$$t_1 = 392, \quad t_2 = 1309, \quad t_3 = 1304.$$

Template Step b   **Build the templates**. Let us fix a particular target signal value and only consider inputs to the cryptographic algorithm that result in traces belonging to the corresponding set A_s (see P-DPA Step 6 from Sect. 4.3.2.1). Let $L_{t,s}$ denote the random variable representing the leakage for such encryption computations at time sample t. Then the random vector

$$\boldsymbol{L}_s := (L_{t_1,s}, L_{t_2,s}, \ldots L_{t_{q_{\mathrm{POI}}},s}) \tag{4.28}$$

can be modeled by a Gaussian random vector. By Definition 1.7.10, to find the PDF of a Gaussian random vector, we need to identify its mean vector and covariance matrix. Using our profiling traces from set A_s, we can compute an approximation for the mean vector, denoted $\boldsymbol{\mu}_s$, using sample means of $L_{t_u,s}$:

$$\boldsymbol{\mu}_s := \left(\overline{l_{t_1,s}}, \overline{l_{t_2,s}}, \ldots, \overline{l_{t_{q_{\mathrm{POI}}},s}} \right).$$

Similarly, an approximation for the covariance matrix is then given by Q_s, where the (u_1, u_2)-entry of Q_s is the sample covariance between $L_{t_{u_1},s}$ and $L_{t_{u_2},s}$ $(1 \leq u_1, u_2, \leq t_{q_{\mathrm{POI}}})$. The pair $(\boldsymbol{\mu}_s, Q_s)$ is called a *template*. With our profiling traces, we can compute M_{signal} templates.

For our illustrations, when the target signal is v, we will have 16 templates. And when the target signal is wt (v), we will have five templates.

Template Step c **Statistical analysis.** In this step, we would like to compute a probability for each key hypothesis given the attack traces. For a fixed key hypothesis $\hat{k}_i$, we divide the M_p attack traces from P-DPA Step 10 into M_{signal} sets, $A_1, A_2, \ldots, A_{M_{signal}}$, depending on the hypothetical target intermediate value $\hat{v}_{ij}$ obtained in P-DPA Step 11. In particular, for an attack trace ℓ_j, let s_{ij} denote the index of the set that it belongs to. Namely

$$\ell_j \in A_{s_{ij}} \quad \text{given key hypothesis } \hat{k}_i.$$

We are only interested in the leakages at the POIs for each attack trace $\ell_j = \left(l_1^j, l_2^j, \ldots, l_q^j \right)$. Define

$$\ell_{j,\text{POI}} := \left(l_{t_1}^j, l_{t_2}^j, \ldots, l_{t_{q_{\text{POI}}}}^j \right). \tag{4.29}$$

With the mean vector $\boldsymbol{\mu}_{s_{ij}}$ and the covariance matrix $Q_{s_{ij}}$ obtained in Template Step b, we can compute the probability of ℓ_j given $\hat{k}_i$ using the PDF of the Gaussian random vector (see Definition 1.7.10) $L_{s_{ij}}$:

$$P(\ell_j|\hat{k}_i) = P(L_{s_{ij}} = \ell_{j,\text{POI}}) = \frac{1}{(2\pi)^{\frac{q_{\text{POI}}}{2}} \sqrt{\det Q_{s_{ij}}}}$$

$$\exp\left(-\frac{1}{2} (\ell_{j,\text{POI}} - \boldsymbol{\mu}_{s_{ij}})^\top Q_{s_{ij}}^{-1} (\ell_{j,\text{POI}} - \boldsymbol{\mu}_{s_{ij}}) \right). \tag{4.30}$$

Furthermore, we can assume the measurements are independent and compute the probability of a set of $\hat{M}_p$ ($1 \leq \hat{M}_p \leq M_p$) traces given the key hypothesis $\hat{k}_i$:

$$P\left(\{\ell_j\}_{j=1}^{\hat{M}_p} \,\middle|\, \hat{k}_i \right) = \prod_{j=1}^{\hat{M}_p} P(\ell_j|\hat{k}_i). \tag{4.31}$$

By the generalized Bayes' theorem (Theorem 1.7.2), the probability of the key hypothesis $\hat{k}_i$ given a set of $\hat{M}_p$ ($\hat{M}_p \leq M_p$) traces is given by

$$P\left(\hat{k}_i \,\middle|\, \{\ell_j\}_{j=1}^{\hat{M}_p}\right) = \frac{P\left(\{\ell_j\}_{j=1}^{\hat{M}_p} \,\middle|\, \hat{k}_i\right) P(\hat{k}_i)}{\sum_{m=1}^{M_k} P\left(\{\ell_j\}_{j=1}^{\hat{M}_p} \,\middle|\, \hat{k}_m\right) P(\hat{k}_m)}.$$

Typically, the key hypothesis follows a uniform distribution in the key space, and we have

$$P(\hat{k}_m) = P(\hat{k}_i)$$

in the above equation, which gives

$$P\left(\hat{k}_i \,\middle|\, \{\ell_j\}_{j=1}^{\hat{M}_p}\right) = \frac{P\left(\{\ell_j\}_{j=1}^{\hat{M}_p} \,\middle|\, \hat{k}_i\right)}{\sum_{m=1}^{M_k} P\left(\{\ell_j\}_{j=1}^{\hat{M}_p} \,\middle|\, \hat{k}_m\right)}. \tag{4.32}$$

For the attack, we expect the correct key hypothesis to have the highest probability. In other words, we are mainly interested in the ordering of the values $P\left(\hat{k}_i \,\middle|\, \{\ell_j\}_{j=1}^{\hat{M}_p}\right)$. Since the denominators are the same for all key hypotheses in Eq. 4.32, we can ignore them. Then Eq. 4.32 is reduced to Eq. 4.31, which can be further simplified by leaving out the common term (see Eq. 4.30)

$$\frac{1}{(2\pi)^{\frac{q_{\text{POI}}}{2}}}.$$

And we get

$$\prod_{j=1}^{\hat{M}_p} \frac{1}{\sqrt{\det Q_{s_{ij}}}} \exp\left(-\frac{1}{2}(\ell_{j,\text{POI}} - \mu_{s_{ij}})^\top Q_{s_{ij}}^{-1}(\ell_{j,\text{POI}} - \mu_{s_{ij}})\right).$$

By taking the natural logarithm, the ordering does not change, and we have

$$-\frac{1}{2}\sum_{j=1}^{\hat{M}_p} \ln\left(\det Q_{s_{ij}}\right) + (\ell_{j,\text{POI}} - \mu_{s_{ij}})^\top Q_{s_{ij}}^{-1}(\ell_{j,\text{POI}} - \mu_{s_{ij}}).$$

Finally, we define the *probability score* of $\hat{k}_i$, denoted $\mathrm{P}_{\hat{k}_i}$, to be

$$P_{\hat{k}_i} = -\sum_{j=1}^{\hat{M}_p} \ln\left(\det Q_{s_{ij}}\right) + (\ell_{j,\text{POI}} - \mu_{s_{ij}})^\top Q_{s_{ij}}^{-1}(\ell_{j,\text{POI}} - \mu_{s_{ij}}).$$

(4.33)

The higher the score, the more likely the hypothesis is equal to the correct key.

Remark 4.3.4 Since the computation of covariances grows quadratically with the number of chosen POIs, in practice, it is also common to assume leakages at different time samples are independent. In this case, the covariance matrix Q_s in Template Step b becomes a diagonal matrix.

First, let us choose the target signal to be the exact value of v. We have built 16 templates. Three POIs (time samples 392, 218, 1328) were chosen as described in Template Step a. Thus for each template, the mean vector has length 3, and the covariance matrix has dimension 3×3. For example, the template for L_1, corresponding to the intermediate value $v = 0$, is given by

$$\mu_1 = (-0.04924, -0.04246, -0.07146),$$

$$Q_1 = \begin{pmatrix} 1.6110 \times 10^{-6} & -6.2968 \times 10^{-9} & -1.0592 \times 10^{-7} \\ -6.2968 \times 10^{-9} & 2.2925 \times 10^{-6} & 3.7191 \times 10^{-7} \\ -1.0592 \times 10^{-7} & 3.7191 \times 10^{-7} & 2.2567 \times 10^{-6} \end{pmatrix}.$$

As another example, the template for L_{12}, corresponding to the intermediate value $v = \text{B}$, is given by

$$\mu_{12} = (-0.04996, -0.05241, -0.07221),$$

$$Q_{12} = \begin{pmatrix} 1.6390 \times 10^{-6} & 1.6328 \times 10^{-7} & 6.3454 \times 10^{-8} \\ 1.6328 \times 10^{-7} & 2.0256 \times 10^{-6} & 1.7985 \times 10^{-7} \\ 6.3454 \times 10^{-8} & 1.7985 \times 10^{-7} & 2.1778 \times 10^{-6} \end{pmatrix}.$$

The probability scores for each key hypothesis are shown in Fig. 4.42, where the blue line corresponds to the correct key hypothesis $\hat{k}_{10} = 9$. We can see that with just a few traces, the correct key hypothesis can be distinguished from the other key hypotheses.

Next, we take the target signal to be the Hamming weight of v, $\text{wt}(v)$. Then we have five templates. The POIs were chosen as described in Template Step a: 392, 1309, 1304. The template for L_1, corresponding to $\text{wt}(v) = 0$, is given by

$$\mu_1 = (-0.04245, 0.08036, -0.03465)$$

$$Q_1 = \begin{pmatrix} 2.2925 \times 10^{-6} & -8.7422 \times 10^{-8} & 1.9156 \times 10^{-7} \\ -8.7422 \times 10^{-8} & 1.4864 \times 10^{-6} & -4.9987 \times 10^{-8} \end{pmatrix}.$$

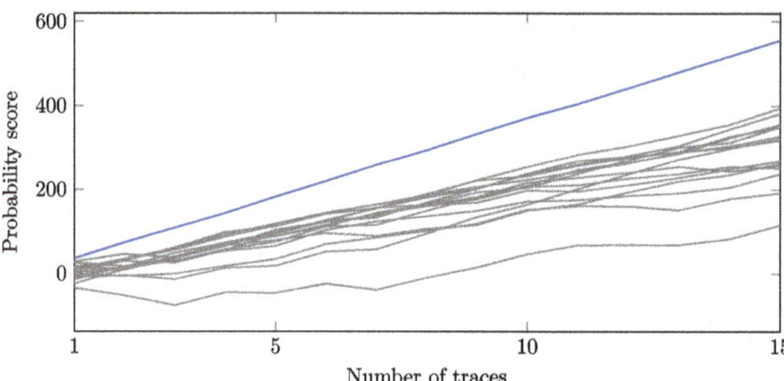

Fig. 4.42 Probability scores (Eq. 4.33) for each key hypothesis computed with different numbers of traces from *Random plaintext dataset*. The target signal is given by the exact value of v, the 0th Sbox output. Three POIs (time samples 392, 218, 1328) were chosen. The blue line corresponds to the correct key hypothesis $\hat{k}_{10} = 9$

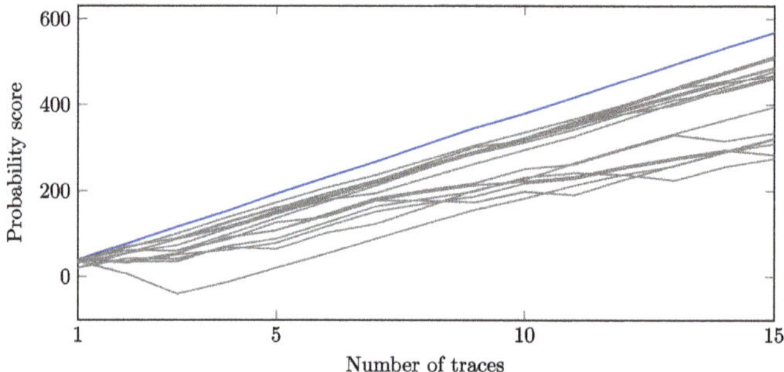

Fig. 4.43 Probability scores (Eq. 4.33) for each key hypothesis computed with different numbers of traces from *Random plaintext dataset*. The target signal is given by wt (v), the Hamming weight of the 0th Sbox output. Three POIs (time samples 392, 1309, 1304) were chosen. The blue line corresponds to the correct key hypothesis $\hat{k}_{10} = 9$

The probability scores for each key hypothesis are shown in Fig. 4.43. Similar to Fig. 4.42, with just 2, 3 traces we can distinguish the correct key hypothesis from the rest.

Attack results on other nibbles For now, we have seen practical demonstrations of how the 0th nibble of the PRESENT first round key can be recovered. As we have mentioned, DPA attacks work in a divide-and-conquer manner, recovering parts of the key in parallel using the same set of traces. As an example, we will detail the attack that recovers the first nibble of the first round key for PRESENT.

In P-DPA Step 1, our target cryptographic implementation is the same as before. The profiling traces from P-DPA Step 2 will still be the *Random dataset*. The chosen part of the key, k, in P-DPA Step 3 is now the first nibble of the first round key. Consequently, the target intermediate value, v, in P-DPA Step 4 will be the first Sbox output. We have the same relation between k, p, and v:

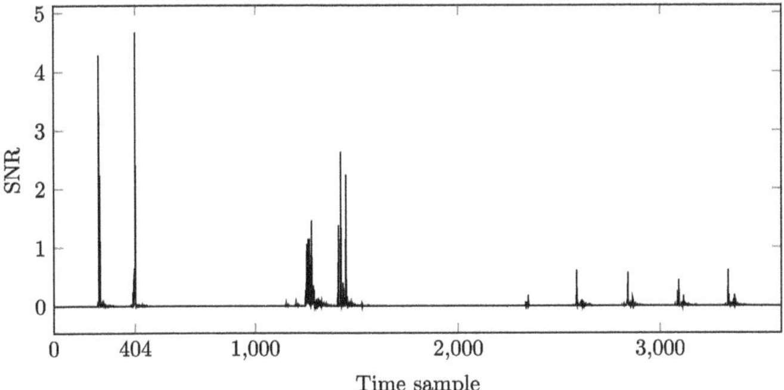

Fig. 4.44 SNR for each time sample, computed using *Random dataset*. The signal is given by the exact value of the 1st Sbox output

$$v = \text{SB}_{\text{PRESENT}}(k \oplus p),$$

where k and p denote the 1st nibble of the first round key and that of the plaintext. For the target signal in P-DPA Step 5, let us choose the exact value of v. Following P-DPA Step 6–P-DPA Step 8, the SNR values are shown in Fig. 4.44. We will choose one POI in Template Step a, which is given by the time sample corresponding to the highest point in the figure-404.

Following Template Step b, 16 templates were computed. For example, the template corresponding to the 1st Sbox output $v = \mathbf{0}$ is given by

$$\mu_1 = -0.039027, \quad \sigma_1^2 = 2.1679112 \times 10^{-6}.$$

As for attack traces in P-DPA Step 10, we use the same traces—*Random plaintext dataset*. Then according to P-DPA Step 11 and Template Step c, the probability scores for each key hypothesis are shown in Fig. 4.45. By Eq. 4.1, the correct value of the 1st nibble of the first round key is given by 8. We can see that similar to the template-based DPA attacks on the 0th key nibble (see Figs. 4.42 and 4.43), with just a few traces, we can recover the correct key nibble value.

As another example, the attack results for attacking the 6th nibble of the first round key are shown in Fig. 4.46, where by profiling, we have identified POI = 464. By Eq. 4.1, the correct value of the 6th nibble of the first round key is given by 3.

4.3.2.4 Success Rate and Guessing Entropy

Comparing Figs. 4.42 and 4.43 to Figs. 4.39 and 4.40, we cannot draw a clear conclusion about which attack method is better. In fact, a different ordering of the traces in *Random plaintext dataset* may affect our attack results. For example, by

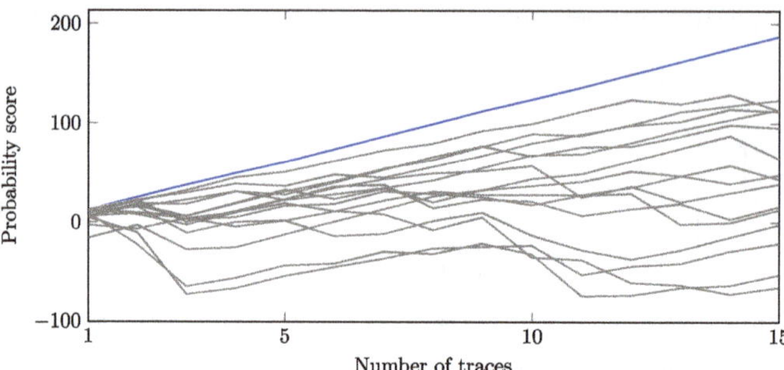

Fig. 4.45 Probability scores (Eq. 4.33) for each key hypothesis computed with different numbers of traces from *Random plaintext dataset*. The target signal is given by the exact value of the 1st Sbox output. One POI (time samples 404) was chosen. The blue line corresponds to the correct key hypothesis 8

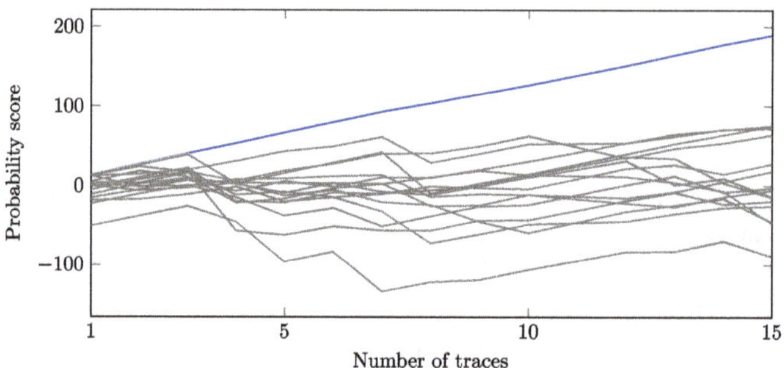

Fig. 4.46 Probability scores (Eq. 4.33) for each key hypothesis computed with different numbers of traces from *Random plaintext dataset*. The target signal is given by the exact value of the 1st Sbox output. One POI (time samples 464) was chosen. The blue line corresponds to the correct key hypothesis 3

arranging the traces in reverse order, we get Figs. 4.47 and 4.48 instead of Figs. 4.39 and 4.40.

To have a fair comparison between different attack methods (e.g., different choices of leakage models, POIs, etc.), we introduce the notion of *success rate* and *guessing entropy* [SMY09].

Note

In this part, our aim is to evaluate the DUT and our implementation against DPA attacks with different settings. Thus, we assume we have the knowledge of the key for the evaluation after the attack.

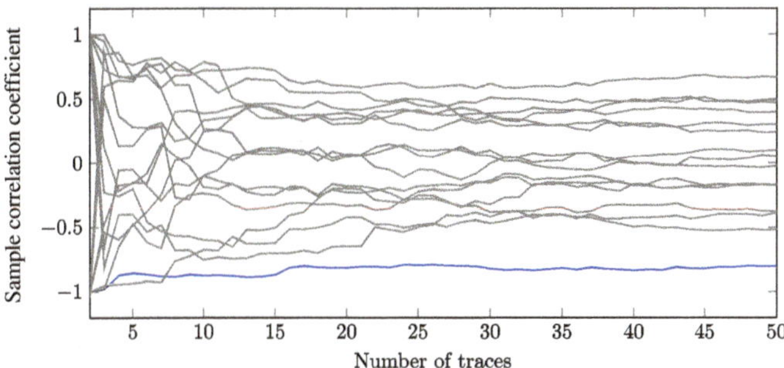

Fig. 4.47 Sample correlation coefficients $r_{i,\text{POI}}^{\hat{M}_p}$ ($i = 1, 2, \ldots, 16$) for POI $= 392$. Computed following Eq. 4.23 with the identity leakage model and the *Random plaintext dataset* arranged in reverse order. The blue line corresponds to the correct key hypothesis $\hat{k}_{10} = 9$

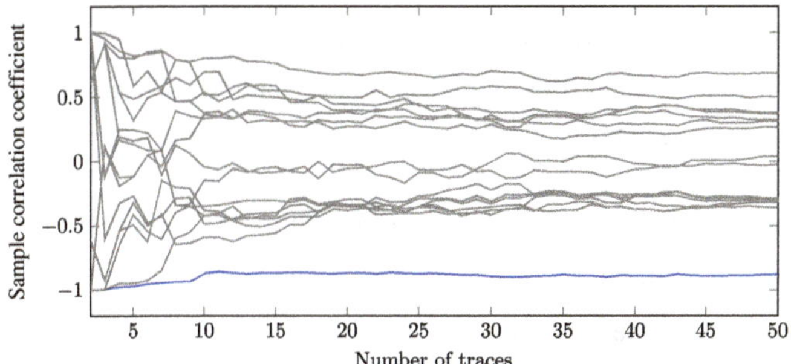

Fig. 4.48 Sample correlation coefficients $r_{i,\text{POI}}^{\hat{M}_p}$ ($i = 1, 2, \ldots, 16$) for POI $= 392$. Computed following Eq. 4.23 with the Hamming weight leakage model and the *Random plaintext dataset* arranged in reverse order. The blue line corresponds to the correct key hypothesis $\hat{k}_{10} = 9$

Fix a number of attack traces $\hat{M}_p$, and for each profiled DPA attack that we have discussed, we can assign a *score* to each key hypothesis after the attack: For leakage model-based DPA attacks, the score of a key hypothesis $\hat{k}_i$ is given by the *absolute value* of the corresponding sample correlation coefficient (Eq. 4.23), and for template-based DPA attacks, the score of a key hypothesis is given by its corresponding probability score (Eq. 4.33). Let $\mathsf{sc}_i^{\hat{M}_p}$ denote the score for the key hypothesis $\hat{k}_i$. We have

$$
\mathsf{sc}_i^{\hat{M}_p} =
\begin{cases}
\left| r_{i,\text{POI}}^{\hat{M}_p} \right| & \text{leakage model-based DPA attack, where } r_{i,\,\text{POI}}^{\hat{M}_p} \text{ is computed} \\
& \text{following Eq. 4.23} \\[2mm]
\mathsf{P}_{\hat{k}_i} & \text{template-based DPA attack, where } \mathsf{P}_{\hat{k}_i} \text{ is computed following} \\
& \text{Eq. 4.33.}
\end{cases}
$$

$$(4.34)$$

We further define $\text{score}^{\hat{M}_p}$ to be a vector consisting of the scores obtained for each key hypothesis with our DPA attack, sorted in descending order:

$$\text{score}^{\hat{M}_p} = \left(\text{sc}_{i_1}^{\hat{M}_p}, \text{sc}_{i_2}^{\hat{M}_p}, \ldots, \text{sc}_{i_{M_k}}^{\hat{M}_p} \right), \quad \text{where } \text{sc}_{i_j}^{\hat{M}_p} \geq \text{sc}_{i_{j+1}}^{\hat{M}_p}$$

$$\text{for } j = 1, 2, \ldots, M_k - 1.$$

The *key rank* of a key hypothesis $\hat{k}_i$, denoted $\text{rank}_{\hat{k}_i}^{\hat{M}_p}$, is given by the index of $\text{sc}_i^{\hat{M}_p}$ in $\text{score}^{\hat{M}_p}$. In particular, let $\hat{k}_c$ denote the correct key hypothesis. We have

$$\text{rank}_{\hat{k}_c}^{\hat{M}_p} = \text{index of } \text{sc}_c^{\hat{M}_p} \text{ in } \text{score}^{\hat{M}_p}. \tag{4.35}$$

With the same number of traces, we may also get different key ranks for the correct key hypothesis due to the different plaintexts/measurements. We consider $\text{rank}_{\hat{k}_c}^{\hat{M}_p}$ as a random variable whose randomness comes from the different plaintexts and measurements.

The ultimate goal of the attack is to achieve $\text{rank}_{\hat{k}_c}^{\hat{M}_p} = 1$[6] so that we can retrieve the correct key hypothesis. Thus, we say that an attack is successful with $\hat{M}_p$ traces if $\text{rank}_{\hat{k}_c}^{\hat{M}_p} = 1$. Then the *success rate* of an attack method with $\hat{M}_p$ traces, denoted $\text{SR}_{\hat{M}_p}$, is defined to be the probability that $\text{rank}_{\hat{k}_c}^{\hat{M}_p} = 1$:

$$\text{SR}_{\hat{M}_p} = P \left(\text{rank}_{\hat{k}_c}^{\hat{M}_p} = 1 \right). \tag{4.36}$$

Empirically, we can estimate the value of $\text{SR}_{\hat{M}_p}$ by computing the frequency of $\text{rank}_{\hat{k}_c}^{\hat{M}_p} = 1$ among a certain number of attacks.

For another metric, the *guessing entropy* for an attack method with $\hat{M}_p$ traces, denoted $\text{GE}_{\hat{M}_p}$, is given by the expectation of the random variable $\text{rank}_{\hat{k}_c}^{\hat{M}_p}$:

$$\text{GE}_{\hat{M}_p} = \text{E} \left[\text{rank}_{\hat{k}_c}^{\hat{M}_p} \right]. \tag{4.37}$$

---

[6] We note that if the key rank is low enough, it is possible to use key enumeration algorithms [VCGRS13] that enable the key recovery even in the case when $\text{rank}_{\hat{k}_c}^{\hat{M}_p} > 1$.

With the terminologies from Sect. 1.8.2, we can approximate $\mathrm{GE}_{\hat{M}_p}$ with a point estimator (see Remark 1.8.4) given by the sample mean of $\mathrm{rank}_{\hat{k}_c}^{\hat{M}_p}$.

Furthermore, when we vary the number of traces $\hat{M}_p$ used for computing $\mathrm{rank}_{\hat{k}_c}^{\hat{M}_p}$, we will get different key ranks for the correct key hypothesis. Thus the probability for the random variable $\mathrm{rank}_{\hat{k}_c}^{\hat{M}_p} = 1$ and its expectation will also vary. To analyze how $\mathrm{SR}_{\hat{M}_p}$ and $\mathrm{GE}_{\hat{M}_p}$ change with increasing values of $\hat{M}_p$, we compute estimations for $\mathrm{SR}_{\hat{M}_p}$ and $\mathrm{GE}_{\hat{M}_p}$ according to Algorithm 4.1.

The input of Algorithm 4.1 takes two user-specified values `max_trace` and `no_of_attack`. `max_trace` is the maximum number of traces (or the biggest value of $\hat{M}_p$) we would like to use for estimating $\mathrm{SR}_{\hat{M}_p}$ and $\mathrm{GE}_{\hat{M}_p}$. In line 3, sizes of S_{sr} and S_{ge} are set to be `max_trace+1` so that the $\hat{M}_p$th entry of each array corresponds to the estimation for $\mathrm{SR}_{\hat{M}_p}$ and $\mathrm{GE}_{\hat{M}_p}$, respectively. For a fixed value of $\hat{M}_p$, `no_of_attack` is the number of attacks to simulate, or equivalently, the number of elements in the sample of $\mathrm{rank}_{\hat{k}_c}^{\hat{M}_p}$ for computing the sample mean (i.e., estimation of $\mathrm{GE}_{\hat{M}_p}$) and frequency of $\mathrm{rank}_{\hat{k}_c}^{\hat{M}_p} = 1$ (i.e., estimation of $\mathrm{SR}_{\hat{M}_p}$). The set of attack traces from P-DPA Step 10 is denoted by `dataset` (line 2). For each value of $\hat{M}_p$ between 2 and `max_trace` (line 4), we simulate `no_of_attack` attacks (line 6). Thus we randomly select $\hat{M}_p \times$`no_of_attack` traces from `dataset`. Those traces are stored in an array A (line 5). Each simulated attack takes $\hat{M}_p$ traces from the array A without repetition (line 7). The key rank of the correct key hypothesis is computed following Eq. 4.35 and the attack steps described in the earlier parts of the section. $S_{ge}[\hat{M}_p]$ stores the sum of the key ranks of the correct key hypothesis for each attack (line 10); then the averaged value is computed as an estimate for the guessing entropy $\mathrm{GE}_{\hat{M}_p}$ (line 14). When the key rank of the correct key hypothesis is 1, $S_{sr}[\hat{M}_p]$ is increased by 1 (line 12). At the end $S_{sr}[\hat{M}_p]$ divided by the number of total simulated attacks gives the frequency of successful attacks (line 13).

As discussed in Sect. 4.2.3, by comparing Figs. 4.12 and 4.13 (or Figs. 4.14 and 4.15), we notice that with more traces, it is more likely for us to capture information about the inputs (or intermediate values) from the side-channel leakages. Naturally, we expect the value of $\mathrm{SR}_{\hat{M}_p}$ to be higher and the value of $\mathrm{GE}_{\hat{M}_p}$ to be lower when $\hat{M}_p$ is bigger. And the attack method that achieves $\mathrm{SR}_{\hat{M}_p} = 1$ or $\mathrm{GE}_{\hat{M}_p} = 1$ with smaller $\hat{M}_p$ is considered to be a better attack.

Now we are ready to compare our attack methods with attack traces from the *Random plaintext dataset*. We have discussed in Example 4.2.19 that by analyzing the *Random dataset*, we identified one POI for the identity leakage model and for the Hamming weight leakage model: 392. For comparison, we also consider the attack with a different POI 1328 for the identity leakage model and 1304 for the Hamming weight leakage model.

Algorithm 4.1: Computation of estimations for guessing entropy and success rate

---

Input: $\texttt{max\_trace}, \texttt{no\_of\_attack}$ // "max_trace" is the maximum number of traces we would like to use for estimating $\text{SR}_{\hat{M}_p}$ and $\text{GE}_{\hat{M}_p}$; for a fixed value of $\hat{M}_p$, "no_of_attack" is the number of attacks, or equivalently, the number of elements in one sample of $\text{rank}_{k_c}^{\hat{M}_p}$.

Output: Estimations of success rate $\text{SR}_{\hat{M}_p}$ and estimations of guessing entropy $\text{GE}_{\hat{M}_p}$ for $\hat{M}_p = 2, 3, \ldots, \texttt{max\_trace}$

1   Follow P-DPA Step 1–P-DPA Step 11 from Sect. 4.3.2.1 to do the profiling and set up the attacks (Template Step a and Template Step b from Sect. 4.3.2.3 apply if we focus on a template-based DPA)

2   Let $\texttt{dataset}$ denote the set of attack traces obtained in P-DPA Step 10

3   **zero array** of size $\texttt{max\_trace+1}$ S_{sr}, S_{ge}// variables to store estimations of success rate and guessing entropy, initialized to zero

4   **for** $\hat{M}_p = 2, \hat{M}_p \leq \texttt{max_trace}, \hat{M}_p + + $ **do**

5   **array** of size $\hat{M}_p \times \texttt{no\_of\_attack}$ $A \xleftarrow{\text{randomly choose}} \texttt{dataset}$// randomly choose "$\hat{M}_p \times \texttt{no\_of\_attack}$" traces from "dataset" and store in A

6   **for** $i = 0, i < \texttt{no_of_attack}, i + + $ **do**

7   **array** of size $\hat{M}_p$ $B = A[i \times \hat{M}_p : (i + 1) \times \hat{M}_p]$// take $\hat{M}_p$ traces from set A without repetition for each ith attack

8   Using the dataset B as attack traces, follow P-DPA Step 12–P-DPA Step 13 from Sect. 4.3.2.1 (Template Step c from Sect. 4.3.2.3 applies if we focus on a template-based DPA) to get the score of each key hypothesis given by Eq. 4.34

9       $rk = \text{rank}_{k_c}^{\hat{M}_p}$ // Key rank of the correct key hypothesis as given in Eq. 4.35

10      $S_{ge}[\hat{M}_p] + = rk$

11   **if** $rk == 1$ **then**

12        $S_{sr}[\hat{M}_p] + = 1$

13     $S_{sr}[\hat{M}_p] = S_{sr}[\hat{M}_p]/\texttt{no\_of\_attack}$// compute the frequency of successful attacks

14     $S_{ge}[\hat{M}_p] = S_{ge}[\hat{M}_p]/\texttt{no\_of\_attack}$// compute the sample mean

15   **return** S_{sr}, S_{ge}// $S_{sr}[M_p]$ (respectively, $S_{ge}[M_p]$) contains the estimation for $\text{SR}_{\hat{M}_p}$ (respectively, $\text{GE}_{\hat{M}_p}$).

As for template-based DPA, we consider two target signals: v and $\text{wt}(v)$. For each target signal, we look at two choices of POIs: one POI (392) and three POIs (392, 218, 1328 for v and 392, 1309, 1304 for $\text{wt}(v)$). When three POIs are chosen, we also analyze the case when leakages at those POIs are assumed to be independent (see Remark 4.3.4).

We note that when just one POI is considered, L_s from Eq. 4.28 becomes a normal random variable L_s. $\ell_{j,\text{POI}}$ (Eq. 4.29) becomes one single point l_{POI}^j. According to the PDF of a normal random variable (Eq. 1.37), Eq. 4.30 becomes

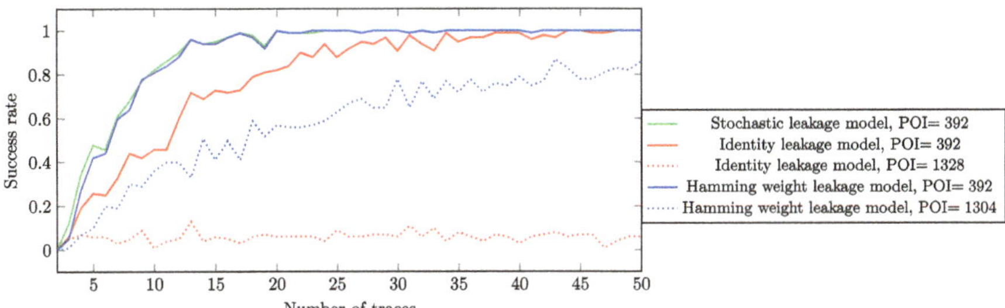

Fig. 4.49 Estimations of success rate computed following Algorithm 4.1 for profiled DPA attacks based on the stochastic leakage model, the identity leakage model, and the Hamming weight leakage model using the *Random plaintext dataset* as attack traces

$$P(\ell_j|\hat{k}_i) = P(L_{s_{ij}} = l_{\mathrm{POI}}^j) = \frac{1}{\sqrt{2\sigma_{s_{ij}}^2 \pi}} \exp\left(-\frac{(l_{\mathrm{POI}}^j - \mu_{s_{ij}})^2}{2\sigma_{s_{ij}}^2}\right),$$

where $\mu_{s_{ij}}$ and $\sigma_{s_{ij}}^2$ are estimations (template) for the mean and variance of $L_{s_{ij}}$. Consequently, the score of $\hat{k}_i$ in Eq. 4.33 is given by

$$P_{\hat{k}_i} = -\sum_{j=1}^{\hat{M}_p} \ln(\sigma_{s_{ij}}^2) + \frac{(l_{\mathrm{POI}}^j - \mu_{s_{ij}})^2}{\sigma_{s_{ij}}^2}.$$

Following Algorithm 4.1, we can compute estimations of $\mathrm{SR}_{\hat{M}_p}$ and $\mathrm{GE}_{\hat{M}_p}$ for our profiled DPA attacks with different settings. We have chosen

$$\texttt{no\_of\_attack} = 100, \quad \texttt{max\_trace} = 50.$$

For a fair comparison, for a given value of $\hat{M}_p$, the same traces are used for all attacks.

The results for leakage model-based profiled DPA are shown in Figs. 4.49 and 4.50. We have seen in Figs. 4.39, 4.40, and 4.41 that with the Hamming weight or the stochastic leakage models, we can distinguish the correct key using fewer traces as compared to using the identity leakage model. As expected, we can see from Fig. 4.49 that fewer traces are needed for SR to reach 1 with the Hamming weight or the stochastic leakage models. Furthermore, we can also see that attack results for the Hamming weight or the stochastic leakage models are similar, with the stochastic leakage model giving slightly better performance. Similarly, Fig. 4.50 shows that fewer traces are needed for GE to reach 1 using the Hamming weight or the stochastic leakage models as compared to the identity leakage model. Moreover, the results also demonstrate that the choice of POI is important for the attack. When the chosen POI has a lower SNR, the attack will need many more traces.

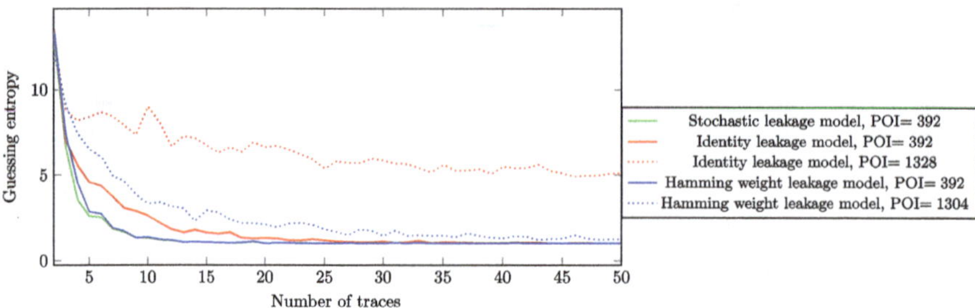

Fig. 4.50 Estimations of guessing entropy computed following Algorithm 4.1 for profiled DPA attacks based on the stochastic leakage model, the identity leakage model, and the Hamming weight leakage model using the *Random plaintext dataset* as attack traces

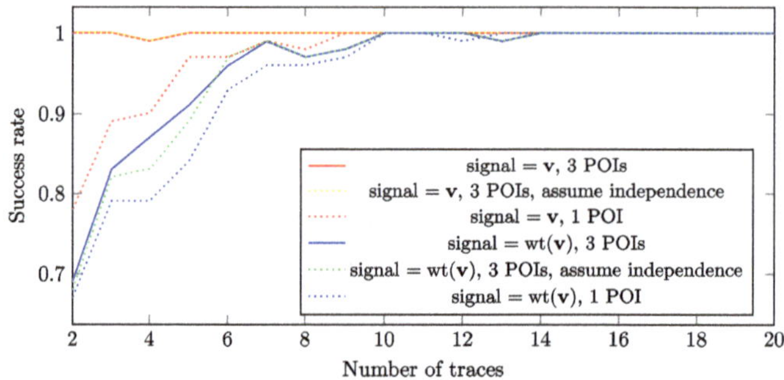

Fig. 4.51 Estimations of success rate computed following Algorithm 4.1 for template-based DPA attacks using the *Random plaintext dataset* as attack traces and the *Random dataset* as profiling traces

The results for template-based DPA are shown in Figs. 4.51 and 4.52. Note that in this case the results are shown for up to 20 traces instead of 50 for leakage model-based DPA attacks, since much fewer traces are needed for a successful attack. We have the following observations:

- When the target signal is given by v, the attack requires fewer traces as compared to the case when the target signal is given by wt (v). This is expected as for the former case we have 16 templates while for the latter we have 5. Of course, the attack results demonstrated that we had enough traces for profiling to get good templates. Without enough profiling traces, different attack results might appear.
- Assuming independence between the leakages at different POIs does not affect the attack results significantly. Especially for the case when the target signal is given by v with three POIs, those two lines are overlapping.
- Using three POIs gives better results than just one POI.
- Compared to Figs. 4.49 and 4.50, template-based DPA, in general, performs better than leakage model-based DPA. This is not surprising as more information is retrieved from the profiling traces using template-based attacks.

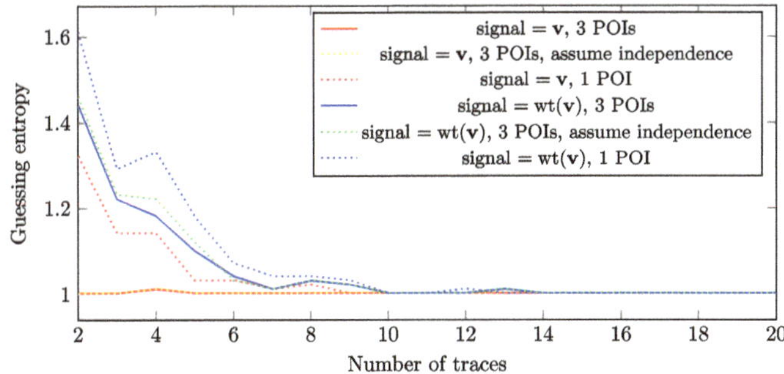

Fig. 4.52 Estimations of guessing entropy computed following Algorithm 4.1 for template-based DPA attacks using the *Random plaintext dataset* as attack traces and the *Random dataset* as profiling traces

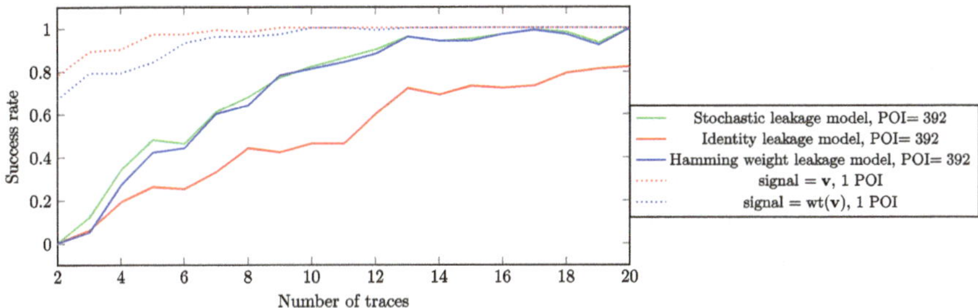

Fig. 4.53 Estimations of success rate computed following Algorithm 4.1 for leakage model-based and template-based DPA attacks with the *Random plaintext dataset* as attack traces

For easy comparison, we have also plotted the results for template-based DPA with one POI and leakage model-based DPA in Figs. 4.53 and 4.54

4.3.3 Side-Channel Assisted Differential Plaintext Attack

Side-channel assisted differential plaintext attack (SCADPA) [BJB18] aims to recover a middle round key of an SPN cipher (see Fig. 3.2) with chosen plaintext and leakages from power traces. The motivation for such an attack is that the developer might choose to protect only the first two/three and the last two/three rounds of a cipher implementation in order to increase the speed (see, e.g., [THM07, SP06]).

Before we continue our discussion on SCADPA, we introduce the notion of *difference distribution table* of an Sbox.

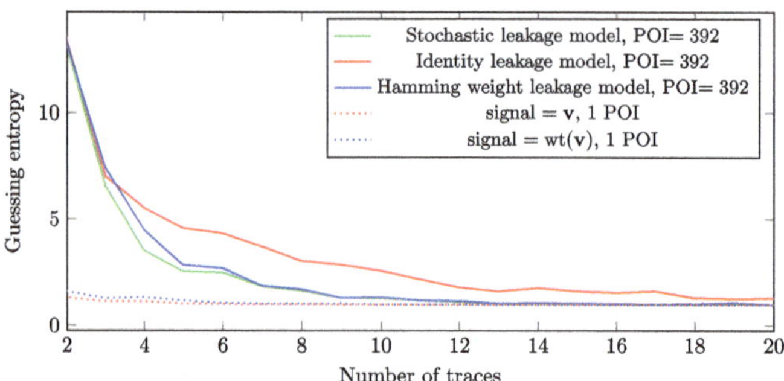

Fig. 4.54 Estimations of guessing entropy computed following Algorithm 4.1 for leakage model-based template-based DPA attacks with the *Random plaintext dataset* as attack traces

Definition 4.3.1 For an Sbox SB: $\mathbb{F}_2^{\omega_1} \to \mathbb{F}_2^{\omega_2}$, the *(extended) difference distribution table (DDT)*[7] of SB is a two-dimensional table T of size $(2^{\omega_1} - 1) \times 2^{\omega_2}$ such that for any $0 < \delta < 2^{\omega_1}$ and $0 \leq \Delta < 2^{\omega_2}$, the entry of T at the Δth row and δth column is given by

$$T[\Delta, \delta] = \left\{ a \mid a \in \mathbb{F}_2^{\omega_1}, \mathrm{SB}(a \oplus \delta) \oplus \mathrm{SB}(a) = \Delta \right\}.$$

We refer to δ as the *input difference* and Δ as the *output difference*.

Example 4.3.6 The difference distribution table for PRESENT Sbox $\mathrm{SB_{PRESENT}}$ (Table 3.11) is detailed in Table 4.1. The row corresponding to output difference $\Delta = 0$ is omitted since it is empty. For example,

$$\mathrm{SB_{PRESENT}}(9 \oplus 3) \oplus \mathrm{SB_{PRESENT}}(9) = \mathrm{SB_{PRESENT}}(A) \oplus E = 1111 \oplus 1110 = 0001 = 1.$$

Hence, 9 is in the entry corresponding to $\delta = 3$ and $\Delta = 1$.

Remark 4.3.5 Suppose we know the input difference and output difference for a particular Sbox input. Then with the DDT we can deduce the possible values of the input. For example, if we know one PRESENT Sbox input a with input difference A gives output difference 2. Then by Table 4.1, $a = 5$ or F. We will utilize such observations for SCADPA attacks and for certain fault attacks in Sect. 5.1.

Attack assumption of SCADPA For SCADPA, we have the following assumptions for the attacker's knowledge and ability:

- The attacker does not have knowledge of the exact details of the implementation. However, the attacker knows certain basic parameters of the implemented

---

[7] In the original definition of DDT [BS12], the entries are $|T[\Delta, \delta]|$, i.e., the cardinalities of $T[\Delta, \delta]$.

Table 4.1 Difference distribution table for PRESENT Sbox (Table 3.11). The columns correspond to input difference δ, and the rows correspond to output difference Δ. The row for $\Delta = 0$ is omitted since it is empty

Δ \ δ	1	2	3	4	5	6	7	8	9	A	B	C	D	E	F
1			9A		36		078F				5E		1C		24BD
2						8E	34		09	5F		1D	67AB	2C	
3	CDEF	46	12					3B		0A			58	79	
4			47		8D				35AC		0B		2F		169E
5		CDEF		0145						2389		67AB			
6		9B	CDEF	37		06	25		18					4A	
7	67AB		03	8C				5D				2E	49	1F	
8						17	AD		6F	4E	2389	0C		5B	
9	0145			9D	BE			2A			7C	3F		68	
A		02	56	BF	9C					7D	1A	48	3E		
B			8B		27	35AC		169E			4F		0D		
C		8a		26	0145	9F	BC		7E					3D	
D	2389	57			AF			4C			1B	6D		0E	
E		13		AE						24BD	6C		59		078F
F						24BD	169E	078F							35AC

algorithm, e.g., whether the implementation is round-based or bit-sliced. Such information may also be deduced by the attacker with visual inspection of the traces.

- The attacker can query encryptions with chosen plaintext and a fixed unknown master key.
- We consider observable leakages in our analysis. Specifically, the adversary can deduce from the side-channel information if a particular intermediate value is different between two distinct encryption operations. Optionally, the attacker may enhance the clarity of side-channel measurements by employing techniques, such as averaging, denoising, filtering, etc.

The goal of the attacker is to recover a middle-round key. SCADPA can be applied to any SPN cipher that has been proposed up to now.

We first give the definition of several basic notations. Suppose our target SPN cipher has in total $\mathtt{Nr}$ rounds. We consider the encryption of two plaintext blocks, denoted by S_0 and S_0'. The corresponding cipher states at the end of round i are represented by S_i and S_i', respectively. A small part (e.g., a bit, a nibble, and a byte) of the $\mathtt{XOR}$ difference between intermediate values of those two encryptions is said to be *active* if it is nonzero. The exact value of this small part is called a *differential value*.

Example 4.3.7 Let us consider AES-128. Fig. 4.55 shows a possible sequence of $\mathtt{XOR}$ differences between the cipher states of two encryptions, where colored squares correspond to active bytes. The two plaintexts S_0 and S_0' differ in the four *main diagonal* bytes. After AddRoundKey and SubBytes operations, those 4 active bytes remain active. Then ShiftRows will move the positions of those 4 active bytes. In this particular case, MixColumns operation changes 4 active bytes to just 1 active byte. Finally, after AddRoundKey, this active byte remains.

Example 4.3.8 In this example, we consider PRESENT. Figure 4.56 shows an example of how the $\mathtt{XOR}$ differences between the cipher states can change in the first three rounds. We adopt terminologies from DDT (Definition 4.3.1) and refer to the value of the active nibble corresponding to the input (respectively, output) of an active Sbox as the input difference (respectively, output difference) of this Sbox.

The plaintext pair S_0 and S_0' differs in the 0th–15th bits, corresponding to the rightmost four Sbox inputs. In this particular case, the output differences of those four Sboxes are all equal to 1. In other words, the differential value of the 0th–15th

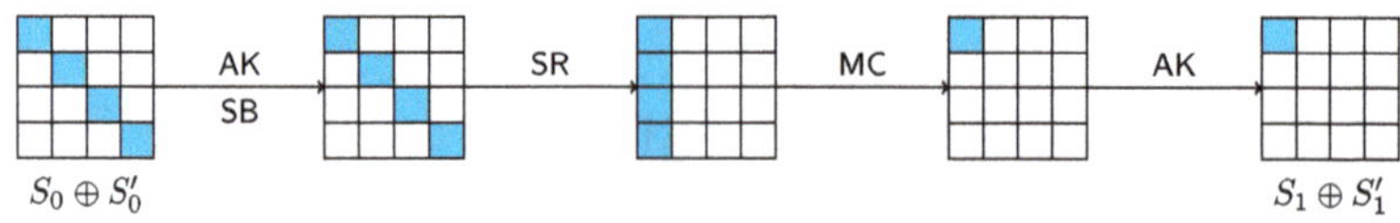

Fig. 4.55 A possible sequence of $\mathtt{XOR}$ differences between the cipher states of two encryptions, where colored squares correspond to active bytes. AK, SB, SR, and MC stand for AddRoundKey, SubBytes, ShiftRows, and MixColumns, respectively

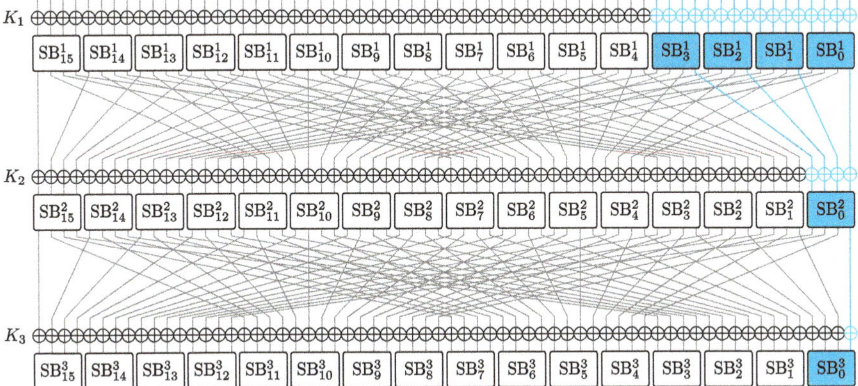

Fig. 4.56 An example of how the XOR differences between the cipher states can change after each round operation of PRESENT. The output differences of the four active Sboxes in round 1 are 1. The output difference of the single active Sbox in round 2 is also 1

bits of sBoxLayer output is 1111. Thus, after the first round, $S_1 \oplus S_1'$ has 4 active bits which correspond to the 0th Sbox input of round 2. Then we again get an output difference 1 for this Sbox, giving us just 1 active bit in $S_2 \oplus S_2'$. Consequently, we have one active nibble after the sBoxLayer operation in round 3.

A cipher state can be written as the concatenation of several small parts of the same bit length ω. In particular, let $\ell = n/\omega$, where n is the block length of the SPN cipher. We have

$$S_i = s_{i0}||s_{i1}||\dots||s_{i\ell-1}, \quad S_i' = s_{i0}'||s_{i1}'||\dots||s_{i\ell-1}', \tag{4.38}$$

where each s_{ij} and s_{ij}' is a binary string of length ω. A *differential characteristic* for round i, denoted ΔS_i, is a binary string of length ℓ:

$$\Delta S_i = (\Delta s_{i0}, \Delta s_{i1}, \dots, \Delta s_{i\ell-1}) \in \mathbb{F}_2^\ell.$$

We say that the intermediate values of two encryptions S_i and S_i' achieve the differential characteristic ΔS_i if

$$s_{ij} \oplus s_{ij}' \begin{cases} = \mathbf{0} & \text{if } \Delta s_{ij} = 0 \\ \neq \mathbf{0} & \text{if } \Delta s_{ij} = 1 \end{cases} \quad \forall j = 0, 1, \dots, \ell - 1.$$

A sequence of ΔS_is

$$\Delta S_0, \Delta S_1, \dots, \Delta S_r, \quad \text{where } r \leq \mathtt{Nr}$$

is called a *differential pattern*. If $\mathrm{wt}\,(\Delta S_r) = 1$, we say that the differential pattern *converges in round r*. A plaintext pair is said to achieve a differential pattern if the corresponding intermediate values achieve each of the differential characteristics in this differential pattern.

Example 4.3.9 [Differential pattern—AES] Continuing Example 4.3.7, we choose $\omega = 8$, and then $\ell = 128/8 = 16$. Figure 4.55 corresponds to the following differential pattern:

$$\Delta S_0, \ \Delta S_1 = 1000010000100001, \ 1000000000000000. \tag{4.39}$$

Since $\mathrm{wt}\,(\Delta S_1) = 1$, this differential pattern converges in round 1.

For example, let us take the following pair of plaintexts:

$$S_0 = \text{4C3C3F54C7AAD34E607110C753C5E990},$$

$$S_0' = \text{033C3F54C725D34E607131C753C5E90F},$$

with the master key

$$3446314634463838341464542413731. \tag{4.40}$$

Then

$$S_0 \oplus S_0' = \text{4F000000008F00000000210000000009F},$$

achieves the differential characteristic ΔS_0 from Eq. 4.39. After one round of AES, we have

$$S_1 = \text{1F1DABAE4071BDD502563FBF63841BAE},$$

$$S_1' = \text{C81DABAE4071BDD502563FBF63841BAE},$$

and

$$S_1 \oplus S_1' = \text{D700000000000000000000000000000000}.$$

Hence S_1 and S_1' achieve the differential characteristic ΔS_1 from Eq. 4.39. The differential value for the 2 active bytes in $S_1 \oplus S_1'$ is D7. We can conclude that the pair of plaintexts S_0 and S_0' achieves the differential pattern given in Eq. 4.39.

Remark 4.3.6

- Following the convention for AES intermediate value representations (see [NIS01]), the string of hexadecimal values is transferred to the 4×4 matrix of bytes (see Eq. 3.2) column by column. For example, $S_0 = \text{4C3C3F54C7AAD34E607110C753C5E990}$ in the matrix format is as follows:

$$\begin{pmatrix} 4C & C7 & 60 & 53 \\ 3C & AA & 71 & C5 \\ 3F & D3 & 10 & E9 \\ 54 & 4E & C7 & 90 \end{pmatrix}.$$

- When PRESENT is considered, we write the indices $j = 0, 1, \ldots, \ell - 1$ in reverse order following the notations for PRESENT cipher (see Sect. 3.1.3).

Example 4.3.10 [Differential pattern—PRESENT] Continuing Example 4.3.8, let $\omega = 1$, then $\ell = 64/1 = 64$. Fig. 4.56 corresponds to the differential pattern:

$$\Delta S_0 = 000000000000\text{FFFF}, \quad \Delta S_1 = 00000000000000\text{0F},$$

$$\Delta S_2 = 0000000000000001. \tag{4.41}$$

Since wt $(\Delta S_2) = 1$, this differential pattern converges in round 2.

For example, let us take the following pair of plaintexts:

$$S_0 = \text{DCFC2D56F32EC070}, \quad S_0' = \text{DCFC2D56F32E3F8F},$$

with the master key

$$1234567812345678. \tag{4.42}$$

Then

$$S_0 \oplus S_0' = 000000000000\text{FFFF},$$

which achieves the differential characteristic ΔS_0 as given in Eq. 4.41. After the first round, we get

$$S_1 = \text{0A93D18CAF9C888B}, \quad S_1' = \text{0A93D18CAF9C8884},$$

which achieves the differential characteristic ΔS_1 from Eq. 4.41 since $4 \oplus \text{B} = \text{F}$. In other words, the differential value for the active nibble in $S_1 \oplus S_1'$ is F. Finally, after the second round, we get

$$S_1 = \text{C09B5DFC8AF48EF3}, \quad S_2' = \text{C09B5DFC8AF48EF2},$$

which achieves the differential characteristic ΔS_2 from Eq. 4.41.

Now, let us fix a differential characteristic ΔS_0. With 2^{M_p} chosen plaintexts, we can construct $2^{2M_p - 1}$ plaintext pairs that achieve the differential characteristic ΔS_0. Suppose the probability for ΔS_0 to result in a differential pattern that converges in round r is $2^{-\text{pr}}$. Then if we would like to get at least one pair of plaintext that achieves a differential pattern starting with ΔS_0, and converging in round r, we should choose M_p plaintexts such that

$$M_p = \frac{\mathrm{pr} + 1}{2}. \tag{4.43}$$

Example 4.3.11 [Probability of convergence—AES] Let us consider AES and the differential characteristic ΔS_0 given by

$$\Delta S_0 = 1000010000100001. \tag{4.44}$$

We would like to compute the probability that ΔS_0 results in a differential pattern that converges in round 1, namely

$$P\left(\mathrm{wt}\left(\Delta S_1\right) = 1 | \Delta S_0 = 1000010000100001\right).$$

If we take any plaintext pair that achieves differential characteristic ΔS_0, after AddRoundKey and SubBytes operations, those 4 active bytes in the main diagonal will remain active. ShiftRows changes their positions to be all in the first column. Then after MixColumns and AddRoundKey, any byte in the first column can be active. Thus, all the possible differential characteristics ΔS_1 following the differential characteristic ΔS_0 are of the form

$$\Delta S_1 = x_0 000 x_1 000 x_2 000 x_3, \tag{4.45}$$

where $x = (x_0, x_1, x_2, x_3) \in \mathbb{F}_2^4$ and $x \neq 0$. There are in total four possible differential characteristics ΔS_1 satisfying $\mathrm{wt}\left(\Delta S_1\right) = 1$, given by four values of x that satisfy $\mathrm{wt}(x) = 1$. Those four differential patterns are shown in Fig. 4.57. We have seen one of them in Fig. 4.55 (see Example 4.3.7).

Furthermore, intermediate values S_1 and S_1' that can achieve ΔS_1 in Eq. 4.45 satisfy

$$S_1 \oplus S_1' = a_0 000 a_1 000 a_2 000 a_3 000,$$

where $a_i \in \mathbb{F}_2^8$ for $i = 0, 1, 2, 3$ and $a_{i_0} \neq 0$ for some $i_0 \in \{0, 1, 2, 3\}$. Then there are in total

$$(2^8)^4 - 1 = 2^{32} - 1$$

possible values for $S_1 \oplus S_1'$. Out of which,

$$4 \times (2^8 - 1) \approx 2^{10} \quad \text{satisfy} \quad \mathrm{wt}\left(\Delta S_1\right) = 1.$$

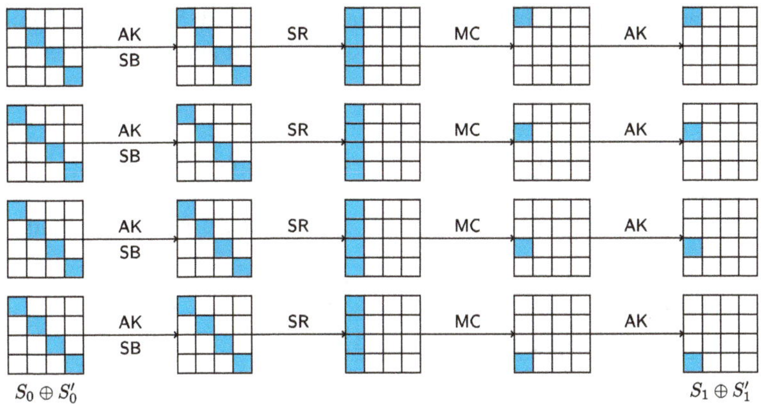

$S_0 \oplus S_0'$ $\qquad\qquad\qquad\qquad\qquad\qquad\qquad\qquad\qquad\qquad\qquad\qquad$ $S_1 \oplus S_1'$

Fig. 4.57 Illustration of how active bytes change for all four differential patterns that start with $\Delta S_0 = 1000010000100001$ and converge in round 1. Blue squares correspond to active bytes. AK, SB, SR, and MC stand for AddRoundKey, SubBytes, ShiftRows, and MixColumns, respectively

There are in total $2^{32} - 1$ possible differential values for the 4 active bytes before MixColumns operation. According to Remark 3.1.3, any value of $S_1 \oplus S_1'$ comes from exactly one differential value for those 4 active bytes. Suppose differential values of those 4 active bytes follow a uniform distribution on $\mathbb{F}_2^{32}$. Then the probability of any value of $S_1 \oplus S_1'$ to occur is $\approx 2^{-32}$. Consequently, we have

$$P\left(\mathrm{wt}\left(\Delta S_1\right) = 1 | \Delta S_0 = 1000010000100001\right) \approx \frac{2^{10}}{2^{32}} = 2^{-22}.$$

In this case, $\mathrm{pr} = 22$. By Eq. 4.43,

$$M_p = \frac{22 + 1}{2} = 11.5.$$

Thus, we need $2^{11.5}$ chosen plaintexts to get a differential pattern that starts with ΔS_0 as given in Eq. 4.44 and converges in round 1.

Example 4.3.12 [Probability of convergence—PRESENT] In this example, we consider PRESENT and the following differential characteristic:

$$\Delta S_0 = 000000000000FFFF. \tag{4.46}$$

Let SB denote the PRESENT Sbox. We would like to compute the probability of a differential pattern that starts with ΔS_0 and converges in round 2, namely

$$P\left(\mathrm{wt}\left(\Delta S_2\right) = 1 | \Delta S_0 = 000000000000FFFF\right).$$

Let SB^i_j denote the jth Sbox in round i. Recall that the 0th Sbox is the right-most Sbox (see Fig. 3.9). Let $\delta_{SB^i_j}$ and $\Delta_{SB^i_j}$ denote the input and output differences of Sbox SB^i_j, respectively.

For ΔS_2 to have Hamming weight 1, we need to have just one active Sbox in round 2 with output difference having Hamming weight 1. Let $SB^2_{j_0}$ be the single active Sbox in round 2.

By the design of pLayer (see Table 3.12), the four active Sboxes in round 1

$$SB^1_0, \quad SB^1_1, \quad SB^1_2, \quad SB^1_3$$

influence the following four Sboxes in round 2:

$$SB^2_0, \quad SB^2_4, \quad SB^2_8, \quad SB^2_{12}.$$

We also notice that the jth bit of all the four Sboxes in round 1 goes to the $(4 * j)$th Sbox in round 2. Since none of the output differences of those four Sboxes in round 1 is equal to 0, to have just one active Sbox in round 2, the output differences of those four active Sboxes in round 1 should all be the same with Hamming weight 1. This implies that the input difference of the single active Sbox in round 2, $SB^2_{j_0}$, is F. Furthermore, by Eq. 4.46, those four active Sboxes in round 1 all have input difference F.

According to Table 4.1, for input difference F, the possible output differences with Hamming weight 1 are 1 and 4. By counting the number of elements in each entry of column F in Table 4.1, we can get that the probability for the output difference to be 1, given that the input difference is F, is $4/16 = 1/4$. The same result holds for output difference 4. The probability that all output differences of the four active Sboxes in round 1 are equal to 1 is then given by

$$P\left(\Delta_{SB^1_0} = \Delta_{SB^1_1} = \Delta_{SB^1_2} = \Delta_{SB^1_3} = 1 \middle| \delta_{SB^1_0} = \delta_{SB^1_1} = \delta_{SB^1_2} = \delta_{SB^1_3} = F\right) = \left(\frac{1}{4}\right)^4$$
$$= 2^{-8}.$$

Similarly, we have

$$P\left(\Delta_{SB^1_0} = \Delta_{SB^1_1} = \Delta_{SB^1_2} = \Delta_{SB^1_3} = 4 \middle| \delta_{SB^1_0} = \delta_{SB^1_1} = \delta_{SB^1_2} = \delta_{SB^1_3} = F\right) = \left(\frac{1}{4}\right)^4$$
$$= 2^{-8}.$$

The probability for the single active Sbox in round 2 to have output difference with Hamming weight 1 is given by

$$P\left(\mathrm{wt}\left(\Delta_{\mathrm{SB}_{j_0}^2}\right) = 1 \middle| \delta_{\mathrm{SB}_{j_0}^2} = \mathrm{F}\right) = P\left(\Delta_{\mathrm{SB}_{j_0}^2} = 1 \middle| \delta_{\mathrm{SB}_{j_0}^2} = \mathrm{F}\right)$$

$$+ P\left(\Delta_{\mathrm{SB}_{j_0}^2} = 4 \middle| \delta_{\mathrm{SB}_{j_0}^2} = \mathrm{F}\right) = \frac{1}{4} + \frac{1}{4} = 2^{-1}.$$

When the output differences of

$$\mathrm{SB}_0^1, \quad \mathrm{SB}_1^1, \quad \mathrm{SB}_2^1, \quad \mathrm{SB}_3^1$$

are all equal to 1 (respectively, 4), the single active Sbox in round 2 is given by SB_0^2 (respectively, SB_8^2). We have

$$P(\mathrm{wt}(\Delta S_2) = 1 | \Delta S_0 = \mathbf{0000000000000FFFF})$$

$$= 2^{-8} P\left(\mathrm{wt}\left(\Delta_{\mathrm{SB}_0^2}\right) = 1 \middle| \delta_{\mathrm{SB}_0^2} = \mathrm{F}\right) + 2^{-8} P\left(\mathrm{wt}\left(\Delta_{\mathrm{SB}_8^2}\right) = 1 \middle| \delta_{\mathrm{SB}_8^2} = \mathrm{F}\right)$$

$$= 2^8 \times 2^{-1} + 2^{-8} \times 2^{-1} = 2^{-9} + 2^{-9} = 2^{-8}.$$

In this case, we have $\mathrm{pr}=8$. By Eq. 4.43,

$$M_p = \frac{8+1}{2} = 4.5.$$

Thus we need $2^{4.5}$ chosen plaintexts to get a differential pattern that starts with ΔS_0 as given in Eq. 4.46 and converges in round 2.

From the discussions above, we can see that there are in total four such differential patterns, corresponding to

$$\Delta_{\mathrm{SB}_0^1} = \Delta_{\mathrm{SB}_1^1} = \Delta_{\mathrm{SB}_2^1} = \Delta_{\mathrm{SB}_3^1} = 1, \Delta_{\mathrm{SB}_0^2} = 1,$$

$$\Delta_{\mathrm{SB}_0^1} = \Delta_{\mathrm{SB}_1^1} = \Delta_{\mathrm{SB}_2^1} = \Delta_{\mathrm{SB}_3^1} = 1, \Delta_{\mathrm{SB}_0^2} = 4,$$

$$\Delta_{\mathrm{SB}_0^1} = \Delta_{\mathrm{SB}_1^1} = \Delta_{\mathrm{SB}_2^1} = \Delta_{\mathrm{SB}_3^1} = 4, \Delta_{\mathrm{SB}_8^2} = 1,$$

$$\Delta_{\mathrm{SB}_0^1} = \Delta_{\mathrm{SB}_1^1} = \Delta_{\mathrm{SB}_2^1} = \Delta_{\mathrm{SB}_3^1} = 4, \Delta_{\mathrm{SB}_8^2} = 4.$$

We have seen the first one in Fig. 4.56 (see Example 4.3.8). The remaining three are shown in Figs. 4.58, 4.59, and 4.60, respectively.

In SCADPA, the attacker queries the encryption with pairs of plaintexts that achieve a *target* differential characteristic ΔS_0 and potentially result in a differential pattern that converges in round r. ΔS_0 and the round number r are chosen so that the probability of convergence is not too small. Then by comparing side-channel

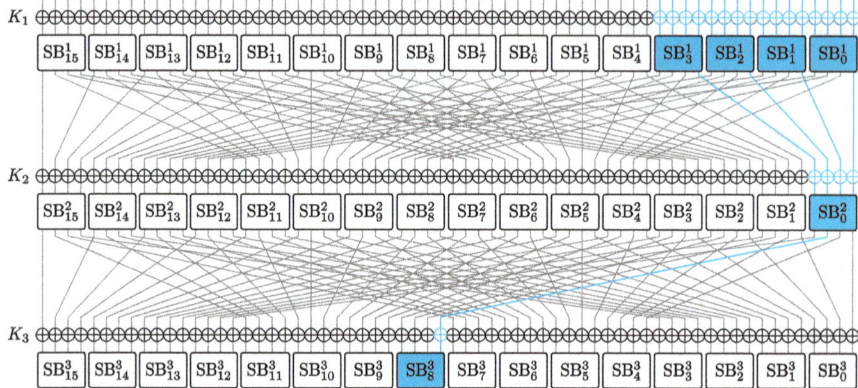

Fig. 4.58 An illustration of how the XOR differences between the cipher states can change after each round operation for PRESENT such that the pair of plaintexts achieves a differential pattern starting with ΔS_0 given in Eq. 4.46 and converging in round 2. The output differences of the four active Sboxes in round 1 are 1. The output difference of the single active Sbox in round 2 is 4

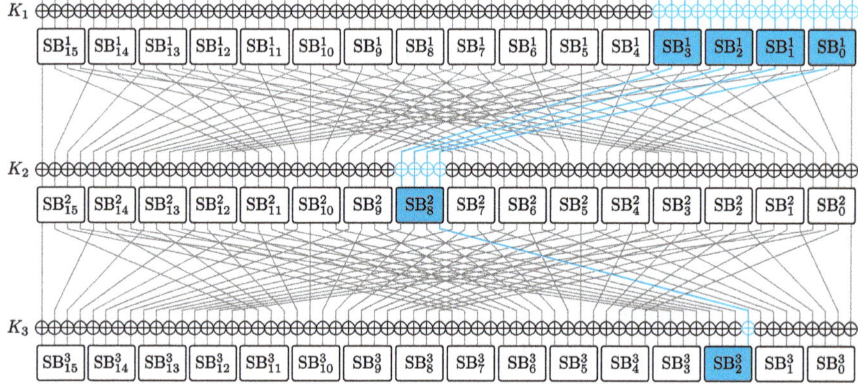

Fig. 4.59 An illustration of how the XOR differences between the cipher states can change after each round operation for PRESENT such that the pair of plaintexts achieves a differential pattern starting with ΔS_0 given in Eq. 4.46 and converges in round 2. The output differences of the four active Sboxes in round 1 are 4. The output difference of the single active Sbox in round 2 is 1

leakages of a middle round from both encryptions for a pair of plaintexts, the attacker tries to confirm if the convergence is achieved and identify the differential characteristic ΔS_r when convergence happens. Thus, we need to choose ΔS_0 and r in a way that we can find a point for side-channel observation so that the leakages can tell us whether the convergence has happened, and if yes, what is the value of ΔS_r.

Example 4.3.13 [Point for side-channel observation—AES] Let us consider AES with $\omega = 8$. As an attacker, we choose the target differential characteristic $\Delta S_0 =$

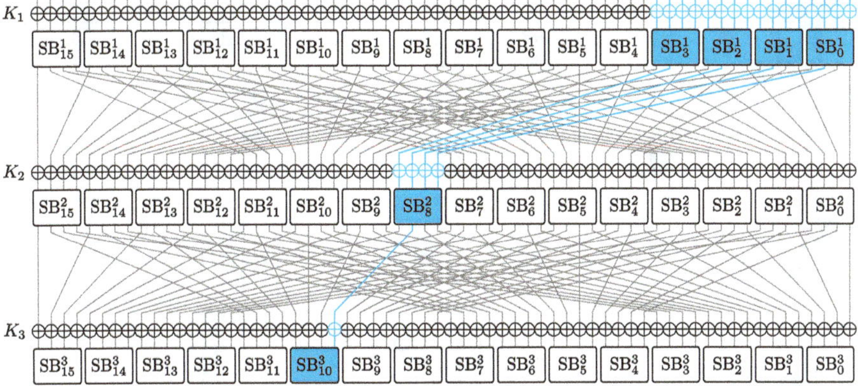

Fig. 4.60 An illustration of how the XOR differences between the cipher states can change after each round operation for PRESENT such that the pair of plaintexts achieves a differential pattern starting with ΔS_0 given in Eq. 4.46 and converges in round 2. The output differences of the four active Sboxes in round 1 are 4. The output difference of the single active Sbox in round 2 is 4

1000010000100001. Then we query the encryption with plaintext pairs that achieve this ΔS_0. For each plaintext, we take, say, N_p traces and use the averaged trace as the leakages for this plaintext. By averaging, the noise can be reduced. Then the difference between averaged traces of each pair of plaintext is computed.

As discussed in Example 4.3.11, there are four differential patterns that start with ΔS_0 and converge in round 1. They are given by the following four values of ΔS_1:

$$1000000000000, \quad 0000100000000, \quad 0000000010000, \quad 0000000000001,$$

corresponding to the single active byte at the end of round 1 being the first, second, third, and fourth bytes in the first column. Figure 4.61 shows how the active bytes change from round 1 to round 3 for all four differential patterns. In the second round, SubBytes does not change the position of this single active byte. ShiftRows changes its position to a different column unless this active byte is the first byte. Due to the property of MixColumns operation (see Remark 3.1.3), this single active byte will influence 4 bytes, leading to 4 active bytes in one single column of the cipher state. Finally, AddRoundKeys in round 2 and SubBytes operation in round 3 will not change the position or number of active bytes.

As discussed in Example 4.3.11, all possible differential characteristics ΔS_1 are of the form as given in Eq. 4.45. In case $\mathrm{wt}\,(\Delta S_1) \neq 1$, we will have more than 1 active byte at the end of round 1, which will be in more than one column after the SubBytes and ShiftRows operations in round 2. Consequently, there will be at least two active columns at the end of round 2. We can then conclude that

$$\Delta S_1 = 1000000000000 \iff \Delta S_2 = 1111000000000000,$$

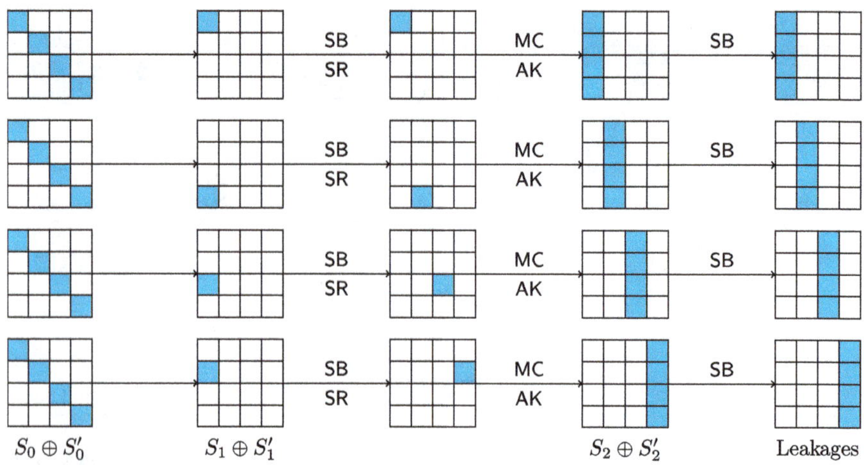

Fig. 4.61 Illustration of how active bytes change from round 1 to round 3 of AES computation, for differential patterns that start with $\Delta S_0 = 1000010000100001$

$$\Delta S_1 = 0000100000000 \Longleftrightarrow \Delta S_2 = 0000111100000000,$$

$$\Delta S_1 = 0000000010000 \Longleftrightarrow \Delta S_2 = 0000000011110000,$$

$$\Delta S_1 = 0000000000001 \Longleftrightarrow \Delta S_2 = 0000000000001111.$$

Suppose the SubBytes operation is implemented column-wise from the first column to the fourth column. Then when we take the trace difference for a pair of plaintexts, we would expect to see peaks around time samples corresponding to active columns and relatively small differences around time samples corresponding to columns that are not active during the SubBytes operation in round 3. By identifying the active columns, we can deduce the value of ΔS_1. In particular, the point of side-channel observation should be SubBytes operation in round 3. Note that we assume using SPA or other methods, the attacker can infer the timing for each operation.

As an example, with the master key from Eq. 4.40 and the experimental setup as described in Sect. 4.1. We adopted the TinyAES[8] implementation for AES, which is widely used for academic purposes. Measurements for the following four pairs of plaintexts were taken:

$$4C3C3F54C7AAD34E607110C753C5E990,$$

$$033C3F54C725D34E607131C753C5E90F; \qquad (4.47)$$

---

[8] https://github.com/kokke/tiny-AES-c

$$3B06201F5EAA0BD6794C249610FBE927,$$

$$5F06201F5E750BD6794CB79610FBE995; \tag{4.48}$$

$$2D2A49F26A79655214056A7B5F35A9E9,$$

$$D12A49F26ACC655214052D7B5F35A9C6; \tag{4.49}$$

$$0EDB19A25C7EF1FDDED31178EE6E7478,$$

$$FADB19A25C06F1FDDED30E78EE6E7415. \tag{4.50}$$

$N_p = 100$ traces were collected for each plaintext. All pairs of plaintexts achieve the same differential characteristic $\Delta S_0 = 1000010000100001$. The ΔS_1 values are given by

$$1000000000000, \quad 0000100000000, \quad 0000000010000, \quad 0000000000001,$$

respectively. Illustrations of the active bytes change for each pair correspond to the four rows of Fig. 4.61.

In Fig. 4.62, the difference between the averaged traces of each pair of the plaintexts is in red, blue, green, and yellow, respectively. We have also plotted the averaged traces for the first plaintext in Eq. 4.47 (in gray), for the purpose of identifying the round operations. Similar to Fig. 4.3, we can find the rough time interval for the SubBytes operation in round 3, which is colored in pink. This is the point for our side-channel observation. After zooming in, we get Fig. 4.63. Recall that the SubBytes operation was implemented column-wise starting from the first column. By the choice of the plaintext pairs, the red, blue, green, and yellow traces correspond to a single active column (see Fig. 4.61) at the first, second, third, and

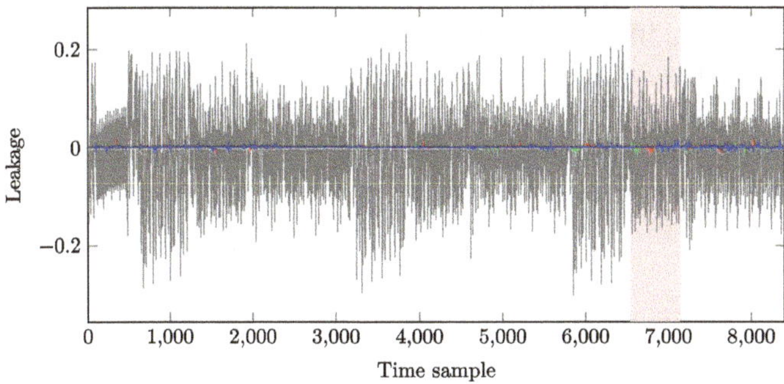

Fig. 4.62 The difference between the averaged traces of plaintext pairs from Eqs. 4.47, 4.48, 4.49, and 4.50 is in red, blue, green, and yellow, respectively. The averaged trace for the first plaintext in Eq. 4.47 is in gray. With this gray plot, similar to Fig. 4.3 we can find the rough time interval for the SubBytes operation in round 3, which is colored in pink

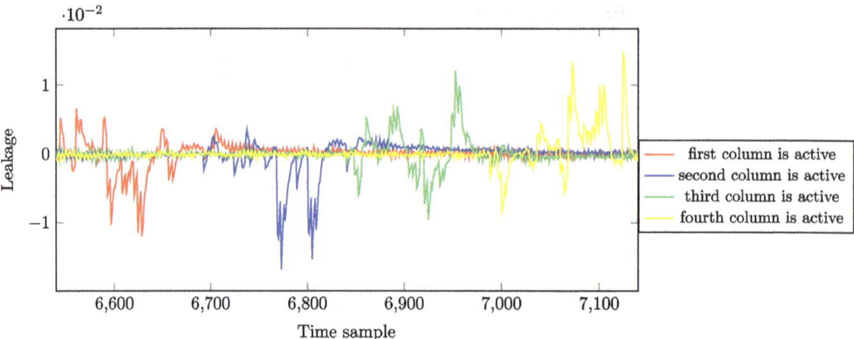

Fig. 4.63 Zoom in to the SubBytes computation (pink area) in Fig. 4.62. The difference between the averaged traces of plaintext pair from Eqs. 4.47, 4.48, 4.49, and 4.50 is in red, blue, green, and yellow, respectively. They correspond to a single active column at the first, second, third, and fourth positions, respectively, during the SubBytes operation in round 3

fourth positions, respectively. This agrees with what we see in Fig. 4.63—the four colored peaks are in sequential order.

Example 4.3.14 [Point for side-channel observation—PRESENT] In this example, we look at PRESENT encryption and let SB denote the PRESENT Sbox. We take $\omega = 1$, and we choose

$$\Delta S_0 = \texttt{0000000000000FFFF}.$$

We aim to find a pair of plaintexts S_0 and S_1 that achieves a differential pattern starting with ΔS_0 and converging in round 2. For each plaintext, we take N_p traces and use the averaged trace as the leakages for this plaintext. Then the difference between averaged traces of each pair of plaintext is computed. We assume that the sBoxLayer operation is implemented nibble-wise, starting from the 0th nibble (right-most) to the 15th nibble (left-most).

Convergence in round 2 means that there is just 1 active bit at the end of round 2. Consequently, we will have just one active Sbox before pLayer in round 3.

On the other hand, suppose there is just one active Sbox in round 3. As discussed in Example 4.3.12, with ΔS_0, the four active Sboxes in round 1 are

$$\text{SB}_0^1, \quad \text{SB}_1^1, \quad \text{SB}_2^1, \quad \text{SB}_3^1.$$

And they will influence four Sboxes in round 2:

$$\text{SB}_0^2, \quad \text{SB}_4^2, \quad \text{SB}_8^2, \quad \text{SB}_{12}^2.$$

By the design of pLayer we know each of those four Sboxes from round 2 will affect four Sboxes in round 3 as shown below:

$$\mathrm{SB}_0^2 : \text{influences } \mathrm{SB}_0^3, \quad \mathrm{SB}_4^3, \quad \mathrm{SB}_8^3, \quad \mathrm{SB}_{12}^3,$$

$$\mathrm{SB}_4^2 : \text{influences } \mathrm{SB}_1^3, \quad \mathrm{SB}_5^3, \quad \mathrm{SB}_9^3, \quad \mathrm{SB}_{13}^3,$$

$$\mathrm{SB}_8^2 : \text{influences } \mathrm{SB}_2^3, \quad \mathrm{SB}_6^3, \quad \mathrm{SB}_{10}^3, \quad \mathrm{SB}_{14}^3,$$

$$\mathrm{SB}_{12}^2 : \text{influences } \mathrm{SB}_3^3, \quad \mathrm{SB}_7^3, \quad \mathrm{SB}_{11}^3, \quad \mathrm{SB}_{15}^3.$$

In particular, they all influence different Sboxes in round 3. Since there is just one active Sbox in round 3, there is just one active Sbox in round 2. We also note that different bits of the output of an Sbox in round 2 go to different Sboxes in round 3, and we can then conclude that there is just 1 active bit at the end of round 2. Moreover, with the position of the active Sbox in round 3, we can further identify the position of the active bit in round 2 with our knowledge of pLayer.

Thus, by observing the leakages around sBoxLayer in round 3, we will be able to see if the convergence has happened and identify the value of ΔS_2.

As an example, let us take the master key to be the one given by Eq. 4.42. We also take the plaintext pair from Example 4.3.10, namely

$$S_0 = \mathrm{DCFC2D56F32EC070}, \quad S_0' = \mathrm{DCFC2D56F32E3F8F}. \tag{4.51}$$

The experimental setup is as described in Sect. 4.1, and measurements were done for three rounds of PRESENT computations. $N_p = 2000$ traces were collected for each plaintext. Recall that this pair of plaintext achieves the following differential pattern:

$$\Delta S_0 = \mathrm{0000000000000FFFF}, \quad \Delta S_1 = \mathrm{000000000000000F},$$

$$\Delta S_2 = \mathrm{0000000000000001}.$$

In particular, there is one single active Sbox SB_0^3 before the pLayer operation of round 3.

For comparison, we also collected 2000 traces for each of the following four plaintexts:

$$\mathrm{8F5F8BD2E7CF5989}, \quad \mathrm{8F5F8BD2E7CFA676} \tag{4.52}$$

and

$$\mathrm{F2DCDC8341D45F79}, \quad \mathrm{F2DCDC8341D4A086}, \tag{4.53}$$

where the first pair of plaintext (Eq. 4.52) achieves the same differential characteristics ΔS_0 and ΔS_1, but at the end of round 2, the differential characteristic is given by

$$\mathrm{0000000100000000}.$$

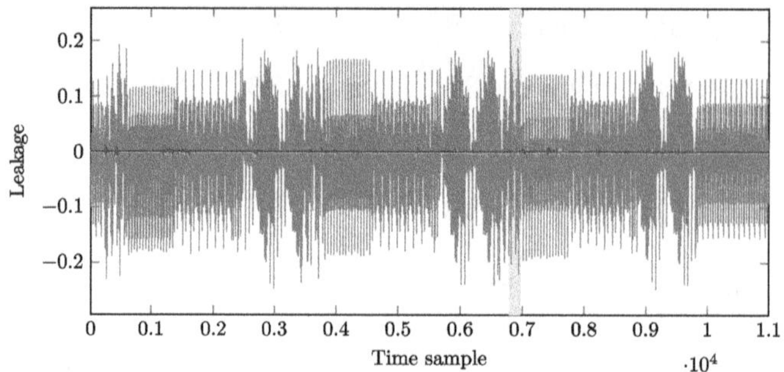

Fig. 4.64 The difference between the averaged traces of S_0 and S_0' from Eq. 4.51 (in red), plaintext pair from Eq. 4.52 (in blue), and plaintext pair from Eq. 4.53 (in green). The averaged trace for S_0 is in gray. With this gray plot, similar to Fig. 4.3 we can find the rough time interval for the sBoxLayer operation in round 3, which is colored in pink

In this case, we have one single active Sbox SB_8^3 before the pLayer operation of round 3.

The second pair of plaintext (Eq. 4.53) also achieves the same ΔS_0 and ΔS_1, while the differential characteristic at the end of round 2 is given by

$$0001000100010000.$$

Then for this pair of plaintext, there are three active Sboxes $(\mathrm{SB}_4^3, \mathrm{SB}_8^3, \mathrm{SB}_{12}^3)$ before the pLayer operation of round 3.

In Fig. 4.64, the difference between the averaged traces of S_0 and S_0' (Eq. 4.51), plaintext pair from Eq. 4.52, and plaintext pair from Eq. 4.53 are in red, blue, and green, respectively. We have also plotted the averaged traces for S_0 (in gray) for the purpose of identifying the round operations. Similar to Fig. 4.3, we can find the rough time interval for the sBoxLayer operation in round 3, which is colored in pink. This time interval corresponds to our point of side-channel observation. After zooming in, we get Fig. 4.65.

Recall that the sBoxLayer is implemented nibble-wise. From the above discussions, we know that the red, blue, and green traces correspond to active Sboxes

$$\mathrm{SB}_0^3; \quad \mathrm{SB}_8^3; \quad \mathrm{SB}_4^3, \mathrm{SB}_8^3, \mathrm{SB}_{12}^3$$

before round 3 pLayer operation, respectively. This agrees with what we see in Fig. 4.65. There is a single peak in the red line and the blue line, while the green line has three peaks. The peak in the red line (SB_0^3) is at the beginning of the sBoxLayer. The first peak of the green line (SB_4^3) is between the peaks of the red (SB_0^3) and blue (SB_8^3) lines. The peak of the blue line coincides with the second peak of the green line (SB_8^3). The last peak of the green line (SB_{12}^3) is in the last quarter of the whole time interval.

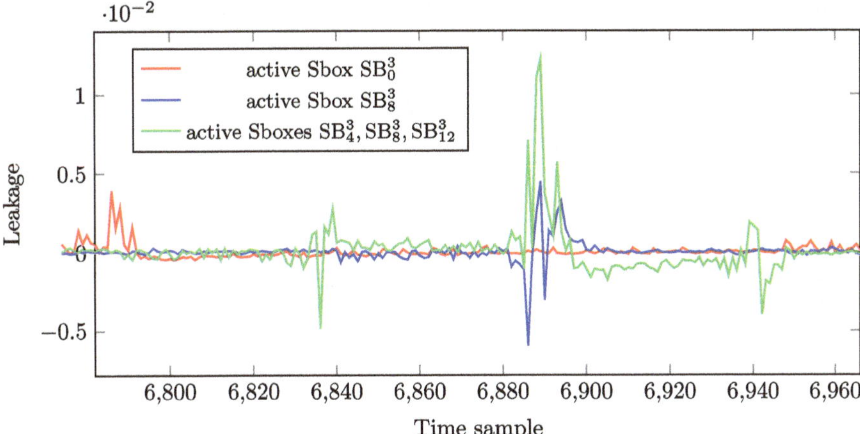

Fig. 4.65 Zoom in to the sBoxLayer computation (pink area) in Fig. 4.64. The difference between the averaged traces of S_0 and S_0' from Eq. 4.51 (in red), plaintext pair from Eq. 4.52 (in blue), and plaintext pair from Eq. 4.53 (in green). They correspond to active Sboxes SB_0^3; SB_8^3; SB_4^3, SB_8^3, SB_{12}^3 before pLayer of round 3

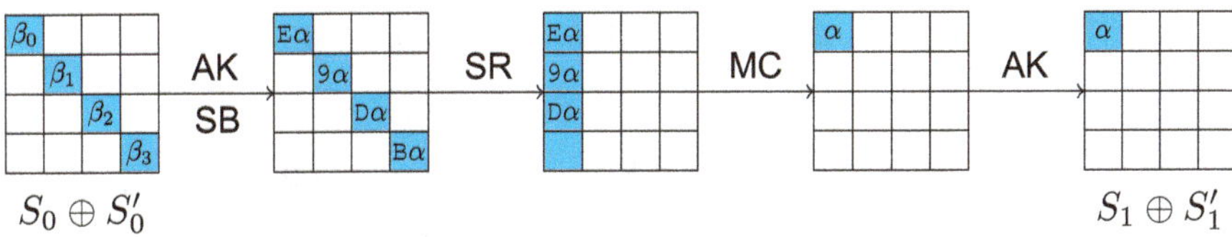

Fig. 4.66 An illustration of differential values for the differential pattern $\Delta S_0 = 1000010000100001$ and $\Delta S_1 = 1000000000000000$

Our ultimate goal is to recover information about the secret keys. Thus, another criterion for choosing ΔS_0 and r is that the possible key hypotheses can be reduced once we find a pair of plaintexts that achieves a converging differential pattern, and we know the value of ΔS_r.

Example 4.3.15 [Reduce key hypotheses—AES] Let us consider AES with $\omega = 8$. As an attacker, we choose the target differential characteristic $\Delta S_0 = 1000010000100001$. Then we query the encryption with plaintext pairs that achieve this ΔS_0. Suppose with the help of side-channel leakages, we have identified a pair of plaintexts S_0 and S_1 that gives a differential pattern converging in round 1 with $\Delta S_1 = 1000000000000000$. Let α be the differential value of the single active byte at the end of round 1. Then, using InvMixColumns (see Eq. 3.7), the differential value of the 4 active bytes right after the SubBytes operation in round 1 is given by

$$\mathtt{0E} \cdot \alpha, \quad \mathtt{09} \cdot \alpha, \quad \mathtt{0D} \cdot \alpha, \quad \mathtt{0B} \cdot \alpha.$$

Let $\beta_0, \beta_1, \beta_2$, and β_0 be the differential values of the 4 active bytes in the main diagonal of the plaintexts. An illustration is shown in Fig. 4.66.

We represent the master key of AES (which is also the whitening key used at the beginning of the encryption) as a matrix:

$$K = \begin{pmatrix} k_{00} & k_{01} & k_{02} & k_{03} \\ k_{10} & k_{11} & k_{12} & k_{13} \\ k_{20} & k_{21} & k_{22} & k_{23} \\ k_{30} & k_{31} & k_{32} & k_{33} \end{pmatrix}.$$

We represent the plaintext S_0 as the following matrix (note that this representation follows the same notation as in Eq. 3.2, which is different from the notations in Eq. 4.38):

$$S_0 = \begin{pmatrix} s_{00} & s_{01} & s_{02} & s_{03} \\ s_{10} & s_{11} & s_{12} & s_{13} \\ s_{20} & s_{21} & s_{22} & s_{23} \\ s_{30} & s_{31} & s_{32} & s_{33} \end{pmatrix}.$$

Then we have

$$\text{SB}_{\text{AES}}(s_{00} \oplus k_{00} \oplus \beta_1) \oplus \text{SB}_{\text{AES}}(s_{00} \oplus k_{00}) = \text{0E} \cdot \alpha$$

$$\text{SB}_{\text{AES}}(s_{11} \oplus k_{11} \oplus \beta_2) \oplus \text{SB}_{\text{AES}}(s_{11} \oplus k_{11}) = \text{09} \cdot \alpha$$

$$\text{SB}_{\text{AES}}(s_{22} \oplus k_{22} \oplus \beta_3) \oplus \text{SB}_{\text{AES}}(s_{22} \oplus k_{22}) = \text{0D} \cdot \alpha$$

$$\text{SB}_{\text{AES}}(s_{33} \oplus k_{33} \oplus \beta_4) \oplus \text{SB}_{\text{AES}}(s_{33} \oplus k_{33}) = \text{0B} \cdot \alpha.$$

Thus,

$$s_{00} \oplus k_{00}, \quad s_{11} \oplus k_{11}, \quad s_{22} \oplus k_{22}, \quad s_{33} \oplus k_{33}$$

are AES Sbox inputs that give output differences

$$\text{0E} \cdot \alpha, \quad \text{09} \cdot \alpha, \quad \text{0D} \cdot \alpha, \quad \text{0B} \cdot \alpha$$

with input differences

$$\beta_1, \quad \beta_2, \quad \beta_3, \quad \beta_4,$$

respectively. Then, by using the difference distribution table for AES Sbox and with the knowledge of the plaintexts, we can reduce the key hypotheses (see Remark 4.3.5).

As an example, let us take the master key to be the one given by Eq. 4.40. Continuing Example 4.3.13, with side-channel leakages, we have identified the following pair of plaintexts that achieves the differential pattern mentioned above, namely,

$$S_0 = \text{4C3C3F54C7AAD34E607110C753C5E990},$$

$$S'_0 = \text{033C3F54C725D34E607131C753C5E90F}.$$

In this case, we have

$$\beta_1 = \text{4C} \oplus \text{03} = \text{4F}$$

$$\beta_2 = \text{AA} \oplus \text{25} = \text{8F}$$

$$\beta_3 = \text{10} \oplus \text{31} = \text{21}$$

$$\beta_4 = \text{90} \oplus \text{0F} = \text{9F}.$$

And

$$s_{00} = \text{4C}, \quad s_{11} = \text{AA}, \quad s_{22} = \text{10}, \quad s_{33} = \text{90}.$$

Thus,

$$\text{4C} \oplus k_{00}, \quad \text{AA} \oplus k_{11}, \quad \text{10} \oplus k_{22}, \quad \text{90} \oplus k_{33}$$

are the AES Sbox inputs that give output differences

$$\text{0E} \cdot \alpha, \quad \text{09} \cdot \alpha, \quad \text{0D} \cdot \alpha, \quad \text{0B} \cdot \alpha$$

with input differences

$$\text{4F}, \quad \text{8F}, \quad \text{21}, \quad \text{9F},$$

respectively. To find the possible values of $k_{00}, k_{11}, k_{22}, k_{33}$, we first find values of α such that the following entries of the AES Sbox DDT are nonempty:

$$(\text{0E} \cdot \alpha, \text{4F}), \quad (\text{09} \cdot \alpha, \text{8F}), \quad (\text{0D} \cdot \alpha, \text{21}), \quad (\text{0B} \cdot \alpha, \text{9F}).$$

There are in total 13 of them, as shown in Table 4.2. Each of those values gives a few hypotheses for $k_{00} \oplus \text{4C}, k_{11} \oplus \text{AA}, k_{22} \oplus \text{10}, k_{33} \oplus \text{90}$.

Consequently, we can find all the possible values for the 4 key bytes, as shown in Table 4.3. The correct master key in Eq. 4.40 and the corresponding correct value of α are marked in blue. We note that the remaining number of key hypotheses is given by

$$2^4 \times 12 + 2^3 \times 4 = 224,$$

while the number of all possible key hypotheses for those 4 bytes is

$$(2^8)^4 = 2^{32}.$$

Table 4.2 In the first column, we list the possible values of α such that the following entries of AES Sbox DDT are nonempty $(0E \cdot \alpha, 4F)$, $(09 \cdot \alpha, 8F)$, $(0D \cdot \alpha, 21)$, $(0B \cdot \alpha, 9F)$. The corresponding hypotheses for $k_{00} \oplus 4C$, $k_{11} \oplus AA$, $k_{22} \oplus 10$, $k_{33} \oplus 90$ are listed in the second, third, and fourth columns, respectively. The correct value of α is marked in blue. A detailed analysis is shown in Example 4.3.15

α	$k_{00} \oplus 4C$	$k_{11} \oplus AA$	$k_{22} \oplus 10$	$k_{33} \oplus 90$
1A	16,59	65,EA	CF,EE	62,FD
29	96,D9	3E,B1	85,A4	78,E7
42	28,67	58,D7	59,78	16,89
5D	AB,E4	40,CF	81,A0	45,DA
66	3,4C	2C,A3	D8,F9	1D,82
71	AF,E0	78,F7	DE,FF	39,A6
74	82,CD	5D,D2	7,26	4E,D1
95	7,48	43,CC	87,A6	65,FA
9C	97,D8	0,3D,8F,B2	44,65	7F,E0
CC	1D,52	37,B8	93,B2	5F,C0
D7	37,78	63,EC	56,77	3E,A1
E7	3A,75	7B,F4	1B,3A	63,FC
EB	BB,F4	34,BB	CD,EC	54,CB

Table 4.3 Possible values of α and the corresponding key hypotheses for $k_{00}, k_{11}, k_{22}, k_{33}$, the main diagonal of the AES master key. The correct key bytes are marked in blue. A detailed analysis is shown in Example 4.3.15

α	k_{00}	k_{11}	k_{22}	k_{33}
1A	5A,15	CF,40	DF,FE	F2,6D
29	DA,95	94,1B	95,B4	E8,77
42	64,2B	F2,7D	49,68	86,19
5D	E7,A8	EA,65	91,B0	D5,4A
66	4F,00	86,9	C8,E9	8D,12
71	E3,AC	D2,5D	CE,EF	A9,36
74	CE,81	F7,78	17,36	DE,41
95	4B,04	E9,66	97,B6	F5,6A
9C	DB,94	AA,97,25,18	54,75	EF,70
CC	51,1E	9D,12	83,A2	CF,50
D7	7B,34	C9,46	46,67	AE,31
E7	76,39	D1,5E	B,2A	F3,6C
EB	F7,B8	9E,11	DD,FC	C4,5B

We can see that the attack can significantly reduce the key hypotheses.

Example 4.3.16 [Reduce key hypotheses—PRESENT] Now we look at PRESENT encryption. Take $\omega = 1$, and let

$$\Delta S_0 = 0000000000000FFFF. \tag{4.54}$$

We aim to find a pair of plaintexts S_0 and S_1 that achieve a differential pattern starting with ΔS_0 and converging in round 2. Suppose by analyzing the side-channel leakages, we have identified such a pair of plaintexts S_0 and S_1 that gives a differential pattern converging in round 2 with

$$\Delta S_2 = 0000000000000001. \tag{4.55}$$

Since there is only 1 active bit (bit 0) at the end of round 2, we know by the design of PRESENT that this means there is only one active Sbox in round 2—Sbox SB_0^2 (see Fig. 4.56). By analyzing the pLayer operation, we know that the output differences of Sboxes $SB_0^1, SB_1^1, SB_2^1, SB_3^1$ are all equal to 1. By our choice of plaintexts, we also know that the input differences of those Sboxes are all equal to F. According to PRESENT Sbox DDT in Table 4.1, the inputs of those four Sboxes are among 2, 4, B, and D. In other words, let

$$S_0 = b_{63}b_{62}\ldots b_1b_0.$$

And let

$$K_1 = \kappa_{63}^1 \kappa_{62}^1 \ldots \kappa_0^1$$

denote the first round key. Then

$$b_{j+3}b_{j+2}b_{j+1}b_j \oplus \kappa_{j+3}^1\kappa_{j+2}^1\kappa_{j+1}^1\kappa_j^1 \in \{2, 4, B, D\}, \quad \text{for } j = 0, 4, 8, 12. \tag{4.56}$$

With the knowledge of the plaintexts, we can reduce the key hypotheses. In particular, the remaining number of key hypotheses for the 0th–15th bits of K_1 is

$$4^4 = 2^8 = 256,$$

while the total number of all possible key hypotheses for those 16 bits is 2^{16}.

As an example, let us take the master key to be the one given by Eq. 4.42. We can compute that the first round key is given by

$$K_1 = \texttt{0000123456781234}. \tag{4.57}$$

Continuing Example 4.3.14, suppose with side-channel leakages, we have identified the following pair of plaintexts that achieves the differential pattern starting with ΔS_0 in Eq. 4.54 and converging in round 2 with ΔS_2 from Eq. 4.55:

$$S_0 = \texttt{DCFC2D56F32EC070}, \quad S_0' = \texttt{DCFC2D56F32E3F8F}.$$

In this case, Eq. 4.56 gives

$$0 \oplus \kappa_3^1\kappa_2^1\kappa_1^1\kappa_0^1 \in \{2, 4, B, D\}, \quad 7 \oplus \kappa_7^1\kappa_6^1\kappa_5^1\kappa_4^1 \in \{2, 4, B, D\},$$

$$0 \oplus \kappa_{11}^1\kappa_{10}^1\kappa_9^1\kappa_8^1 \in \{2, 4, B, D\}, \quad C \oplus \kappa_{15}^1\kappa_{14}^1\kappa_{13}^1\kappa_{12}^1 \in \{2, 4, B, D\}.$$

We can then reduce all the possible key hypotheses for the 0th–15th bits of K_1:

$$\kappa_3^1\kappa_2^1\kappa_1^1\kappa_0^1 \in \{2, 4, B, D\}, \quad \kappa_7^1\kappa_6^1\kappa_5^1\kappa_4^1 \in \{5, 3, C, A\},$$

$$\kappa_{11}^1 \kappa_{10}^1 \kappa_9^1 \kappa_8^1 \in \{2, 4, B, D\}, \ \kappa_{15}^1 \kappa_{14}^1 \kappa_{13}^1 \kappa_{12}^1 \in \{E, 8, 7, 1\},$$

where the correct key nibbles given by Eq. 4.57 are marked in blue.

Up to now, we have seen how SCADPA can reduce the key hypotheses on 4 bytes of AES master key and 4 nibbles of the first round key for PRESENT. In general, the steps for SCADPA are as follows:

SCADPA Step 1 **Choose the target cryptographic implementation**. SCADPA applies to all SPN ciphers that have been proposed so far. As running examples, we will continue to discuss the attacks on AES-128 and PRESENT.

SCADPA Step 2 **Choose the value** ω. Based on our chosen cipher, we need to decide the value of ω for our attack. This value is highly dependent on the cipher design. In general, for AES-like ciphers, we would choose ω to be the same as the size of the Sbox. And for bit permutation-based ciphers (e.g., PRESENT), we choose ω to be 1.

SCADPA Step 3 **Identify a target differential characteristic ΔS_0, a round number r for convergence, and a point for side-channel observation**. We would like to look for plaintext pairs that achieve a differential pattern starting with ΔS_0 and converging in round r. We also need to decide on a point for side-channel leakage analysis during the computation after round r. The choice of ΔS_0, r, and the point for side-channel observation should satisfy the following conditions:

- The probability of convergence is not too small. In particular, if the probability is $2^{-\text{pr}}$, we will need 2^{M_p} chosen plaintexts for the attack, where $M_p = 0.5\text{pr} + 0.5$.
- Using side-channel leakages at the chosen point of measurement, we should be able to confirm if the convergence has appeared for the differential pattern between a pair of plaintexts. Furthermore, it is possible to identify the value of ΔS_r in case the convergence appears.
- The possible key hypotheses can be reduced once we find a pair of plaintexts that achieves a converging differential pattern and obtain the value of ΔS_r.

SCADPA Step 4 **Choose plaintexts**. We choose 2^{M_p} distinct plaintexts so that each pair of them achieves the target differential characteristic ΔS_0.

SCADPA Step 5 **Side-channel measurement and observation**. With each plaintext, we measure N_p traces. The average trace of those N_p traces is computed for each plaintext. For each pair of plaintexts, we take the difference of the corresponding average traces and analyze the difference trace at the chosen point of observation. Once we find one difference trace that indicates the convergence has occurred, we deduce the value of ΔS_r from the measurements and carry on to the next step.

SCADPA Step 6 **Reduce key hypotheses**. Once we identify a pair of plaintexts that archives a converging differential pattern, we can reduce the key hypotheses using the knowledge of ΔS_r and the plaintexts.

Example 4.3.17 In summary, a SCADPA attack on AES-128 starts with choosing

$$\omega = 8, \quad \Delta S_0 = 1000010000100001, \quad r = 1,$$

and the point for side-channel observation being the SubBytes operation in round 3. Then we query AES encryption with $2^{11.5}$ (see Example 4.3.11) chosen plaintexts such that each pair of them achieves the differential characteristic ΔS_0. With side-channel leakages, we can deduce if convergence has happened, and if yes, we record the value of ΔS_1 (see Example 4.3.13). Finally with a similar computation as in Example 4.3.15, we reduce the key hypotheses for the 4 bytes in the main diagonal of the master key.

Similar attacks can be carried out on the other *"diagonals"* of the master key to reduce the key hypotheses of the whole master key. In particular, the other values of ΔS_0 can be

$$0100001000011000, \quad 0010000110000100, \quad 0001100001000010.$$

The possible differential patterns for each ΔS_0 are shown in Fig. 4.67, where each figure represents four different differential patterns starting with the same ΔS_0. The blue-colored squares represent active bytes, and only one of those 4 colored bytes is active in the last two cipher states (so that the differential pattern converges in round 1).

Example 4.3.18 As for SCADPA attack on PRESENT, we start by choosing

$$\omega = 1, \quad \Delta S_0 = 0000000000000FFFF, \quad r = 2,$$

and point for side-channel observation being the sBoxLayer operation in round 3. Then we query PRESENT encryption with $2^{4.5}$ (see Example 4.3.12) chosen plaintexts such that each pair of them achieves the differential characteristic ΔS_0. With side-channel leakages, we can deduce if convergence has happened, and if yes, we record the value of ΔS_2 (see Example 4.3.14). Finally with a similar computation as in Example 4.3.16, we reduce the key hypotheses for the 0th–15th bits of the first round key. We have also computed that the remaining number of key hypotheses will be 2^8 instead of the original 2^{16}.

Similar attacks can be carried out on the other bits of the first round key to reduce the key hypotheses of the whole round key. In particular, the other values of ΔS_0 can be

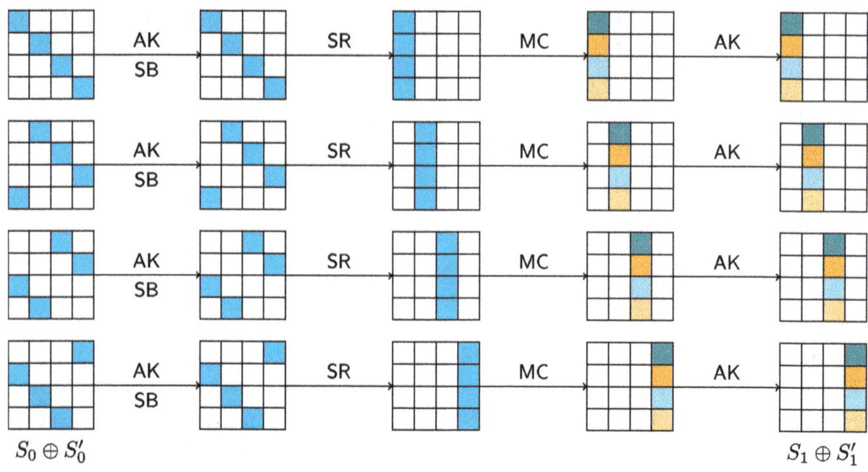

Fig. 4.67 The possible differential patterns for AES encryption with $\triangle S_0$ equal to 1000010000100001, 0100001000011000, 0010000110000100, 0001100001000010, respectively. Each figure represents four different differential patterns starting with the same $\triangle S_0$. The blue-colored squares represent active bytes and only one of those 4 colored bytes is active in the last two cipher states

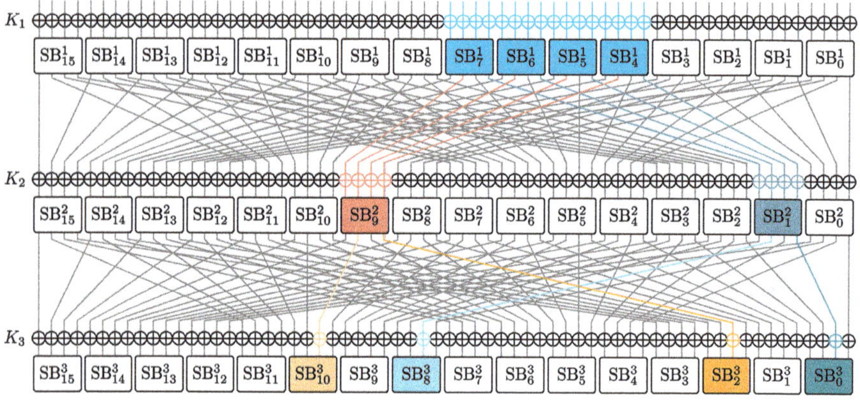

Fig. 4.68 The possible differential patterns for PRESENT encryption that start with $\triangle S_0 = $ 00000000FFFF0000 and converge in round 2. There are in total four patterns—the single active bit at the end of round 2 can be the 4th, 6th, 32nd, or 34th bit

00000000FFFF0000,   0000FFFF00000000,   FFFF000000000000.

The possible differential patterns for each of the three values of $\triangle S_0$ are shown in Figs. 4.68, 4.69, and 4.70. Each figure shows four differential patterns that converge in round 2. Each differential pattern has 1 active nibble at the end of round 1 and a single active bit at the end of round 2.

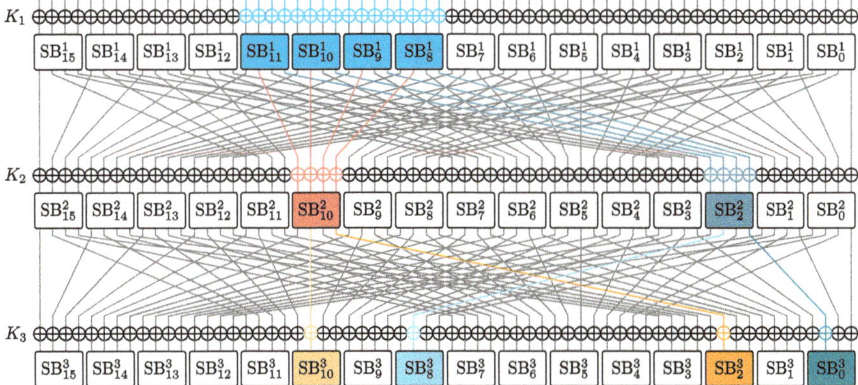

Fig. 4.69 The possible differential patterns for PRESENT encryption that start with $\Delta S_0 = $ 0000FFFF00000000 and converge in round 2. There are in total four patterns—the single active bit at the end of round 2 can be the 8th, 10th, 36th, or 38th bit

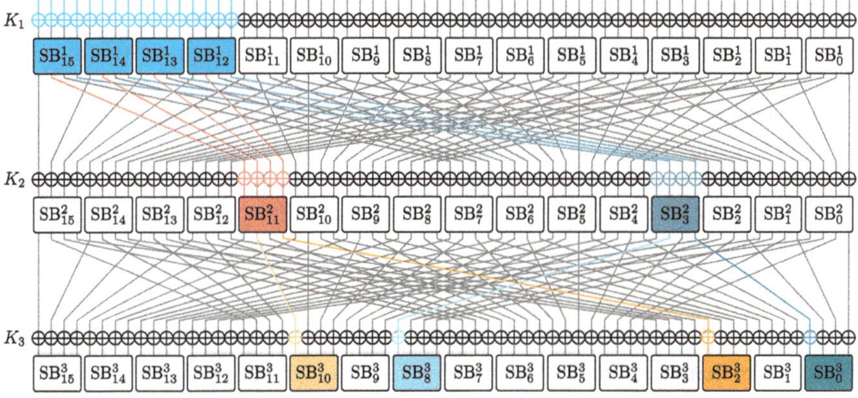

Fig. 4.70 The possible differential patterns for PRESENT encryption that start with $\Delta S_0 = $ FFFF000000000000 and converge in round 2. There are in total four patterns—the single active bit at the end of round 2 can be the 12th, 14th, 40th, or 42nd bit

Remark 4.3.7 As mentioned in Example 4.3.13, for the attack on AES, we assume the SubBytes operation is implemented column-wise from the first column to the fourth column. We note that a different ordering of the columns in the implementation is also vulnerable to the attack, provided the attacker knows the ordering of the columns. Similarly, for our attack on PRESENT, we have mentioned in Example 4.3.14 that the sBoxLayer operation is implemented nibble-wise from the 0th nibble to the 15th nibble. A different ordering of the nibbles still can be attacked as long as the attacker has the knowledge of the specific ordering.

4.4 Side-Channel Analysis Attacks on RSA and RSA Signatures

In this section, we will discuss one SPA and one DPA attack on implementations of RSA and RSA signatures.

Following the same notations from Sect. 3.3, let p, q be two distinct odd primes. $n = pq$ and $e \in \mathbb{Z}_{\varphi(n)}^*$ are the public keys. $d = e^{-1} \bmod \varphi(n)$ is the private key. Furthermore, let

$$d_{\ell_d-1} d_{\ell_d-2} \ldots d_1 d_0$$

be the binary representation of d.

We will show how SPA and DPA can be used to recover the value of d during the computation of

$$a^d \bmod n \tag{4.58}$$

for some $a \in \mathbb{Z}_n$. For both attacks, we focus on one particular method for implementing the modular exponentiation—the left-to-right square and multiply algorithm (Algorithm 3.8). A similar SPA attack can also be applied to the right-to-left square and multiply algorithm (Algorithm 3.7). We note that the attacks can be carried out during either the decryption of RSA or the signature signing procedure of RSA signatures.

For the experiments, we have set the values of the parameters as given in Examples 3.3.2:[9]

$$p = 29, \quad q = 41, \quad n = 1189, \quad \varphi(n) = 1120, \quad e = 3, \quad d = 747. \tag{4.59}$$

Then our implementation of Algorithm 3.8 can be described by Algorithm 4.2.

Remark 4.4.1 We note that since $\varphi(n) = (p-1)(q-1)$ is even and $\gcd(d, \varphi(n)) = 1$, d is odd. In particular $d_0 = 1$.

4.4.1 Simple Power Analysis

We have seen that DPA exploits the relationship between leakages at specific time samples and the data being processed in the DUT. SPA, on the other hand, analyzes leakages along the time axis, exploiting relationships between leakages

---

[9] Note that for easy illustration, the values we choose for p and q are much smaller than practical values.

Algorithm 4.2: Left-to-right square and multiply algorithm for computing modular exponentiation (see Algorithm 3.8) with parameters from Eq. 4.59

Input: a // $a \in \mathbb{Z}_{1189}$
Output: $a^{747} \bmod 1189$
1 $n = 1189$
2 $dbin = [1, 1, 0, 1, 0, 1, 1, 1, 0, 1]$// binary representation of $d = 747$, $d_0 = 1$,
    $d_1 = 1$
3 $\ell_d = $ length of $dbin$// bit length of d
4 $t = 1$
5 **for** $i = \ell_d - 1, i \geq 0, i - - $ **do**
6 $\quad$ $t = t * t \bmod n$
   $\quad$ // ith bit of d is 1
7 $\quad$ **if** $d_i = 1$ **then**
8 $\quad\quad$ $t = a * t \bmod n$

9 **return** t

---

and operations. Similar to profiled DPA, SPA requires knowledge of the exact implementation.

We have seen in the analysis of Fig. 4.3 that different operations can be deduced from observing the power traces. An SPA attack on the square and multiply algorithm works with a similar method—we examine the traces to figure out if both square and multiplication are executed in one loop from line 5 (the corresponding bit of d is 1) or not (the corresponding bit of d is 0). Following Kerckhoffs' principle (see Definition 2.1.3), we assume the attacker has the knowledge of Algorithm 4.2 except for the values of bits of d in line 2.

With the experimental setting as described in Sect. 4.1, we measured one power trace for the computation of Algorithm 4.2 on our DUT. The trace is shown in Fig. 4.71. We can see ten similar patterns. By examining Algorithm 4.2, we have two guesses:

Guess a Each pattern corresponds to one modular operation (modular square from line 6 or modular multiplication from line 8).
Guess b Each pattern corresponds to one loop from line 5.

Let S denote the modular square operation from line 6 and M the modular multiplication from line 8. We observe that the loop in line 5 contains either one square operation (S) or one square followed by one multiplication operation (SM). We also have the following correspondence between operations in loop i and the ith bit of the secret key d:

$$\text{loop } i \text{ contains only S} \iff d_i = 0, \quad \text{loop } i \text{ consists of SM} \iff d_i = 1.$$

We further notice that there are mainly two types of patterns in Fig. 4.71, one with a single cluster of peaks and one with more than one cluster of peaks. They are colored in green and blue in Fig. 4.72, respectively.

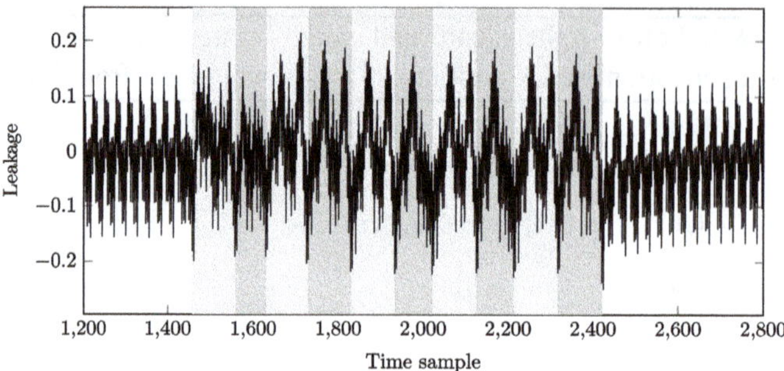

Fig. 4.71 One trace corresponding to the computation of Algorithm 4.2. We can see ten similar patterns

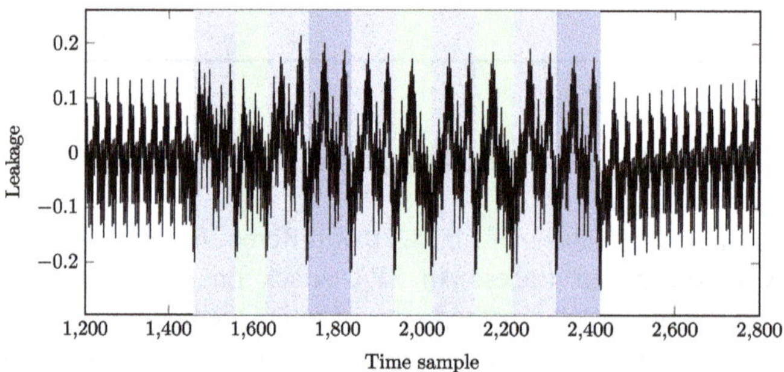

Fig. 4.72 Highlighted two types of patterns from Fig. 4.71. One pattern with a single cluster of peaks (colored in green) and one with more than one cluster of peaks (colored in blue)

Let us first assume that Guess a is correct. Based on the above observations, we have two possibilities to consider:

- The (green colored) single peaked patterns correspond to modular square operation (S), and the (blue colored) multiple peaked patterns correspond to modular multiplication operation (M).
- The (green colored) single peaked patterns correspond to modular multiplication operation (M), and the (blue colored) multiple peaked patterns correspond to modular square operation (S).

We know that $d_0 = 1$ (see Remark 4.4.1). Then we can deduce that the last blue-colored pattern in Fig. 4.72 does not represent a single modular square operation (S). On the other hand, the start of the computation will always be a modular square operation, which then indicates that the first blue-colored pattern corresponds to S. We have reached a contradiction, and we conclude that Guess a is not correct.

Next, we assume Guess b is correct. Similarly, we have two possibilities to consider:

- The (green colored) single peaked patterns represent a single modular square operation (S), i.e., the corresponding bit of d is 0, and the (blue colored) multiple peaked patterns represent SM and the corresponding bit of d is 1.
- The green-colored patterns correspond to SM, and the blue-colored patterns correspond to S.

As discussed above, the end of the computation does not stop with $d_0 = 0$, and thus the blue-colored patterns represent SM, i.e., the corresponding bit of d is 1. Consequently, the green-colored patterns correspond to loops with the bit of d being 0. We can then read out the value of bits d_i $(i = \ell_d - 1, \ldots, 0, 1)$ from Fig. 4.72:

$$1 \quad 0 \quad 1 \quad 1 \quad 1 \quad 0 \quad 1 \quad 0 \quad 1 \quad 1.$$

Finally, we recover the secret key

$$d = 1011101011_2 = 747.$$

One might argue that the first green pattern in Fig. 4.72 may also be a multiple peaked blue pattern. We note that this pattern is shorter than the other blue patterns. Hence it is more likely to correspond to one operation instead of two. Nevertheless, in a realistic attack, one could use brute force to recover this bit.

Remark 4.4.2 By the design of the Montgomery powering ladder (Algorithm 3.9), there is always a multiplication followed by a square operation, making it safe against our SPA attack presented above.

4.4.2   Differential Power Analysis

For DPA attacks on RSA implementations, we focus on Montgomery's method for implementing modular multiplication MonPro (see Algorithm 3.17). Following Eq. 4.59 and Example 3.5.17, we have

$$\begin{aligned} p = 29, \quad q &= 41, \quad n = 1189, \quad \varphi(n) = 1120, \, e = 3, \\ d = 747, \, r &= 2048, \, r^{-1} = 717, \, \hat{n} = 1235. \end{aligned} \tag{4.60}$$

Then our implementation of Montgomery left-to-right square and multiply algorithm (Algorithm 3.20) can be described by Algorithm 4.4. Also, our implementation of MonPro (Algorithm 3.17) becomes Algorithm 4.3.

We have implemented Algorithm 4.4 in our DUT. With experimental settings as described in Sect. 4.1, one trace is shown in Fig. 4.73. This trace is different from Fig. 4.72—we cannot see two distinct types of patterns. If we take a closer look at the computation of MonPro in Algorithm 4.3, we can see that the main difference between a square and a multiply is in line 3, which does not involve modular n as

---

Algorithm 4.3: `MonPro`, Montgomery product algorithm with parameters from Eq. 4.59

Input: a, b// $a, b \in \mathbb{Z}_{1189}$
Output: $717ab \bmod 1189$
1 $\hat{n} = 1235$
2 $n = 1189$
3 $t = ab$
4 $m = t\hat{n}\mathrm{AND}2047$
5 $u = (t + mn) >> 11$
6 **if** $u \geq n$ **then**
7 $\quad \lfloor \quad u = u - n$
8 **return** u

Algorithm 4.4: Montgomery left-to-right square and multiply algorithm with parameters from Eq. 4.59. `MonPro` is given by Algorithm 4.3

Input: a// $a \in \mathbb{Z}_{1189}$;
Output: $a^{747} \bmod 1189$
1 $n = 1189, \quad r = 2048$
2 $dbin = [1, 0, 1, 1, 1, 0, 1, 0, 1, 1]$// binary representation of $d = 747$, $d_0 = 1$, $d_1 = 1$
3 $\ell_d = $ length of $dbin$// bit length of d
4 $t_r = r \bmod n$
5 $a_r = ar \bmod n$
6 **for** $i = \ell_d - 1, i \geq 0, i - - $ **do**
7 $\quad \mid \quad t_r = \mathrm{MonPro}(t_r, t_r)$// $t_r = t_r \times_{\mathrm{Mon}} t_r$.
8 $\quad \mid \quad$ **if** $dbin[i] = 1$ **then**
9 $\quad \mid \quad \lfloor \quad t_r = \mathrm{MonPro}(t_r, a_r)$// $t_r = t_r \times_{\mathrm{Mon}} a_r$.
10 $t = \mathrm{MonPro}(t_r, 1)$// $t = t_r \times_{\mathrm{Mon}} 1 = t_r * r^{-1} \bmod n$.
11 **return** t

---

compared to lines 6 and 8 in Algorithm 4.2. This missing modular n operation might be the main reason for the missing pattern structure in Fig. 4.73.

Nevertheless, we can still gain important information from the trace. First, we note that there are 18 similar patterns in Fig. 4.73. By examining Algorithm 4.4, similar to Guess a and Guess b from Sect. 4.4.1, we can assume each of those 18 patterns corresponds to either one execution of `MonPro` or one loop from line 6. Since there is one extra `MonPro` operation in line 10, we know that the last pattern will not represent a loop. If each of the other patterns corresponds to one loop, we will have a secret key of bit length 17, which is longer than the bit length of n (bit length of 1189 is 10) and hence impossible. We conclude that there is a high possibility that each pattern corresponds to one execution of `MonPro`. Since when $d_i = 1$ there are two executions of `MonPro` and when $d_i = 0$ there is one execution of `MonPro`, our observations reveal that

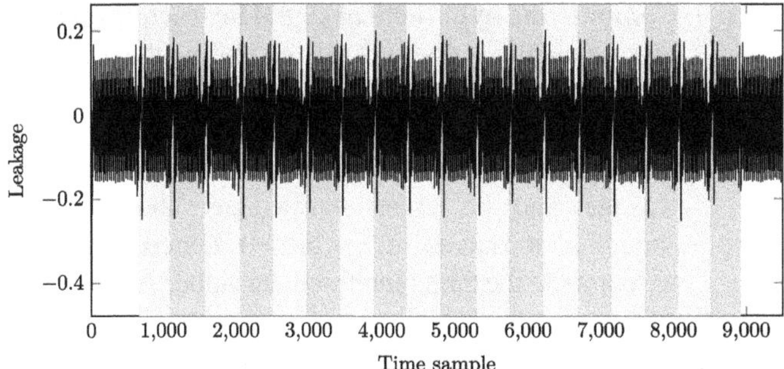

Fig. 4.73 One trace corresponding to the computation of Algorithm 4.4. We can see 18 similar patterns

$$\ell_d + \mathrm{wt}\,(d) = 17.$$

We follow similar attack steps as DPA attacks on symmetric block ciphers presented in Sect. 4.3.1.1. However, we will only describe one particular attack on RSA (originally proposed in [AFV07]), while Sect. 4.3.1.1 outlines attack steps for a generic DPA attack on any symmetric block ciphers.

DPA-RSA Step 1 **Identify the target cryptographic implementation**. As mentioned above, we focus on the left-to-right square and multiply algorithm with Montgomery's method for modular multiplication. In particular, our attack will be on an implementation of Algorithm 4.4. We remark that to have a better signal, most part of Algorithm 4.3 was implemented in ARM assembly.

DPA-RSA Step 2 **Experimental setup and measure leakages**. With the same experimental setting as in Sect. 4.1, we have measured $M = 10,000$ traces, each for a random input $a \in \mathbb{Z}_{1189}$. Let a_j ($j = 1, 2, \ldots, M$) denote the jth input with corresponding power trace $\ell_j = (l_1^j, l_2^j, \ldots, l_q^j)$, where the total number of time samples in one trace is $q = 9500$.

DPA-RSA Step 3 **Choose the part of the key to recover**. In this attack, we aim to recover the full secret key d.

DPA-RSA Step 4 **Choose the target intermediate value**. Our target intermediate value is the 0th byte of the value a_r, defined in line 5 of Algorithm 4.4. We note that a_r is only used in the algorithm when $d_i = 1$ (line 9), and thus we expect the correlation between the leakages and information related to a_r to be higher when line 9 is executed. Consequently, we will know that $d_i = 1$ for the corresponding loop. Since in practice a_r is a big integer, it is more reasonable to focus on just part of a_r. For our experiments, $a_r \in \mathbb{Z}_{1189}$ has bit length at most 11. We will focus on the 0th byte (bits $0, 1, 2 \ldots, 7$) of a_r.

DPA-RSA Step 5 **Compute the hypothetical signal for each target intermediate value.** Our attack does not rely on finding the best key hypothesis that achieves the highest absolute correlation coefficient as in DPA attacks on symmetric block ciphers. The information we exploit is that when the absolute correlation coefficient between leakages and the target intermediate value is high, the corresponding loop has secret key bit $= 1$. For each of the M inputs a_j, we compute the target intermediate value, denoted $\boldsymbol{v}_j$, as follows:

$$\boldsymbol{v}_j = \text{bits } 0, 1, 2, \ldots, 7 \text{ of } a_j r \bmod n, \quad j = 1, 2, \ldots, 10{,}000. \tag{4.61}$$

As we have seen in Sect. 4.3.2.1, the Hamming weight leakage model (Eq. 4.4) is a good estimate for leakages of our DUT. We compute the hypothetical signal corresponding to a_j, denoted $\mathcal{H}_j$, as follows:

$$\mathcal{H}_j = \text{wt}\left(\boldsymbol{v}_j\right), \quad j = 1, 2, \ldots, 10{,}000. \tag{4.62}$$

DPA-RSA Step 6 **Statistical analysis.** We view the hypothetical signal as a random variable $\mathcal{H}$ that varies when the input a changes. For a fixed time sample t, we also consider the leakage at t as a random variable L_t. Then our computations from DPA-RSA Step 5 and our traces from 11 give us a sample for this pair of random variables $(\mathcal{H}, L_t)$:

$$\left\{ (\mathcal{H}_j, l_t^j) \mid j = 1, 2, \ldots, 10{,}000 \right\}.$$

To see at what time samples the leakages are correlated to $\mathcal{H}$, the same as in DPA Step 7, we adopt the notion of correlation coefficient (Definition 1.7.11). And for each time sample t, we compute the sample correlation coefficient (Example 1.8.1), denoted by r_t, of $\mathcal{H}$ and L_t:

$$r_t := \frac{\sum_{j=1}^{M}(\mathcal{H}_j - \overline{\mathcal{H}})(l_t^j - \overline{l_t})}{\sqrt{\sum_{j=1}^{M}(\mathcal{H}_j - \overline{\mathcal{H}})^2}\sqrt{\sum_{j=1}^{M}(l_t^j - \overline{l_t})^2}},$$

$$M = 10{,}000, \quad t = 1, 2, \ldots, 9500. \tag{4.63}$$

Example 4.4.1 For our experiments, we have

$$a_1 = 900, \quad a_2 = 1083, \quad a_3 = 881, \quad a_4 = 852.$$

Then

$$a_1 r \bmod n = 900 \times 2048 \bmod 1189 = 250 = \text{FA},$$

$$a_2 r \bmod n = 1083 \times 2048 \bmod 1189 = 499 = \text{1F3},$$

$$a_3 r \bmod n = 881 \times 2048 \bmod 1189 = 575 = \text{23F},$$

$$a_4 r \bmod n = 852 \times 2048 \bmod 1189 = 633 = 279.$$

According to Eq. 4.61, we have

$$v_1 = \text{FA}, \quad v_2 = \text{F3}, \quad v_3 = \text{3F}, \quad v_4 = 79.$$

Then the hypothetical signals from DPA-RSA Step 5 are given by (Eq. 4.62)

$$\mathcal{H}_1 = \text{wt}\,(\text{FA}) = \text{wt}\,(11111010) = 6,$$

$$\mathcal{H}_2 = \text{wt}\,(\text{F3}) = \text{wt}\,(11110011) = 6,$$

$$\mathcal{H}_3 = \text{wt}\,(\text{3F}) = \text{wt}\,(00111111) = 6,$$

$$\mathcal{H}_3 = \text{wt}\,(79) = \text{wt}\,(01111001) = 5.$$

The sample correlation coefficients for all time samples are shown in Fig. 4.74. We can see a sequence of 18 patterns. To recover the secret key, we need the help of SPA. We have discussed before that there are 18 patterns in Fig. 4.73, and each of them most likely corresponds to one execution of `MonPro`.

If we put Figs. 4.73 and 4.74 together, we get Fig. 4.75. We can see that the 18 patterns corresponding to sample correlation coefficients and those corresponding to leakages coincide. Thus, we can assume each pattern in Fig. 4.74 represents one execution of `MonPro`.

Let us then take a closer look at Fig. 4.74. We can see there are mainly two types of patterns: one with a lower peak and one with a higher peak and a small high peak

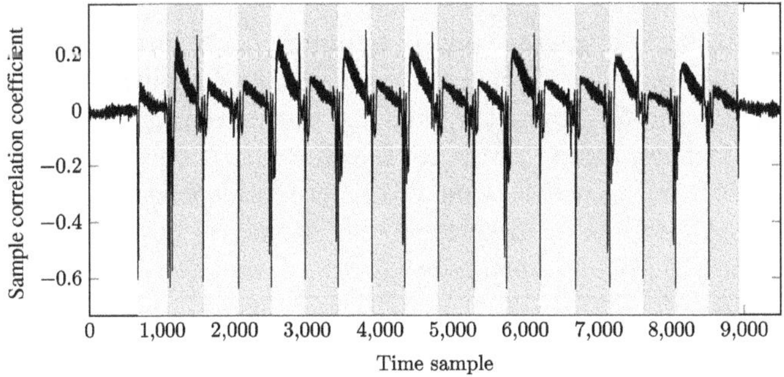

Fig. 4.74 Sample correlation coefficients r_t (Eq. 4.63) for time samples $t = 1, 2, \ldots, 9500$. We can see a sequence of 18 patterns

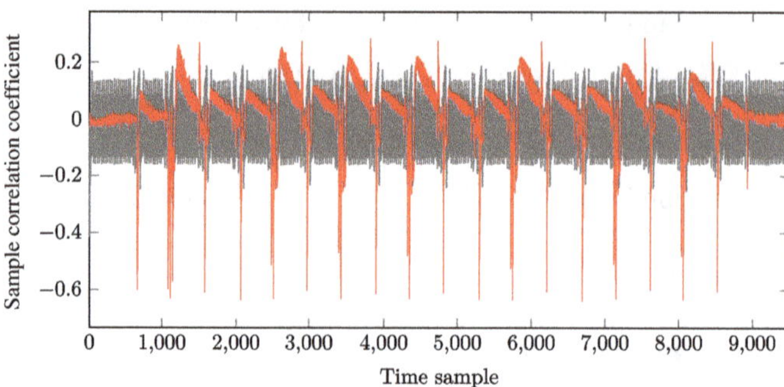

Fig. 4.75 Sample correlation coefficients from Fig. 4.73 (in red) with one power trace from Fig. 4.74 in gray. We can see that the 18 patterns corresponding to sample correlation coefficients and those corresponding to leakages coincide

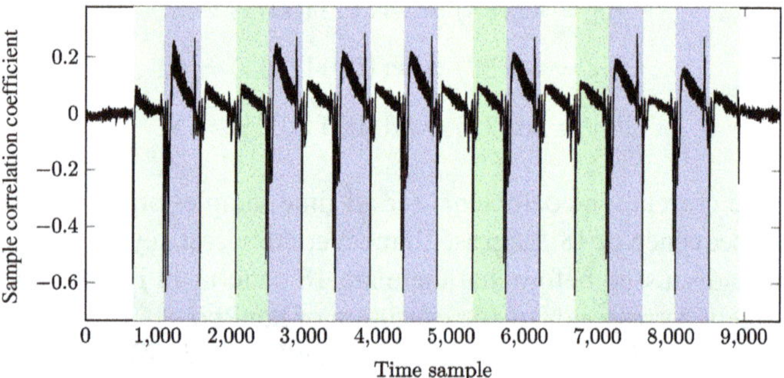

Fig. 4.76 There are mainly two types of patterns in Fig. 4.74: one with a lower peak and one with a higher peak and a small high peak at the end of the pattern. In this figure, they are highlighted in green and blue, respectively

at the end of the pattern. They are highlighted in green and blue, respectively, in Fig. 4.76.

We know that the last pattern in Fig. 4.76 corresponds to line 10 in Algorithm 4.4. Then each of the remaining 17 patterns represents the computation of either line 9 or line 7. Let S and M denote the modular square and modular multiplication computations in lines 7 and 9, respectively. Since a_r is only used in M, we can assume that a higher peaked (blue-colored) pattern corresponds to M. Consequently, a lower peaked (green-colored) pattern corresponds to S. Using Fig. 4.76, we can deduce the sequence of square and multiply operations in one execution of Algorithm 4.4:

$$SMSSMSMSMSSMSSMSM.$$

A loop in Algorithm 4.4 contains either a single S or SM. We can then map this sequence of operations into different loops (separated by spaces)

$$\text{SM S SM SM SM S SM S SM SM.}$$

Furthermore, a loop containing a single S corresponds to the secret bit $= 0$, while a loop containing SM corresponds to the secret bit $= 1$. We can then read out the bits of the secret key d:

$$1 \quad 0 \quad 1 \quad 1 \quad 1 \quad 0 \quad 1 \quad 0 \quad 1 \quad 1,$$

and we can reconstruct the key

$$d = 1011101011 = 747.$$

4.5 Countermeasures Against Side-Channel Analysis Attacks

In this section, we will discuss a few implementation-level SCA countermeasures for both symmetric block cipher and RSA implementations. We have seen how the dependency of a device's leakages (power consumption) on data and operations can be exploited to recover the secret keys of a cryptographic implementation. The goal of the countermeasures that we will see is to make the leakage of the DUT independent of the operations or the intermediate values of the executed cryptographic implementation. We will detail two types of countermeasures—hiding and masking/blinding.

The goal of a hiding-based countermeasure is to remove the operation or data dependency of leakages. This can be done by changing the leakage of the DUT in a way that every operation requires a similar (balance the leakages) or a random (randomize the leakages) amount of energy. On the other hand, the goal of a masking/blinding-based countermeasure is to remove the data dependency of the leakages by randomizing the intermediate values that the DUT is processing. The rationale is that since the value being processed in the DUT is randomized and independent of the intermediate value of the cryptographic computation, we cannot capture information on the actual intermediate value from the leakages. In practice, both types of countermeasures are used.

4.5.1 Hiding

As mentioned above, hiding-based countermeasure aims to either randomize or balance the leakages of the DUT for different operations or data. In this section, we will discuss two countermeasures that aim to balance the leakages—one for symmetric block ciphers and one for RSA implementations.

4.5.1.1 Encoding-Based Countermeasure for Symmetric Block Ciphers

In Sect. 4.3.2.2, we have discussed the stochastic leakage model. In this part, we will show a countermeasure that is based on analyzing the stochastic leakage of the DUT [MSB16].

Recall that with the stochastic leakage model, we can characterize the leakage at a single time sample. For the countermeasure, we also focus on one time sample. The coefficients (see Eq. 4.24) of the stochastic leakage model will be estimated using the measured traces. Based on the estimated leakage model, we choose a binary code (Definition 1.6.1) that results in a lower SNR at this particular time sample and makes the attacks require more effort.

In Sects. 4.3.1 and 4.3.2 we have seen attacks based on the Hamming weight leakage model on PRESENT implementations. Thus, to provide more protection, we further require the codewords in our code to have the same Hamming weight, as shown in [HBK23]. In this way, attacks based on the Hamming weight of the intermediate values will not be possible.

The steps for the countermeasure are as follows:

Code-SCA Step 1 **Identify the target instruction and target intermediate value.** As the stochastic leakage model is specific to one time sample, we need to first decide what is the most vulnerable instruction and which intermediate value needs to be protected the most. Let $v = v_{m_v-1}v_{m_v-2}\ldots v_1 v_0$ denote the target intermediate value of bit length at most m_v. In general, it is recommended that the implementation is done in assembly to identify the most vulnerable instruction.

For our illustrations, we will choose the instruction MOV for our microcontroller, and we focus on the PRESENT Sbox output. Hence $m_v = 4$. The operation we implemented is then

$$\text{MOV}\quad \text{r0}\quad a, \tag{4.64}$$

where r0 represents a register and a is the input of the target instruction.

Code-SCA Step 2 **Choose the code length n_C and the Hamming weight w_H of each codeword.** We would like to choose a code to represent our secret intermediate value v such that instead of processing v with our DUT, the corresponding codewords will be used. Clearly, the size of the binary code will be 2^{m_v} in order to represent all values of v. The length of the binary code should be at least $m_v + 1$ so that it allows us to choose which word to use as our codewords. Longer length in general not only gives us more freedom but also causes more overhead. Let n_C denote our chosen code length.

As mentioned before, we would also like the codewords in our binary code to have the same Hamming weight, making attacks based on the Hamming weight leakage model impossible. Note that there are in total $\binom{n_C}{w_H}$ words in $\mathbb{F}_2^{2^{n_C}}$ that have Hamming weight w_H. One criterion for the choice of n_C and w_H is then

$$\binom{n_C}{w_H} > 2^{m_v}, \tag{4.65}$$

so that we will have enough codewords to represent all values of v.

In summary, we are looking for an $(n_C, 2^{m_v})$-binary code such that each codeword has Hamming weight w_H. And n_C and w_H should satisfy Eq. 4.65.

For our experiments, we choose $n_C = 8$ and $w_H = 6$. We are interested in $(8, 16)$-binary codes such that each codeword has Hamming weight 6.

Code-SCA Step 3   **Experimental setup and trace measurement**. In this step, we will collect two datasets, denoted $\mathcal{T}_1$ and $\mathcal{T}_2$. Using our DUT, we repeatedly run the target instruction with random inputs. One trace is measured for each input. We first take M_1 inputs with random values from $\mathbb{F}_2^{m_v}$, which give us the dataset $\mathcal{T}_1$. Then using M_2 inputs with values from $\mathbb{F}_2^{n_C}$, we get $\mathcal{T}_2$. Suppose each trace contains q time samples. Note that we assume the traces in those two datasets are well aligned.

$\mathcal{T}_1$ represents how leakages behave when random values of v are being processed by the DUT. It will be used to identify the POI in Code-SCA Step 4, which is the time of the computation that is supposed to be the most vulnerable.

To choose a good binary code, we would like to profile leakages when the input is a codeword. Let $x = x_{n_C-1}x_{n_C-2}\ldots x_1 x_0$ be a word from $\mathbb{F}_2^{2^{n_C}}$ of bit length at most n_C. Recall from Sect. 4.3.2.2 (Eq. 4.24) that the stochastic leakage model specifies the leakage is related to the value x being processed in the device as follows:

$$\mathcal{L}(x) = \sum_{s=0}^{n_C-1} \alpha_s x_s + \text{noise}, \tag{4.66}$$

where noise $\sim \mathcal{N}(0, \sigma^2)$ denotes the noise with mean 0 and variance σ^2. Estimations for the coefficients α_s ($s = 0, 1, \ldots, n_C - 1$) will be computed with profiling traces from $\mathcal{T}_2$. We can see that it is important for traces in $\mathcal{T}_1$ and $\mathcal{T}_2$ to be aligned so that the profiling of $\mathcal{T}_2$ is carried out with the correct POI.

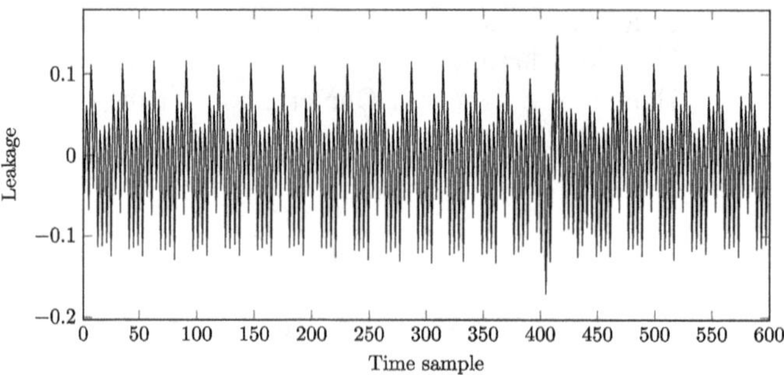

Fig. 4.77 An example of a trace from dataset $\mathcal{T}_1$, obtained in Code-SCA Step 3, which corresponds to MOV instruction surrounded by NOPs

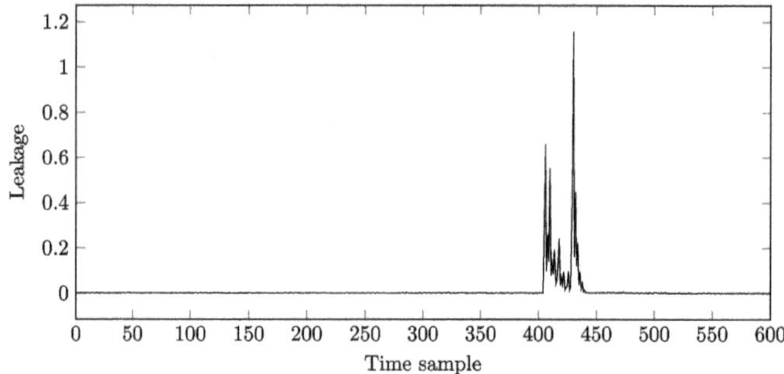

Fig. 4.78 SNR values for each time sample computed with dataset $\mathcal{T}_1$ obtained in Code-SCA Step 3. The highest point is our POI = 430

For our experiment, we measured $M_1 = 10{,}000$ traces for the MOV instruction with inputs between 0 and F and $M_2 = 10{,}000$ traces with inputs between 00 and FF. Each trace contains $q = 600$ time samples. The MOV instruction was surrounded with NOP operations. Figure 4.77 shows what one trace from $\mathcal{T}_1$ looks like. We can see that the operation MOV happens between time samples 400 and 450. Traces from $\mathcal{T}_2$ look very similar.

Code-SCA Step 4 **Identity the POI**. With the traces from $\mathcal{T}_1$ obtained in Code-SCA Step 3, we compute the SNR for each time sample following similar methods as in P-DPA Step 6–P-DPA Step 8 from Sect. 4.3.2.1. Our target signal is the exact value of $\boldsymbol{v}$, thus $M_{signal} = 2^{m_v}$ (P-DPA Step 6). The number of time samples is $q_{pf} = q$ (P-DPA Step 7). The POI is taken to be the time sample with the highest SNR.

With our 10,000 traces from $\mathcal{T}_1$, the SNR values for each time sample are shown in Fig. 4.78. And our POI is 430.

Code-SCA Step 5 **Estimate coefficients for the stochastic leakage model**. Following SLM Step a–SLM Step c in Sect. 4.3.2.2, we compute the estimations for α_s in Eq. 4.66 using dataset $\mathcal{T}_2$. With the notations from Sect. 4.3.2.2, we have $M_{pf} = M_2$. The POI was identified in Code-SCA Step 4. Note that the target intermediate value is not the value $\boldsymbol{v}$ from Code-SCA Step 1, but words from $\mathbb{F}_2^{2^n_C}$. Let $\hat{\alpha}_s$ denote the estimated value for α_s ($s = 0, 1, \ldots, n_C - 1$).

Using our dataset $\mathcal{T}_2$ for MOV instruction, POI $= 430$, and $n_C = 8$, we get the following estimations $\hat{\alpha}_s$ for α_s:

$$\hat{\alpha}_0 \approx -0.00245761, \ \hat{\alpha}_1 \approx -0.00130026, \ \hat{\alpha}_2 \approx -0.00135884,$$
$$\hat{\alpha}_3 \approx -0.00122801, \ \hat{\alpha}_4 \approx -0.00131569, \ \hat{\alpha}_5 \approx -0.00213467,$$
$$\hat{\alpha}_6 \approx -0.00209748, \ \hat{\alpha}_7 \approx -0.00221288.$$

$$(4.67)$$

Code-SCA Step 6 **Compute the estimated signal for each word with Hamming weight w_H**. Using the estimated values, $\hat{\alpha}_s$, for the coefficients, according to the stochastic leakage model (Eq. 4.66), we can compute the estimated signal of each word $\boldsymbol{x} = x_{n_C-1}x_{n_C-2}\ldots x_1 x_0$ from $\mathbb{F}_2^{2^{n_C}}$, denoted SG($\boldsymbol{x}$),

$$\mathrm{SG}(\boldsymbol{x}) = \sum_{s=0}^{n_c-1} \hat{\alpha}_s x_s. \qquad (4.68)$$

We can identify each integer between 0 and $2^{n_C} - 1$ with a unique binary string from $\mathbb{F}_2^{2^{n_C}}$ using the integer's binary representation and compute its estimated signal with Eq. 4.68. Let Words be the table of integers between 0 and $2^{n_C} - 1$ whose binary representation has Hamming weight w_H. Then the table T_{SG} of estimated signals is constructed such that

$$T_{\mathrm{SG}}[a] = \mathrm{SG}(\mathrm{Words}[a]), \quad a = 0, 1, \ldots, 2^{n_C} - 1. \qquad (4.69)$$

For our experiments,

$$\mathrm{Words} = [3F, 5F, 6F, 77, 7B, 7D, 7E, 9F, AF, B7, BB, BD, BE, CF,$$
$$D7, DB, DD, DE, E7, EB, ED, EE, F3, F5, F6, F9, FA, FC].$$

$$(4.70)$$

For example, with $\hat{\alpha}_s$ from Eq. 4.67, we have

$$T_{SG}[0] = SG(3F) = SG(00111111) = \sum_{i=0}^{5} \hat{\alpha}_i \approx -0.009795.$$

The table T_{SG} can be found in Table E.1 (Appendix E).

Code-SCA Step 7 **Find the optimal code.** Finally, we search for an optimal $(n_C, 2^{m_v})$-binary code whose codewords all have Hamming weight w_H using Algorithm 4.5 (see [MSB16, Algorithm 1]). The input of the algorithm consists of m_v, the maximum bit length of the target intermediate value from Code-SCA Step 1; n_C and w_H, the code length and the Hamming weight for each codeword chosen in Code-SCA Step 2; Words, the table of integers between 0 and $2^{n_C} - 1$ with Hamming weight w_H obtained in Code-SCA Step 6; and T_{SG}, the table of estimated signals for each integer from Words as specified in Eq. 4.69. As discussed in Code-SCA Step 2, the total number of codewords is 2^{m_v} (line 1), and the total number of binary strings of length n_C and Hamming weight w_H is $\binom{n_C}{w_H}$ (line 2). Firstly, we sort the values in T_{SG} in ascending order and save the sorted values in T_{sorted}, where $T_{\text{sorted}}[0]$ contains the lowest value from T_{SG} (line 6). The array I records the corresponding integer in Words for each estimated signal in T_{sorted} (lines 7 and 8). Next, the difference between the values in $T_{\text{sorted}}[j + \text{code\_size} - 1]$ and $T_{\text{sorted}}[j]$ is stored in the jth entry of the array D, for $j = 0, 1, \ldots, \text{total\_word-code\_size}$ (lines 9 and 10). The index of the smallest value in D is denoted by ind (line 11). Finally, the binary code consists of codewords that correspond to estimated signals in the range $D[\text{ind}]$ and $D[\text{ind}+\text{code\_size}-1]$ (lines 12 and 13).

With $m_v = 4$, $n_C = 8$, $w_H = 6$, Words from Eq. 4.70, and T_{SG} in Table E.1, we get the following $(8, 16)$-binary code

$$C_{(8,16)} := [\text{B7, D7, BD, AF, DD, 77, CF, BB,}$$

$$\text{DB, 7D, 6F, 7B, F6, FC, EE, FA}]. \tag{4.71}$$

The sorted table T_{sorted} (line 6 of Algorithm 4.5) can be found in Table E.2 (Appendix E), where the codewords are highlighted in blue.

We will argue that Algorithm 4.5 indeed outputs an optimal $(n_C, 2^{m_v})$-binary code that achieves a lower SNR, according to the stochastic leakage model with coefficients $\hat{\alpha}_s$ obtained in Code-SCA Step 5. First, let $A = \{a_1, a_2, \ldots, a_\beta\}$ be a set of β $(\beta \geq 2)$ real numbers. Define

---

Algorithm 4.5: Finding the optimal code for encoding countermeasure against SCA

Input: m_v, n_C, w_H, Words, T_{SG} // m_v is the maximum bit length of the target intermediate value identified in Code-SCA Step 1; n_C is the code length and w_H is the Hamming weight for each codeword chosen in Code-SCA Step 2; Words is the table of integers between 0 and $2^{n_C} - 1$ with Hamming weight w_H as discussed in Code-SCA Step 6; T_{SG} is the table of estimated signals for each integer from Words as specified in Eq. 4.69.

Output: An $(n_C, 2^{m_v})$-binary code with each codeword having Hamming weight w_H

1 code_size $= 2^{m_v}$ // number of codewords in our code

2 total_word $= \binom{n_C}{w_H}$ // total number of words of length n_C and Hamming weight w_H

3 array of size total_word$-$code_size$+1$ D

4 array of size total_word I
// C will store the codewords

5 array of size code_size C
// $T_{\text{sorted}}[0]$ contains the lowest value from T_{SG}

6 $T_{\text{sorted}} = T_{\text{SG}}$ sorted in ascending order

7 for $j = 0$, $j <$ total_word, $j++$ **do**
// I records the corresponding word in Words for each estimated signal in T_{sorted}

8 $\quad I[j] =$ Words [index of $T_{\text{sorted}}[j]$ in T_{SG}]

9 for $j = 0$, $j \leq$ total_word $-$ code_size, $j++$ **do**
// the jth entry of D is given by the difference between the value in $T_{\text{sorted}}[j + \text{code\_size} - 1]$ and $T_{\text{sorted}}[j]$

10 $\quad D[j] = T_{\text{sorted}}[j + \text{code\_size} - 1] - T_{\text{sorted}}[j]$

// ind is the index of the smallest value in D

11 ind $= \arg\min_j D[j]$

// the code consists of codewords that correspond to estimated signals in the range $D[\text{ind}]$ and $D[\text{ind} + \text{code\_size} - 1]$

12 for $j = 0$, $j <$ code_size, $j++$ **do**

13 $\quad C[j] = I[\text{ind} + j]$

14 return C

$$d(A) := \max \left\{ |a_i - a_j| \mid a_i, a_j \in A \right\}$$

to be the largest absolute difference between elements in A. We also define the variance of values in A, denoted Var(A), by (see Eq. 1.35)

$$\text{Var}(A) := \frac{1}{\beta} \sum_{i=1}^{\beta} (a_i - \overline{a})^2,$$

where $\overline{a}$ is the average of a_i given by

$$\overline{a} = \frac{1}{\beta} \sum_{i=1}^{\beta} a_i.$$

It is easy to see that

$$(a_i - \overline{a})^2 \leq d(A)^2,$$

and hence

$$\mathrm{Var}(A) \leq d(A)^2.$$

Now, let C be an $(n_C, 2^{m_v})$-binary code, and define

$$A(C) := \{\mathrm{SG}(c) \mid c \in C\}$$

to be the set of estimated signals for codewords in C. When C is used for encoding the target intermediate value, the variance of the signal at POI is then given by

$$\mathrm{Var}(X_{\mathrm{POI}}) = \mathrm{Var}(A(C)).$$

The goal of Algorithm 4.5 is to find a C such that $d(A(C))$ is the minimum among all $(n_C, 2^{m_v})$-binary codes whose codewords have Hamming weight w_H. According to the above discussions, we can conclude that the SNR of the code found by the algorithm is also relatively small. Even though this code may not be the one that achieves the lowest SNR, another code with a lower SNR will have a bigger $d(A(C))$, which might be exploited to improve the attack results.

To see how effective is the countermeasure, we have simulated template-based DPA attacks (see Sect. 4.3.2.3) on the 0th Sbox output of PRESENT. The dataset $\mathcal{T}_1$ obtained in Code-SCA Step 3 is used as the profiling traces to build templates for unprotected implementation. A total of $10,000$ traces were collected with random inputs from $C_{(8,16)}$ (Eq. 4.71) as profiling traces to build templates for attacks on protected implementations.

To get the attack traces for the unprotected implementation, for each plaintext nibble p and a fixed key nibble 9 (the same as the 0th nibble of the key in Eq. 4.1), we precomputed the Sbox output (see Table 3.11)

$$v = \mathrm{SB}_{\mathrm{PRESENT}}(p \oplus 9).$$

Then we carried out the measurement for the operation described in Eq. 4.64 with v as the input a. $100,000$ traces were collected for random plaintext nibbles p. Attack traces for the protected implementation were obtained in a similar manner. Instead of v, we pass the corresponding codeword from $C_{(8,16)}$, $C_{(8,16)}[v]$, as input a in Eq. 4.64. We have also measured $100,000$ traces with random plaintext nibbles.

The attacks follow steps from Sect. 4.3.2.3, where we have set the target signal to be the exact value of v (or the corresponding codeword). Since we only focus on one POI, according to Template Step b, we have computed a mean leakage for each value of v for unprotected implementation and for each value of codeword in $C_{(8,16)}$ for the protected implementation. The mean leakage values for different v are given by

$$\{-0.01055, -0.00943, -0.00680, -0.00772, -0.00698, -0.00778, -0.00656,$$
$$-0.00748, -0.00677, -0.00764, -0.00641, -0.00732, -0.00649, -0.00732,$$
$$-0.00619, -0.00716\}.$$

The mean leakage values for different codewords are

$$\{-0.01064, -0.01054, -0.01036, -0.01037, -0.01032, -0.01058, -0.01038,$$
$$-0.01031, -0.01023, -0.01027, -0.01036, -0.01034, -0.01021, -0.01008,$$
$$-0.01014, -0.00998\}.$$

It is easy to see that the differences between mean leakages in the first set are bigger compared to those in the second set. If we compute the variance between mean leakages in those two sets, we get

$$1.2067 \times 10^{-6} \text{ and } 2.8459 \times 10^{-8}.$$

This shows that it is more difficult to distinguish between the leakages of codewords in $C_{(8,16)}$ than that of different values of v. Since DPA attacks rely on exploiting the difference between leakages for different data, we expect the protected implementation to be more challenging to attack with DPA.

The attack results are shown in Figs. 4.79 and 4.80. Computations of estimations for success rates and guessing entropy followed Algorithm 4.1, where we have set

$$\mathtt{max\_trace} = 1000, \quad \mathtt{no\_of\_attack} = 100.$$

We can see that the unprotected implementation can be broken with about 150 traces, while the protected implementation cannot be broken with even 1000 traces.

We note that the number of traces required for a successful attack on unprotected implementation is more than what we have obtained in Sect. 4.3.2.3 (see Figs. 4.51 and 4.52). This is expected as the highest SNR we have for MOV instruction (Fig. 4.78) is much less than that for one round of PRESENT (Fig. 4.20).

For comparison, we have also repeated the same steps for the proposed countermeasure with different values of $w_H = 2, 3, 4, 5$. The template-based DPA attack results are shown in Figs. 4.81 and 4.82. We can see that all the codes increase the number of traces needed for a successful attack. And the code with $w_H = 4$ behaves the best.

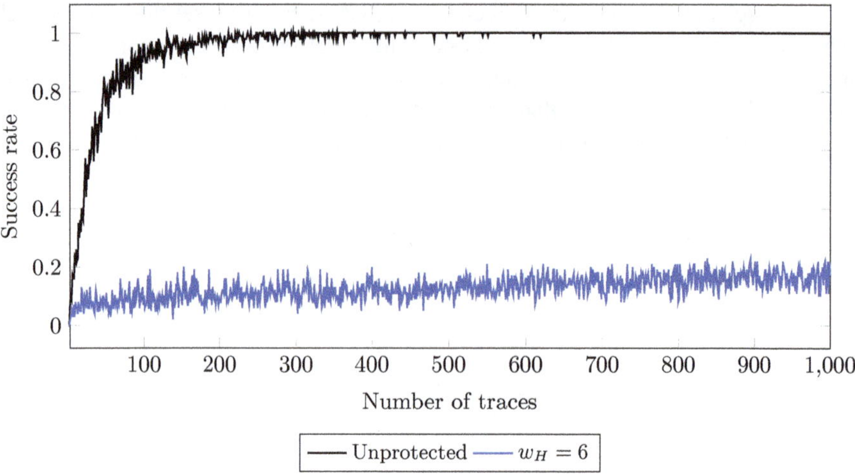

Fig. 4.79 Estimations of success rate computed following Algorithm 4.1 for template-based DPA attack on the MOV instruction taking the PRESENT Sbox output as an input. The black line corresponds to unprotected intermediate values. The blue line corresponds to encoded intermediate values with the binary code $C_{(8,16)}$ (Eq. 4.71), where all codewords have Hamming weight 6

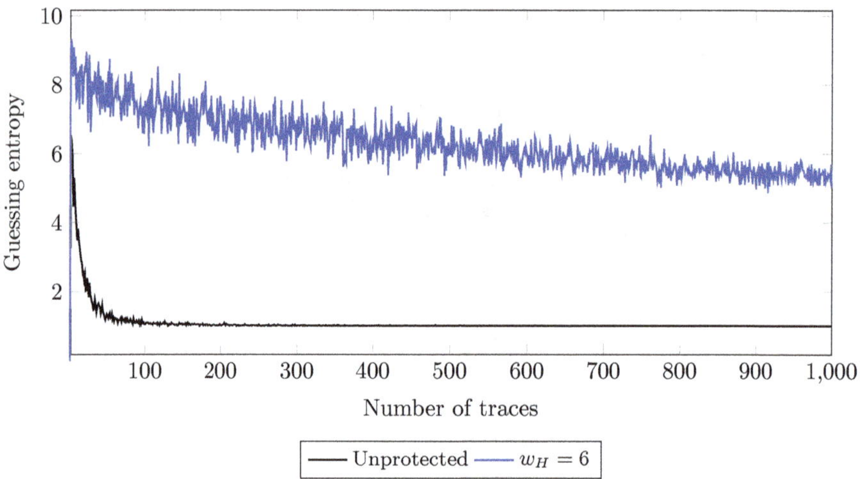

Fig. 4.80 Estimations of guessing entropy computed following Algorithm 4.1 for template-based DPA attack on the MOV instruction taking the PRESENT Sbox output as an input. The black line corresponds to unprotected intermediate values. The blue line corresponds to encoded intermediate values with the binary code $C_{(8,16)}$ (Eq. 4.71), where all codewords have Hamming weight 6

We note that the presented countermeasure focuses on one instruction; therefore the chosen binary code is only optimal for the target instruction and target intermediate value. Nevertheless, the whole cipher state can be encoded, giving a certain level of protection to all the other instructions. A method of encoding

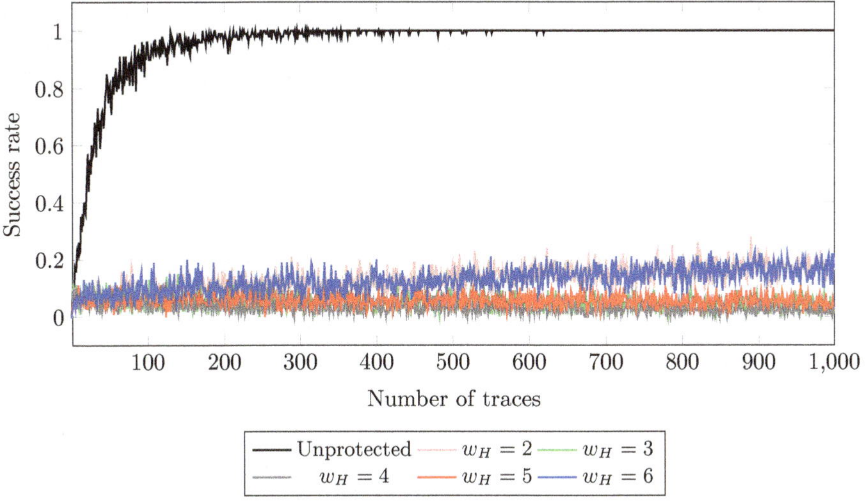

Fig. 4.81 Estimations of success rate computed following Algorithm 4.1 for template-based DPA attack on the MOV instruction taking the PRESENT Sbox output as an input. The black line corresponds to unprotected intermediate values. The other lines correspond to encoded intermediate values with (8, 16)-binary codes obtained following Code-SCA Step 1–Code-SCA Step 7, where we have set $w_H = 2, 3, 4, 5, 6$

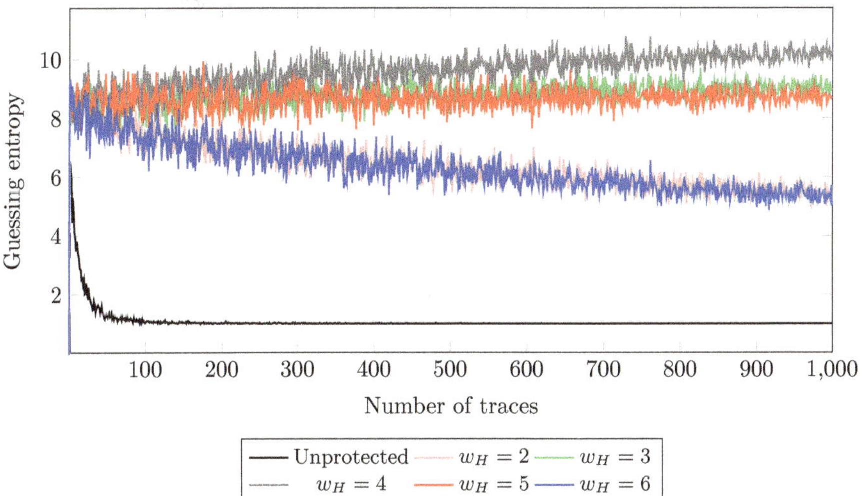

Fig. 4.82 Estimations of guessing entropy computed following Algorithm 4.1 for template-based DPA attack on the MOV instruction taking the PRESENT Sbox output as an input. The black line corresponds to unprotected intermediate values. The other lines correspond to encoded intermediate values with (8, 16)-binary codes obtained following Code-SCA Step 1—Code-SCA Step 1, where we have set $w_H = 2, 3, 4, 5, 6$

the whole encryption computation will be discussed in Sect. 5.2.1. Different codes might work for different devices, but as implementers, we would have access to the device we want to protect and can choose the best code that is suitable for the device.[10]

4.5.1.2 Square and Multiply Always

In Sect. 4.4.1 we have seen one SPA attack on RSA implementations that exploits the part of the square and multiply algorithm where multiplication is carried out only when the secret key bit is 1. A natural countermeasure is that we always compute multiplication no matter what the value of the secret key bit is. Such an algorithm is called the *square and multiply always algorithm* [Cor99].

We keep the notations from Sect. 3.3. Let $n = pq$ be the product of two distinct odd primes. Let $d \in \mathbb{Z}^*_{\varphi(n)}$ be the secret key of RSA/RSA signatures. We would like to compute

$$a^d \bmod n$$

for some $a \in \mathbb{Z}_n$.

Recall that we have presented the right-to-left (Algorithm 3.7) and left-to-right (Algorithm 3.8) square and multiply algorithms. Correspondingly, we have the right-to-left and left-to-right square and multiply always algorithms, detailed in Algorithms 4.6 and 4.7, respectively. In both algorithms, the modular multiplication computation is always carried out. And when the secret bit is 0, the result is discarded (line 6 in Algorithm 4.6 and line 7 in Algorithm 4.7).

As an illustration, let us consider our attack presented in Sect. 4.4.1. With the square and multiply always countermeasure, Algorithm 4.2 becomes Algorithm 4.8. With the same experimental setting as described in Sect. 4.1, we have measured one trace for the computation of Algorithm 4.8 with our DUT. To make sure line 10 will be executed, we have turned off the compiler optimization. The trace is shown in Fig. 4.83. We can see that we still observe ten patterns, the same as in Sect. 4.4.1. But in this case, all of them have more than one peak cluster. We know from the discussions in Sect. 4.4.1 that this is because each of the patterns corresponds to one loop from line 5 and each loop contains one modular square (line 6) and one modular multiplication operation (line 8 or line 10). Thus, we cannot repeat the same attack as presented in Sect. 4.4.1. However, we can deduce that the secret key has bit length 10. In practical settings, the bit length will be much bigger. To the best of our knowledge, this information alone cannot reveal the secret key.

On the other hand, we will show that the square and multiply always algorithm is still vulnerable to the DPA attack presented in Sect. 4.4.2. With square and multiply always countermeasure, Algorithm 4.4 becomes Algorithm 4.9.

---

[10] Naturally, creating a different code for every device would be impractical for serial production.

Algorithm 4.6: Right-to-left square and multiply always algorithm for computing modular exponentiation. A hiding-based countermeasure against SCA attacks

---

Input: n, a, d // $n \in \mathbb{Z}, n \geq 2$; $a \in \mathbb{Z}_n$; $d \in \mathbb{Z}_{\varphi(n)}$ has bit length ℓ_d

Output: $a^d \bmod n$

1 result $= 1, t = a$
2 **for** $i = 0, i < \ell_d, i + +$ **do**
      // ith bit of d is 1
3   **if** $d_i = 1$ **then**
         // mutiply by a^{2^i}
4   result $=$ result $* t \bmod n$
5   **else**
         // ith bit of d is 0, compute multiplication and discard the result
6   tmp $=$ result $* t \bmod n$
    // $t = a^{2^{i+1}}$
7      $t = t * t \bmod n$
8 **return** result

Algorithm 4.7: Left-to-right square and multiply always algorithm for computing modular exponentiation. A hiding-based countermeasure against SCA attacks

---

Input: n, a, d // $n \in \mathbb{Z}, n \geq 2$; $a \in \mathbb{Z}_n$; $d \in \mathbb{Z}_{\varphi(n)}$

Output: $a^d \bmod n$

1 $t = 1$
2 **for** $i = \ell_d - 1, i \geq 0, i - -$ **do**
3      $t = t * t \bmod n$
    // ith bit of d is 1
4   **if** $d_i = 1$ **then**
5          $t = a * t \bmod n$
6   **else**
         // ith bit of d is 0, compute multiplication and discard the result
7   tmp $= a * t \bmod n$
8 **return** t

With the same experimental setting as in Sects. 4.1 and 4.4.2, one trace for computation of Algorithm 4.9 is shown in Fig. 4.84. We note that there are 21 similar patterns in the figure. By examining Algorithm 4.9, we can guess that each of them might correspond to one loop from line 6 or one execution of MonPro. If the former is true, we will have a private key d with a bit length bigger than the bit length of n. Thus, we can conclude that most likely each of them corresponds to one execution of MonPro. Then the last one corresponds to line 12, and the remaining 20 tells us that $\ell_d = 10$.

Algorithm 4.8: Protected implementation of Algorithm 4.2. Left-to-right square and multiply always algorithm for computing modular exponentiation (see Algorithm 3.8) with parameters from Eq. 4.59

Input: a// $a \in \mathbb{Z}_{1189}$
Output: $a^{747} \bmod 1189$
1 $n = 1189$
2 $dbin = [1, 1, 0, 1, 0, 1, 1, 1, 0, 1]$// binary representation of $d = 747$, $d_0 = 1$,
    $d_1 = 1$
3 ℓ_d = length of $dbin$// bit length of d
4 $t = 1$
5 for $i = \ell_d - 1, i \geq 0, i - - $ **do**
6 $t = t * t \bmod n$
      // ith bit of d is 1
7   **if** $d_i = 1$ **then**
8 $t = a * t \bmod n$
9   **else**
        // ith bit of d is 0, compute multiplication and discard the
        result
10   tmp $= a * t \bmod n$

11 return t

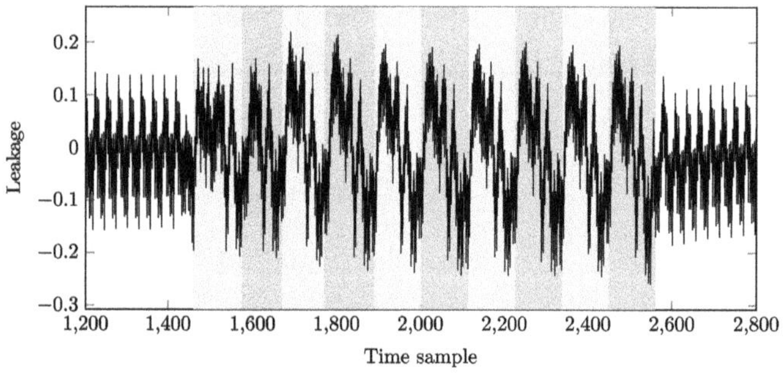

Fig. 4.83 One trace corresponding to the computation of Algorithm 4.8. We can see ten similar patterns

Following the attack steps as in Sect. 4.4.2, we have collected $M = 10,000$ traces, each with $q = 10,800$ time samples (11). The sample correlation coefficients (see DPA-RSA Step 6) are shown in Fig. 4.85, where the trace from Fig. 4.84 is in gray in the background. We can see that there are 21 patterns in the sample correlation coefficient plot which coincide with those from the trace plot in Fig. 4.84—each corresponds to one execution of MonPro.

We also see mainly two different patterns, one with a higher positive peak cluster and one with a lower positive peak cluster. They are colored in blue and green, respectively, in Fig. 4.86. Clearly, the one with a higher positive peak cluster corresponds to the computation of line 9 or line 11 since it results in bigger

---

Algorithm 4.9: Montgomery left-to-right square and multiply always algorithm with parameters from Eq. 4.59. `MonPro` is given by Algorithm 4.3

Input: a // $a \in \mathbb{Z}_n$;
Output: $a^{747} \bmod 1189$
1 $n = 1189, \quad r = 2048$
2 $dbin = [1, 0, 1, 1, 1, 0, 1, 0, 1, 1]$ // binary representation of $d = 747$, $d_0 = 1$, $d_1 = 1$
3 $\ell_d = $ length of $dbin$ // bit length of d
4 $t_r = r \bmod n$
5 $a_r = ar \bmod n$
6 **for** $i = \ell_d - 1, i \geq 0, i - - $ **do**
7 $\quad$ $t_r = \mathtt{MonPro}(t_r, t_r)$ // $t_r = t_r \times_{\mathrm{Mon}} t_r$.
8 $\quad$ **if** $dbin[i] = 1$ **then**
9 $\quad\quad$ $t_r = \mathtt{MonPro}(t_r, a_r)$ // $t_r = t_r \times_{\mathrm{Mon}} a_r$.
10 $\quad$ **else**
$\quad\quad$ // ith bit of d is 0, compute multiplication and discard the result
11 $\quad\quad$ $\mathtt{tmp} = \mathtt{MonPro}(t_r, a_r)$

12 $t = \mathtt{MonPro}(t_r, 1)$ // $t = t_r \times_{\mathrm{Mon}} 1 = t_r * r^{-1} \bmod n$.
13 **return** t

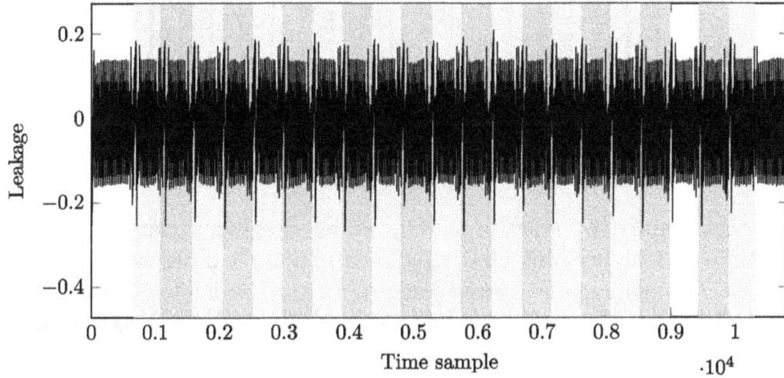

Fig. 4.84 One trace corresponding to the computation of Algorithm 4.9. We can see 21 similar patterns. Each of them corresponds to one execution of `MonPro`

correlation coefficient with a_r. The green-colored patterns correspond to line 7, except for the last one, which corresponds to line 12. Our attack will work if we can distinguish between line 9 and line 11. We take a closer look at those blue-colored patterns. We can see that some of them have a high peak at the end—they are colored in lighter blue. The others do not have this high peak—they are colored in darker blue. We know that the first secret bit $d_{\ell-1} = 1$; thus we deduce that those lighter blue-colored patterns correspond to line 9 and the darker blue-colored ones correspond to line 11. Consequently, each lighter blue-colored pattern indicates a secret bit $= 1$, and each darker blue-colored pattern indicates a secret bit $= 0$. We can write down the bits of the secret key as follows:

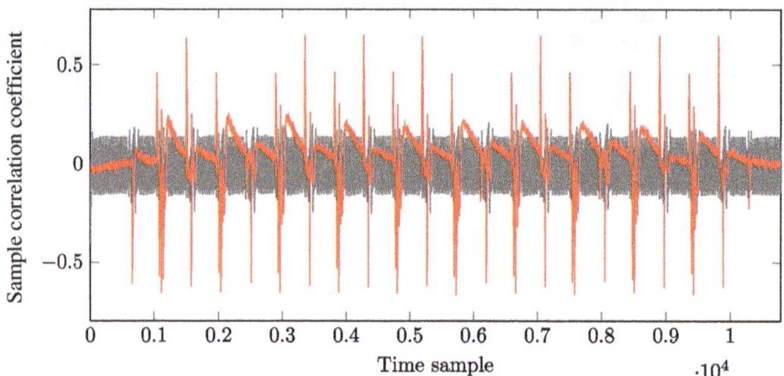

Fig. 4.85 Sample correlation coefficients computed following attack steps from Sect. 4.4.2 with 10, 000 traces for the computation of Algorithm 4.9. The trace from Fig. 4.84 is gray in the background. We can see that there are 21 patterns in the sample correlation coefficient plot that coincide with those from Fig. 4.84—each corresponds to one execution of `MonPro`

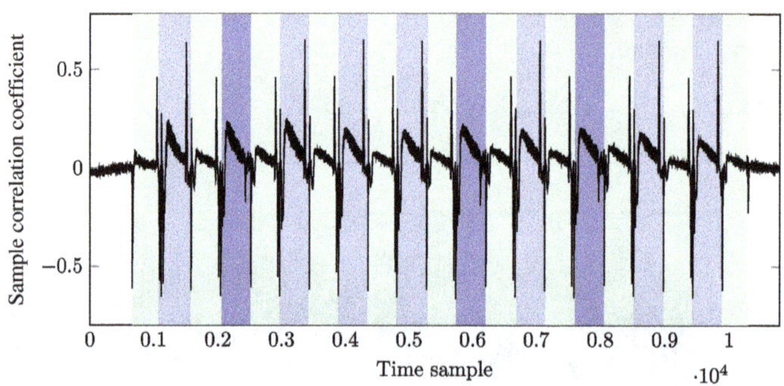

Fig. 4.86 There are mainly two types of patterns in the sample correlation coefficient plot from Figure 4.85—one with a higher peak cluster (colored in blue) and one with a lower peak cluster (colored in green). Among the blue-colored patterns, we further divide them into two types—one with a high peak at the end (in lighter blue) and one without this peak (in darker blue)

$$1 \quad 0 \quad 1 \quad 1 \quad 1 \quad 0 \quad 1 \quad 0 \quad 1 \quad 1,$$

and the secret key d is given by

$$d = 1011101011 = 747.$$

4.5.2 Masking and Blinding

As mentioned before, the goal of masking/blinding is to randomize the intermediate values being processed in the DUT. When such a countermeasure is applied to symmetric block ciphers, we refer to it as *masking*. And when it is applied to public key cryptosystem implementations, following the convention, it is called *blinding* instead.

Let v be the secret intermediate value that we would like to mask. The masked value, denoted v_{m}, is concealed by a random value m, called a *mask*, with a binary operation $\cdot$ such that

$$v_{\mathrm{m}} = v \cdot \mathrm{m}.$$

When the binary operation $\cdot$ is given by bitwise XOR, we have a *Boolean masking*. When $\cdot$ is a modular addition or modular multiplication, we have an *arithmetic masking*. We will discuss Boolean masking for symmetric block ciphers in Sects. 4.5.2.1–4.5.2.3 and arithmetic masking for RSA implementations in Sect. 4.5.2.4.

4.5.2.1   Introduction to Boolean Masking

As one can imagine, the cryptographic algorithm needs to be changed a bit for us to carry out computations with the masked intermediate values and keep track of all the masks. So that at the end of the encryption, we can remove the masks to output the original ciphertext. In general, a *masking scheme* specifies how masks are applied to the plaintext and intermediate values, as well as how they are removed from the ciphertext. There are a few principles we follow for a masking scheme design:

- All intermediate values should be masked during the computation. In particular, we would apply masks to the plaintext (and the key).
- We assume the attacker does not have knowledge of the masks—otherwise, the attacker can carry out similar DPA attacks by making hypotheses about the key values as in Sects. 4.3.1 and 4.3.2.
- When some intermediate values are to be XOR-ed with each other (e.g., in AES MixColumns operation), different masks should be applied to each of them. Otherwise, the same valued masks will cancel out.
- Each encryption has a different set of randomly generated masks.

For any function f, the mask that is applied to an input of f is called the *input mask* of f. The corresponding mask for the output is called the *output mask* of f.

Definition 4.5.1 Let $f : \mathbb{F}_2^{m_1} \to \mathbb{F}_2^{m_2}$ be a function, where m_1 and m_2 are positive integers. f is said to be *linear* (w.r.t. $\oplus$) if for any $x, y \in \mathbb{F}_2^{m_1}$, we have

$$f(x \oplus y) = f(x) \oplus f(y).$$

f is *nonlinear* if it is not linear.

Example 4.5.1

- AddRoundKey operation in AES (Sect. 3.1.2) round function is a linear function. In fact, bitwise XOR with a round key is a linear function in general.
- DES (Sect. 3.1.1) Sboxes are nonlinear functions. Any Sbox proposed so far for symmetric block ciphers is nonlinear.
- pLayer in PRESENT (Sect. 3.1.3) round function is linear.
- MixColumns operation in AES is linear (see Remark 3.1.3).

With Boolean masking, it is easy to keep track of the masks with linear operations. Let f be a linear function, and take any input of f, v, with a corresponding mask m; we have

$$f(v \oplus m) = f(v) \oplus f(m).$$

Thus, when the input mask is m, the output mask is given by $f(m)$. One of the main challenges in designing a masking scheme is to find ways to keep track of masks for nonlinear operations.

4.5.2.2 Boolean Masking for AES-128

In this part, we will discuss a masking scheme for AES-128. The scheme was first proposed in [HOM06], see also [MOP08, Section 9.2.1].

The only nonlinear operation in AES encryption is SubBytes. Let SB denote AES Sbox. We will consider a table lookup implementation (see Sect. 3.2.1) for the SubBytes operation. We choose an input mask $m_{\text{in, SB}}$ and an output mask $m_{\text{out, SB}}$ for SB. Then we generate a table that implements the masked Sbox, denoted SB_m, such that

$$SB_m(v \oplus m_{\text{in, SB}}) = SB(v) \oplus m_{\text{out, SB}}. \tag{4.72}$$

The masking scheme works as follows. Firstly, at the beginning of each encryption:

- We randomly generate six independent masks with values from $\mathbb{F}_2^8$, denoted by

$$m_{\text{in, SB}}, \quad m_{\text{out, SB}}, \quad m_0, \quad m_1, \quad m_2, \quad m_3.$$

  $m_{\text{in, SB}}$ and $m_{\text{out, SB}}$ will be the input and output masks for AES Sbox computation. m_0, m_1, m_2, m_3 will be used as input masks for MixColumns operation.
- Compute the lookup table for masked Sbox as given in Eq. 4.72.
- Calculate m_0', m_1', m_2', m_3' from m_0, m_1, m_2, m_3 using the MixColumns operation (see Eq. 3.6):

$$\begin{pmatrix} m'_0 \\ m'_1 \\ m'_2 \\ m'_3 \end{pmatrix} = \begin{pmatrix} 02 & 03 & 01 & 01 \\ 01 & 02 & 03 & 01 \\ 01 & 01 & 02 & 03 \\ 03 & 01 & 01 & 02 \end{pmatrix} \begin{pmatrix} m_0 \\ m_1 \\ m_2 \\ m_3 \end{pmatrix}. \tag{4.73}$$

Let us keep the matrix representation of the AES cipher state as in Eq. 3.2. During the encryption, the masking scheme continues as follows. We apply masks m'_0, m'_1, m'_2, m'_3 to the plaintext such that the 4 bytes in row $i + 1$ are masked with m'_i. Then the cipher state before the initial AddRoundKey is of the format

$$\begin{pmatrix} s_{00} \oplus m'_0 & s_{01} \oplus m'_0 & s_{02} \oplus m'_0 & s_{03} \oplus m'_0 \\ s_{10} \oplus m'_1 & s_{11} \oplus m'_1 & s_{12} \oplus m'_1 & s_{13} \oplus m'_1 \\ s_{20} \oplus m'_2 & s_{21} \oplus m'_2 & s_{22} \oplus m'_2 & s_{23} \oplus m'_2 \\ s_{30} \oplus m'_3 & s_{31} \oplus m'_3 & s_{32} \oplus m'_3 & s_{33} \oplus m'_3 \end{pmatrix}. \tag{4.74}$$

We will not detail the masking scheme for the key schedule. For a round key K, we use the following matrix representation:

$$\begin{pmatrix} k_{00} & k_{01} & k_{02} & k_{03} \\ k_{10} & k_{11} & k_{12} & k_{13} \\ k_{20} & k_{21} & k_{22} & k_{23} \\ k_{30} & k_{31} & k_{32} & k_{33} \end{pmatrix}.$$

We assume that the round keys, except for the last round key, are all masked such that the bytes in row $i + 1$ are masked with $m'_i \oplus m_{\text{in, SB}}$. Then for a round key K, the representation of its masked value in the matrix format will be

$$\begin{pmatrix} k_{00} \oplus m'_0 \oplus m_{\text{in, SB}} & k_{01} \oplus m'_0 \oplus m_{\text{in, SB}} & k_{02} \oplus m'_0 \oplus m_{\text{in, SB}} & k_{03} \oplus m'_0 \oplus m_{\text{in, SB}} \\ k_{10} \oplus m'_1 \oplus m_{\text{in, SB}} & k_{11} \oplus m'_1 \oplus m_{\text{in, SB}} & k_{12} \oplus m'_1 \oplus m_{\text{in, SB}} & k_{13} \oplus m'_1 \oplus m_{\text{in, SB}} \\ k_{20} \oplus m'_2 \oplus m_{\text{in, SB}} & k_{21} \oplus m'_2 \oplus m_{\text{in, SB}} & k_{22} \oplus m'_2 \oplus m_{\text{in, SB}} & k_{23} \oplus m'_2 \oplus m_{\text{in, SB}} \\ k_{30} \oplus m'_3 \oplus m_{\text{in, SB}} & k_{31} \oplus m'_3 \oplus m_{\text{in, SB}} & k_{32} \oplus m'_3 \oplus m_{\text{in, SB}} & k_{33} \oplus m'_3 \oplus m_{\text{in, SB}} \end{pmatrix}. \tag{4.75}$$

After the initial AddRoundKey, according to Eqs. 4.74 and 4.75, the cipher state becomes

$$\begin{pmatrix} s_{00} \oplus m_{\text{in, SB}} & s_{01} \oplus m_{\text{in, SB}} & s_{02} \oplus m_{\text{in, SB}} & s_{03} \oplus m_{\text{in, SB}} \\ s_{10} \oplus m_{\text{in, SB}} & s_{11} \oplus m_{\text{in, SB}} & s_{12} \oplus m_{\text{in, SB}} & s_{13} \oplus m_{\text{in, SB}} \\ s_{20} \oplus m_{\text{in, SB}} & s_{21} \oplus m_{\text{in, SB}} & s_{22} \oplus m_{\text{in, SB}} & s_{23} \oplus m_{\text{in, SB}} \\ s_{30} \oplus m_{\text{in, SB}} & s_{31} \oplus m_{\text{in, SB}} & s_{32} \oplus m_{\text{in, SB}} & s_{33} \oplus m_{\text{in, SB}} \end{pmatrix}, \tag{4.76}$$

where each byte is masked with $m_{\text{in, SB}}$.

For round 1–round 9, the changes in cipher states after each operation of AES-128 with masked implementation are detailed below:

- **SubBytes**. The SubBytes operation is performed using the table designed for SB_m. By Eq. 4.72, after the SubBytes operation; each byte of the cipher state is masked by $m_{out,\,SB}$:

$$
\begin{pmatrix}
s_{00} \oplus m_{out,\,SB} & s_{01} \oplus m_{out,\,SB} & s_{02} \oplus m_{out,\,SB} & s_{03} \oplus m_{out,\,SB} \\
s_{10} \oplus m_{out,\,SB} & s_{11} \oplus m_{out,\,SB} & s_{12} \oplus m_{out,\,SB} & s_{13} \oplus m_{out,\,SB} \\
s_{20} \oplus m_{out,\,SB} & s_{21} \oplus m_{out,\,SB} & s_{22} \oplus m_{out,\,SB} & s_{23} \oplus m_{out,\,SB} \\
s_{30} \oplus m_{out,\,SB} & s_{31} \oplus m_{out,\,SB} & s_{32} \oplus m_{out,\,SB} & s_{33} \oplus m_{out,\,SB}
\end{pmatrix}. \tag{4.77}
$$

- **ShiftRows**. ShiftRows does not change the masks, each byte of the cipher state is still masked by $m_{out,\,SB}$.
- **MixColumns**. Before MixColumns, we change the masks of the cipher state by XOR-ing the four bytes in row $i + 1$ with $m'_i \oplus m_{out,\,SB}$. In this way, the input of MixColumns is of the format

$$
\begin{pmatrix}
s_{00} \oplus m_0 & s_{01} \oplus m_0 & s_{02} \oplus m_0 & s_{03} \oplus m_0 \\
s_{10} \oplus m_1 & s_{11} \oplus m_1 & s_{12} \oplus m_1 & s_{13} \oplus m_1 \\
s_{20} \oplus m_2 & s_{21} \oplus m_2 & s_{22} \oplus m_2 & s_{23} \oplus m_2 \\
s_{30} \oplus m_3 & s_{31} \oplus m_3 & s_{32} \oplus m_3 & s_{33} \oplus m_3
\end{pmatrix}.
$$

By the choice of m'_i (see Eq. 4.73), the cipher state at the output of MixColumns is the same as in Eq. 4.74:

$$
\begin{pmatrix}
s_{00} \oplus m'_0 & s_{01} \oplus m'_0 & s_{02} \oplus m'_0 & s_{03} \oplus m'_0 \\
s_{10} \oplus m'_1 & s_{11} \oplus m'_1 & s_{12} \oplus m'_1 & s_{13} \oplus m'_1 \\
s_{20} \oplus m'_2 & s_{21} \oplus m'_2 & s_{22} \oplus m'_2 & s_{23} \oplus m'_2 \\
s_{30} \oplus m'_3 & s_{31} \oplus m'_3 & s_{32} \oplus m'_3 & s_{33} \oplus m'_3
\end{pmatrix}.
$$

- **AddRoundKey**. After the AddRoundKey of the round, the cipher state becomes the same as the input of this round, as given in Eq. 4.76:

$$
\begin{pmatrix}
s_{00} \oplus m_{in,\,SB} & s_{01} \oplus m_{in,\,SB} & s_{02} \oplus m_{in,\,SB} & s_{03} \oplus m_{in,\,SB} \\
s_{10} \oplus m_{in,\,SB} & s_{11} \oplus m_{in,\,SB} & s_{12} \oplus m_{in,\,SB} & s_{13} \oplus m_{in,\,SB} \\
s_{20} \oplus m_{in,\,SB} & s_{21} \oplus m_{in,\,SB} & s_{22} \oplus m_{in,\,SB} & s_{23} \oplus m_{in,\,SB} \\
s_{30} \oplus m_{in,\,SB} & s_{31} \oplus m_{in,\,SB} & s_{32} \oplus m_{in,\,SB} & s_{33} \oplus m_{in,\,SB}
\end{pmatrix}.
$$

We can repeat the above for every round from round 1 to round 9. Finally, the input of round 10 is in the form of Eq. 4.76. After SubBytes and ShiftRows in round 10, the cipher state will be the same as in Eq. 4.77. Thus we require that each byte of the last round key is masked by $m_{out,\,SB}$. In this way, we will get unmasked ciphertext.

Table 4.4 Relation between the output bits of Sboxes from the Quotient group Qj^i and the input bits of Sboxes from the corresponding Remainder group Rj^{i+1}. For example, the 0th input bit of SB^{i+1}_{j+4} in Rj^{i+1} comes from the first output bit of SB^i_{4j} in Qj^i

Rj^{i+1} \ Qj^i	SB^i_{4j}	SB^i_{4j+1}	SB^i_{4j+2}	SB^i_{4j+3}
SB^{i+1}_{j}	$(0,0)$	$(1,0)$	$(2,0)$	$(3,0)$
SB^{i+1}_{j+4}	$(0,1)$	$(1,1)$	$(2,1)$	$(3,1)$
SB^{i+1}_{j+8}	$(0,2)$	$(1,2)$	$(2,2)$	$(3,2)$
SB^{i+1}_{j+12}	$(0,3)$	$(1,3)$	$(2,3)$	$(3,3)$

4.5.2.3 Boolean Masking for PRESENT

We will present two methods for masking PRESENT encryption. Let SB denote the PRESENT Sbox (Table 3.11) for the rest of this part.

Before we go into details of the masking scheme, we introduce the notion of Quotient group and Remainder group. We number the Sboxes in the ith round of PRESENT as $\mathrm{SB}^i_0, \mathrm{SB}^i_1, \ldots, \mathrm{SB}^i_{15}$, where SB^i_0 is the right-most Sbox in Fig. 3.9. Those Sboxes can be grouped in two different ways: the *Quotient group* and the *Remainder group*:

$$Qj^i := \left\{ \mathrm{SB}^i_{4j}, \mathrm{SB}^i_{4j+1}, \mathrm{SB}^i_{4j+2}, \mathrm{SB}^i_{4j+3} \right\}, \quad Rj^i := \left\{ \mathrm{SB}^i_j, \mathrm{SB}^i_{j+4}, \mathrm{SB}^i_{j+8}, \mathrm{SB}^i_{j+12} \right\},$$

where $j = 0, 1, 2, 3$. Such a grouping allows us to relate the bits for each Sbox output in round i to bits of each Sbox input in round $i + 1$ in a certain way through pLayer, as shown in Table 4.4. In particular, we observe that:

- Bits of the 0th Sbox (SB^i_{4j}) output in Quotient group Qj^i are permuted to the 0th bits of Sbox inputs in the corresponding Remainder group Rj^{i+1};
- Bits of the first Sbox (SB^i_{4j+1}) output in Qj^i are permuted to the first bits of Sbox inputs in Rj^{i+1}.
- Bits of the second Sbox (SB^i_{4j+2}) output in Qj^i are permuted to the second bits of Sbox inputs in Rj^{i+1}.
- Bits of the third Sbox (SB^i_{4j+3}) output in Qj^i are permuted to the third bits of Sbox inputs in Rj^{i+1}.

An illustration is shown in Fig. 4.87.

Hence pLayer can be considered as four identical parallel bitwise operations where each is a function $\mathsf{p} : \mathbb{F}_2^{16} \to \mathbb{F}_2^{16}$ that takes one Quotient group output and permutes it to the corresponding Remainder group input.

The first masking scheme follows a similar methodology as the masking scheme for AES presented in Sect. 4.5.2.2. Given an input mask m_{in} and an output mask $\mathsf{m}_{\mathrm{out}}$ for the PRESENT Sbox, we compute a table T that implements the masked Sbox such that

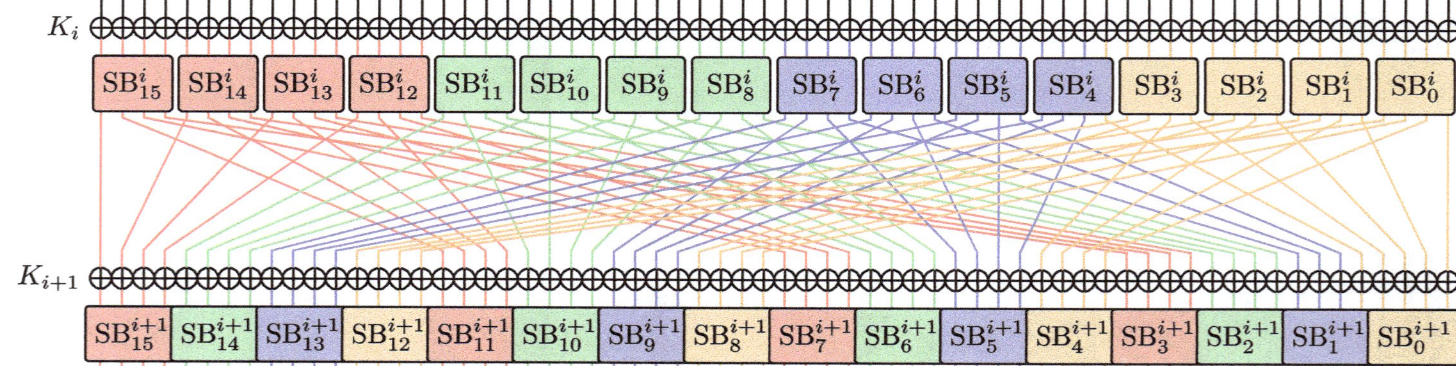

Fig. 4.87 An illustration of the relation between Sbox outputs in a Quotient group to Sbox inputs in the corresponding Remainder group. Sboxes in Quotient groups $Q0^i$, $Q1^i$, $Q2^i$, $Q3^i$ and their corresponding Remainder groups $R0^{i+1}$, $R1^{i+1}$, $R2^{i+1}$, $R3^{i+1}$ are in orange, blue, green, and red colors, respectively

$$T[v \oplus m_{in}] = SB(v) \oplus m_{out}. \tag{4.78}$$

At the beginning of each encryption:

- We randomly generate two independent masks m_{in}, m_{out} with values from $\mathbb{F}_2^4$.
- Compute lookup table T (given in Eq. 4.78) for the masked Sbox.
- Calculate m_3, m_2, m_1, m_0 from m_{out} with the pLayer operation

$$m_3, m_2, m_1, m_0 = p(m_{out}, m_{out}, m_{out}, m_{out}). \tag{4.79}$$

Let us represent the intermediate values of PRESENT encryption as

$$b_{15}, b_{14}, \ldots, b_1, b_0, \tag{4.80}$$

where each b_j denotes a nibble of the cipher state. At the start of the encryption, we mask the ith four nibbles of the plaintext with m_i, m_i, m_i, m_i ($i = 0, 1, 2, 3$). This means the cipher state at the input of round 1 is given by

$$b_{15} \oplus m_3, \ldots, b_{12} \oplus m_3, b_{11} \oplus m_2, \ldots, b_8 \oplus m_2, b_7 \oplus m_1, \ldots, b_4 \oplus m_1,$$

$$b_3 \oplus m_0, \ldots, b_0 \oplus m_0. \tag{4.81}$$

The cipher state changes for each round of PRESENT are as follows:

- **addRoundKey**. We assume the key schedule is changed so that the ith ($i = 0, 1, 2, 3$) four nibbles of each round key, except for the last round key, are masked by

$$m_i \oplus m_{in}, m_i \oplus m_{in}, m_i \oplus m_{in}, m_i \oplus m_{in}.$$

  Then after the addRoundKey operation, the cipher state is of the following format:

$$b_{15} \oplus m_{in}, b_{14} \oplus m_{in}, \ldots, b_1 \oplus m_{in}, b_0 \oplus m_{in},$$

  where each nibble is masked by m_{in}.
- **sBoxLayer**. By our design of the masked Sbox lookup table (Eq. 4.78), after sBoxLayer, each nibble of the cipher state will be masked by m_{out}:

$$b_{15} \oplus m_{out}, b_{14} \oplus m_{out}, \ldots, b_1 \oplus m_{out}, b_0 \oplus m_{out}.$$

- **pLayer**. After the pLayer computation, according to our discussion above about Quotient group, Remainder group, and Eq. 4.79, the cipher state will become (see Fig. 4.87)

Table 4.5 An example of T2, which specifies the output mask $m_{out,SB}$ for each input mask $m_{in,SB}$ of PRESENT Sbox [SBM18] such that all possible values of $m_{in} \oplus m_{out}$ appear

$m_{in,SB}$	0	1	2	3	4	5	6	7	8	9	A	B	C	D	E	F
$m_{out,SB} = T2[m_{in,SB}]$	E	4	F	9	0	3	D	5	7	8	A	2	B	1	6	C
$m_{in,SB} \oplus m_{out,SB}$	E	5	D	A	4	6	B	2	F	1	0	9	7	C	8	3

$$b_{15} \oplus m_3, \ldots, b_{12} \oplus m_3, b_{11} \oplus m_2, \ldots, b_8 \oplus m_2, b_7 \oplus m_1, \ldots, b_4 \oplus m_1,$$

$$b_3 \oplus m_0, \ldots, b_0 \oplus m_0,$$

which is the same as in Eq. 4.81. Thus the above can be repeated for all 31 rounds.

We assume the last round key has the same masks as the plaintext. Then after the final addRoundKey operation, we will get unmasked ciphertext.

The second masking scheme for PRESENT is detailed in [SBM18]. Different from the masked AES Sbox lookup table, this time we compute a lookup table, denoted T1, such that for any $v \in \mathbb{F}_2^4$, any input mask $m_{in} \in \mathbb{F}_2^4$, and the corresponding output mask $m_{out} \in \mathbb{F}_2^4$ for PRESENT Sbox,

$$T1[v \oplus m_{in}, m_{in}] = SB(v) \oplus m_{out}. \tag{4.82}$$

We also need another table T2 that helps us to keep track of the masks

$$T2[m_{in}] = m_{out}, \quad m_{in} = 0, 1, \ldots, F. \tag{4.83}$$

In this way, we do not need to generate a masked Sbox lookup table whenever the input mask for the Sbox changes. The size of T1 is 8×4, and the storage required is $2^8 \times 2^4 = 2^{12}$ bits or 2^9 bytes. The table T2 requires 16 bits of memory. It is suggested that T2 should be designed such that all possible values of $m_{in} \oplus m_{out}$ appear. For example, one possible choice of T2 is given in Table 4.5, originally presented in [SBM18].

In fact, in general, we have the following observations:

Remark 4.5.1 Let f be a function, and let $m_{in,f}$ denote its input mask with corresponding output mask $m_{out,f}$. For any input x of f, we have

$$(x \oplus f(x)) \oplus (m_{in,f} \oplus m_{out,f}) = (x \oplus m_{in,f}) \oplus (f(x) \oplus m_{out,f}).$$

Thus, when choosing the input mask $m_{in,f}$ and its corresponding output mask $m_{out,f}$, we need to ensure that all possible values of $m_{in,f} \oplus m_{out,f}$ appear. Otherwise, the distribution induced by $(x \oplus f(x)) \oplus (m_{in,f} \oplus m_{out,f})$ will not be uniform, and the signal corresponding to the value of $x \oplus f(x)$ cannot be properly concealed, making it vulnerable to DPA attacks.

Since the pLayer operation is linear, we can simply apply pLayer to the masks to keep track of their changes. We use the same notation as in Eq. 4.80 for PRESENT cipher state. At the beginning of one encryption, we randomly generate 16 masks, each is applied to one nibble of the plaintext. Suppose the cipher state at the input of round i is of the following format:

$$b_{15} \oplus m^{i-1}_{15,\text{in}}, b_{14} \oplus m^{i-1}_{14,\text{in}}, \ldots, b_1 \oplus m^{i-1}_{1,\text{in}}, b_0 \oplus m^{i-1}_{0,\text{in}}.$$

The changes in cipher states of PRESENT for round i are as follows:

- **addRoundKey**. We do not apply masks to the round keys. Consequently, after the addRoundKey operation, each nibble of the cipher state still has the same mask:

$$b_{15} \oplus m^{i-1}_{15,\text{in}}, b_{14} \oplus m^{i-1}_{14,\text{in}}, \ldots, b_1 \oplus m^{i-1}_{1,\text{in}}, b_0 \oplus m^{i-1}_{0,\text{in}}.$$

- **sBoxLayer**. Let

$$m^{i-1}_{j,\text{out}} = T2\left[m^{i-1}_{j,\text{in}} \right], \quad j = 0, 1, \ldots, 15,$$

denote the output mask for PRESENT Sbox corresponding to the input mask $m^{i-1}_{j,\text{in}}$. Then after sBoxLayer, the cipher state is of the following format:

$$b_{15} \oplus m^{i-1}_{15,\text{out}}, b_{14} \oplus m^{i-1}_{14,\text{out}}, \ldots, b_1 \oplus m^{i-1}_{1,\text{out}}, b_0 \oplus m^{i-1}_{0,\text{out}},$$

- **pLayer**. We apply the pLayer operation to both the cipher state and the mask for the whole cipher state. The mask for the whole cipher state is the string obtained by concatenating all 16 masks $m^{i-1}_{j,\text{out}}$:

$$m^{i-1}_{15,\text{out}}, m^{i-1}_{14,\text{out}}, \ldots, m^{i-1}_{1,\text{out}}, m^{i-1}_{0,\text{out}}.$$

After pLayer, masks for each nibble of the cipher state will be changed and the cipher state will become

$$b_{15} \oplus m^{i}_{15,\text{in}}, b_{14} \oplus m^{i}_{14,\text{in}}, \ldots, b_1 \oplus m^{i}_{1,\text{in}}, b_0 \oplus m^{i}_{0,\text{in}},$$

where

$$m^{i}_{15,\text{in}}, m^{i}_{14,\text{in}}, \ldots, m^{i}_{1,\text{in}}, m^{i}_{0,\text{in}} = \text{pLayer}(m^{i-1}_{15,\text{out}}, m^{i-1}_{14,\text{out}}, \ldots, m^{i-1}_{1,\text{out}}, m^{i-1}_{0,\text{out}}).$$

Consequently, $m^{i}_{j,\text{in}}$ will be the input mask for the jth Sbox in round $i+1$.

Finally, after 31 rounds, we have another addRoundKey operation, which does not change the masks of the cipher state since the round keys are not masked. The cipher state will be

$$b_{15} \oplus \mathrm{m}^{31}_{15,\mathrm{in}},\; b_{14} \oplus \mathrm{m}^{31}_{14,\mathrm{in}},\; \ldots,\; b_1 \oplus \mathrm{m}^{31}_{1,\mathrm{in}},\; b_0 \oplus \mathrm{m}^{31}_{0,\mathrm{in}}.$$

To get the unmasked ciphertext, we remove the masks by XOR-ing the cipher state with

$$\mathrm{m}^{31}_{15,\mathrm{in}},\; \mathrm{m}^{31}_{14,\mathrm{in}},\; \ldots,\; \mathrm{m}^{31}_{1,\mathrm{in}},\; \mathrm{m}^{31}_{0,\mathrm{in}}.$$

An algorithmic description for masked PRESENT computation is given in Algorithm 4.10. The changes in masks are recorded in the variable masks, and we remove them at the end of the computation (line 12).

Algorithm 4.10: Masked implementation of PRESENT

Input: p, T1, T2, K_i ($i = 1, 2, \ldots, 32$)// p is the plaintext for encryption; T1 is the table for masked Sbox as given in Eq. 4.82; T2 specifies the output mask given the input mask for PRESENT Sbox as defined in Eq. 4.83; K_i are round keys for PRESENT encryption
Output: ciphertext
1 randomly generate 16 masks $\mathrm{m}_0, \mathrm{m}_1, \ldots, \mathrm{m}_{15}$
2 **array** of size 16   state = $p \oplus \mathrm{m}_{15}, \mathrm{m}_{14}, \ldots, \mathrm{m}_1, \mathrm{m}_0$// mask the jth nibble of the plaintext with m_j, each entry of the array is one masked nibble
3 **array** of size 16   masks = $\mathrm{m}_{15}, \mathrm{m}_{14}, \ldots, \mathrm{m}_1, \mathrm{m}_0$
4 **for** $i = 0, i < 31, i + +$ **do**
5   state = addRoundKey(state, K_i)
6   **for** $j = 0, j < 16, j + +$ **do**
        // for each nibble
7   state[j] = T1[state[j], masks[j]]// masked Sbox computation
8   masks[j] = T2[masks[j]]// record the output masks of Sbox computation
9   state = pLayer(state)// apply pLayer to the cipher state
10   masks = pLayer(masks)// apply pLayer to the masks
11 state = addRoundKey(state, K_i)
12 state = state $\oplus$ masks
13 **return** state

As an illustration, we have implemented masked PRESENT following Algorithm 4.10, where we used Table 4.5 to choose the output mask $\mathrm{m}_{\mathrm{out,SB}}$ given the input mask $\mathrm{m}_{\mathrm{in,SB}}$ for PRESENT Sbox. With the experimental setup described in Sect. 4.1, we have collected four datasets, with similar settings as those datasets in Sect. 4.1. All the datasets contain traces that capture one round of masked software implementation of PRESENT encryption.

- *Masked fixed dataset A*: This dataset contains 100 traces with a fixed round key FEDCBA0123456789 and a fixed plaintext ABCDEF1234567890.
- *Masked fixed dataset B*: This dataset contains 100 traces with a fixed round key FEDCBA0123456789 and a fixed plaintext 84216BA484216BA4.

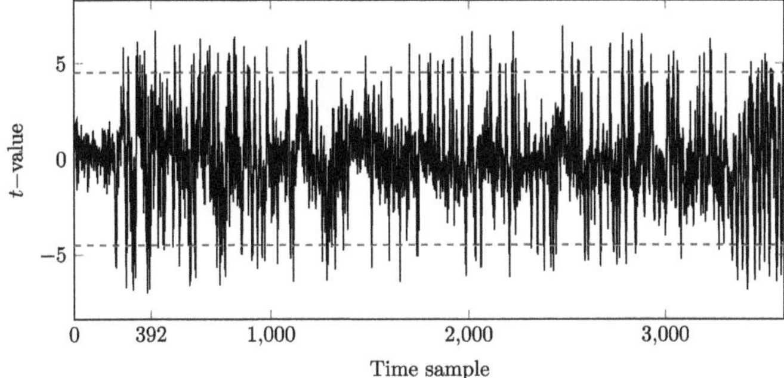

Fig. 4.88 t-Values (Eq. 4.17) for all time samples $1, 2, \ldots, 3600$ computed with 50 traces from *Masked fixed dataset A* and 50 traces from *Masked fixed dataset B*. The signal is given by the plaintext value, and the fixed versus fixed setting is chosen. Blue dashed lines correspond to the threshold 4.5 and −4.5

- *Masked random plaintext dataset*: This dataset contains 20000 traces with a fixed round key

$$FEDCBA0123456789 \tag{4.84}$$

and a random plaintext for each trace.
- *Masked random dataset*: This dataset contains 10,000 traces with a random round key and a random plaintext for each trace.

In each case, the execution of the cipher is surrounded by nop instructions so that the round operation patterns can be clearly distinguished from the provided plots. While the raw traces are all 5000 time samples long, for plotting and analysis purposes, we shorten them to 3600 time samples as the later parts correspond to nop instructions and do not contain any useful information. We also note that for these datasets, we reduced the number of collected time samples by a factor of 3.

Following the TVLA steps from Sect. 4.2.3, we have computed the t-values using 50 traces from *Masked fixed dataset A* and 50 traces from *Masked fixed dataset B*. The intermediate value v (TVLA Step 2) is chosen to be the plaintext value. Fixed versus fixed setting (TVLA Step 3) is used. The results are shown in Fig. 4.88. We can see that compared to Fig. 4.13, the t-values are much lower, and there are no points with very high peaks to stand out from the rest. Even though some time samples have t-values outside of the threshold, they are not far from it. This indicates that the implementation should exhibit less leakage as compared to the unprotected one.

To see how the implementation is resistant to DPA attacks, we have adopted the template-based DPA as described in Sect. 4.3.2.3. Following steps from Sect. 4.3.2.1, we take *Masked random dataset* as our profiling traces. The same as in Sect. 4.3.2.3, the target part of the key (P-DPA Step 3) is the 0th nibble of the first round key. The target intermediate value (P-DPA Step 4) is the 0th Sbox

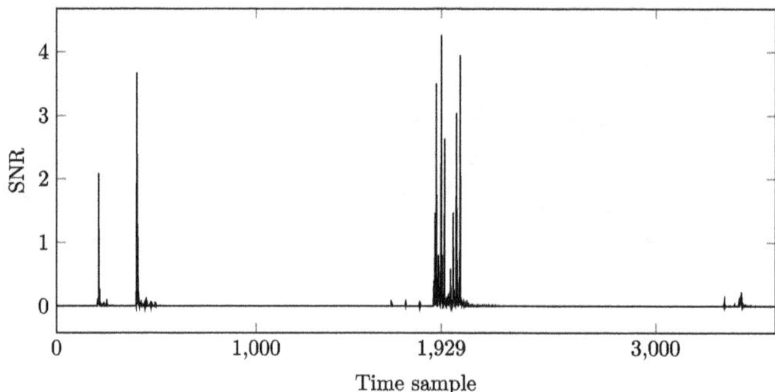

Fig. 4.89 SNR computed with *Masked random dataset*. The signal is given by the exact value of the 0th Sbox output

output of the first round. We consider the target signal (P-DPA Step 5) to be the exact value of $\boldsymbol{v}$, since in Sect. 4.3.2.3 we have seen that this is a better choice than taking $\mathrm{wt}(\boldsymbol{v})$ to be the target signal (see Fig. 4.52). Consequently, we group our profiling traces *Masked random dataset* into 16 sets (P-DPA Step 6). Using the methodology from P-DPA Step 7 and P-DPA Step 8, we have computed the SNR values for all 3600 time samples using *Masked random dataset*. The results are shown in Fig. 4.89.

The time sample achieving the highest SNR is $t = 1929$, which will be our POI (Template Step a from Sect. 4.3.2.3). Following Template Step b from Sect. 4.3.2.3, we have built the template for this POI. In particular, the mean values μ_s for $s = 0, 1, \ldots, 15$ (corresponding to $v = 0, 1, \ldots, 16$) are as follows:

$$\{-0.04463, -0.04415, -0.04443, -0.04401, -0.03993, -0.03977, -0.03977,$$

$$-0.03959, -0.04437, -0.04397, -0.04419, -0.04374, -0.03958,$$

$$-0.03948, -0.03947, -0.03932\}\,.$$

We take the *Masked random plaintext dataset* as our attack traces (P-DPA Step 10). There are 16 key hypotheses $\hat{k}_i = i - 1$ ($i = 0, 1, \ldots, 15$). Based on our implementation, the hypothetical intermediate value should be given by

$$\hat{\boldsymbol{v}}_{ij} = \mathrm{SB}_{\mathrm{PRESENT}}(\hat{k}_i \oplus p_j \oplus \mathrm{m}_{0,j}), \quad i = 1, 2, \ldots, 16, \quad j = 1, 2, \ldots, \hat{M}_p,$$

where p_j is the 0th nibble of the plaintext corresponding to the jth trace, $\mathrm{m}_{0,j}$ is the input mask applied to this nibble, and $\hat{M}_p \leq 20000$ is the number of traces used for the attack. However, as mask values are unknown to the attacker, we will only compute the unmasked hypothetical intermediate value (P-DPA Step 11)

$$\hat{\boldsymbol{v}}_{ij} = \mathrm{SB}_{\mathrm{PRESENT}}(\hat{k}_i \oplus p_j), \quad i = 1, 2, \ldots, 16, \quad j = 1, 2, \ldots, \hat{M}_p.$$

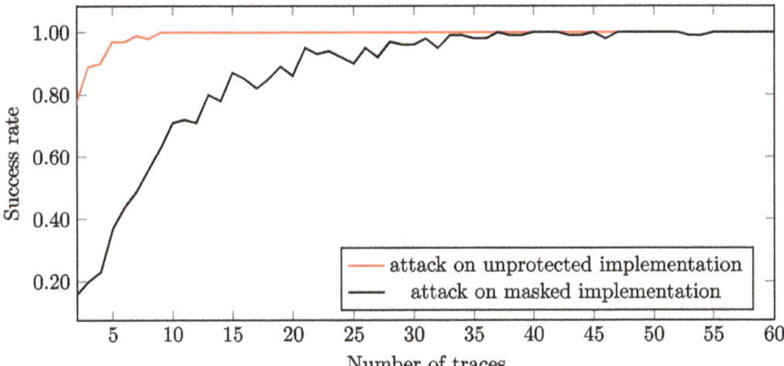

Fig. 4.90 Estimations of guessing entropy computed following Algorithm 4.1 for template-based DPA attacks on the *Masked random plaintext dataset* (in black) and on the *Random plaintext dataset* (in red)

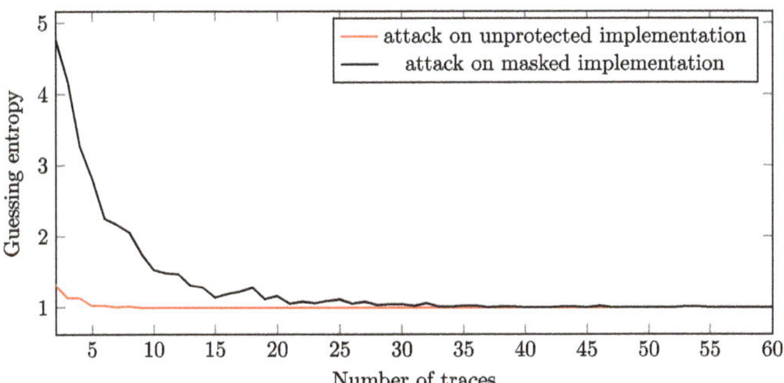

Fig. 4.91 Estimations of guessing entropy computed following Algorithm 4.1 for template-based DPA attacks on the *Masked random plaintext dataset* (in black) and on the *Random plaintext dataset* (in red)

Since we chose the signal to be the exact value of v, our leakage model will be the identity leakage model. The hypothetical signals are given by (P-DPA Step 12)

$$\mathcal{H}_{ij} = \hat{v}_{ij}, \quad i = 1, 2, \ldots, 16, \quad j = 1, 2, \ldots, \mathring{M}_p.$$

Following Template Step c, we can compute the probability score for each key hypothesis. Then with Algorithm 4.1, we can calculate the estimations for guessing entropy and success rate of the attack. We have set

$$\texttt{max\_trace} = 60, \quad \texttt{no\_of\_attack} = 100.$$

The results are shown in Figs. 4.90 and 4.91.

For comparison, we have also plotted the results for template-based DPA attack on unprotected implementations with one POI (in red in both figures). We can see that compared to the attacks on unprotected implementations, the same attack on the masked PRESENT requires more traces for the attack to be successful.

In practice, more than one mask will be applied to provide better protection, leading us to *higher order masking*. We will briefly introduce this notion in Sect. 4.6.

4.5.2.4 Blinding for RSA and RSA Signatures

As mentioned before, the application of masking in the context of a public cryptosystem is called *blinding*. Normally an arithmetic mask is applied.

Let p, q be two distinct odd primes, $n = pq$ be the RSA modulus, $d \in \mathbb{Z}^*_{\varphi(n)}$ be the private key for RSA, and $e = d^{-1} \mod \varphi(n)$ be the public key. The attacks we have seen in Sect. 4.4 exploit leakages during the computation of

$$a^d \mod n$$

for some $a \in \mathbb{Z}_n$. Those attacks can be during the RSA signature signing process or RSA decryption. More attacks will be discussed in Sect. 4.6.

Given those attacks, it is recommended to blind the secret values during the computation. It is also required that the masks and blinded values should be updated frequently or even during the computations. In this case, it will be difficult for the attacker to combine whatever partial information obtained from the leakages of the previously blinded value and the newly leaked information.

In this part, we will discuss a few methods, including exponent blinding, message blinding, and modulus blinding. The countermeasures are mostly designed against DPA attacks. In particular, the message blinding method will be effective against the DPA attack we have presented in Sect. 4.4.2. The original proposals can be found in [BCDG10, KJJR11].

Exponent blinding First, we consider how we can randomize the secret exponent d. One method is that we generate a random number $\lambda \in [0, 2^\ell - 1]$. Then instead of computing

$$a^d \mod n,$$

we compute

$$a^{d+\lambda\varphi(n)} \mod n. \tag{4.85}$$

Note it follows from Corollary 1.4.5 that

$$a^{d+\lambda\varphi(n)} \equiv a^d \mod n.$$

Example 4.5.2 Let $p = 3$, $q = 5$, then $n = 15$ and $\varphi(n) = 2 \times 4 = 8$. The same as in Example 3.3.1, we choose $e = 3$, and we get $d = 3$. Take $a = 8$ and $\lambda = 2$. We have computed in Example 3.3.1 that

$$a^d \mod n = 8^3 \mod 15 = 2.$$

With the above countermeasure (Eq. 4.85), we have

$$a^{d+\lambda\varphi(n)} \bmod n = 8^{3+2\times8} \bmod 15 = 8^{19} \bmod 15 = 8 \times (8^2)^9 \bmod 15$$

$$= 8 \times 4^9 \bmod 15 = 8 \times 4 \times (4^2)^4 \bmod 15 = 32 \bmod 15 = 2.$$

Typically, for RSA modulus of bit length 1024, we take $\ell = 20$ or 30 to guarantee a reasonable overhead [BCDG10].

The next method takes a random number λ and calculates

$$a^{\lambda} \times a^{d-\lambda} \bmod n.$$

It is easy to see that

$$a^{\lambda} \times a^{d-\lambda} \bmod n = a^d \bmod n.$$

A third method generates a random number λ such that $\gcd(\lambda, \varphi(n)) = 1$ and calculates

$$(a^{\lambda})^{d(\lambda^{-1} \bmod \varphi(n))} \bmod n. \tag{4.86}$$

Since

$$\lambda(d(\lambda^{-1} \bmod \varphi(n))) \equiv d \bmod \varphi(n),$$

it follows from Corollary 1.4.5 that

$$(a^{\lambda})^{d(\lambda^{-1} \bmod \varphi(n))} \bmod n = a^d \bmod n.$$

Example 4.5.3 The same as in Example 4.5.2, let $p = 3, q = 5, n = 15, d = 3, a = 8$. We have computed that

$$a^d \bmod n = 2.$$

Choose $\lambda = 3$, which is coprime to $\varphi(n) = 8$. Then by the extended Euclidean algorithm,

$$8 = 3 \times 2 + 2, \quad 3 = 2 + 1 \Longrightarrow 1 = 3 - 2 = 3 - (8 - 3 \times 2) = 3 \times 3 - 8,$$

we have

$$\lambda^{-1} \bmod \varphi(n) = 3^{-1} \bmod 8 = 3.$$

According to Eq. 4.86,

$$(a^\lambda)^{d(\lambda^{-1} \bmod \varphi(n))} \bmod n = (8^3)^{3\times 3} \bmod 15 = (512)^9$$

$$\bmod 15 = 2^9 \bmod 15 = 512 \bmod 15 = 2.$$

Another exponent blinding method considers a CRT-based RSA implementation. Recall from Sect. 3.5.1.3 that to compute

$$a^d \bmod n,$$

following CRT-based RSA, we can first calculate

$$a_p := a^{d \bmod (p-1)} \bmod p, \quad a_q := a^{d \bmod (q-1)} \bmod q. \tag{4.87}$$

Then $a^d \bmod n$ is given by

$$a_p y_q q + a_q y_p p \bmod n, \quad \text{or equivalently} \quad a_p + ((a_q - a_p)y_p \bmod q)p,$$

where

$$y_q = q^{-1} \bmod p, \quad y_p = p^{-1} \bmod q.$$

The countermeasure takes two random numbers λ_1, λ_2, and instead of computing a_p, a_q with Eq. 4.87, we calculate

$$a_p = a^{d+\lambda_1(p-1)} \bmod p, \quad a_q = a^{d+\lambda_2(q-1)} \bmod q.$$

It follows from Corollary 1.4.3 that

$$a^{d+\lambda_1(p-1)} \bmod p = a^{d \bmod (p-1)} \bmod p, \quad a^{d+\lambda_2(q-1)} \bmod q = a^{d \bmod (q-1)} \bmod q.$$

Example 4.5.4 The same as in Example 4.5.2, let

$$p = 3, \quad q = 5, \quad n = 15, \quad d = 3, \quad a = 8.$$

We have computed in Example 3.5.5 that

$$a_p = a^{d \bmod (p-1)} \bmod p = 8^{3 \bmod 2} \bmod 3 = 2,$$

$$a_q = a^{d \bmod (q-1)} \bmod q = 8^{3 \bmod 4} \bmod 5 = 2.$$

Take $\lambda_1 = 2$, $\lambda_2 = 3$, with our countermeasure, we have

$$a_p = a^{d+\lambda_1(p-1)} \bmod p = 8^{3+2\times2} \bmod 3 = 8^7 \bmod 3 = 2^7 \bmod 3 = 128 \bmod 3 = 2$$

$$a_q = a^{d+\lambda_2(q-1)} \bmod q = 8^{3+3\times4} \bmod 5 = 8^{15} \bmod 5 = 3^{15} \bmod 5 = 3\times(3^2)^7 \bmod 5$$

$$= 3 \times 4^7 \bmod 5 = 3\times4\times(4^2)^3 \bmod 5 = 12 \bmod 5 = 2.$$

Our SCA attacks from Sect. 4.4 rely on exploiting the leakages to get the value of each bit of the secret exponent d. We can see that for all the exponent blinding methods above, assuming an attack on one RSA decryption (or RSA signatures singing) execution, with the same methods, we can only recover the value of d+a random number or $d\times$a random number, making the real value of d concealed from the attacker. On the other hand, if two computations with different masks are attacked, the secret key can be recovered. For example, with the first countermeasure, if we know the values for

$$d + \lambda_1\varphi(n), \quad d + \lambda_2\varphi(n),$$

then we can get

$$(\lambda_1 - \lambda_2)\varphi(n).$$

Since λ_1 and λ_2 have bit length 20 or 30, we can factorize $(\lambda_1 - \lambda_2)\varphi(n)$ by trying all possible values of λ_1 and λ_2.

Message blinding We can also mask the value a. In this way, DPA attacks (e.g., the attack in Sect. 4.4.2) that rely on knowing certain intermediate values related to a cannot be carried out.

Take a random number λ such that $\gcd(\lambda, n) = 1$, and compute

$$a_1 = \lambda^e \bmod n, \quad a_2 = \lambda^{-1} \bmod n.$$

To get

$$a^d \bmod n,$$

we calculate

$$(((aa_1)^d \bmod n)a_2) \bmod n.$$

Since

$$ed \equiv 1 \bmod \varphi(n),$$

by Corollary 1.4.5,

$$\lambda^{ed} \equiv \lambda \bmod n \implies \lambda^{ed-1} \bmod n = 1.$$

Then

$$(((aa_1)^d \bmod n)a_2) \bmod n = (((a\lambda^e \bmod n)^d \bmod n)(\lambda^{-1} \bmod n)) \bmod n$$
$$= ((a^d \bmod n)(\lambda^{ed-1} \bmod n)) \bmod n = a^d \bmod n.$$

The first mask a_1 randomizes the input of the computation, and the second mask a_2 corrects the output to the expected result.

Example 4.5.5 Keep the same parameters as in Example 4.5.2:

$$p = 3, \quad q = 5, \quad n = 15, \quad e = 3, \quad d = 3, \quad a = 8, \quad \varphi(n) = 8.$$

We know that $a^d \bmod n = 2$. Take $\lambda = 4$, which is coprime with n. Then with the message blinding countermeasure above, we have

$$a_1 = \lambda^e \bmod n = 4^3 \bmod 15 = 64 \bmod 15 = 4.$$

By the extended Euclidean algorithm,

$$15 = 4 \times 3 + 3, \quad 4 = 3 + 1 \implies 1 = 4 - 3 = 4 - (15 - 4 \times 3) = 4 \times 4 - 15$$

and

$$a_2 = \lambda^{-1} \bmod n = 4^{-1} \bmod 15 = 4.$$

Finally,

$$(((aa_1)^d \bmod n)a_2) \bmod n = (((8 \times 4)^3 \bmod 15) \times 4) \bmod 15$$
$$= ((2^3 \bmod 15) \times 4) \bmod 15$$
$$= 32 \bmod 15 = 2.$$

Modulus blinding When the modulus is random during the computations, similar to random values of a, DPA attacks such as the one in Sect. 4.4.2 cannot be carried out as the attacker does not know the modulus to derive the target intermediate values.

For blinding the modulus n, we generate a random number λ and compute

$$(a^d \bmod (\lambda n)) \bmod n.$$

It is easy to see that

$$(a^d \bmod (\lambda n)) \bmod n = a^d \bmod n.$$

Example 4.5.6 Keep the same parameters as in Example 4.5.2:

$$p = 3, \quad q = 5, \quad n = 15, \quad e = 3, \quad d = 3, \quad a = 8, \quad \varphi(n) = 8.$$

We know that $a^d \bmod n = 2$. Let $\lambda = 4$, then we have

$$(a^d \bmod (\lambda n)) \bmod n = (8^3 \bmod (4 \times 15))$$

$$\bmod 15 = (512 \bmod 60) \bmod 15 = 32 \bmod 15 = 2.$$

Remark 4.5.2 Note that the message blinding and the modulus blinding methods we have presented can also be used in a similar way to protect the computation of

$$a^d \bmod p, \quad a^d \bmod q,$$

in CRT-based RSA implementations.

4.6 Further Reading

Leakage model We note that the Hamming distance, Hamming weight, identity (Sect. 4.2.1), and stochastic leakage (Sect. 4.3.2.2) models all assume there are no differences in the leakage when the value in a bit switches from 0 to 1 or from 1 to 0. Improved models can be found in, e.g., [PSQ07, GHP04].

Leakage assessment TVLA (see Sect. 4.2.3) was first proposed in 2011 [GGJR+11]. More discussions on how to set the threshold 4.5 can be found in [DZD+18]. Another prominent leakage assessment method is Person's χ^2-test [SM15], which is normally used as a replacement for TVLA when analyzing multivariate and horizontal leakages.

Simple power analysis We have seen that by visual inspection of the power traces, the attacker can gain information about the operations being executed on the device. SPA was first introduced in [KJJ99], which is also the very first proposal of power analysis attacks. The authors mentioned that programs involving conditional branch operations depending on secret parameters are at risk. Later this idea was applied to develop an SPA attack on RSA [MDS99b] (see Sect. 4.4.1).

[Nov02] (see also [KJJR11, Section 3.3]) proposes an attack that exploits vulnerability in Garner's algorithm for CRT-based RSA. The authors demonstrate that with SPA, we can identify if $a \bmod p > a \bmod q$. Then with adaptive chosen ciphertext and binary search, the value of p can be recovered. [FMP03] shows that with only known messages, assuming p and q have different lengths, in case $q < p/2^\ell$, p and q can be recovered by performing $60 \times 2^\ell$ signatures on average. A lower bound of ℓ is specified in the paper.

SPA has also been used to obtain the Hamming weight of operands [MS00] or attack AES key schedule [Man03]. Similar to profiled DPA, we can carry out a profiled SPA attack, see, e.g., [Man03, Section 5.3].

Differential power analysis A DPA attack on DES can be found in, e.g., [MDS99a]. For AES, detailed descriptions are given in [MOP08, Chapter 6].

For DPA attacks on RSA, [MDS99b] lists different variants of DPA on RSA, where some can be considered as extended SPA attacks. [dBLW03] proposes a DPA attack on CRT-based implementation using Garner's algorithm. The target intermediate value is the remainder after the modular reduction with one of the primes. [AFV07] studies more attacks on other intermediate values of CRT-based RSA. We have elaborated one of the methods in Sect. 4.4.2.

We also refer the readers to [MOP08, Sta10, KJJR11] for more discussions on SPA and DPA.

Template attacks The idea of template attacks was first introduced in [CRR03]. In Sect. 4.3.2.3, we discussed how templates can be used for DPA on symmetric block ciphers. In a similar manner, template-based attacks can also be applied to SPA on symmetric block ciphers [MOP08, Section 5.3], and SCA on RSA [VEW12, XLZ$^+$18].

We note that the template attacks we have described used normal distributions to approximate the distributions induced by leakages. One might refer to this as a Gaussian template attack. A more generic method, MIA, can be found in [GBTP08], where the authors aim to approximate the mutual information between the hypothetical leakages and the actual measured leakages without making assumptions on the leakage distribution.

SCADPA Side-channel assisted differential plaintext attack (SCADPA) was first proposed in [BJB18] for PRESENT and in [BJHB19] for GIFT implementations. It was later generalized to all SPN block ciphers in [BBH$^+$20]. The attack presented in Sect. 4.3.3 is based on this generalized attack. We refer the readers to the original paper [BBH$^+$20] for attacks on more ciphers and analysis of attack complexity.

More attacks Other side-channel attack methods exist for symmetric block ciphers. For example, collision attacks [SWP03] identify the collision of intermediate values between two encryptions using power traces to recover the secret key. Algebraic side-channel attacks [RS09] express both the target algorithm and its leakages as equations to achieve successful attacks with unknown plaintext/ciphertext. Soft analytical SCA [VCGS14] constructs a graph for the implementation and uses the belief propagation algorithm on this graph to efficiently combine the information of all leakage points. DCSCA (differential ciphertext SCA) [HBB21] targets GIFT cipher. The attack analyzes the statistical distribution of intermediate values with the help of side-channel leakages to recover the last round key. The authors also demonstrated the extension of the attack to GIFT-based AEAD schemes.

Preprocessing of traces During the measurements, it can happen that the traces contain too much noise. Or if there are certain countermeasures in place, the traces can also be misaligned. There are various classical methods for preprocessing traces. For example, *moving average* computes the average of the leakages from a few time samples to smooth the signal. *Principal component analysis* [BHvW12] aims to reduce the noise in the traces by projecting high-dimensional data to a lower dimensional subspace while preserving the data variance. *Elastic alignment* [vWWB11] aligns the traces by focusing on the synchronization of trace shape and generating artificial samples. It can be used to counter jitter-based countermeasures. The method is based on the dynamic time warping algorithm designed for speech recognition [SC78].

Hiding-based countermeasure A hiding-based countermeasure aims to make the leakage random or constant independent of the operation/data.

To randomize the leakage, we can insert random delays (jitters) [CK09] or shuffle the execution order of independent operations. For example, shuffle Sboxes in AES implementations [HOM06] randomize the sequence of square and multiply operations in RSA [Wal02]. Another approach to randomizing the leakage proposes to use residue number systems to allow randomizing the representation of finite field elements for computing exponentiation [BILT04].

To make the leakage constant, different methodologies have been proposed on different levels. For the cell level (or logic design level), we have, for example, dual-rail precharge logic (DPL) [TV06] and dynamic and differential logic styles [TAV02]. DPL has two phases: in the precharge phase, values in the wires are set to a precharge value (either 0 or 1); then during the evaluation phase, one wire carries the signal 0, and the other wire carries the signal 1. We note that this is equivalent to using the binary code {01, 10} for encoding 0, 1. For the software level, we have seen encoding-based countermeasures for symmetric block ciphers in Sect. 4.5.1.1 and square and multiply always algorithm for RSA in Sect. 4.5.1.2. The original proposal of the square and multiply always algorithm can be found in [Cor99]. More on encoding-based countermeasures can be found in, e.g., [CG16]. [CG16] uses linear complementary dual code—a code C is a complementary dual code if $C \cap C^{\perp} = \{\mathbf{0}\}$ (see Definition 1.6.9). Another example of software level countermeasure can be found in [HDD11], where the authors propose to use DPL in software for symmetric block ciphers. See also [RGN13] for a DPL in software countermeasure with provable security for bitsliced implementation of PRESENT.

Masking-based countermeasures Those countermeasures are designed to make the leakage dependent on some random value. Masking was first proposed by Goubin and Patarin [GP99] and Chari et al. [CJRR99] independently. It has been proven that masking-based countermeasure is secure given that the source of randomness is truly random [PR13]. Due to this sound mathematical basis, it has become the most adopted countermeasure for symmetric block ciphers.

Instead of a naive lookup table implementation of a masked Sbox as presented in Sects. 4.5.2.2 and 4.5.2.3, many other methods have also been proposed. We refer the readers to [DCRB$^+$16, OS05] for masked AES Sbox and [PMK$^+$11, SBM18] for masked PRESENT Sbox.

In Sects. 4.5.2.1–4.5.2.3 we have seen how *Boolean masking*, i.e., the intermediate value is concealed by $\oplus$ with the mask(s), can be implemented for AES and PRESENT. Section 4.5.2.4 focuses on arithmetic masking, where the intermediate value is concealed by an arithmetic operation (modular addition or modular multiplication). There are many other ways of applying the masks for example, affine masking [VW01], polynomial masking [GM11], inner product masking [BFGV12], etc. The exact operation for applying a mask is typically chosen depending on the operations that are used in the cryptographic algorithm that we would like to protect. Some cryptographic algorithms (e.g., AES) contain both Boolean and arithmetic operations,[11] and they can be protected using both Boolean and arithmetic masking. But switching from one type of masking to another is not a trivial task. We refer the readers to [Mes00, Gou01, CT03, AG01] for more discussions on this topic.

Masking-based countermeasures can also be implemented on the hardware level, for example, masking buses [BGM$^+$03], Boolean masking of DLP [PM05], random precharging [BGLT04]. However, it has been shown that masked gates in hardware are vulnerable to DPA attacks due to glitches in CMOS circuits [MPG05]. This leads to the development of threshold implementations [NRS11], which are based on multiparty computation [CD$^+$15], as a way to realize secure Boolean masking in hardware.

Higher order masking In Sects. 4.5.2.1–4.5.2.3 we focused on masking schemes with one mask only. In particular, the masked value v_m is related to the original value v and the mask m through a binary operation $v_m = v \cdot$ m. In the language of *secret sharing* [Bei11], we can say that the secret value v is represented by two *shares*, v_m and m. We can see that given only one of the two shares, no information about v can be revealed. Instead of two shares (or one mask), we can also use several shares, resulting in a higher order masking. In particular, a *dth-order masking* applies $d - 1$ masks to the secret value v.

Similarly, in Sects. 4.3.1 and 4.3.2, we only focused on one target intermediate value. Such a DPA attack is also called a *first-order DPA*. When leakages of several different intermediate values (e.g., at different time samples) are analyzed, we have a *higher order DPA* [CJRR99]. The number of traces needed for a higher order DPA to succeed is exponential in the standard deviation of the noise. The exponent is given by $d + 1$, where $d + 1$ is the order of the masking (i.e., d masks are applied) [PR13].

For the security of a Boolean masking. Let us take a secret value v of bit length at most m_v. The masked value is given by $v \oplus$ m, where m $\in \mathbb{F}_2^{m_v}$. We can consider

---

[11] AES Sbox is based on modular computations in a specific field—see Eq. 3.3.

the value of m as a discrete random variable. In case the distribution induced by this random variable is uniform on $\mathbb{F}_2^{mv}$, the distribution induced by the value of $v \oplus m$ is also uniform on $\mathbb{F}_2^{mv}$, regardless of the value of v. Thus, we expect the leakage to be independent of v when only first-order DPA is carried out. The security proof for first-order Boolean masking against first-order DPA can be found in [BGK04]. Results for higher order Boolean masking are given in [RP10]. However, the proofs rely on the masks to be truly random, which is not easy to achieve in practice. For example, our masked implementation from Sect. 4.5.2.3 can still be attacked with first-order DPA (see Figs. 4.90 and 4.91). We also note that the choice of masks should follow certain rules so that the masking scheme is more secure (see, e.g., [BGN$^+$15])

Blinding Blinding was first suggested in [Koc96]. It was then later formalized by J. S. Coron [Cor99]. It is worth noting that several patents have been published about masking [KJJ10] and blinding [KJ01].

Various attacks on blinding have also been published. For example, [FV03] proposes an attack on the left-to-right square and multiply algorithm that recovers a blinded secret exponent with SPA. [WvWM11] discusses a DPA attack on the square and multiply always algorithm and message blinding. [FRVD08] exploits the leakage during the computation of the random exponent.

More about countermeasures For SCA countermeasures, except for those introduced in this chapter, there are also many other techniques. In general, we can divide them according to the levels of protection.

Protocol level countermeasures aim to design cryptographic protocols to survive leakage analysis. For example, by limiting the number of communications that can be performed with any given key, fewer measurements can be done by the attacker for the same key or by rekeying [MSGR10].

Cryptographic primitive level countermeasures are proposals of new cipher designs that are resistant to side-channel attacks.

Implementation level countermeasures were the focus of this chapter, where we discussed some hiding and masking/blinding techniques in Sect. 4.5. There are also other implementation-level countermeasures, for example, time randomization [MMS01b] and encryption of the buses [BHT01].

Architecture level countermeasures refer to techniques that modify the architecture of the computation device. For example, [MMS01a] proposes to use a nondeterministic processor to randomly change the sequence of the executed program during each execution; [SVK$^+$03] integrates secure instructions into a nonsecure processor.

Hardware level countermeasures protect the implementations through external means, for example, conforming glues [AK96], protective coating [TSS$^+$06], and detachable power supplies [Sha00].

Attacks on post-quantum cryptographic implementations Several papers propose SCA on post-quantum cryptosystems.

For example, in [UXT$^+$22], side-channel leakage during the execution of a pseudorandom function in the re-encryption of key encapsulation mechanism decapsulation is exploited. With the leakage, the attacker gains information on whether the public key decryption result is equivalent to the reference plaintext. The authors in [GJJ22] propose to submit special ciphertexts to the decryption oracle that correspond to cases of single errors. Through leakage in the additive Fast Fourier Transform step used to evaluate the error locator polynomial, a single entry of the secret key can be determined. A survey on SCA and FIA on Kyber and Dilithium post-quantum schemes with novel countermeasures was presented in [RCDB22].

Attacks on neural networks SCA techniques have also been adopted for attacking neural network implementations. In such cases, normally a black-box scenario is assumed and the attacker's goal is to recover the secret parameters of the target neural network. [BBJP19] demonstrated how a timing-based attack can recover the architecture information. Then with DPA techniques, weights for each neuron can be recovered. The provided experiments were done on the ARM Cortex-M3 microcontroller. [YMY$^+$20] experimented on a hardware implementation of neural networks on an FPGA. For a more comprehensive overview of the topic, we refer the interested reader to [BBB$^+$22].

Correspondingly, SCA countermeasures for cryptographic implementations have also been adopted for protecting neural network implementations. For example, masking methods have been utilized in [DCA20, DAP$^+$22, DCSA22]. Threshold implementation was proposed in [MBFC22] with a Trivium stream cipher to generate the randomness. From the area of hiding countermeasures, shuffling was implemented in [NY21] and desynchronization by adding a random jitter in [BJHB23].

4.6.1 AI-Assisted SCA

AI-based methods have been applied for side-channel analysis in the past few years. If we look at DPA (Sects. 4.3.1, 4.3.2, and 4.4.2), the key recovery is essentially a classification problem. In particular, in a profiled setting, the profiling phase corresponds to the training phase of an AI-based algorithm. During the attack phase, the analysis of the leakage traces can be seen as a classification problem where the goal of an attacker is to classify those traces based on the related data (e.g., a specific Sbox output value). Various AI-based techniques have been adopted for SCA, e.g., k-nearest neighbor algorithm [MZMM16], random forest [LBM15], support vector machines [HZ12], multilayer perception (MLP) [GHO15], and convolutional neural networks (CNNs) [ZBHV20]. It has also been shown that, with neural networks, protected implementations can be broken. For example, [WP20] used autoencoder to break hiding countermeasures, while in [MPP16], the authors successfully broke masking countermeasures with deep learning techniques.

As an example, let us consider the case of a neural network used for the classification problem in a DPA attack on AES implementations (see Sect. 4.3.1). The input of the network will then be (part of) the traces. The output layer will have a softmax activation function, and each class corresponds to one possible value of the target Sbox output, hence leading to one key byte hypothesis with the knowledge of the plaintext. Then during the inference, for each input data, the network output indicates the possibilities of the 256 values for the Sbox output, which gives a possibility of each of the corresponding key byte hypotheses.

Success rate and guessing entropy Given a few, say $\hat{M}_p$, data (trace), we can compute a score for each key hypothesis by summing up the corresponding probabilities predicted using each data. Then we can rank the key hypotheses according to their scores with the one ranked the first having the highest score. Let us denote the rank of the correct key hypothesis by $rk_{AI}^{\hat{M}_p}$. It is easy to see that we can consider $rk_{AI}^{\hat{M}_p}$ as a random variable whose randomness comes from different plaintexts/measurements.

Recall that in Eq. 4.36, we have defined the success rate for a DPA attack. For AI-based SCA attacks, we have an equivalent definition of success rate, namely the probability that $rk_{AI}^{\hat{M}_p} = 1$. Similarly, we can also define the guessing entropy (see Eq. 4.37) to be the expectation of the random variable $rk_{AI}^{\hat{M}_p}$. Same as for DPA attacks, we can estimate success rate with the frequency of successful attacks among a number of trials and estimate guessing entropy using the sample mean of $rk_{AI}^{\hat{M}_p}$.

In particular, for a fixed $\hat{M}_p$, we randomly select $\hat{M}_p$ data from the test set and carry out an attack with $\hat{M}_p$ traces, and then we compute $rk_{AI}^{\hat{M}_p}$. We repeat this procedure for, e.g., 100 times, which gives us a sample of $rk_{AI}^{\hat{M}_p}$. Its mean is then an estimation for the guessing entropy. An estimation for the success rate is the frequency of $rk_{AI}^{\hat{M}_p} = 1$ among those 100 simulated attacks.

In most cases, the goal of AI-based SCA is to achieve a low guessing entropy or a high success rate with as few traces as possible after training.

Different research topics in AI-assisted SCA Many different aspects of AI-assisted SCA have been analyzed by researchers.

Firstly, there are a few publications on public datasets, which are used to evaluate novel proposals of AI-based techniques. To name a few, ASCAD dataset [BPS+20, BPS+21] contains power traces for software implementations of AES with masking countermeasures and artificially introduced random jitters. AES_HD [BJP20] dataset is EM traces corresponding to unprotected AES hardware implementation on FPGA. AES_RD [CK09, CK10, CK18] dataset consists of power traces of software implementations of AES with random delay.

The most studied direction is of course to achieve high success rates or low guessing entropy. By examining the similarity of side-channel traces to time series data (e.g., audio signals), [KPH$^+$19] proposed a VGG15-like network together with a regularization method achieved by adding noise to the traces. Zaid et al. [ZBHV20] introduced a methodology for the design of CNNs in the SCA context. The paper analyzed several datasets and constructed an optimal CNN for each dataset. [WJB20] showed an improvement of [ZBHV20] using data oversampling. [PCP20] used ensemble models to achieve good generalization from the training set to the validation set for a given dataset. On the other hand, Won et al. [WHJ$^+$21] utilized Multi-scale Convolutional Neural Networks for SCA to achieve the goal of integrating classical trace preprocessing techniques and attacking several datasets without changing the network architectures.

Hyperparameter tuning, an important problem in AI algorithm development in general, naturally attracted attention in the domain of SCA. Various methods have been proposed, for example, Bayesian optimization and random search [WPP22], reinforcement learning [RWPP21], and genetic algorithm for choosing architectures [MPP16] or for choosing all hyperparameters [AGF21].

It has been shown that test accuracy in machine learning cannot properly assess SCA performance [PHJ$^+$19]. Because of this observation, many training strategies are studied, for example, stopping criteria based on success rate [RZC$^+$21] or based on mutual information [PBP21].

Recently, non-profiled AI-based SCA has also gained attention in the research community. For example, in [Tim19], the authors propose to train a neural network for each key hypothesis. To do this, the attacker splits the traces based on the key hypothesis, just like when carrying out DPA. The network that achieves the best training metrics then reveals the actual key byte. This method was titled *Differential Deep Learning Analysis (DDLA)*.

Stream ciphers were targeted by a combination of machine learning, mixed integer linear programming, and satisfiability modulo theory methods [KDB$^+$22].

Furthermore, AI-based methods have also been adopted for the identification of points of interest [LZC$^+$21] and leakage assessment [MWM21].

Chapter 5
Fault Attacks and Countermeasures

Fault attacks are active attacks where the attacker tries to perturb the internal computations by external means. Such attacks exploit a scenario where the attacker has access to the device and can tamper with it.

Fault attacks can be achieved with different techniques, ranging from simple clock/voltage glitches to sophisticated optical fault injections (see Sect. 6.2 for more details).

The attacker's goal is to recover the secret master key of the cryptographic algorithm. The attack methodologies are normally developed on the algorithmic level. But implementation-specific vulnerabilities also exist (Sect. 5.1.4).

There are different effects that a fault injection can achieve. *Instruction skip* and *instruction change* perturbs the instruction being executed by modifying the opcode of the instruction. *Bit flip* flips the bits in the data. The number of bits affected is normally limited by the register size (although, technically it is possible to affect a few registers at once). We use the notation m-bit flip to indicate how many bits are flipped by the fault attack. This notion is consistent with our previous definition of bit flip (see Definition 1.2.17). *Bit set/rest* fixes the bit value to be 1 (set) or 0 (reset). *Random byte fault* changes the byte value to a random number. *Stuck-at fault* permanently changes the value of one bit to 0 (stuck-at-0) or 1 (stuck-at-1). We refer to those different effects as *fault models*.

If the fault injected in an intermediate value x results in a faulty value x', we refer to $\varepsilon := x \oplus x'$ as the *fault mask*, which represents the change in the faulted value.

We can divide the faults into two types depending on how long the effects last. A *permanent fault* is a destructive fault that changes the value of a memory cell permanently and hence affects data during the computations. Whereas when a *transient fault* is injected, the circuit recovers its original behavior after the fault stimulus ceases (usually just one instruction) or after the device reset. A transient fault can perturb both data and instruction. In this chapter, we only consider transient faults.

353

X. Hou, J. Breier, *Cryptography and Embedded Systems Security*,
https://doi.org/10.1007/978-3-031-62205-2_5

After the fault injection, there are two possible scenarios. The output (ciphertext) is faulty, or the fault is ineffective and the ciphertext is not changed. We will see that both scenarios can be exploited.

In the rest of this chapter, we will discuss fault attacks and countermeasures for symmetric block ciphers (Sects. 5.1 and 5.2) and for RSA and RSA signatures (Sects. 5.3 and 5.4).

5.1 Fault Attacks on Symmetric Block Ciphers

This section presents a few fault attack methods on symmetric block ciphers. By convention (see Kerckhoffs' principle in Definition 2.1.3), we assume that the specifications of round functions and key schedules are public. The master key, and hence also the round keys, is secret. We also assume that throughout the attack, the same master key is used, and the goal of the attacker is normally to recover certain round key(s). The methodologies presented can be applied to an unprotected implementation of any symmetric block cipher proposed up to now.

Fault attacks normally aim to recover the last/first round key(s) and then use the inverse key schedule to find the master key. As mentioned in Remarks 3.1.1, 3.1.4, and 3.1.5, for DES and PRESENT-80, the knowledge of any round key gives 48 and 64 bits of the master key, respectively, and the rest of the key bits can be brute forced, while for AES, the value of any round key reveals the value of the full master key.

5.1.1 Differential Fault Analysis

Differential Fault Analysis (DFA) was first introduced by Biham et al. [BS97] in 1997. It has been studied by numerous researchers in different settings and is one of the most popular fault attack analysis methods for symmetric block ciphers.

DFA considers a fault injection into the intermediate state of the cipher, normally in the last few rounds. Then the difference between correct and faulty ciphertexts is analyzed to recover the round key(s).

Before going into details of DFA, we recall the notion of differential distribution table of an Sbox from Definition 4.3.1.

Example 5.1.1 Let us consider one of the DES Sboxes, $\mathrm{SB}_{\mathrm{DES}}^1 : \mathbb{F}_2^6 \rightarrow \mathbb{F}_2^4$ (Table 3.3). We note that since the maximum bit length of the input, 6, is longer than that of the output, 4, the output difference can be zero for some cases. The size of the table is $2^6 \times (2^4 - 1)$. Part of it can be found in Table 5.1.

For example,

$$\mathrm{SB}_{\mathrm{DES}}^1(5 \oplus 8) \oplus \mathrm{SB}_{\mathrm{DES}}^1(5) = \mathrm{SB}_{\mathrm{DES}}^1(\mathrm{D}) \oplus \mathrm{SB}_{\mathrm{DES}}^1(000101)$$

$$= \mathrm{SB}_{\mathrm{DES}}^1(001101) \oplus 7 = 13 \oplus 7 = \mathrm{A}.$$

Table 5.1 Part of the difference distribution table for SB_{DES}^1 (Table 3.3)

Δ \ δ	1	2	...	7	8	...
0			...	13,14		...
1			...	1,6,30,37		...
2			...	3,4,A,D,23,24,31,33,34,36		...
3	1C,1D,2C,2D,3C,3D	5,7,C,E,21,23,30,32	...	2,5,8,F	11,12,13,17,19,1A, 1B,1F,26,2E,37,3F	...
4			...			...
5	6,7	20,22,3C,3E	...	1B,1C,3B,3C	7,F,23,27,2B,2F,35,3D	...
6	C,D,24,25	24,26,2D,2F	...	9,E,11,12,15,16,20,27	4,C,10,14,18,1C,32,3A	...
7	16,17,32,33	15,17,1C,1E	...	21,26,2A,2D	22,2A,36,3E	...
8			...	10,17		...
9	E,F,10,11,28,29,36,37,38,39	10,12,2C,2E,38,3A	...	B,C,22,25	6,E,20,25,28,2D	...
A	4,5,14,15,26,27,30,31,34,35,3A,3B	0,2,14,16,25,27,39,3B	...	0,7,1A,1D,28,2F,39,3E	5,D	...
⋮	⋮	⋮	⋮	⋮	⋮	⋱

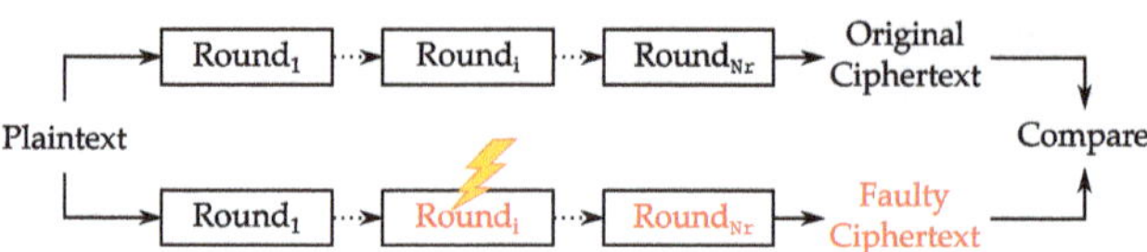

Fig. 5.1 An illustration of DFA

An illustration of DFA is shown in Fig. 5.1. The attacker injects a fault in a chosen round of the algorithm to get the desired fault propagation at the end of the encryption. By examining the differences between a correct and a faulty ciphertext, the possible values of the secret key can be narrowed down, and we also say that the *key hypotheses are reduced*. Another important concept needed for DFA is *nonlinear* functions (see Definition 4.5.1). The fault is usually injected at the input of a nonlinear function of the algorithm.

Example 5.1.2 (How DFA works on a simple example) Let us consider the AND operation (see Example 1.2.14) that takes inputs $a, b \in \mathbb{F}_2$ and outputs

$$c = a \,\&\, b.$$

All possible values of a, b, c are given by

a	b	$c = a \,\&\, b$
0	0	0
0	1	0
1	0	0
1	1	1

Suppose the output c can be observed by the attacker and a, b are unknown. The goal of the attacker is to recover the value of a. This can be achieved by DFA—during the computation, the attacker injects a fault in b by flipping it. By the knowledge of the faulty and the correct outputs, the attacker can easily recover the value of a: If the output stays the same, then $a = 0$; otherwise $a = 1$.

Next, we will detail how DFA works on an Sbox. Let SB: $\mathbb{F}_2^{\omega_1} \rightarrow \mathbb{F}_2^{\omega_2}$ be an Sbox, and let $a \in \mathbb{F}_2^{\omega_1}$, $b \in \mathbb{F}_2^{\omega_2}$ be fixed secret values. Define

$$f : \mathbb{F}_2^{\omega_1} \rightarrow \mathbb{F}_2^{\omega} \tag{5.1}$$

$$x \mapsto \mathrm{SB}(x \oplus a) \oplus b.$$

We will show how to recover the values of a and b with DFA.

Let us consider faults injected in the input of f. We use x' to denote the faulty value of x. The same as in Example 5.1.2, we assume a bit-flip fault model. Let ε denote the fault mask, i.e., $\varepsilon = x \oplus x'$.

Suppose the attacker has the knowledge of the Sbox design, inputs and outputs of f, and the fault mask ε. Furthermore, the attacker can repeat the computation with the same input (not chosen by the attacker). With details of the Sbox, the attacker can compute the DDT, denoted by T, of SB.

Let Δ denote the difference between the correct and the faulty output; we have

$$\Delta = (\mathrm{SB}(x \oplus a) \oplus b) \oplus (\mathrm{SB}(x' \oplus a) \oplus b) = \mathrm{SB}(x \oplus a) \oplus \mathrm{SB}(x' \oplus a)$$
$$= \mathrm{SB}(x \oplus a) \oplus \mathrm{SB}(x \oplus a \oplus \varepsilon).$$
$$(5.2)$$

Then the value $x \oplus a$ is in the entry of T corresponding to input difference $\delta = \varepsilon$ and output difference Δ. Thus, the possible values for $x \oplus a$ can be reduced to those in $T[\Delta, \varepsilon]$. With the knowledge of x, the attacker can narrow down the possible values of a. With the knowledge of the input and output of f, each value of a gives a unique value of b. The attacker can repeat the attack until the value of a (and hence b) is recovered or until brute force is possible to try the remaining values.

Example 5.1.3 (How DFA works on PRESENT Sbox) Let us consider the case when the Sbox in the definition of f (Eq. 5.1) is the PRESENT Sbox (Table 3.11). Suppose the attacker fixes the input to be $x = 0$, and they know that the correct output of f is 0.

When the attacker injects fault in x with fault mask $\varepsilon_1 = 3$, they get a faulty output 1. By Eq. 5.2, we have

$$\Delta_1 = 0 \oplus 1 = 1.$$

Thus $x \oplus a$ is in the entry of DDT of PRESENT Sbox corresponding to input difference 3 and output difference 1. By Table 4.1, the possible values for $x \oplus a$ are given by 9 and A.

When the attacker injects another fault with fault mask $\varepsilon_2 = 2$, they get a faulty output 6. We have $\Delta_2 = 6$. Again by Table 4.1, the possible values for $x \oplus a$ are given by 9 and B.

Thus, the attacker can conclude that

$$x \oplus a = 9.$$

Since $x = 0$, we know $a = 9$. With the knowledge that the correct output is 0, we have

$$\mathrm{SB}_{\mathrm{PRESENT}}(0 \oplus 9) \oplus b = 0 \implies b = \mathrm{SB}_{\mathrm{PRESENT}}(9).$$

Table 3.11 gives $b = \mathrm{E}$.

We can check that for $\varepsilon_1 = 3$,

$$\Delta_1 = f(0 \oplus 3) = \mathrm{SB}_{\mathrm{PRESENT}}(3 \oplus 9) \oplus \mathrm{E} = \mathrm{SB}_{\mathrm{PRESENT}}(\mathrm{A}) \oplus \mathrm{E} = \mathrm{F} \oplus \mathrm{E} = 1,$$

and for $\varepsilon_2 = 2$,

$$\Delta_2 = f(0 \oplus 2) = \text{SB}_{\text{PRESENT}}(2 \oplus 9) \oplus E = \text{SB}_{\text{PRESENT}}(B) \oplus E = 8 \oplus E = 6.$$

One might ask, how many faults are needed to recover the values of a and b. If we take a closer look at Table 4.1, we can see that in case the attacker can choose the fault mask, they only need two faults. For example, fault masks 3 and 5 can uniquely determine the Sbox input—any two distinct elements that appear in the same entry in column $\delta = 3$ are in two different entries in column $\delta = 5$. When a random fault mask is considered, a brute force analysis can show that at most four different fault masks are needed.

5.1.1.1   DFA on DES

Now, we will discuss how DFA can break implementations of DES (Sect. 3.1.1).

Recall that DES is a Feistel cipher. Its cipher state at the end of round i can be denoted as L_i and R_i, where L stands for left and R stands for right. The DES round function F satisfies

$$(L_i, R_i) = F(L_{i-1}, R_{i-1}), \text{ where } L_i = R_{i-1}, \ R_i = L_{i-1} \oplus f(R_{i-1}, K_i).$$
$$(5.3)$$

Before the first round function, the encryption starts with an initial permutation (IP). The inverse of IP, called the final permutation (IP^{-1}), is applied to the cipher state after the last round before outputting the ciphertext. In our analysis, we ignore the final permutation and consider the value before that as the ciphertext. Otherwise, the attacker can easily obtain this value by applying IP to the ciphertext.

At the ith round, the function f in the round function (Eq. 5.3) of DES takes input $R_{i-1} \in \mathbb{F}_2^{32}$ and round key $K_i \in \mathbb{F}_2^{48}$ and outputs a 32-bit intermediate value as follows:

$$f(R_{i-1}, K_i) = P_{\text{DES}}(\text{Sboxes}(E_{\text{DES}}(R_{i-1}) \oplus K_i)).$$
$$(5.4)$$

First, R_{i-1} is passed to an expansion function $E_{\text{DES}} : \mathbb{F}_2^{32} \to \mathbb{F}_2^{48}$ (Table 3.2). Then the output $E_{\text{DES}}(R_{i-1})$ is XOR-ed with the round key K_i, producing a 48-bit intermediate value. This 48-bit value is divided into eight 6-bit subblocks. Eight distinct Sboxes, $\text{SB}_{\text{DES}}^j : \mathbb{F}_2^6 \to \mathbb{F}_2^4$ ($1 \le j \le 8$), are applied to each of the 6 bits. Finally, the resulting 32-bit intermediate value goes through a permutation function $P_{\text{DES}} : \mathbb{F}_2^{32} \to \mathbb{F}_2^{32}$ (Table 3.4).

For $j = 1, 2, \ldots, 8$, let $E_{\text{DES}}(R_i)^j$ denote the jth 6 bits of $E_{\text{DES}}(R_i)$. For example, $E_{\text{DES}}(R_i)^1$ are bits at positions 1, 2, 3, 4, 5, 6 of $E_{\text{DES}}(R_i)$ (see also Note in Sect. 3.1.1). Similarly, let K_i^j denote the jth 6 bits of K_i and $P_{\text{DES}}^{-1}(R_i \oplus L_{i-1})^j$ be the jth 4 bits of $P_{\text{DES}}^{-1}(R_i \oplus L_{i-1})$. By Eqs. 5.3 and 5.4, we have

$$P_{\text{DES}}^{-1}(R_i \oplus L_{i-1})^j = \text{SB}_{\text{DES}}^j(E_{\text{DES}}(R_{i-1})^j \oplus K_i^j). \tag{5.5}$$

We consider a fault injection at the right half of the cipher state at the beginning of the 16th round, i.e., fault in R_{15}. Suppose the fault model is 1-bit flip. In other words, the fault mask $\varepsilon \in \mathbb{F}_2^{32}$ satisfies $\text{wt}\,(\varepsilon) = 1$ and

$$R'_{15} = R_{15} \oplus \varepsilon.$$

We assume the attacker has the knowledge of the output of DES (correct and faulty ciphertexts), fault model, and fault location. They can also repeat the computation with the same plaintext, not chosen by the attacker. The attacker's goal is to recover K_{16}, the last round key.

Let L'_{16} and R'_{16} denote the left and right parts of the faulty ciphertext, respectively. By our assumption, the attacker has the knowledge of L'_{16} and L_{16}. Since $R_{15} = L_{16}$, we have

$$L'_{16} \oplus L_{16} = R'_{15} \oplus R_{15} = \varepsilon \quad \text{(fault mask).} \tag{5.6}$$

Define

$$\Delta R_{16} := R'_{16} \oplus R_{16}. \tag{5.7}$$

By Eq. 5.5,

$$P_{\text{DES}}^{-1}(R_{16} \oplus L_{15})^j = \text{SB}_{\text{DES}}^j(E_{\text{DES}}(L_{16})^j \oplus K_{16}^j),$$

$$P_{\text{DES}}^{-1}(R'_{16} \oplus L_{15})^j = \text{SB}_{\text{DES}}^j(E_{\text{DES}}(L'_{16})^j \oplus K_{16}^j)$$

$$= \text{SB}_{\text{DES}}^j(E_{\text{DES}}(L_{16} \oplus \varepsilon)^j \oplus K_{16}^j).$$

Since P_{DES} and E_{DES} are linear, we have

$$P_{\text{DES}}^{-1}(\Delta R_{16})^j$$

$$= \text{SB}_{\text{DES}}^j(E_{\text{DES}}(L_{16})^j \oplus K_{16}^j \oplus E_{\text{DES}}(\varepsilon)^j) \oplus \text{SB}_{\text{DES}}^j(E_{\text{DES}}(L_{16})^j \oplus K_{16}^j).$$

Thus, $E_{\text{DES}}(L_{16})^j \oplus K_{16}^j$ is an input for the jth DES Sbox such that with input difference $E_{\text{DES}}(\varepsilon)^j$, the output difference is $P_{\text{DES}}^{-1}(\Delta R_{16})^j$. With the knowledge of ε, ΔR_{16}, and L_{16}, the attacker can reduce the key hypotheses for K_{16}^j.

We note that if $E_{\text{DES}}(\varepsilon)^j = 0$, the input for the jth Sbox is not changed, and the output will also not change. In this case, we say that this Sbox is *inactive*. Otherwise, we say the Sbox is *active*. For an inactive Sbox, a different fault mask will be needed to activate this Sbox. Since we consider a 1-bit flip, by the design of E_{DES} (Table 3.2), 16 bits of the input are repeated in the output; thus only one or two Sboxes will be active for one fault mask.

Example 5.1.4 Let

$$L_{15} = 00000000, \quad R_{15} = 00000000, \quad K_{16} = 14D8F55DAA7A.$$

By Eq. 5.4,

$$f(R_{15}, K_{16}) = P_{\text{DES}}(\text{Sboxes}(E_{\text{DES}}(R_{15}) \oplus K_{16})) = 832ABB8E.$$

By Eq. 5.3,

$$L_{16} = R_{15} = 00000000, \quad R_{16} = L_{15} \oplus 832ABB8E = 832ABB8E.$$

Suppose fault mask $\varepsilon = 40000000$, then

$$R'_{15} = 40000000,$$

and

$$L'_{16} = R'_{15} = 40000000, \; R'_{16} = L_{15} \oplus f(R'_{15}, K_{16}) = 83AAB98E.$$

We note that the values agree with Eq. 5.6:

$$\varepsilon = L'_{16} \oplus L_{16} = 40000000.$$

By Eq. 5.7,

$$\Delta R_{16} = R'_{16} \oplus R_{16} = 00800200.$$

Since the bit flip is in the second bit of input for E_{DES}, according to Table 3.2, the third bit of the output of E_{DES} will be changed. Consequently, $\text{SB}_{\text{DES}}^{j}$ is active for $j = 1$ and inactive otherwise. We have

$$E_{\text{DES}}(\varepsilon)^1 = 8, \quad E_{\text{DES}}(\varepsilon)^j = 0 \text{ for } j \neq 1.$$

By Table 3.4, the first 4 bits of the output of P_{DES} are given by the 9th, 17th, 23rd, and 31st bits of the input, hence

$$P_{\text{DES}}^{-1}(\Delta R_{16})^1 = 1010 = A.$$

Consequently, the input of SB_{DES}^1, which is

$$E_{\text{DES}}(R_{15})^1 \oplus K_{16}^1 = K_{16}^1,$$

gives output difference A when the input difference is 8. By Table 5.1, K_{16}^1 is equal to one of the two possible values: 5 and D, where 5 agrees with the first 6 bits of K_{16}.

In [BS97], the authors reported that with exhaustive search, they found that, on average, four possible 6-bit key hypotheses remain for each active Sbox. An improved attack that considers fault injection in the earlier rounds can be found in [Riv09]

5.1.1.2   Diagonal DFA on AES-128

In this part, we discuss a DFA attack on AES-128 implementations. Recall that AES cipher state can be represented as a 4×4 matrix of bytes (see Eq. 3.2):

$$\begin{pmatrix} s_{00} & s_{01} & s_{02} & s_{03} \\ s_{10} & s_{11} & s_{12} & s_{13} \\ s_{20} & s_{21} & s_{22} & s_{23} \\ s_{30} & s_{31} & s_{32} & s_{33} \end{pmatrix}. \tag{5.8}$$

Let us represent those bytes by squares as in Fig. 3.6 for visual illustration. Suppose a fault is injected at the beginning of one round (except for the last round) in byte s_{00}. Then the fault propagation in this round can be represented by Fig. 5.2, where blue squares correspond to bytes that might be affected by the fault. Since SubBytes and ShiftRows only affect 1 byte and the first row does not change in ShiftRows operation, in the first three states, the blue squares stay in the same position. MixColumns takes one column as input and outputs one column. AddRoundKey does not change the fault effects. Hence in the last state, the whole first column can be affected by the fault. Similarly, if the fault is injected at the beginning of one round in any combination of bytes $s_{00}, s_{11}, s_{22}, s_{33}$, at the end of this round, the whole first column might be affected by the fault. Some cases are shown in Fig. 5.3.

Let us refer to the bytes $s_{00}, s_{11}, s_{22}, s_{33}$ as a *diagonal* of AES state. We consider a fault attack where a random byte fault is injected in this diagonal of the AES state at the end of round 7. By the above discussion, we know that at the end of round 8, the whole first column might be affected by the fault. Similarly, we can study the fault propagation in round 9. Let δ_i $(i = 1, 2, 3, 4)$ denote the differences between the four correct and faulty bytes in the first column of the cipher state after SubBytes in round 9. An illustration is shown in Fig. 5.4, where S_8 (respectively, S_9) denotes

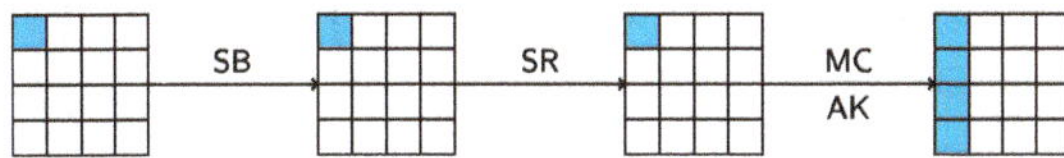

Fig. 5.2 Visual illustration of how the fault propagates when a fault is injected at the beginning of one AES round (not the last round) in byte s_{00}. Blue squares correspond to bytes that can be affected by the fault

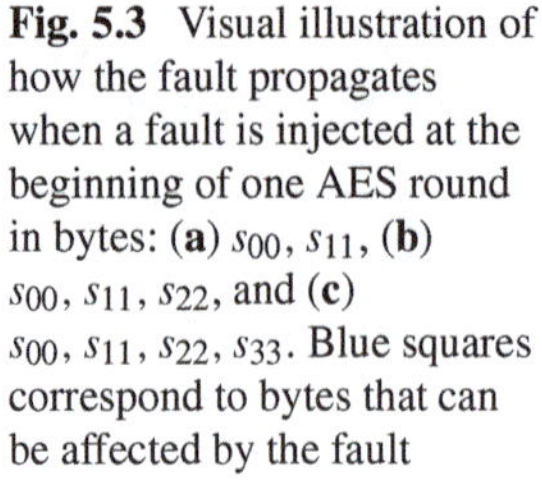

Fig. 5.3 Visual illustration of how the fault propagates when a fault is injected at the beginning of one AES round in bytes: (**a**) s_{00}, s_{11}, (**b**) s_{00}, s_{11}, s_{22}, and (**c**) $s_{00}, s_{11}, s_{22}, s_{33}$. Blue squares correspond to bytes that can be affected by the fault

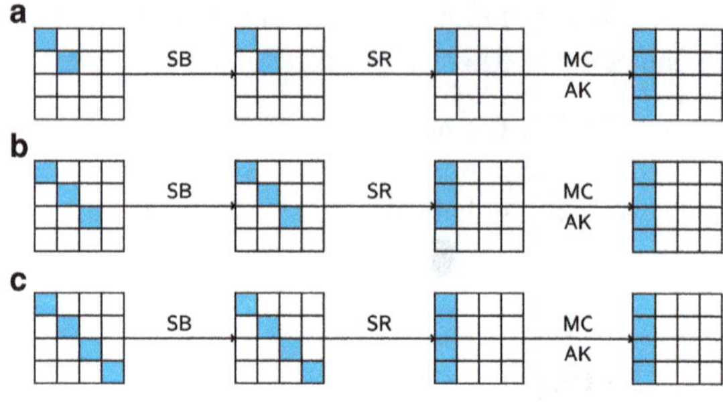

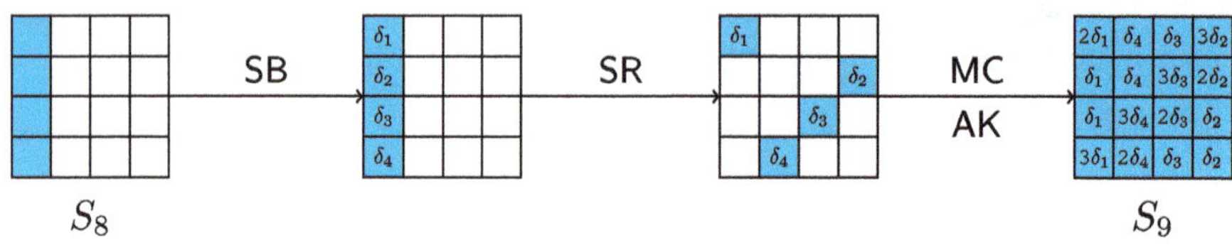

Fig. 5.4 Visual illustration of fault propagation in the 9th round of AES when the fault was injected in the diagonal $s_{00}, s_{11}, s_{22}, s_{33}$ of the AES cipher state at the end of round 7

the cipher state at the end of round 8 (respectively, round 9). After ShiftRows, those four δ_is move to different positions as shown in the third cipher state in the figure.

Recall that MixColumns multiplies one column by the following matrix (see Eq. 3.6):

$$\begin{pmatrix} 02 & 03 & 01 & 01 \\ 01 & 02 & 03 & 01 \\ 01 & 01 & 02 & 03 \\ 03 & 01 & 01 & 02 \end{pmatrix}.$$

Since this is a linear operation, the differences will also be multiplied by the corresponding coefficients in the matrix. Consequently, we get the last state S_9 as shown in Fig. 5.4.

Let us represent the cipher state at the end of round 9 S_9, the correct ciphertext c, and the last round key K_{10} with the following matrices:

$$S_9 = \begin{pmatrix} a_{00} & a_{01} & a_{02} & a_{03} \\ a_{10} & a_{11} & a_{12} & a_{13} \\ a_{20} & a_{21} & a_{22} & a_{23} \\ a_{30} & a_{31} & a_{32} & a_{33} \end{pmatrix}, \quad c = \begin{pmatrix} c_{00} & c_{01} & c_{02} & c_{03} \\ c_{10} & c_{11} & c_{12} & c_{13} \\ c_{20} & c_{21} & c_{22} & c_{23} \\ c_{30} & c_{31} & c_{32} & c_{33} \end{pmatrix}, \quad K_{10} = \begin{pmatrix} k_{00} & k_{01} & k_{02} & k_{03} \\ k_{10} & k_{11} & k_{12} & k_{13} \\ k_{20} & k_{21} & k_{22} & k_{23} \\ k_{30} & k_{31} & k_{32} & k_{33} \end{pmatrix}.$$

In Round 10, we have SubBytes, ShiftRows, and AddRoundKey operations. In particular, we have

$$c_{00} = \text{SB}_{\text{AES}}(a_{00}) \oplus k_{00}, \quad c_{13} = \text{SB}_{\text{AES}}(a_{10}) \oplus k_{13},$$

$$c_{22} = \text{SB}_{\text{AES}}(a_{20}) \oplus k_{22}, \quad c_{31} = \text{SB}_{\text{AES}}(a_{30}) \oplus k_{31},$$

which gives

$$a_{00} = \text{SB}^{-1}_{\text{AES}}(c_{00} \oplus k_{00})$$

$$a_{10} = \text{SB}^{-1}_{\text{AES}}(c_{13} \oplus k_{13})$$

$$a_{20} = \text{SB}^{-1}_{\text{AES}}(c_{22} \oplus k_{22})$$

$$a_{30} = \text{SB}^{-1}_{\text{AES}}(c_{31} \oplus k_{31}).$$

Let us denote the faulty ciphertext by c', we write

$$c' = \begin{pmatrix} c'_{00} & c'_{01} & c'_{02} & c'_{03} \\ c'_{10} & c'_{11} & c'_{12} & c'_{13} \\ c'_{20} & c'_{21} & c'_{22} & c'_{23} \\ c'_{30} & c'_{31} & c'_{32} & c'_{33} \end{pmatrix}.$$

Then

$$a'_{00} = \text{SB}^{-1}_{\text{AES}}(c'_{00} \oplus k_{00})$$

$$a'_{10} = \text{SB}^{-1}_{\text{AES}}(c'_{13} \oplus k_{13})$$

$$a'_{20} = \text{SB}^{-1}_{\text{AES}}(c'_{22} \oplus k_{22})$$

$$a'_{30} = \text{SB}^{-1}_{\text{AES}}(c'_{31} \oplus k_{31}).$$

Let $\delta = \delta_1$. By observing the first column of S_9 in Fig. 5.4, we have

$$2\delta = a_{00} \oplus a'_{00} = \text{SB}^{-1}_{\text{AES}}(c_{00} \oplus k_{00}) \oplus \text{SB}^{-1}_{\text{AES}}(c'_{00} \oplus k_{00})$$

$$\delta = a_{10} \oplus a'_{10} = \text{SB}^{-1}_{\text{AES}}(c_{13} \oplus k_{13}) \oplus \text{SB}^{-1}_{\text{AES}}(c'_{13} \oplus k_{13})$$

$$\delta = a_{20} \oplus a'_{20} = \text{SB}^{-1}_{\text{AES}}(c_{22} \oplus k_{22}) \oplus \text{SB}^{-1}_{\text{AES}}(c'_{22} \oplus k_{22})$$

$$3\delta = a_{30} \oplus a'_{30} = \text{SB}^{-1}_{\text{AES}}(c_{31} \oplus k_{31}) \oplus \text{SB}^{-1}_{\text{AES}}(c'_{31} \oplus k_{31}).$$

Then for each value of δ, the possible values for $k_{00}, k_{13}, k_{22}, k_{31}$ will be restricted by the above four equations. In particular,

$$a_{00} = \text{SB}^{-1}_{\text{AES}}(c_{00} \oplus k_{00})$$

can be considered as an AES Sbox input that corresponds to input difference 2δ and output difference $c_{00} \oplus c'_{00}$. Similarly,

$$a_{10} = \text{SB}^{-1}_{\text{AES}}(c_{13} \oplus k_{13}), \quad a_{20} = \text{SB}^{-1}_{\text{AES}}(c_{22} \oplus k_{22}), \quad a_{30} = \text{SB}^{-1}_{\text{AES}}(c_{31} \oplus k_{31})$$

are AES Sbox inputs that give output differences

$$c_{13} \oplus c'_{13}, \quad c_{22} \oplus c'_{22}, \quad c_{31} \oplus c'_{31},$$

when the input differences are

$$\delta, \quad \delta, \quad 3\delta,$$

respectively. It was shown [SMR09] that, on average, the key hypotheses for $(k_{00}, k_{13}, k_{22}, k_{31})$ can be reduced to 2^8.

Example 5.1.5 Suppose the master key is

$$\begin{pmatrix} 00\ 04\ 08\ 0C \\ 01\ 05\ 09\ 0D \\ 02\ 06\ 0A\ 0E \\ 03\ 07\ 0B\ 0F \end{pmatrix}$$

and the plaintext is

$$\begin{pmatrix} 00\ 44\ 88\ CC \\ 11\ 55\ 99\ DD \\ 22\ 66\ AA\ EE \\ 33\ 77\ BB\ FF \end{pmatrix}.$$

By AES encryption and key schedule (Sect. 3.1.2), we can find that (see [NIS01] Appendix C)

$$S_7 = \begin{pmatrix} D1\ 79\ B4\ D6 \\ 87\ C4\ 55\ 6F \\ 6C\ 30\ 94\ F4 \\ 0F\ 0A\ AD\ 1F \end{pmatrix},$$

$$K_8 = \begin{pmatrix} 47\ A4\ E0\ AE \\ 43\ 1C\ 16\ BF \\ 87\ 65\ BA\ 7A \\ 35\ B9\ F4\ D2 \end{pmatrix}, \quad K_9 = \begin{pmatrix} 54\ F0\ 10\ BE \\ 99\ 85\ 93\ 2C \\ 32\ 57\ ED\ 97 \\ D1\ 68\ 9C\ 4E \end{pmatrix}, \quad K_{10} = \begin{pmatrix} 13\ E3\ F3\ 4D \\ 11\ 94\ 07\ 2B \\ 1D\ 4A\ A7\ 30 \\ 7F\ 17\ 8B\ C5 \end{pmatrix}.$$

Then the intermediate values in round 8 are as follows:

$$S_7 \xrightarrow{\text{SB}} \begin{pmatrix} \text{3E B6 8D F6} \\ \text{17 1C FC A8} \\ \text{50 04 22 BF} \\ \text{76 67 95 C0} \end{pmatrix} \xrightarrow{\text{SR}} \begin{pmatrix} \text{3E B6 8D F6} \\ \text{1C FC A8 17} \\ \text{22 BF 50 04} \\ \text{C0 76 67 95} \end{pmatrix} \xrightarrow{\text{MC}} \begin{pmatrix} \text{BA A1 D5 5F} \\ \text{A0 F9 51 41} \\ \text{3D B5 2C 4D} \\ \text{E7 6E BA 23} \end{pmatrix}$$

$$\xrightarrow{\text{AK}} \begin{pmatrix} \text{FD 05 35 F1} \\ \text{E3 E5 47 FE} \\ \text{BA D0 96 37} \\ \text{D2 D7 4E F1} \end{pmatrix} = S_8,$$

where SB, SR, MC, and AR stand for SubBytes (Table 3.9), ShiftRows, Mix-Columns, and AddRoundKey, respectively. The operations in round 9 compute

$$S_8 \xrightarrow{\text{SB}} \begin{pmatrix} \text{54 6B 96 A1} \\ \text{11 D9 A0 BB} \\ \text{F4 70 90 9A} \\ \text{B5 0E 2F A1} \end{pmatrix} \xrightarrow{\text{SR}} \begin{pmatrix} \text{54 6B 96 A1} \\ \text{D9 A0 BB 11} \\ \text{90 9A F4 70} \\ \text{A1 B5 0E 2F} \end{pmatrix} \xrightarrow{\text{MC}} \begin{pmatrix} \text{E9 02 1B 35} \\ \text{F7 30 F2 3C} \\ \text{4E 20 CC 21} \\ \text{EC F6 F2 C7} \end{pmatrix}$$

$$\xrightarrow{\text{AK}} \begin{pmatrix} \text{BD F2 0B 8B} \\ \text{6E B5 61 10} \\ \text{7C 77 21 B6} \\ \text{3D 9E 6E 89} \end{pmatrix} = S_9.$$

In round 10 we have

$$S_9 \xrightarrow{\text{SB}} \begin{pmatrix} \text{7A 89 2B 3D} \\ \text{9F D5 EF CA} \\ \text{10 F5 FD 4E} \\ \text{27 0B 9F A7} \end{pmatrix} \xrightarrow{\text{SR}} \begin{pmatrix} \text{7A 89 2B 3D} \\ \text{D5 EF CA 9F} \\ \text{FD 4E 10 F5} \\ \text{A7 27 0B 9F} \end{pmatrix} \xrightarrow{\text{AK}} \begin{pmatrix} \text{69 6A D8 70} \\ \text{C4 7B CD B4} \\ \text{E0 04 B7 C5} \\ \text{D8 30 80 5A} \end{pmatrix} = c.$$

Suppose a fault is injected in byte s_{00} of S_7 with fault mask D8. We have

$$s'_{00} = \text{D1} \oplus \text{D8} = \text{09}.$$

The computations in round 8 become

$$S'_7 = \begin{pmatrix} \text{09 79 B4 D6} \\ \text{87 C4 55 6F} \\ \text{6C 30 94 F4} \\ \text{0F 0A AD 1F} \end{pmatrix} \xrightarrow{\text{SB}} \begin{pmatrix} \text{01 B6 8D F6} \\ \text{17 1C FC A8} \\ \text{50 04 22 BF} \\ \text{76 67 95 C0} \end{pmatrix} \xrightarrow{\text{SR}} \begin{pmatrix} \text{01 B6 8D F6} \\ \text{1C FC A8 17} \\ \text{22 BF 50 04} \\ \text{C0 76 67 95} \end{pmatrix}$$

$$\xrightarrow{\text{MC}} \begin{pmatrix} \text{C4 A1 D5 5F} \\ \text{9F F9 51 41} \\ \text{02 B5 2C 4D} \\ \text{A6 6E BA 23} \end{pmatrix} \xrightarrow{\text{AK}} \begin{pmatrix} \text{83 05 35 F1} \\ \text{DC E5 47 FE} \\ \text{85 D0 96 37} \\ \text{93 D7 4E F1} \end{pmatrix} = S_8'.$$

Round 9 then calculates

$$S_8' \xrightarrow{\text{SB}} \begin{pmatrix} \text{EC 6B 96 A1} \\ \text{86 D9 A0 BB} \\ \text{97 70 90 9A} \\ \text{DC 0E 2F A1} \end{pmatrix} \xrightarrow{\text{SR}} \begin{pmatrix} \text{EC 6B 96 A1} \\ \text{D9 A0 BB 86} \\ \text{90 9A 97 70} \\ \text{A1 DC 0E 2F} \end{pmatrix}$$

$$\xrightarrow{\text{MC}} \begin{pmatrix} \text{82 6B 78 97} \\ \text{4F 59 57 09} \\ \text{F6 9B 0A B6} \\ \text{3F 24 91 50} \end{pmatrix} \xrightarrow{\text{AK}} \begin{pmatrix} \text{D6 9B 68 29} \\ \text{D6 DC C4 25} \\ \text{C4 CC E7 21} \\ \text{EE 4C 0D 1E} \end{pmatrix} = S_9'.$$

And, in round 10, we have

$$S_9' \xrightarrow{\text{SB}} \begin{pmatrix} \text{F6 14 45 A5} \\ \text{F6 86 1C 3F} \\ \text{1C 4B 94 FD} \\ \text{28 29 D7 72} \end{pmatrix} \xrightarrow{\text{SR}} \begin{pmatrix} \text{F6 14 45 A5} \\ \text{86 1C 3F F6} \\ \text{94 FD 1C 4B} \\ \text{72 28 29 D7} \end{pmatrix} \xrightarrow{\text{AK}} \begin{pmatrix} \text{E5 F7 B6 E8} \\ \text{97 88 38 DD} \\ \text{89 B7 BB 7B} \\ \text{0D 3F A2 12} \end{pmatrix} = c'.$$

The attacker obtains the following equations:

$$2\delta = \text{SB}_{\text{AES}}^{-1}(69 \oplus k_{00}) \oplus \text{SB}_{\text{AES}}^{-1}(\text{E5} \oplus k_{00})$$

$$\delta = \text{SB}_{\text{AES}}^{-1}(\text{B4} \oplus k_{13}) \oplus \text{SB}_{\text{AES}}^{-1}(\text{DD} \oplus k_{13})$$

$$\delta = \text{SB}_{\text{AES}}^{-1}(\text{B7} \oplus k_{22}) \oplus \text{SB}_{\text{AES}}^{-1}(\text{BB} \oplus k_{22})$$

$$3\delta = \text{SB}_{\text{AES}}^{-1}(30 \oplus k_{31}) \oplus \text{SB}_{\text{AES}}^{-1}(\text{3F} \oplus k_{31}).$$

Thus the possible values of

$$\text{SB}_{\text{AES}}^{-1}(69 \oplus k_{00}), \quad \text{SB}_{\text{AES}}^{-1}(\text{B4} \oplus k_{13}), \quad \text{SB}_{\text{AES}}^{-1}(\text{B7} \oplus k_{22}), \quad \text{SB}_{\text{AES}}^{-1}(30 \oplus k_{31})$$

are inputs of AES Sbox that with input differences

$$2\delta, \quad \delta, \quad \delta, \quad 3\delta$$

produce output differences

$$69 \oplus \text{E5} = \text{8C}, \quad \text{B4} \oplus \text{DD} = 69, \quad \text{B7} \oplus \text{BB} = \text{0C}, \quad 30 \oplus \text{DD} = \text{ED},$$

Table 5.2 Part of the difference distribution table for AES Sbox (Table 3.9) corresponding to output differences 0C, 69, 8C, and ED

Δ $\diagdown$ δ	1	2	3	4	5	6	7	8	9	...
0C	2,3	35,37	7D,7E		E2,E7	0,6,A3,A5		4,C	D2,DB	...
69	48,49					70,76				...
8C		D9,DB		42,47				E5,ED		...
ED	52,53		49,4A	41,45			68,6F	70,78	C5,CC	...

respectively.

Part of the AES Sbox difference distribution table corresponding to output differences 8C, 69, 0C, and ED is shown in Table 5.2. We can see that $\delta \neq 02$ since for input difference 02, the entry for row 69 is empty. In other words, there are no inputs that have output difference 69 for input difference 02. Thus δ can only take values that give nonempty entries in the DDT for columns 2δ, δ, δ, 3δ and corresponding rows 8C, 69, 0C, ED. By searching the rows 8C, 69, 0C, ED, we can find all possible values of δ:

$$01, \quad 06, \quad 0B, \quad 28, \quad 3D, \quad 49, \quad 6B, \quad 76, \quad 8F, \quad 90, \quad A6, \quad B2, \quad B8, \quad D0, \quad EE,$$

in total 15 choices. In most of the entries of AES Sbox DDT, there are only two values; thus the remaining number of key hypotheses is roughly

$$2^4 \times 15 \approx 2^8.$$

We can also check that for the correct values $k_{00} = 13$, $k_{13} = 2B$, $k_{22} = A7$, $k_{31} = 17$ (see Table 3.10 for $\mathrm{SB}_{\mathrm{AES}}^{-1}$),

$$2\delta = \mathrm{SB}_{\mathrm{AES}}^{-1}(7A) \oplus \mathrm{SB}_{\mathrm{AES}}^{-1}(F6) = BD \oplus D6 = 6B$$

$$\delta = \mathrm{SB}_{\mathrm{AES}}^{-1}(9F) \oplus \mathrm{SB}_{\mathrm{AES}}^{-1}(F6) = 6E \oplus D6 = B8$$

$$\delta = \mathrm{SB}_{\mathrm{AES}}^{-1}(10) \oplus \mathrm{SB}_{\mathrm{AES}}^{-1}(1C) = 7C \oplus C4 = B8$$

$$3\delta = \mathrm{SB}_{\mathrm{AES}}^{-1}(27) \oplus \mathrm{SB}_{\mathrm{AES}}^{-1}(28) = 3D \oplus EE = D3.$$

According to Examples 1.5.17 and 1.5.18,

$$B8 \times 02 = 10111000 \times 02 = 01110000 \oplus 1B = 6B, \quad B8 \times 03 = 6B \oplus B8 = D3.$$

The other three columns of S_9 in Fig. 5.4 can provide similar results, reducing the key hypotheses for other key bytes of K_{10}. Consequently, with just *one pair* of correct and faulty ciphertext, the key hypotheses for K_{10} can be reduced to 2^{32} as opposed to the original 2^{128}.

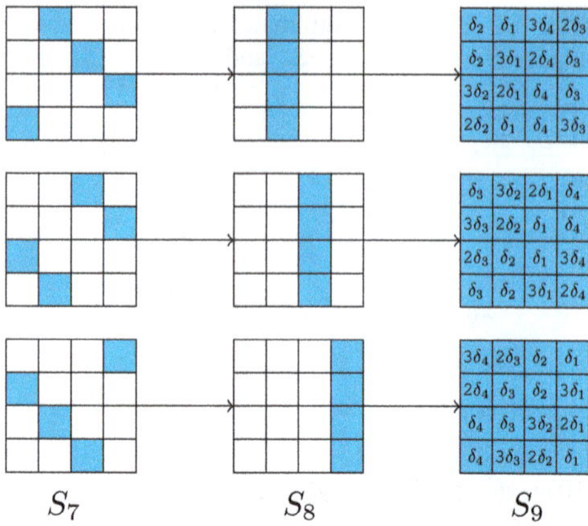

Fig. 5.5 Fault propagation for random byte fault injected in the "diagonals" of the cipher state at the end of round 7

We note that in this attack, we assume the attacker has the knowledge of the fault location (diagonal of cipher state at the end of round 7), fault model (random byte), and output of AES (correct and faulty ciphertext). Since the attack is on the diagonal of the cipher state, it is also called the *diagonal DFA*. Similar attacks can be carried out if the fault is injected in the other three "diagonals" of the cipher state at the end of round 7. The corresponding fault propagations are depicted in Fig. 5.5, where S_i denotes the cipher state at the end of round i.

5.1.2 Statistical Fault Analysis

Statistical Fault Analysis (SFA) [FJLT13] assumes no knowledge of plaintext or correct ciphertext for the attacker. Only knowledge of faulty ciphertext and a nonuniform fault model is required.

We will provide more details on the definition of a nonuniform fault model. We consider fault models that change an intermediate value x to x'. We can model these two intermediate values as random variables X and X'. Based on the fault properties, we can draw a table with probabilities for the value x to be changed to x', i.e., $P(X' = x' | X = x)$. Such a table is called a *fault distribution table*. We say that the fault model is *nonuniform* if

$$P(X' = x' | X = x) \neq \frac{1}{2^b}$$

for some x and x', where b is the maximum bit length of x.

Example 5.1.6 Let us consider the case when x is just 1 bit. A stuck-at-0 fault changes x to 0 with probability 1. A bit-flip fault model changes x to $x \oplus 1$ with probability 1. A random fault changes x to $x \oplus 1$ with probability 0.5. The fault distribution tables for those three fault models are shown in Table 5.3. In this case,

Table 5.3 Fault distribution tables for fault models: (a) stuck-at-0, (b) bit flip, and (c) random fault

		x'				x'				x'	
		0	1			0	1			0	1
x	0	1	0		0	0	1		0	0.5	0.5
	1	1	0		1	1	0		1	0.5	0.5
		(a)				(b)				(c)	

Table 5.4 Fault distribution tables for fault models: (a) stuck-at-0 with probability 0.5 and (b) random-AND with δ, where δ follows a uniform distribution

		x'				x'	
		0	1			0	1
x	0	1	0		0	1	0
	1	0.5	0.5		1	0.5	0.5
		(a)				(b)	

the bit length of x is 1, and a fault model is nonuniform if

$$P(X' = x' | X = x) \neq \frac{1}{2}$$

for some x and x'. Thus both stuck-at-0 and bit-flip fault models are nonuniform.

Example 5.1.7 We again consider the case when x is 1 bit. We discuss two more complicated nonuniform fault models. Stuck-at-0 with probability 0.5 changes x to 0 with probability 0.5. The corresponding fault distribution table is shown in Table 5.4 (a). Random-AND with δ, where δ follows a uniform distribution, has the same fault distribution table. For example,

$$P(x' = 1 | x = 1) = P(\delta = 1) = 0.5.$$

5.1.2.1 SFA Attack on AES-128 Round 9

In this part, we will discuss an SFA attack on AES-128. We represent the cipher state at the end of round 9 S_9, the correct ciphertext c, and the last round key K_{10} with the following matrices:

$$S_9 = \begin{pmatrix} s_{00} & s_{01} & s_{02} & s_{03} \\ s_{10} & s_{11} & s_{12} & s_{13} \\ s_{20} & s_{21} & s_{22} & s_{23} \\ s_{30} & s_{31} & s_{32} & s_{33} \end{pmatrix}, \quad c = \begin{pmatrix} c_{00} & c_{01} & c_{02} & c_{03} \\ c_{10} & c_{11} & c_{12} & c_{13} \\ c_{20} & c_{21} & c_{22} & c_{23} \\ c_{30} & c_{31} & c_{32} & c_{33} \end{pmatrix}, \quad K_{10} = \begin{pmatrix} k_{00} & k_{01} & k_{02} & k_{03} \\ k_{10} & k_{11} & k_{12} & k_{13} \\ k_{20} & k_{21} & k_{22} & k_{23} \\ k_{30} & k_{31} & k_{32} & k_{33} \end{pmatrix}.$$

According to the round operations, SubBytes, ShiftRows, and AddRoundKey, in round 10, we have

$$c_{00} = \mathrm{SB}_{\mathrm{AES}}(s_{00}) \oplus k_{00} \implies s_{00} = \mathrm{SB}_{\mathrm{AES}}^{-1}(c_{00} \oplus k_{00}). \tag{5.9}$$

We consider a fault in s_{00} with a nonuniform fault model. Let S_{00} and S_{00}' denote the random variables corresponding to s_{00} and its faulty value s_{00}', respectively. Suppose the attacker has the knowledge of the fault location and the fault distribution table, i.e., the probabilities

$$P(S_{00}' = s_{00}' \mid S_{00} = s_{00})$$

for all s_{00} and s_{00}' from $\mathbb{F}_2^8$. The goal of the attacker is to recover k_{00}.

We assume S_{00} follows a uniform distribution, i.e.,

$$P(S_{00} = s_{00}) = \frac{1}{256}, \quad \forall s_{00} \in \mathbb{F}_2^8.$$

Then by Lemma 1.7.2,

$$P(S_{00}' = s_{00}') = \sum_{s_{00}=0}^{255} P(S_{00}' = s_{00}' \mid S_{00} = s_{00}) P(S_{00} = s_{00})$$

$$= \frac{1}{256} \sum_{s_{00}=0}^{255} P(S_{00}' = s_{00}' \mid S_{00} = s_{00}). \tag{5.10}$$

To carry out the attack, the attacker injects fault in s_{00} above and collects a set of m faulty ciphertexts $\{c'^1, c'^2, \ldots, c'^m\}$. Let $\hat{k}_{00}$ denote a key hypothesis for k_{00}. Then for each c'^i, we can compute a hypothetical value for s_{00}' using Eq. 5.9, denoted $\hat{s}_{00}^i$, as follows:

$$\hat{s}_{00}^i = \mathrm{SB}_{\mathrm{AES}}^{-1}(c'^i_{00} \oplus \hat{k}_{00}). \tag{5.11}$$

The probability that the faulty value of s_{00} in the ith encryption equals $\hat{s}_{00}^i$ can be found using the fault distribution table with Eq. 5.10:

$$P(S_{00}' = \hat{s}_{00}^i) = \frac{1}{256} \sum_{s_{00}=0}^{255} P(S_{00}' = \hat{s}_{00}^i \mid S_{00} = s_{00}).$$

Define $\ell(\hat{k}_{00})$ to be the probability that the faulty value of s_{00} in the ith encryption equals the hypothetical value $\hat{s}_{00}^i$ for all i, i.e.,

$$\ell(\hat{k}_{00}) := \prod_{i=1}^{m} P(S_{00}' = \hat{s}_{00}^i). \tag{5.12}$$

Then the correct key can be found using the *maximum likelihood* approach

$$k_{00} = \arg\max_{\hat{k}_{00}} \ell(\hat{k}_{00}).$$

Example 5.1.8 Let us consider a stuck-at-0 fault model, i.e.,

$$P(S'_{00} = s'_{00}|S_{00} = s_{00}) = \begin{cases} 1 & s'_{00} = \mathbf{00} \\ 0 & \text{Otherwise,} \end{cases}$$

for all $s_{00} \in \mathbb{F}_2^8$.

In this case, one faulty ciphertext is enough to recover k_{00}. Since the attacker knows that the faulty value s'_{00} is always $\mathbf{00}$, they can recover k_{00} by computing

$$k_{00} = c'_{00} \oplus \mathrm{SB}_{\mathrm{AES}}(\mathbf{00}) = c'_{00} \oplus 63.$$

Example 5.1.9 In this example, we consider a random-AND fault model such that

$$P(S'_{00} = s'_{00}|S_{00} = s_{00}) = \begin{cases} 1 & s'_{00} = s_{00}\mathrm{AND}\delta \\ 0 & \text{Otherwise,} \end{cases}$$

where

$$P(\delta = x) = \frac{1}{256}, \quad \forall x \in \mathbb{F}_2^8.$$

By Eq. 5.10,

$$P(S'_{00} = s'_{00}) = \frac{1}{256} \sum_{s_{00}=0}^{255} P(S'_{00} = s'_{00}|S_{00} = s_{00})$$

$$= \frac{1}{256} \sum_{s_{00}=0}^{255} \left(\sum_{x=1}^{255} P(S'_{00} = s'_{00}|S_{00} = s_{00}, \delta = x)P(\delta = x) \right)$$

$$= \frac{1}{256^2} \sum_{s_{00}=0}^{255} |\{\delta \mid s'_{00} = \delta\mathrm{AND}s_{00}\}|$$

$$= \frac{|\{(\delta, s_{00}) \mid s'_{00} = \delta\mathrm{AND}s_{00}, \quad s_{00} \in \mathbb{F}_2^8, \quad \delta \in \mathbb{F}_2^8\}|}{256^2} = \frac{3^{8-\mathrm{wt}(s'_{00})}}{256^2},$$

$$(5.13)$$

where $\mathrm{wt}\left(s'_{00}\right)$ denotes the Hamming weight of s'_{00} (See Definition 1.6.10). To derive the last equality, we note that if 1 bit of s'_{00} is 0, then the corresponding bit for δ and s_{00} can be either 0 or 1 but not both 1, giving us three choices. If 1 bit of s'_{00} is 1, then the corresponding bit in δ and s_{00} must both be 1.

Let $s_{00} = \text{AB}$, $k_{00} = \text{00}$. Then (see Table 3.9 for SB_{AES})

$$c_{00} = \text{SB}_{\text{AES}}(s_{00} \oplus k_{00}) = 62.$$

Suppose five injected faults result in values of $\delta = \text{0F}, \text{F0}, \text{FF}, \text{54}, \text{CD}$, respectively. Then the corresponding faulty values of s''^{i}_{00} are $\text{0B}, \text{A0}, \text{AB}, \text{00}, \text{89}$. And the faulty ciphertext bytes c''^{i}_{00} are $\text{2B}, \text{E0}, 62, 63, \text{A7}$.

Take $\hat{k}_{00} = \text{1A}$, and by Eq. 5.11, we have (see Table 3.10 for $\text{SB}_{\text{AES}}^{-1}$)

$$\hat{s}^{1}_{00} = \text{SB}_{\text{AES}}^{-1}(\text{2B} \oplus \text{1A}) = \text{SB}_{\text{AES}}^{-1}(31) = \text{C7},$$

$$\hat{s}^{2}_{00} = \text{SB}_{\text{AES}}^{-1}(\text{E0} \oplus \text{1A}) = \text{SB}_{\text{AES}}^{-1}(\text{FA}) = \text{2D},$$

$$\hat{s}^{3}_{00} = \text{SB}_{\text{AES}}^{-1}(62 \oplus \text{1A}) = \text{SB}_{\text{AES}}^{-1}(78) = \text{BC},$$

$$\hat{s}^{4}_{00} = \text{SB}_{\text{AES}}^{-1}(63 \oplus \text{1A}) = \text{SB}_{\text{AES}}^{-1}(79) = \text{B6},$$

$$\hat{s}^{5}_{00} = \text{SB}_{\text{AES}}^{-1}(\text{A7} \oplus \text{1A}) = \text{SB}_{\text{AES}}^{-1}(\text{BD}) = \text{7A}.$$

By Eqs. 5.12 and 5.13,

$$\ell(\text{1A}) = \prod_{i=1}^{5} P(S'_{00} = \hat{s}^{i}_{00}) = P(S'_{00} = \text{C7})P(S'_{00} = \text{2D})$$

$$P(S'_{00} = \text{BC})P(S'_{00} = \text{B6})P(S'_{00} = \text{7A})$$

$$= \frac{1}{256^{10}} \times 3^{8 \times 5 - \text{wt(C7)} - \text{wt(2D)} - \text{wt(BC)} - \text{wt(B6)} - \text{wt(7A)}}$$

$$= \frac{1}{256^{10}} \times 3^{40-5-4-5-5-5} = \frac{3^{16}}{256^{10}}.$$

Take $\hat{k}_{00} = \text{00}$, we have

$$\hat{s}^{1}_{00} = \text{SB}_{\text{AES}}^{-1}(\text{2B}) = \text{0B},$$

$$\hat{s}^{2}_{00} = \text{SB}_{\text{AES}}^{-1}(\text{E0}) = \text{A0},$$

$$\hat{s}^{3}_{00} = \text{SB}_{\text{AES}}^{-1}(62) = \text{AB},$$

$$\hat{s}^{4}_{00} = \text{SB}_{\text{AES}}^{-1}(63) = \text{00},$$

$$\hat{s}^{5}_{00} = \text{SB}_{\text{AES}}^{-1}(\text{A7}) = 89.$$

And

$$\ell(\mathtt{00}) = \prod_{i=1}^{5} P(S'_{00} = \hat{s}^i_{00}) = P(S'_{00} = \mathtt{0B})\,P(S'_{00} = \mathtt{A0})\,P(S'_{00} = \mathtt{AB})$$

$$P(S'_{00} = \mathtt{00})\,P(S'_{00} = \mathtt{89})$$

$$= \frac{1}{256^{10}} \times 3^{8\times 5 - \mathrm{wt}(\mathtt{0B}) - \mathrm{wt}(\mathtt{A0}) - \mathrm{wt}(\mathtt{AB}) - \mathrm{wt}(\mathtt{00}) - \mathrm{wt}(\mathtt{89})}$$

$$= \frac{1}{256^{10}} \times 3^{40-3-2-5-3} = \frac{3^{27}}{256^{10}}.$$

We can see that $\ell(\mathtt{00}) > \ell(\mathtt{1A})$.

It was shown in [FJLT13] that with high probability, the correct key byte can be found with only a few faults. The same method can recover other bytes of K_{10}. We note that each byte can be recovered in parallel; hence the number of faults required to get the full round key depends on the number of bytes that can be faulted with one fault injection.

In case the attacker only knows that the fault model is nonuniform, without the knowledge of its fault distribution table, a metric based on the *Square Euclidean Imbalance (SEI)* can be used. Define

$$\mathrm{SEI}(\hat{k}_{00}) := \sum_{j=0}^{255} \left(\frac{|\{i \mid \hat{s}^i_{00} = j\}|}{m} - \frac{1}{256} \right)^2.$$

We can see that by definition, SEI measures a certain distance between the obtained hypothetical distribution of S'_{00} and the uniform distribution. Since we know that the fault model is nonuniform, we expect the distribution induced by S'_{00} to be far from the uniform distribution. Thus we take the correct key to be

$$k_{00} = \arg\max_{\hat{k}_{00}} \ \mathrm{SEI}(\hat{k}_{00}).$$

5.1.2.2   SFA on AES-128 Round 8

In this part, we consider the fault to be injected in the output of round 8, S_8. Similar to before, we represent the cipher state at the end of round 8 S_8, the correct ciphertext c, and the last round key K_{10} with the following matrices:

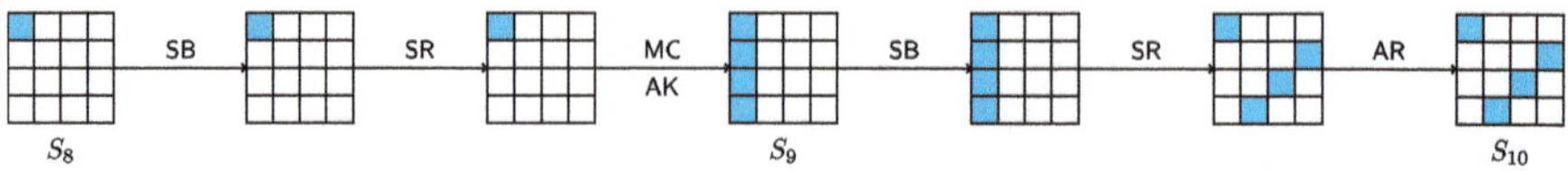

Fig. 5.6 Illustration of fault propagation for a fault injected in the first byte of S_8 (the cipher state at the end of round 8)

$$
S_8 = \begin{pmatrix} s_{00} & s_{01} & s_{02} & s_{03} \\ s_{10} & s_{11} & s_{12} & s_{13} \\ s_{20} & s_{21} & s_{22} & s_{23} \\ s_{30} & s_{31} & s_{32} & s_{33} \end{pmatrix}, \quad
c = \begin{pmatrix} c_{00} & c_{01} & c_{02} & c_{03} \\ c_{10} & c_{11} & c_{12} & c_{13} \\ c_{20} & c_{21} & c_{22} & c_{23} \\ c_{30} & c_{31} & c_{32} & c_{33} \end{pmatrix}, \quad
K_{10} = \begin{pmatrix} k_{00} & k_{01} & k_{02} & k_{03} \\ k_{10} & k_{11} & k_{12} & k_{13} \\ k_{20} & k_{21} & k_{22} & k_{23} \\ k_{30} & k_{31} & k_{32} & k_{33} \end{pmatrix}.
$$

We further represent the output of InvMixColumns operation on the second last round key, K_9, as follows:

$$
\mathrm{InvMixColumns}(K_9) = \begin{pmatrix} a_{00} & a_{01} & a_{02} & a_{03} \\ a_{10} & a_{11} & a_{12} & a_{13} \\ a_{20} & a_{21} & a_{22} & a_{23} \\ a_{30} & a_{31} & a_{32} & a_{33} \end{pmatrix}.
$$

Suppose a fault is injected in s_{00} with a nonuniform fault model. The fault propagation is shown in Fig. 5.6.

We can see that s_{00} is related to $c_{00}, c_{13}, c_{22}, c_{31}$ and $k_{00}, k_{13}, k_{22}, k_{31}$ as follows:

$$
s_{00} = \mathrm{SB}_{\mathrm{AES}}^{-1}(a_{00} \oplus \mathrm{InvMixColumns\ for\ the\ first\ column}
$$

$$
(\mathrm{SB}_{\mathrm{AES}}^{-1}(c_{00} \oplus k_{00}), \mathrm{SB}_{\mathrm{AES}}^{-1}(c_{13} \oplus k_{13}), \mathrm{SB}_{\mathrm{AES}}^{-1}(c_{22} \oplus k_{22}), \mathrm{SB}_{\mathrm{AES}}^{-1}(c_{31} \oplus k_{31}))).
$$

As discussed in Sect. 3.1.2, InvMixColumns computation is equivalent to multiplication with the following matrix:

$$
\begin{pmatrix} \mathtt{0E} & \mathtt{0B} & \mathtt{0D} & \mathtt{09} \\ \mathtt{09} & \mathtt{0E} & \mathtt{0B} & \mathtt{0D} \\ \mathtt{0D} & \mathtt{09} & \mathtt{0E} & \mathtt{0B} \\ \mathtt{0B} & \mathtt{0D} & \mathtt{09} & \mathtt{0E} \end{pmatrix}.
$$

We have

$$
s_{00} = \mathrm{SB}_{\mathrm{AES}}^{-1}(a_{00} \oplus \mathtt{0E} \cdot \mathrm{SB}_{\mathrm{AES}}^{-1}(c_{00} \oplus k_{00}) \oplus \mathtt{0B} \cdot \mathrm{SB}_{\mathrm{AES}}^{-1}(c_{13} \oplus k_{13})
$$

$$
\oplus\, \mathtt{0D} \cdot \mathrm{SB}_{\mathrm{AES}}^{-1}(c_{22} \oplus k_{22}) \oplus \mathtt{09} \cdot \mathrm{SB}_{\mathrm{AES}}^{-1}(c_{31} \oplus k_{31})).
$$

With a set of m faulty ciphertexts $\{c'^1, c'^2, \ldots, c'^m\}$, the attacker can make hypothesis on the values of $k_{00}, k_{13}, k_{22}, k_{31}$, and a_{00}, denoted by $\hat{k}_{00}, \hat{k}_{13}, \hat{k}_{22}, \hat{k}_{31},$

and $\hat{a}_{00}$. Then they can compute the corresponding hypothetical values for s'_{00}, denoted $\hat{s}^i_{00}$, as follows:

$$\hat{s}^i_{00} = \mathrm{SB}^{-1}_{\mathrm{AES}}(\hat{a}_{00} \oplus \mathtt{0E} \cdot \mathrm{SB}^{-1}_{\mathrm{AES}}(c'^{i}_{00} \oplus \hat{k}_{00}) \oplus \mathtt{0B} \cdot \mathrm{SB}^{-1}_{\mathrm{AES}}(c'^{i}_{13} \oplus \hat{k}_{13})$$

$$\oplus \mathtt{0D} \cdot \mathrm{SB}^{-1}_{\mathrm{AES}}(c'^{i}_{22} \oplus \hat{k}_{22}) \oplus \mathtt{09} \cdot \mathrm{SB}^{-1}_{\mathrm{AES}}(c'^{i}_{31} \oplus \hat{k}_{31})).$$

The correct key bytes can be recovered with either maximum likelihood (when the fault distribution table is known) or SEI (when the fault distribution table is unknown) as discussed above.

We refer the reader to the original paper [FJLT13] for other methods of obtaining the correct key hypothesis and attacks in even earlier rounds of AES.

The main advantage of SFA is that only faulty ciphertexts are required, and there is no need for repeated plaintexts. However, the attack assumes that each fault injection is successful.

5.1.3   *Persistent Fault Analysis*

Persistent Fault Analysis (PFA) [ZLZ$^+$18] considers a fault in the memory, normally where the Sbox lookup table is stored. As we do not expect the table to be rewritten during the computation, the fault would stay until the device is reset; hence the name "persistent" is used in the attack method.

We will use AES-128 as a running example to show how the attack works. The methodology also applies to other block ciphers.

We consider a random byte fault model. Suppose the fault location is in the first byte of the Sbox lookup table. Then the output for $\mathrm{SB}_{\mathrm{AES}}(\mathtt{00}) = v$ is changed to v', the fault mask $\varepsilon \in \mathbb{F}_2^8$ is given by

$$\varepsilon = v \oplus v'.$$

By Table 3.9, we know that $v = 63$. We assume the attacker has the knowledge of the output of AES (correct and faulty ciphertexts), fault model, and fault location. However, the attacker does not know the fault mask. The attacker aims to recover the last round key K_{10}.

Recall that in round 10, the operations for AES encryption include Sub-Bytes, ShiftRows, and AddRoundKey. We represent the cipher state right before AddRoundKey in round 10, denoted S, the ciphertext c, and K_{10} with the following matrices:

$$S = \begin{pmatrix} s_{00} & s_{01} & s_{02} & s_{03} \\ s_{10} & s_{11} & s_{12} & s_{13} \\ s_{20} & s_{21} & s_{22} & s_{23} \\ s_{30} & s_{31} & s_{32} & s_{33} \end{pmatrix}, \quad c = \begin{pmatrix} c_{00} & c_{01} & c_{02} & c_{03} \\ c_{10} & c_{11} & c_{12} & c_{13} \\ c_{20} & c_{21} & c_{22} & c_{23} \\ c_{30} & c_{31} & c_{32} & c_{33} \end{pmatrix}, \quad K_{10} = \begin{pmatrix} k_{00} & k_{01} & k_{02} & k_{03} \\ k_{10} & k_{11} & k_{12} & k_{13} \\ k_{20} & k_{21} & k_{22} & k_{23} \\ k_{30} & k_{31} & k_{32} & k_{33} \end{pmatrix}.$$

We have $c_{00} = s_{00} \oplus k_{00}$.

Since the fault affects the encryption starting from the first round, here we only consider the values of S and c with fault present in the Sbox lookup table and do not look into the original values of S and c. We omit the superscript $'$ in S' and c' which were used to indicate the faulty value in our previous discussions. We note that it is possible that for some encryptions the fault does not affect the result, e.g. when 00 is never used as an input for the Sbox computations. However, such cases do not affect the attack method.

Let S_{00} denote the random variable corresponding to the value of s_{00}. Due to the diffusion and confusion layers in AES, it is reasonable to assume

$$P(S_{00} = s_{00}) \begin{cases} \approx \frac{2}{256} & s_{00} = v \oplus \varepsilon \\ = 0 & s_{00} = v \\ \approx \frac{1}{256} & \text{otherwise.} \end{cases}$$

Since $c_{00} = s_{00} \oplus k_{00}$, given c_{00}, we know

$$k_{00} \neq c_{00} \oplus v.$$

Thus the attacker can collect a set of m (faulty) ciphertexts $\{c^1, c^2, \ldots, c^m\}$ and eliminate key hypotheses for k_{00} that are equal to

$$c_{00}^i \oplus v.$$

Example 5.1.10 Let the key byte $k_{00} = 45$ and the fault mask $\varepsilon = 12$. Then the faulty output of AES Sbox for input 00 becomes

$$63 \oplus 12 = 71.$$

In this case, no matter what input the AES Sbox gets during the computations, the output will never be 63. In particular, we have $s_{00} \neq 63$. Equivalently,

$$k_{00} \oplus 63 = 45 \oplus 63 = 26$$

would never appear in the first byte of the ciphertexts. Otherwise, we would have

$$s_{00} = 26 \oplus k_{00} = 26 \oplus 45 = 63,$$

which is not possible.

Suppose the attacker collects the following values for c_{00} with the fault present in the Sbox lookup table: 00, 12, FE. Then the attacker can eliminate

$$00 \oplus 63 = 63, \quad 12 \oplus 63 = 71, \quad FE \oplus 63 = 9D$$

from the key hypotheses.

Next, we consider the case when the attacker has the knowledge of the fault mask ε. Let Y denote the random variable corresponding to the value of c_{00}, and we have

$$P(Y = c_{00}) \begin{cases} \approx \frac{2}{256} & c_{00} = v \oplus \varepsilon \oplus k_{00} \\ = 0 & c_{00} = v \oplus k_{00} \\ \approx \frac{1}{256} & \text{otherwise.} \end{cases}$$

Given a set of ciphertexts $\{c^1, c^2, \ldots, c^m\}$, if we look at the first byte of those ciphertexts, we expect $v \oplus \varepsilon \oplus k_{00}$ to appear with the highest frequency. Thus the attacker computes

$$y_{\max} := \arg \max_y |\{c_{00}^i \mid c_{00}^i = y\}|.$$

Then a candidate for the correct key is given by $y_{\max} \oplus v \oplus \varepsilon$.

Alternatively, we can also calculate the empirical probabilities for each y, denoted $\hat{P}(Y = y)$, as follows:

$$\hat{P}(Y = y) = \frac{|\{c_{00}^i \mid c_{00}^i = y\}|}{m}.$$

For a large enough sample, we expect

$$\hat{P}(y_{\max}) \approx \frac{2}{256}.$$

The simulated results in [ZLZ$^+$18] show that with about 4000 faulty ciphertexts, the empirical probability of $y_{\max}$ is high enough to be distinguished from that of other values of y.

5.1.4 Implementation-Specific Fault Attack

In this subsection, we discuss an implementation-specific DFA attack on PRESENT [BHL18]. We consider the implementation of PRESENT with the second method discussed in Sect. 3.2.2.1 (Algorithm 3.5). Recall that the method combines sBoxLayer and pLayer by using four 8×8 tables—Table 1, Table 2, Table 3, and Table 4. Each table extracts certain bits of the cipher state, and the final output will be obtained by combining those bits using bitwise OR. The particular implementation we target is from [PV13, AV13], which was written in AVR assembly (see [Atm16] for 8-bit AVR Instruction Set Manual). Part of the implementation is listed in Algorithm 5.1.

Algorithm 5.1: Part of an implementation for PRESENT encryption that combines sBoxLayer and pLayer in AVR assembly [PV13, AV13]. A pseudocode can be found in Algorithm 3.5

```
 1 ...
 2 ldi ZH, 0x06     // load Table one address
 3 mov ZL, r0       // r0 contains input to Table one
   // lookup program memory at address Z, store in r21. This is
       equivalent to storing the Table one output corresponding to input
       from r0 in r21. Thus in this step, we lookup bits
       0, 1, 16, 17, 32, 33, 48, 49.
 4 lpm r21, Z
 5 andi r21, 0xC0   // extract output bits 0, 1
 6 ...              // load Table Two address
 7 lpm r23, Z       // lookup Table two (bits 50, 51, 2, 3, 18, 19, 34, 35) and
       store in r23
 8 andi r23, 0x30   // extract output bits 2, 3
 9 or r21, r23      // combine bits 0, 1, 2, 3
10 ...              // lookup Table three (bits 36, 37, 52, 53, 4, 5, 20, 21)
       and store in r23
11 andi r23, 0x0C   // extract output bits 4, 5
12 or r21, r23      // combine bits 0, 1, 2, 3, 4, 5
13 ...              // lookup Table four (bits 22, 23, 38, 39, 54, 55, 6, 7) and
       store in r23
14 andi r23, 0x03   // extract output bits 6, 7
15 or r21, r23      // combine bits 0, 1, 2, 3, 4, 5, 6, 7
16 ...
```

In line 2, we load the address of Table 1, then line 3 looks up the table, and finally, line 4 stores the table output in the register r21. These three lines implement line 1 in Algorithm 3.5. Afterward, in line 5, the leftmost 2 bits are extracted from r21 and stored in r21. This corresponds to line 5 in Algorithm 3.5. As we have explained in Sect. 3.2.2.1, these 2 bits correspond to bits at positions 0 and 1 of pLayer output. Similarly, lines 6–8 extract the 2nd and 3rd bits of pLayer output using Table 2 and store them in r23. Then line 9 combines bits 0, 1, 2, and 3 with bitwise OR. Then the implementation continues to extract pLayer output bits at positions 4 and 5 with Table 3 and at positions 6 and 7 with Table 4. Those bits are all combined through bitwise OR to register r21 (lines 12 and 15).

The fault attack on this implementation injects fault in register r23 between lines 14 and 15 in the final round of PRESENT. The fault model used is a bit flip.

Recall that in this round of PRESENT, we have operations addRoundKey, sBoxLayer, and pLayer, which will be followed by another addRoundKey (see Sect. 3.1.3). Let $\kappa_7\kappa_6\ldots\kappa_1\kappa_0$ denote the 0th byte of the round key K_{32}, which is the round key used right before outputting the ciphertext. Let $b_7b_6b_5\ldots b_0$ denote the intermediate value contained in register r21 right after line 15. Then the 0th byte of the correct ciphertext, denoted $c_7c_6c_5\ldots c_0$, is given by

$$c_7c_6c_5\ldots c_0 = b_7b_6b_5\ldots b_0 \oplus \kappa_7\kappa_6\ldots\kappa_1\kappa_0.$$

And the value in register r23 between lines 14 and 15 is $000000b_1b_0$.

We assume the attacker has the knowledge of the outputs of PRESENT (ciphertext) and can repeat the computation with the same plaintext (not chosen by the attacker). Furthermore, we consider a relatively strong attacker model where the attacker can choose the fault mask ε. In particular, we let $\varepsilon = 11111100$. Consequently, we inject a 6-bit flip in register r23 between lines 14 and 15. The faulty value in register r23 will be

$$000000b_1b_0 \oplus \varepsilon = 000000b_1b_0 \oplus 11111100 = 111111b_1b_0.$$

After line 15, the value in register r21 will then become $111111b_1b_0$. And the faulty ciphertext byte $c_7'c_6'c_5'\ldots c_0'$ is given by

$$c_7'c_6'c_5'\ldots c_0' = 111111b_1b_0 \oplus \kappa_7\kappa_6\ldots\kappa_1\kappa_0.$$

Since the faulty ciphertext byte $c_7'c_6'c_5'\ldots c_0'$ is known, the attacker can recover 6 bits of K_{32} by computing

$$\kappa_7\kappa_6\kappa_5\kappa_4\kappa_3\kappa_2 = c_7'c_6'c_5'c_4'c_3'c_2' \oplus 111111.$$

Similar methods can be used to recover all other bits of K_{32}.

We note that this attack is specific to the implementation considered. It shows that even with theoretically secure countermeasures in place, the programmer should verify its implementation, for example, by using an automated tool (see [BHL18, HBZL19, HSP20] for automated evaluation of SW implementations and [BGE$^+$17, PGP$^+$19] for HW implementations).

5.2   Fault Countermeasures for Symmetric Block Ciphers

A simple countermeasure one might consider to protect against certain fault attacks would be to repeat the encryption, compare the two outputs, and only return the ciphertext if those two outputs are equal [BECN$^+$06]. For example, for DFA attacks described in Sect. 5.1.1, such a countermeasure will be successful since those attacks require the knowledge of the faulty ciphertext. However, an easy attack on this countermeasure would be fault injection on both encryption computations or skipping the instruction for checking the outputs [SHS16].

In this section, we will discuss in detail two more sophisticated countermeasures against fault attacks on symmetric block ciphers.

5.2.1   Encoding-Based Countermeasure

We recall from Definition 1.6.3 that the *(minimum) distance* of a binary code C, denoted $\operatorname{dis}(C)$, is given by

$$\operatorname{dis}(C) = \min\{\operatorname{dis}(c_1, c_2) \mid c_1, c_2 \in C, c_1 \neq c_2\}.$$

We have seen that a binary code with minimum distance $\operatorname{dis}(C)$ can detect $\operatorname{dis}(C) - 1$ bit flips (see Definition 1.6.5 and Theorem 1.6.1). Thus a natural choice for fault countermeasure is to consider encoding the intermediate values during the computation. The question is, which code to choose and how to implement it?

As an example of what kind of code to use, we will discuss one proposal of using anticode (see Definition 1.6.12) for the countermeasure against bit flips and instruction skips [BHL19]. Recall that a binary (n, M, d, δ)-anticode has length n, cardinality M, minimum distance d, and maximum distance δ, where the *maximum distance* of a binary code C (see Definition 1.6.12) is given by

$$\operatorname{maxdis}(C) = \max\{\operatorname{dis}(c_1, c_2) \mid c_1, c_2 \in C\}.$$

For example, $\{10, 01\}$ is a binary $(2, 2, 2, 2)$-anticode.

Following Kerckhoffs' principle (see Definition 2.1.3), we assume the code used for the countermeasure is public. In particular, the attacker has the knowledge of all the codewords and how the information is encoded.

Intuitively if the minimum distance of the code is too small, we know that the code cannot detect a large number of bit flips. On the other hand, let us consider a code of length n and size M that contains at least two codewords, say c_1, c_2, with $\operatorname{dis}(c_1, c_2) = n$. If an n-bit flip is injected when c_1 or c_2 is used for the computation, then the resulting faulty value is still a codeword and cannot be detected. Since there are in total M codewords, the possibility for the fault to go undetected is at least $2/M$. Thus, a very big maximum distance is also not desirable.

We refer the reader to the original paper [BHL19] for the formalization of encoding-based countermeasures for symmetric block ciphers and calculations of the probability of detecting any m-bit flips and instruction skips given a binary code. The authors also provide a theoretical analysis which concludes that to have overall good protection against all possible bit flips, it is better to use code with not too small minimum distance and not too big maximum distance.

Such an observation leads us to the notion of anticode (see Definition 1.6.12). The paper [BHL19] also demonstrated the effectiveness of using anticodes with simulated results.

In the rest of this part, we would like to focus on how encoding countermeasures can be implemented in software for PRESENT encryption. The implementation we present has the following properties:

- Each operation is implemented as a table lookup from memory.
- Before the table lookup, the destination register of an operation is precharged to **0**.
- When any of the inputs is **0**, the output is **0**.
- When an error is detected, the output is **0** (error message).

Furthermore, we also assume the registers are precharged to **0** before the program starts, and this process cannot be faulted. Such a design can protect the implementation from single instruction skips.

For example, Algorithm 5.2 implements the computation of a binary operation through a table lookup. The two inputs a and b are loaded to registers r0 and r1 in instructions 1 and 2. The binary operation is computed by table lookup, and the result is stored in register r2 (instruction 4). Note that instruction 3 puts the error message **0** in r2 before it is used. Since the registers are supposed to be precharged to **0**, skipping instruction 1 or 2 will result in the input of the table lookup being **0**, and by our design, the final output will be **0**. Skipping instruction 3 will not change the output or the program flow. Skipping instruction 4 will make the final output to be **0**. Thus a single instruction skip of any instruction of Algorithm 5.2 will either make no change to the output or result in outputting **0**, which is the error message.

Algorithm 5.2: A simple program to demonstrate protection against single instruction skip attacks

```
1 LDI r0 a// load input a
2 LDI r1 b// load input b
3 EOR r2 r2// precharge register r2 to zero
4 LPM r2 r0 r1// execution of an operation by table lookup
```

Clearly, we need to choose codes that do not contain **0** as a codeword. Of course, the error message can be changed to a different value, allowing the usage of codes containing **0**, e.g., linear codes (see Definition 1.6.7). But for the implementation technique we are going to discuss, the structure of linear codes is not important.

In case the fault changes some encoded intermediate value to a word that is not a codeword, the table lookup will produce **0**, which indicates an error. In the subsequent instructions, when the input of a table is **0**, the output will always be **0** since **0** is not a codeword. In such cases, we say that the fault is *detected*. Otherwise, when a successful fault injection does not result in **0** output, we say the fault is *undetected*.

Table 5.5 Lookup table for carrying out XOR between a, b ($a, b \in \mathbb{F}_2$) using 01 as the codeword for 0 and 10 as the codeword for 1

	00	01	10	11
00	00	00	00	00
01	00	01	10	00
10	00	10	01	00
11	00	00	00	00

Example 5.2.1 As a simple example, let us consider $\{01, 10\}$, a binary $(2, 2, 2, 2)$-anticode. Since there are two codewords, it can be used to encode 1 bit of information. Let 01 be the codeword for 0 and 10 be the codeword for 1. The lookup table for carrying out XOR between a, b ($a, b \in \mathbb{F}_2$) is shown in Table 5.5. As mentioned before, 00 indicates an error. Thus the table outputs 00 if one input is not a codeword.

Example 5.2.2 Let us consider bit flip attacks on the inputs of XOR operation from Example 5.2.1. We can see that any 1-bit flip will be detected: If the fault is injected in input 01, with 1-bit flip, we get either 00 or 11; both will give output 00. Similarly, if 1-bit flip is injected in input 10, we will have 00 or 11, and the output will again be 00.

On the other hand, a 2-bit flip will be undetected. For example, suppose we would like to compute $0 \oplus 0$. Then the inputs for the table lookup will be 01 and 01, and the output will be 01, which corresponds to 0. If a 2-bit flip is injected in the first input, we get 10 and 01 for the table lookup. The result will be 10. Such a fault will not be detected and can successfully change the output of the operation.

We recall the notion of *Quotient group* and *Remainder group* for PRESENT Sboxes from Sect. 4.5.2.3. We have discussed that pLayer can be considered as four identical parallel bitwise operations where each is a function $\mathsf{p} : \mathbb{F}_2^{16} \to \mathbb{F}_2^{16}$ that takes one Quotient group output and permutes it to the corresponding Remainder group input. Furthermore, we have seen in Sect. 3.1.3 that addRoundKey is a function $\mathbb{F}_2^{64} \to \mathbb{F}_2^{64}$. Each Sbox in the sBoxLayer is a function SB: $\mathbb{F}_2^4 \to \mathbb{F}_2^4$. Thus, one convenient code choice would be those with cardinality 16, encoding 4 bits of information. In particular, we are looking for a binary $(n, 16, d, \delta)$-anticode, where d is the minimum distance of the code and δ is the maximum distance of the code.

We refer the readers to [BHL19] for an algorithm for finding anticodes that achieve a low probability of undetected faults with given length, minimum distance, and maximum distance. In the rest of this subsection, we will use the following binary $(8, 16, 2, 7)$-anticode as a running example

$$\{01, 08, 02, 0B, 04, 1D, 1E, 30, 07, 65, 6A, AD, B3, CE, D9, F6\}. \qquad (5.14)$$

In particular, 01 is the codeword for 0000, 08 in the codeword for 0001, etc. And we write

$$01 = \text{encode}(0000).$$

Given an anticode C, the addRoundKey operation can be implemented using an XOR table similar to the one shown in Example 5.2.1. The size of the table will be $2^8 \times 2^8$. Let $\widetilde{\oplus}$ denote this table lookup operation.

Example 5.2.3 Using the anticode given in Eq. 5.14, the table entry corresponding to 01 and 08 will be

$$\text{encode}(0000 \oplus 0001) = \text{encode}(0001) = 08.$$

And we write

$$01\widetilde{\oplus}08 = 08.$$

The implementation of sBoxLayer and pLayer is based on four 16×64 lookup tables, $T0, T1, T2, T3$. Let $x = x_3x_2x_1x_0$ be an element in $\mathbb{F}_2^4$. We write

$$\text{SB}(x_3x_2x_1x_0) = x_3^s x_2^s x_1^s x_0^s.$$

Example 5.2.4 Take $D = 1101$, then

$$x_3 = 1, \quad x_2 = 1, \quad x_1 = 0, \quad x_0 = 1.$$

Since $\text{SB}(D) = 7 = 0111$ (see Table 3.11), we have

$$x_3^s = 0, \quad x_2^s = 1, \quad x_1^s = 1, \quad x_0^s = 1.$$

The design of tables $T0, T1, T2,$ and $T3$ is as follows:

$$T0 : C \to C \times C \times C \times C$$

$$\text{encode}(x_3x_2x_1x_0) \mapsto \text{encode}(000x_3^s), \text{encode}(000x_2^s),$$

$$\text{encode}(000x_1^s), \text{encode}(000x_0^s)$$

$$T1 : C \to C \times C \times C \times C$$

$$\text{encode}(x_3x_2x_1x_0) \mapsto \text{encode}(00x_3^s0), \text{encode}(00x_2^s0),$$

$$\text{encode}(00x_1^s0), \text{encode}(00x_0^s0)$$

$$T2 : C \rightarrow C \times C \times C \times C$$

$$\text{encode}(x_3 x_2 x_1 x_0) \mapsto \text{encode}(0x_3^s 00), \text{encode}(0x_2^s 00),$$

$$\text{encode}(0x_1^s 00), \text{encode}(0x_0^s 00)$$

$$T3 : C \rightarrow C \times C \times C \times C$$

$$\text{encode}(x_3 x_2 x_1 x_0) \mapsto \text{encode}(x_3^s 000), \text{encode}(x_2^s 000),$$

$$\text{encode}(x_1^s 000), \text{encode}(x_0^s 000).$$

Thus, each table extracts the bits of the Sbox output, permutes them, and outputs the corresponding codeword. It is easy to see that each entry of the outputs of each table can be either encode(0000) or encode(0001) for $T0$, encode(0010) for $T1$, encode(0100) for $T2$, and encode(1000) for $T3$.

Example 5.2.5 Suppose the input is $01 = \text{encode}(0000)$. The corresponding Sbox output would be $C = 1100$ (see Table 3.11), i.e., $x_3^s x_2^s x_1^s x_0^s = 1100$. Using the anticode given in Eq. 5.14, the output of $T0$ will be

$$\text{encode}(0001)= 08, \text{encode}(0001)= 08, \text{encode}(0000)= 01, \text{encode}(0000)= 01.$$

The output of $T1$ is

$$\text{encode}(0010)= 02, \text{encode}(0010)= 02, \text{encode}(0000)= 01, \text{encode}(0000)= 01.$$

$T2$ gives

$$\text{encode}(0100)= 04, \text{encode}(0100)= 04, \text{encode}(0000)= 01, \text{encode}(0000)= 01.$$

Finally, $T3$ produces

$$\text{encode}(1000)= 07, \text{encode}(1000)= 07, \text{encode}(0000)= 01, \text{encode}(0000)= 01.$$

Example 5.2.6 Suppose the input is $08 = \text{encode}(0001)$. The corresponding Sbox output would be $5 = 0101$, i.e., $x_3^s x_2^s x_1^s x_0^s = 0101$. Using the anticode given in Eq. 5.14, the output of $T0$ will be

$$\text{encode}(0000)= 01, \text{encode}(0001)= 08, \text{encode}(0000)= 01, \text{encode}(0001)= 08.$$

The output of $T1$ is

$$\text{encode}(0000)= 01, \text{encode}(0010)= 02, \text{encode}(0000)= 01, \text{encode}(0010)= 02.$$

$T2$ gives

$\mathrm{encode}(0000) = 01$, $\mathrm{encode}(0100) = 04$, $\mathrm{encode}(0000) = 01$, $\mathrm{encode}(0100) = 04$.

Finally, $T3$ produces

$\mathrm{encode}(0000) = 01$, $\mathrm{encode}(1000) = 07$, $\mathrm{encode}(0000) = 01$, $\mathrm{encode}(1000) = 07$.

Now, let the original cipher state at sBoxLayer input be $b_{63}b_{62}\ldots b_0$. For the encoding-based implementation, the corresponding cipher state will be

$$\mathrm{encode}(b_{63}b_{62}b_{61}b_{60})\mathrm{encode}(b_{59}b_{58}b_{57}b_{56})\ldots$$

$$\mathrm{encode}(b_7b_6b_5b_4)\mathrm{encode}(b_3b_2b_1b_0).$$

Each codeword in this cipher state will be passed to tables $T0$, $T1$, $T2$, $T3$, and the outputs will be recorded. Then the output of pLayer will be computed by combining those table outputs through $\widetilde{\oplus}$.

Example 5.2.7 By Table 3.12, the pLayer output bits at positions $0, 1, 2, 3$ come from the bits at positions $0, 4, 8, 12$ of the input of pLayer. Thus, we first get $\mathrm{encode}(000b_0^s)$ from $T0$ output, $\mathrm{encode}(00b_4^s0)$ from $T1$, $\mathrm{encode}(0b_8^s00)$ from $T2$, and $\mathrm{encode}(b_{12}^s000)$ from $T3$, and then the 0th nibble of pLayer output will be

$$\mathrm{encode}(000b_0^s)\widetilde{\oplus}\mathrm{encode}(00b_4^s0)\widetilde{\oplus}\mathrm{encode}(0b_8^s00)\widetilde{\oplus}\mathrm{encode}(b_{12}^s000).$$

As another example, the third nibble (bits 16, 17, 18, 19) of pLayer output is given by

$$\mathrm{encode}(000b_1^s)\widetilde{\oplus}\mathrm{encode}(00b_5^s0)\widetilde{\oplus}\mathrm{encode}(0b_9^s00)\widetilde{\oplus}\mathrm{encode}(b_{13}^s000).$$

Remark 5.2.1 By the design of our implementation, when the faulty intermediate value is not a codeword, the table lookup returns 0, and the attacker will not be able to tell what the original faulty ciphertext is. Since both DFA and SFA require analysis of the faulty ciphertexts, they can be prevented when the fault model is bit flip, and the number of bit flips is lower than the minimum distance of the binary code.

We have also seen that binary codes can *correct* error. According to Theorem 1.6.2, if m bits are flipped during the computation, a binary code C used for encoding-based countermeasure can correct this fault as long as $m \leq \lfloor (d-1)/2 \rfloor$, where d is the minimum distance of C. Note that to realize the incomplete decoding rule, we need an error message to indicate more than one codeword is at the same smallest distance from the input word.

For example, let us consider the 3-repetition code $C_{[3,1,3]} = \{000, 111\}$, which is a $[3, 1, 3]$-linear code (see Example 1.6.8). Since $C_{[3,1,3]}$ contains two codewords, it can be used to encode 1 bit of information. As 000 is a codeword of $C_{[3,1,3]}$, we cannot use it as the error message. On the other hand, we note that no word in $\mathbb{F}_2^3$ is at the same distance from 000 and 111, which means we will always be able to find a codeword using the minimum distance decoding rule. $C_{[3,1,3]}$ has minimum distance

Table 5.6 Lookup table for error-correcting code based computation of AND between a, b ($a, b \in \mathbb{F}_2$), using the 3-repetition code {000, 111}. 000 is the codeword for 0, and 111 is the codeword for 1

&	000	001	010	011	100	101	110	111
000	000	000	000	000	000	000	000	000
001	000	000	000	000	000	000	000	000
010	000	000	000	000	000	000	000	000
011	000	000	000	111	000	111	111	111
100	000	000	000	000	000	000	000	000
101	000	000	000	111	000	111	111	111
110	000	000	000	111	000	111	111	111
111	000	000	000	111	000	111	111	111

3. Then we know that an implementation based on encoding countermeasure with $C_{[3,1,3]}$ will be able to correct errors caused by 1-bit flip attacks.

Let 000 be the codeword for 0 and 111 be the codeword for 1. The lookup table for computation of AND between a, b ($a, b \in \mathbb{F}_2$) with error correction is shown in Table 5.6. For example, if the inputs are 0 (000) and 1 (111), the correct output should be 0, which corresponds to codeword 000.

We can also see that if there are more bit flips, the faulty output might be corrected to a wrong codeword. For example, if the inputs are 111 and 111, but the second 111 is faulted to 001 with a 2-bit flip attack, then the table lookup gives output 000. However, since 1 & 1 = 1, the output should be 111. Thus, it is better to only use error-correcting code-based countermeasure when we know at most $\lfloor (d-1)/2 \rfloor$ bits can be flipped, where d is the minimum distance of the binary code.

We refer the readers to [BKHL20] for an encoding-based hardware implementation of PRESENT using the 3-repetition code $C_{[3,1,3]}$.

5.2.2 Infective Countermeasure

The idea of infective countermeasure is to process the ciphertext in a way that the output becomes useless for an attacker when faults are injected during the computations. We will take the proposal from [TBM14] and only focus on the case for AES-128. The protection for other AES variants can be done in a similar way; we refer the readers to [TBM14] for more details.

The main methodology of the countermeasure is to compute each round of AES encryption twice before moving to the next round. The results of the two computations of the same round will be compared, if a fault is detected, the rest of the computation should produce random values. Computations of dummy rounds are also randomly added in between the AES rounds so that the attacker would not know where the fault was actually injected.

As mentioned in Sect. 3.1.2, AES-128 has key size 128 bits and round number Nr= 10. We define F_i ($i = 0, 1, 2, \ldots, 10$) as follows: F_0 denotes the initial AddRoundKey operation in AES; for $i = 1, 2, \ldots, 9$, F_i denotes the AES round

---

Algorithm 5.3: Infective Countermeasure for AES-128

Input: p, β, keys, t // p is a plaintext block; β is a random number; keys contains the AES round keys K_i and the dummy round keys κ_i (Eq. 5.15) for $i = 0, 1, 2, \ldots, 10$, see Eq. 5.16; t is a user-specified security parameter.

Output: ciphertext or infected ciphertext

1 $R_0 = p$ // cipher state
2 $R_1 = p$ // redundant cipher state
3 $R_2 = \beta$ // dummy round state
4 Generate $\text{rstr} \in \mathbb{F}_2^t$ // contains 22 of 1s corresponding to AES rounds and $t - 22$ of 0s corresponding to dummy rounds
5 $j = 0$
6 $\text{idx} = 1$
7 **while** $\text{idx} \leq t$ **do**
8     $i = \lfloor j/2 \rfloor$ // i is the round counter
9     $\lambda = \text{rstr}[\text{idx}]$ // λ is given by the idxth bit of rstr, $\lambda = 0$ implies a dummy round
10   $a = ((\text{LSB of } j) \mathbin{\&} \lambda) \oplus 2(\neg\lambda)$ // LSB stands for the least significant bit, $\&$ is bitwise AND (see Definition 1.3.6), $\neg$ is logical negation
11   $R_a = F_i(R_a, \text{keys}[\lambda][i])$
12   $\gamma = \lambda \mathbin{\&} (\text{LSB of } j) \mathbin{\&} (\neg 1_0(R_0 \oplus R_1))$ // if j is odd and $\lambda = 1$, detect fault injection in AES
13   $\delta = (\neg\lambda) \mathbin{\&} (\neg 1_0(R_2 \oplus \beta))$ // detect fault injection in dummy round when $\lambda = 0$
14   $R_0 = (\neg(\gamma \vee \delta) \cdot R_0) \oplus ((\gamma \vee \delta) \cdot R_2)$
15   $j = j + \lambda$
16   $\text{idx} = \text{idx} + 1$
17 **return** R_0

---

function, in particular, F_i consists of the following operations: SubBytes, ShiftRows, MixColumns, and AddRoundKey; F_{10} denotes the AES round function for the last round. It consists of SubBytes, ShiftRows, and AddRoundKey. Let K_i ($i = 0, 1, 2, \ldots, 10$) denote the round keys for AES. Each F_i takes as input the cipher state at the end of round $i - 1$ and K_i, and outputs the cipher state at the end of round i.

Correspondingly, we also generate a random number β and the round keys for the dummy rounds, denoted κ_i ($i = 0, 1, 2, \ldots, 10$), such that

$$F_i(\beta, \kappa_i) = \beta \tag{5.15}$$

for $i = 0, 1, 2 \ldots, 10$. We note that since F_0 is an AddRoundKey operation,

$$\kappa_0 = 00000000000000000.$$

Furthermore,

$$\kappa_i = \beta \oplus \text{MixColumns}(\text{ShiftRows}(\text{SubBytes}(\beta))), \quad \text{for } i = 1, 2, \ldots, 9$$

and

$$\kappa_{10} = \beta \oplus \text{ShiftRows}(\text{SubBytes}(\beta)).$$

We set an array of keys of size 2×11, denoted **keys** as

$$\text{keys}[0][i] = \kappa_i, \quad \text{keys}[1][i] = K_i. \tag{5.16}$$

The details of the countermeasure are shown in Algorithm 5.3. As mentioned before, each AES round is computed twice. The user-specified number t determines how many dummy rounds will be added during the computation. The cipher state for the first AES computation is stored in R_0, and the cipher state in the redundant AES computation is stored in R_1. Both are initialized to be the plaintext (lines 1 and 2). The dummy round state is stored in R_2 and initialized to be the random number β (line 3). j (line 5) counts the total number (including the redundant ones) of AES rounds computed, and $i = \lfloor j/2 \rfloor$ (line 8) is the actual round counter. The random string **rstr** contains 22 of 1s corresponding to two computations of each F_i for $i = 0, 1, \ldots, 10$ and $t - 22$ bits of 0 corresponding to dummy rounds. In each loop, we go through the **idx**th bit of **rstr** (line 9), and the value is stored in λ. The value of **idx** is increased by 1 (line16) at the end of each loop so that in the next loop we will go to the next bit of **rstr**.

In line 10, we note that if j is even (respectively, odd), the least significant bit (LSB) of j is 0 (respectively, 1), thus

$$a = ((\text{LSB of } j) \,\&\, \lambda) \oplus 2(\neg\lambda) = \begin{cases} 0 \oplus 2 = 2, & \text{if } \lambda = 0 \\ (0 \,\&\, 1) \oplus 0 = 0, & \text{if } \lambda = 1 \text{ and } j \text{ is even} \\ (1 \,\&\, 1) \oplus 0 = 1, & \text{if } \lambda = 1 \text{ and } j \text{ is odd.} \end{cases}$$

Then, in line 11, when $\lambda = 0$, $a = 2$, we compute a dummy round i with

$$R_2 = F_i(R_2, \text{keys}[0][i]) = F_i(R_2, \kappa_i).$$

When $\lambda = 1$, j is even, and $a = 0$, we compute AES round i with

$$R_0 = F_i(R_0, \text{keys}[1][i]) = F_i(R_0, K_i).$$

When $\lambda = 1$, j is odd, and $a = 1$, we compute a redundant AES round i with

$$R_1 = F_i(R_1, \text{keys}[1][i]) = F_i(R_1, K_i).$$

The total round counter j is increased by λ at the end of each loop (line 15). When $\lambda = 0$, only a dummy round is computed.

Up to now, we have seen how the AES rounds and dummy rounds are computed. Next, we discuss how fault is handled in the algorithm.

First, we recall the notion of indicator function from Definition 3.2.2. We consider the indicator function for $\mathbf{0}$ with domain $\mathbb{F}_2^{128}$:

$$1_{\mathbf{0}} : \mathbb{F}_2^{128} \to \mathbb{F}_2$$

$$\mathbf{x} \mapsto \prod_i (1 - x_i).$$

In other words,

$$1_{\mathbf{0}}(\mathbf{x}) = \begin{cases} 1 & \text{if } \mathbf{x} = 0 \\ 0 & \text{otherwise.} \end{cases}$$

Then, with logical negation, $\neg 1_{\mathbf{0}}(\mathbf{x}) : \mathbb{F}_2^{128} \to \mathbb{F}_2$ and

$$\neg 1_{\mathbf{0}}(\mathbf{x}) = \begin{cases} 0 & \text{if } \mathbf{x} = 0 \\ 1 & \text{otherwise.} \end{cases}$$

Consequently, in line 12, we have

$$\gamma = \lambda \ \& \ (\text{LSB of } j) \ \& \ (\neg 1_{\mathbf{0}}(R_0 \oplus R_1)) = \begin{cases} 0 & \text{if } \lambda = 0 \text{ or } j \text{ is even} \\ \neg 1_{\mathbf{0}}(R_0 \oplus R_1) & \text{otherwise} \end{cases}$$

$$= \begin{cases} 0 & \text{if } \lambda = 0 \text{ or } j \text{ is even or } R_0 = R_1 \\ 1 & \lambda = 1, j \text{ is odd, and } R_0 \neq R_1. \end{cases}$$

Thus, when j is odd and $\lambda = 1$ (i.e., in the loop when the redundant AES round is computed), γ indicates if the cipher state in the AES round computation, R_0, is equal to the redundant cipher state, R_1, or equivalent, whether fault happened in AES round or in the redundant round computation. If there was no fault, $\gamma = 0$; otherwise, $\gamma = 1$.

Algorithm 5.4: Computation of AES round in the infective Countermeasure for AES-128 from Algorithm 5.3

---

1 j is even
2 $i = \lfloor j/2 \rfloor$ // i is the round counter
3 $\lambda = 1$
4 $a = 0$
5 $R_0 = F_i(R_0, \text{keys}[1][i])$ // $\text{keys}[1][i] = K_i$ is the ith round key for AES
6 $\gamma = 0$
7 $\delta = 0$
8 $R_0 = R_0$

Algorithm 5.5: Computation of redundant AES round in the infective Countermeasure for AES-128 from Algorithm 5.3

```
1  j is odd
2  i = ⌊j/2⌋// i is the round counter
3  λ = 1
4  a = 1
5  R₁ = Fᵢ(R₁, keys[1][i])// keys[1‖i] = Kᵢ is the ith round key for AES
6  γ = ¬1₀(R₀ ⊕ R₁)// detect fault injection in AES
7  δ = 0
8  R₀ = ((¬γ) · R₀) ⊕ (γ · R₂)// if there is fault in AES computation, R₀ = R₂
        becomes a random number
```

Algorithm 5.6: Computation of the dummy round in the infective Countermeasure for AES-128 from Algorithm 5.3

```
1  λ = 0
2  a = 2
3  R₂ = Fᵢ(R₂, keys[0][i])// i is the round counter, keys[0‖i] = κᵢ is the ith
        round key for the dummy rounds
4  γ = 0
5  δ = ¬1₀(R₂ ⊕ β)// detect fault injection in dummy round
6  R₀ = ((¬δ) · R₀) ⊕ (δ · R₂)// if there is fault in the dummy round
        computation, R₀ = R₂ becomes a random number
```

Similarly, in line 13, we have

$$\delta = \begin{cases} 0 & \text{if } \lambda = 1 \\ \neg 1_0(R_2 \oplus \beta) & \text{if } \lambda = 0 \end{cases} = \begin{cases} 0 & \text{if } \lambda = 1 \text{ or } R_2 = \beta \\ 1 & \text{if } \lambda = 0 \text{ and } R_2 \neq \beta. \end{cases}$$

Thus, when $\lambda = 0$, i.e., in the loop when the dummy round is computed, δ indicates if there is a fault injected in the computation of the dummy round state R_2. By the design of dummy round keys and β (see Eq. 5.15), if there are no faults, $R_2 = \beta$ and $\delta = 0$. Otherwise, $R_2 \neq \beta$ and $\delta = 1$.

Finally, in line 14, we have

$$R_0 = (\neg(\gamma \vee \delta) \cdot R_0) \oplus ((\gamma \vee \delta) \cdot R_2) = \begin{cases} R_0 & \text{if } \gamma = 0 \text{ and } \delta = 0 \\ R_2 & \text{otherwise.} \end{cases}$$

This line guarantees that R_0 will be changed to a random number R_2 if a fault is detected in any of the computations. Consequently, the output will be a random number or *infected ciphertext*.

The computations for the AES round, the redundant round, and the dummy round are shown in Algorithms 5.4, 5.5, and 5.6.

5.3   Fault Attacks on RSA and RSA Signatures

As discussed in Sect. 2.1.2, a public key cryptosystem has a public key and a private key. For fault attacks that will be discussed in this section, the attacker's goal will be the recovery of the secret key.

Unlike fault attacks on symmetric block ciphers, attacks on public key ciphers depend on the underlying intractable problem, and we do not have a systematic methodology. However, the general attack concept can be applied to ciphers based on similar intractable problems. This section will focus on fault attacks on implementations of RSA signatures. We will discuss a few fault attacks during the signature signing procedure to recover the private key.

> **Note**
> We note that the attacks on RSA signature signing procedure can also be applied to RSA decryption process.

For the rest of this section, let p and q be two distinct odd primes. Let $n = pq$ and $e \in \mathbb{Z}^*_{\varphi(n)}$ be the public key for RSA signatures. $d = e^{-1} \bmod \varphi(n)$ denotes the private key. The goal of the attacker is to recover d. As discussed in Sect. 3.5.1.3, the signature is computed on the hash value, $h(m)$, of the intended message m, where h is a fast public hash function (see Sect. 2.1.1). For simplicity, we will use m to denote the hash value $h(m)$.

Let ℓ_d and ℓ_n denote the bit length of d and n, respectively. We have the following binary representation (see Theorem 1.1.1) of d:

$$d = \sum_{i=0}^{\ell_d - 1} d_i 2^i.$$

We recap here the CRT-based implementation for RSA signatures. Following the discussions in Sect. 3.5.1.3, to sign the signature for m, the owner of the private key, say Alice, computes

$$s_p := m^{d \bmod (p-1)} \bmod p, \quad s_q := m^{d \bmod (q-1)} \bmod q, \tag{5.17}$$

and the signature s is given by Gauss's algorithm,

$$s = s_p y_q q + s_q y_p p \bmod n,$$

or by Garner's algorithm,

$$s = s_p + ((s_q - s_p) y_p \bmod q) p,$$

where

$$y_q = q^{-1} \bmod p, \quad y_p = p^{-1} \bmod q. \tag{5.18}$$

Alice sends s and m to Bob. To verify the signature, Bob computes and checks if

$$s^e \bmod n = m.$$

If the equality holds, Bob considers the signature valid.

5.3.1   Bellcore Attack

We first describe an attack that recovers the private key of RSA signatures by exploiting a faulty signature. The attack was first introduced by Boneh et al. [BDL97]. The name "Bellcore" comes from the company the authors were working for at the time of the publication. This paper is also the very first paper that introduced fault attacks to cryptographic implementations.

As mentioned in Sect. 3.5.1.3, y_q and y_p (Eq. 5.18) can be precomputed. We assume there are no faults in their computations.

By the design of s_p, s_q, y_p, and y_q, we have

$$s \equiv s_q \bmod q, \quad s \equiv s_p \bmod p, \tag{5.19}$$

which gives

$$m \equiv s^e \equiv s_q^e \bmod q, \quad m \equiv s^e \equiv s_p^e \bmod p. \tag{5.20}$$

Suppose a malicious fault was induced during the signing of the signature and the computation of s_p or s_q (Eq. 5.17), but not both, is corrupted. Let us assume that s_p is faulty and s_q is computed correctly. A similar attack applies if s_q is faulty and s_p is correct. Let s' denote the faulty signature. By Eq. 5.19,

$$s' \equiv s \equiv s_q \bmod q, \quad s' \not\equiv s \bmod p.$$

In other words,

$$q \mid (s' - s), \quad p \nmid (s' - s).$$

Recall that n and e are public. If the attacker further has the knowledge of s and s', then they can compute

$$q = \gcd(s' - s, n), \quad p = \frac{n}{q}.$$

As mentioned in Sect. 3.3, after factorizing n, the attacker can compute

$$\varphi(n) = (p-1)(q-1)$$

and eventually recover the private key

$$d = e^{-1} \bmod \varphi(n)$$

by the extended Euclidean algorithm (Algorithm 1.2)

For a different attack [Len96], we assume the attacker does not have the knowledge of the correct signature s. Instead, the attacker can obtain the faulty signature s' and the original message hash value m. For example, the attacker can request Alice for the signature of a chosen message. By Eq. 5.20,

$$s'^{e} \equiv m \bmod q, \quad s'^{e} \not\equiv m \bmod p,$$

i.e.,

$$q \mid (s'^{e} - m), \quad p \nmid (s'^{e} - m).$$

Thus the attacker can factorize n by computing

$$q = \gcd(s'^{e} - m, n), \quad p = \frac{n}{q}.$$

Example 5.3.1 Let $p = 5$, $q = 7$, and $e = 5$. We have calculated that $d = 5$ in Example 3.4.1 and $y_q = 3$, $y_p = 3$ in Example 3.5.8. Suppose $m = 6$. By Eq. 5.17, to calculate the signature, Alice computes

$$s_p = m^{d \bmod (p-1)} \bmod p = 6^{5 \bmod 4} \bmod 5 = 1,$$
$$s_q = m^{d \bmod (q-1)} \bmod q = 6^{5 \bmod 6} \bmod 7 = 6.$$

And the signature is

$$s = s_p + ((s_q - s_p)y_p \bmod q)p = 1 + ((6-1) \times 3 \bmod 7) \times 5 = 6.$$

We can verify that

$$s^{e} \bmod n = 6^{5} \bmod 35 = 6 = m.$$

Now suppose the computation of s_p is faulty and $s'_p = 3$. Then we have

$$s' = s'_p + ((s_q - s'_p)y_p \bmod q)p = 3 + ((6-3) \times 3 \bmod 7) \times 5 = 3 + 2 \times 5 = 13.$$

If the attacker has the knowledge of $s = 6$ and $s' = 13$, they can compute

$$q = \gcd(s' - s, n) = \gcd(13 - 6, 35) = \gcd(7, 35) = 7.$$

If the attacker has the knowledge of $s' = 13$ and $m = 6$, they can compute

$$q = \gcd(s'^e - m, n) = \gcd(13^5 - 6, 35) = \gcd(371287, 35).$$

By the Euclidean algorithm,

$$371287 = 35 \times 10608 + 7, \quad \gcd(371287, 35) = \gcd(35, 7),$$
$$35 = 7 \times 5, \quad\quad\quad\quad\quad\quad \gcd(35, 7) = 7,$$

and $q = 7$.

Similarly, suppose the computation of s_q is faulty and $s'_q = 2$. Then

$$s' = s_p + ((s'_q - s_p)y_p \bmod q)p = 1 + ((2 - 1) \times 3 \bmod 7) \times 5 = 16.$$

If the attacker has the knowledge of $s = 6$ and $s' = 16$, they can compute

$$p = \gcd(s' - s, n) = \gcd(16 - 6, 35) = \gcd(10, 35) = 5.$$

If the attacker has the knowledge of $s' = 16$ and $m = 6$, they can compute

$$p = \gcd(s'^e - m, n) = \gcd(16^5 - 6, 35) = \gcd(1048570, 35).$$

By the Euclidean algorithm,

$$1048570 = 35 \times 29959 + 5, \quad \gcd(1048570, 35) = \gcd(35, 5),$$
$$35 = 5 \times 7, \quad\quad\quad\quad\quad\quad \gcd(35, 5) = 5.$$

Hence $p = 5$.

Example 5.3.2 Let $p = 11, q = 13$. Then $n = 143$,

$$\varphi(n) = 10 \times 12 = 120.$$

Choose $e = 11$, which is coprime with $\varphi(n)$. By the extended Euclidean algorithm,

$$120 = 11 \times 10 + 10, \ 11 = 10 \times 1 + 1 \implies 1 = 11 - (120 - 11 \times 10) = 11 \times 11 - 120,$$

and we have $d = 11^{-1} \bmod 120 = 11$. Again, by the extended Euclidean algorithm,

$$13 = 11 \times 1 + 2, \ 11 = 2 \times 5 + 1 \implies 1 = 11 - 2 \times 5 = 11 - 5 \times (13 - 11) = 11 \times 6 - 13 \times 5,$$

and we have

$$y_q = q^{-1} \bmod p = 13^{-1} \bmod 11 = -5 \bmod 11 = 6,$$

$$y_p = p^{-1} \bmod q = 11^{-1} \bmod 13 = 6.$$

Let $m = 2$. By Eq. 5.17, to calculate the signature, Alice computes

$$s_p = m^{d \bmod (p-1)} \bmod p = 2^{11 \bmod 10} \bmod 11 = 2 \bmod 11 = 2,$$

$$s_q = m^{d \bmod (q-1)} \bmod q = 2^{11 \bmod 12} \bmod 13 = 2048 \bmod 13 = 7.$$

Using Garner's algorithm, the signature is

$$s = s_p + ((s_q - s_p)y_p \bmod q)p = 2 + ((7 - 2) \times 6 \bmod 13) \times 11 = 2 + 4 \times 11 = 46.$$

We have

$$46^2 \bmod 143 = 114, \qquad\qquad 46^3 \bmod 143 = 114 \times 46 \bmod 143 = 96,$$

$$46^5 \bmod 143 = 114 \times 96 \bmod 143 = 76, \; 46^{10} \bmod 143 = 76^2 \bmod 143 = 56.$$

We can then verify that

$$s^e \bmod n = 46^{11} \bmod 143 = 56 \times 46 \bmod 143 = 2 = m.$$

Now suppose the computation of s_p is faulty and $s_p' = 7$. Then we have

$$s' = s_p' + ((s_q - s_p')y_p \bmod q)p = 7 + ((7 - 7) \times 6 \bmod 13) \times 11 = 7.$$

If the attacker has knowledge of $s = 46$ and $s' = 7$, they can compute

$$q = \gcd(s' - s, n) = \gcd(7 - 46, 143) = \gcd(-39, 143) = \gcd(39, 143).$$

By the Euclidean algorithm,

$$
\begin{aligned}
143 &= 39 \times 3 + 26, & \gcd(39, 143) &= \gcd(39, 26), \\
39 &= 26 + 13, & \gcd(39, 26) &= \gcd(26, 13), \\
26 &= 13 \times 2, & \gcd(26, 13) &= 13.
\end{aligned}
$$

Hence $q = 13$.

If the attacker has knowledge of $s' = 7$ and $m = 2$, they can compute

$$q = \gcd(s'^e - m, n) = \gcd(7^{11} - 2143) = \gcd(1977326741, 143).$$

By the Euclidean algorithm,

$$1977326741 = 143 \times 13827459 + 104, \quad \gcd(1977326741, 143) = \gcd(143, 104),$$
$$143 = 104 + 39, \quad\quad\quad\quad\quad\quad\quad\quad \gcd(143, 104) = \gcd(104, 39),$$
$$104 = 39 \times 2 + 26, \quad\quad\quad\quad\quad\quad\quad \gcd(104, 39) = \gcd(39, 26),$$
$$39 = 26 + 13, \quad\quad\quad\quad\quad\quad\quad\quad\quad \gcd(39, 26) = \gcd(26, 13),$$
$$26 = 13 \times 2, \quad\quad\quad\quad\quad\quad\quad\quad\quad\quad q = \gcd(26, 13) = 13.$$

Similarly, suppose the computation of s_q is faulty and $s_q' = 2$. Then

$$s' = s_p + ((s_q' - s_p)y_p \bmod q)p = 2 + ((2 - 2) \times 6 \bmod 13) \times 11 = 2.$$

If the attacker has knowledge of $s = 46$ and $s' = 2$, they can compute

$$p = \gcd(s' - s, n) = \gcd(2 - 46, 143) = \gcd(-44, 143) = \gcd(44, 143).$$

By the Euclidean algorithm,

$$143 = 44 \times 3 + 11, \quad \gcd(44, 143) = \gcd(44, 11),$$
$$44 = 11 \times 4, \quad\quad\quad q = \gcd(44, 11) = 11.$$

If the attacker has knowledge of $s' = 2$ and $m = 2$, they can compute

$$p = \gcd(s'^e - m, n) = \gcd(2^{11} - 2, 143) = \gcd(2046, 143).$$

By the Euclidean algorithm,

$$2046 = 143 \times 14 + 44, \quad \gcd(2046, 143) = \gcd(143, 44),$$
$$143 = 44 \times 3 + 11, \quad\quad \gcd(143, 44) = \gcd(44, 11),$$
$$44 = 11 \times 4, \quad\quad\quad\quad\quad p = \gcd(44, 11) = 11.$$

5.3.2 *Attack on the Square and Multiply Algorithm*

In this subsection, we will look at fault attacks on the square and multiply algorithm. We will first detail the bit flip attack proposed in [BDH+97], and then we will discuss an improved version proposed in [JQBD97].

Instead of a CRT-based implementation, we assume the implementation computes the signature with the right-to-left square and multiply algorithm. Following Algorithm 3.7, to compute $m^d \bmod n$, we have Algorithm 5.7, where ℓ_d is the bit length of d.

For the attack, we inject a bit-flip fault model so that 1 bit of d, say d_i, is flipped. Let d' denote the faulty value of d. Then the faulty signature is given by $s' = m^{d'} \bmod n$. From Algorithm 5.7, lines 4 and 5, we can see that the computations

---

Algorithm 5.7: Computing RSA signature with the right-to-left square and multiply algorithm

---

Input: n, m, d // n is the RSA modulus; m is hash value of the message; d is the private key of bit length ℓ_d

Output: $s = m^d \bmod n$

1 $s = 1$

2 $t = m$

3 **for** $i = 0$, $i < \ell_d$, $i + +$ **do**

      // ith bit of d is 1

4   **if** $d_i = 1$ **then**

          // multiply by m^{2^i}

5          $s = s * t \bmod n$

      // $t = m^{2^{i+1}}$

6      $t = t * t \bmod n$

7 **return** s

---

of s and s' will differ by the multiplication of m^{2^i}. In particular, we have

$$\frac{s'}{s} \equiv \begin{cases} m^{-2^i} \bmod n, & \text{if } d_i = 1, d_i' = 0 \\ m^{2^i} \bmod n, & \text{if } d_i = 0, d_i' = 1. \end{cases} \tag{5.21}$$

Suppose the attacker has the knowledge of s, s', and m; then they can compute

$$\frac{s'}{s} \bmod n$$

and compare it with

$$m^{2^i} \bmod n$$

to recover the value of d_i.

To improve the attack, we loosen the assumption on the attacker and assume that they only have the knowledge of s' and m (not knowing s). In this case, we note that

$$\frac{s'^e}{s^e} \equiv \frac{s'^e}{m} \bmod n.$$

Then it follows from Eq. 5.21 that

$$\frac{s'^e}{m} \equiv \begin{cases} m^{-e2^i} \bmod n, & \text{if } d_i = 1, d_i' = 0 \\ m^{e2^i} \bmod n, & \text{if } d_i = 0, d_i' = 1. \end{cases} \tag{5.22}$$

Thus the attacker can compute

$$\frac{s'^e}{m} \bmod n$$

and compare with

$$m^{e2^i} \bmod n$$

to recover the value of d_i.

Both attacks can be repeated for different bits of d to recover the whole private key.

Example 5.3.3 Let $p = 3$, $q = 5$. We have $n = 15$ and $\varphi(n) = 2 \times 4 = 8$. Suppose $d = 3 = 11_2$ and $m = 2$. We have computed in Example 3.5.1 that

$$s = m^d \bmod n = 8.$$

The intermediate values for the computation with Algorithm 5.7 will be

i	d_i	t	result
0	1	4	2
1	1	1	8

By the extended Euclidean algorithm, we get

$$e = d^{-1} \bmod \varphi(n) = 3^{-1} \bmod 8 = 3.$$

Suppose d_0 is flipped, then $d' = 2 = 10_2$. The resulting computation following Algorithm 5.7 will then have the intermediate values as follows:

i	d_i	t	result
0	0	4	1
1	1	1	4

Thus $s' = 4$.

With the knowledge of $s = 8$, $s' = 4$, and $m = 2$, the attacker computes

$$\frac{s'}{s} \equiv \frac{4}{8} \equiv 2^{-1} \bmod 15, \quad m^{2^i} \equiv 2^1 \equiv 2 \bmod 15 \quad \Longrightarrow \quad \frac{s'}{s} \equiv m^{-2^i} \bmod n.$$

By Eq. 5.21, $d_0 = 1$.

In case the attacker does not have the knowledge of s, they can compute

$$\frac{s'^e}{m} \equiv \frac{4^3}{2} \equiv 32 \equiv 2 \bmod 15.$$

By the extended Euclidean algorithm,

$$15 = 2 \times 7 + 1 \implies 2^{-1} \bmod 15 = -7 \bmod 15 = 8.$$

And we have

$$m^{e2^i} \equiv 2^{3 \times 2^0} \equiv 2^3 \equiv 8 \bmod 15, \quad m^{-e2^i} \equiv 2^{-3 \times 2^0} \equiv 2^{-3} \equiv 8^3 \equiv 512 \equiv 2 \bmod 15.$$

Thus

$$\frac{s'^e}{m} \equiv m^{-e2^i} \bmod n.$$

By Eq. 5.22, $d_0 = 1$.

5.3.3 Attack on the Public Key

In this subsection, we will discuss an attack [BCG08] that injects faults into the RSA public key n, during the signature singing, and recovers the private key d. Since the value n is big, it will be stored in a few registers. The fault can be injected during loading or preparing n. The attack is specific to the right-to-left square and multiply algorithm.

The RSA signature computation with the right-to-left square and multiply algorithm is detailed in Algorithm 5.7. Let n' denote the faulty RSA modulus and

$$\varepsilon := n \oplus n'$$

be the fault mask. Suppose the fault is injected in round j ($1 \le j \le \ell_d - 2$), resulting in a faulty square computation in line 6

$$t = t * t \bmod n',$$

and this faulty n' is also used for the rest of the computation. Then the faulty signature is given by

$$s' = \left[\left(\prod_{i=0}^{j-1} m^{2^i d_i} \bmod n \right) \prod_{i=j}^{\ell_d - 1} \left(m^{2^{j-1}} \bmod n \right)^{2^{i-j+1} d_i} \right] \bmod n'. \tag{5.23}$$

We note that if the fault is injected in round $j = \ell_d - 1$ for the computation of the square, the output will not be affected, and hence the faulty signature will not be useful for recovery of the secret key. If the fault is injected in round 0, since $m \in \mathbb{Z}_n$, the computation result will be $m^d \bmod n'$, and the attacker would need to brute force all possible values of d to find out which one gives the faulty signature. Hence we assume $j \geq 1$.

Recall that the correct signature is given by

$$s = \prod_{i=0}^{\ell_d-1} m^{2^i d_i} \bmod n.$$

Define

$$d_{(j)} := d_{\ell_d-1}\ldots d_{j+1}d_j 00\ldots 00,$$

then

$$m^{d_{(j)}} = \prod_{i=j}^{\ell_d-1} m^{2^i d_i} \bmod n,$$

and

$$s' = \left[(sm^{-d_{(j)}} \bmod n) \prod_{i=j}^{\ell_d-1} \left(m^{2^{j-1}} \bmod n \right)^{2^{i-j+1} d_i} \right] \bmod n'.$$

There are 2^{ℓ_d-j} possible values for $d_{(j)}$.

Suppose the attacker has knowledge of ε (hence n'), the message hash value m, the correct signature s, and the faulty signature s'. For each guessed value of $d_{(j)}$, denoted

$$\hat{d}_{(j)} = \hat{d}_{\ell_d-1}\ldots \hat{d}_{j+1}\hat{d}_j 00\ldots 0,$$

the attacker computes

$$\hat{s}' = \left[(sm^{-\hat{d}_{(j)}} \bmod n) \prod_{i=j}^{\ell_d-1} \left(m^{2^{j-1}} \bmod n \right)^{2^{i-j+1} \hat{d}_i} \right] \bmod n' \tag{5.24}$$

and compares it with s'. Then they record values of $\hat{d}_{(j)}$ that satisfy

$$\hat{s}' = s',$$

which reduces the hypotheses for the jth—$(\ell_d - 1)$th bits of d.

The attack can be repeated for other bits of d to reduce the key hypotheses further.

Example 5.3.4 Let $n = 15$, $m = 2$. Then $\varphi(n) = 8$. Let $d = 5 = 101_2$. Computing

$$s = m^d \bmod n = 2^5 \bmod 15$$

with Algorithm 5.7, we have the following intermediate values in each loop:

i	d_i	t	s
0	1	4	2
1	0	1	2
2	1	1	2

and the correct signature $s = 2$.

Suppose a fault is injected in n when line 6 is executed in the iteration $i = 1$, resulting in $n' = 13$. The intermediate values will be

i	d_i	t	s
0	1	4	2
1	0	3	2
2	1	9	6

and the faulty signature $s' = 6$, which agrees with Eq. 5.23:

$$
\begin{aligned}
s' &= \left[\left(\prod_{i=0}^{j-1} m^{2^i d_i} \bmod n \right) \prod_{i=j}^{\ell_d-1} \left(m^{2^{j-1}} \bmod n \right)^{2^{i-j+1} d_i} \right] \bmod n' \\
&= \left[\left(m^{2^0 d_0} \bmod n \right) \prod_{i=1}^{2} \left(m^{2^0} \bmod n \right)^{2^i d_i} \right] \bmod n' \\
&= \left[(m^{d_0} \bmod n)(m \bmod n)^{2 d_1 + 2^2 d_2} \right] \bmod n' \\
&= (2 \bmod 15)(2 \bmod 15)^4 \bmod 13 = 2^5 \bmod 13 = 6.
\end{aligned}
$$

To recover the secret key d, the attacker takes all possible values for $d_{(1)} = d_2 d_1 0$ and computes the corresponding possible faulty signatures with Eq. 5.24:

$$
\begin{aligned}
\hat{s}' &= \left[(s m^{-\hat{d}_{(j)}} \bmod n) \prod_{i=j}^{\ell_d-1} \left(m^{2^{j-1}} \bmod n \right)^{2^{i-j+1} \hat{d}_i} \right] \bmod n' \\
&= \left[(2 m^{-\hat{d}_{(1)}} \bmod n)(m \bmod n)^{2\hat{d}_1 + 2^2 \hat{d}_2} \right] \bmod n' \\
&= \left[(2^{1-\hat{d}_{(1)}} \bmod 15) \times 2^{2\hat{d}_1 + 2^2 \hat{d}_2} \right] \bmod n'.
\end{aligned}
$$

For $\hat{d}_{(1)} = 000$, we have

$$
\hat{s}' = \left[(2^{1-\hat{d}_{(1)}} \bmod 15) \times 2^{2\hat{d}_1 + 2^2 \hat{d}_2} \right] \bmod n' = 2 \times 1 \bmod 13 = 2.
$$

For $\hat{d}_{(1)} = 010$,

$$\hat{s}' = \left[(2^{1-\hat{d}_{(1)}} \bmod 15) \times 2^{2\hat{d}_1+2^2\hat{d}_2}\right] \bmod n' = (2^{-1} \bmod 15) \times 2^2$$
$$\bmod n' = 8 \times 4 \bmod 13 = 6.$$

For $\hat{d}_{(1)} = 100$,

$$\hat{s}' = \left[(2^{1-\hat{d}_{(1)}} \bmod 15) \times 2^{2\hat{d}_1+2^2\hat{d}_2}\right] \bmod n' = (2^{-3} \bmod 15) \times 2^4$$
$$\bmod n' = 2 \times 16 \bmod 13 = 6.$$

For $\hat{d}_{(1)} = 110$,

$$\hat{s}' = \left[(2^{1-\hat{d}_{(1)}} \bmod 15) \times 2^{2\hat{d}_1+2^2\hat{d}_2}\right] \bmod n' = (2^{-5} \bmod 15) \times 2^6$$
$$\bmod n' = 8 \times 64 \bmod 13 = 5.$$

Thus the attacker can conclude that $d_{(1)} = 010$ or 100, i.e., $d_1 d_2 = 01$ or 10.

In case the attacker does not have the knowledge of the exact fault mask ε (and hence n'), but instead, they know the range for ε. Then they can brute force all possible values of ε and $\hat{d}_{(j)}$ to reduce the key candidate. We refer the readers to [BCG08] for more details.

5.3.4 Safe Error Attack

This part looks into implementations that are either based on the right-to-left square and multiply algorithm (Sect. 3.5.1.1) or on the Montgomery powering ladder (Sect. 3.5.1.2). We further require that the modular multiplication is implemented with Blakely's method (Sect. 3.5.2.1). We will discuss a fault attack that is specific to such a setting.

The attack exploits the knowledge of whether an intermediate faulty value is used or not by observing whether the final output is changed, thus the name *safe error attack* [YJ00]. Since only knowing whether the output is changed or not is enough, if we implement a countermeasure that repeats the computation, compares the final results, and outputs an error when a fault is detected, the safe error attack still applies.

Let ω be the computer's word size (see Sect. 2.1.2). Take $\kappa = \lceil \ell_n/\omega \rceil$, i.e.,

$$(\kappa - 1)\omega < \ell_n \leq \kappa\omega,$$

where ℓ_n is the bit length of n.

5.3.4.1 Safe Error Attack on the Montgomery Powering Ladder

With the Montgomery powering ladder and Blakley's method, the signature $s = m^d \bmod n$ is computed with Algorithm 5.8 (see Algorithm 3.15).

Since ℓ_n is the bit length of n, the bit lengths of the variables R_0 and R_1 are at most ℓ_n. We can write

$$R_0 = \sum_{i=0}^{\kappa-1} R_{0i}(2^\omega)^i, \qquad R_1 = \sum_{i=0}^{\kappa-1} R_{1i}(2^\omega)^i.$$

We can also assume each of R_{0i} and R_{1i} is stored in one register.

Suppose $d_j = 0$, and a fault is injected during the jth iteration of the outer loop, when $i < i_0$ in the loop starting from line 6, in the variable R_{0i_0}, for some i_0 such that $0 \leq i_0 \leq \kappa - 1$. Then the value in R_1 in line 9 will not be affected since R_{0i_0} is used when $i = i_0$. However, the value in R_0 in line 14 will be faulty. Hence the final output will be faulty.

On the other hand, suppose $d_j = 1$, and a fault is injected during the jth iteration of the outer loop and when $i < i_0$ in the loop starting from line 17, in the variable R_{0i_0}, for some i_0 such that $0 \leq i_0 \leq \kappa - 1$. Then the fault will go unnoticed since R_{0i_0} is used when $i = i_0$ and the value in R_0 will be rewritten in line 20. Thus the final output will be correct.

We assume the attacker has the knowledge of the correct signature, and they can rerun the algorithm with the same inputs, inject fault, and observe the final output. To recover the value of d_j, the attacker fixes an i_0, estimates the time for i to be less than i_0 in loop j, and injects fault in R_{0i_0}. If the signature is faulty, then $d_j = 0$, and if the signature is correct, then $d_j = 1$. We note that the computation times for one loop starting from line 6 and one loop starting from line 17 are similar since they both involve two multiplications and one modular reduction. The attack can be repeated for different bits of d to recover the full private key.

Example 5.3.5 Let us repeat the computations in Examples 5.3.3 and 3.5.1 with Algorithm 5.8. We have

$$p = 3, \quad q = 5, \quad n = 15, \quad \varphi(n) = 2 \times 4 = 8, \quad d = 3 = 11_2, \quad m = 2.$$

And $\ell_n = 4$, $\ell_d = 2$. Suppose $\omega = 2$, then

$$\kappa = \left\lceil \frac{\ell_n}{\omega} \right\rceil = \left\lceil \frac{4}{2} \right\rceil = 2.$$

With Algorithm 5.8, lines 1 and 2 give

$$R_0 = 1, \; R_{00} = 01, \; R_{01} = 00. \quad R_1 = 2, \; R_{10} = 10, \; R_{11} = 00.$$

The intermediate values are

Algorithm 5.8: RSA signature computation with Montgomery powering ladder and Blakely's method

---

Input: n, m, d // n is the RSA modulus of bit length ℓ_n; m is the hash value of the message; d is the private key of bit length ℓ_d

Output: $m^d \bmod n$

1 $R_0 = 1$
2 $R_1 = m$
3 **for** $j = \ell_d - 1, j \geq 0, j - - $ **do**
4   **if** $d_j = 0$ **then**
         // lines 5 - 9 implement $R_1 = R_0 R_1 \bmod n$
5          $R = 0$
6   **for** $i = \kappa - 1, i \geq 0, i - -$ **do**
           // $\kappa = \lceil \ell_n / \omega \rceil$, where ω is the word size of the computer
7             $R = 2^\omega R + R_{0i} R_1$
8             $R = R \bmod n$
9          $R_1 = R$
         // lines 10 - 14 implement $R_0 = R_0^2 \bmod n$
10        $R = 0$
11   **for** $i = \kappa - 1, i \geq 0, i - -$ **do**
12          $R = 2^\omega R + R_{0i} R_0$
13          $R = R \bmod n$
14        $R_0 = R$
15   **else**
         // lines 16 - 20 implement $R_0 = R_0 R_1 \bmod n$
16        $R = 0$
17   **for** $i = \kappa - 1, i \geq 0, i - -$ **do**
18          $R = 2^\omega R + R_{0i} R_1$
19          $R = R \bmod n$
20        $R_0 = R$
         // lines 21 - 25 implement $R_1 = R_1^2 \bmod n$
21        $R = 0$
22   **for** $i = \kappa - 1, i \geq 0, i - -$ **do**
23          $R = 2^\omega R + R_{1i} R_1$
24          $R = R \bmod n$
25        $R_1 = R$
26 **return** R_0

$j = 1 \; d_1 = 1$

    loop line 17 $i = 1$ $R = 2^\omega R + R_{01} R_1 \bmod n = 0$
                  $i = 0$ $R = 2^\omega R + R_{00} R_1 \bmod n = 2 \bmod 15 = 2$
    line 20 $R_0 = 2 \; R_{00} = 10, \; R_{01} = 00$
    loop line 22 $i = 1$ $R = 2^\omega R + R_{11} R_1 \bmod n = 0$
                  $i = 0$ $R = 2^\omega R + R_{10} R_1 \bmod n = 2 \times 2 \bmod 15 = 4$
    line 25 $R_1 = 4 \; R_{10} = 00, \; R_{11} = 01$

$j = 0 \; d_0 = 1$

    loop line 17 $i = 1$ $R = 2^\omega R + R_{01} R_1 \bmod n = 0$
                  $i = 0$ $R = 2^\omega R + R_{00} R_1 \bmod n = 2 \times 4 \bmod 15 = 8$
    line 20 $R_0 = 8 \; R_{00} = 00, \; R_{01} = 10$

Hence the output is 8.

Suppose the attacker would like to find out what is d_0. They estimate the time for $j = 0$ in the outer loop and $i = 0$ in the loop starting from either line 6 or line 17. Then they inject fault into R_{01} at this time. We note that R_{01} is used (blue R_{01} in the above equations) before $i = 0$ and reassigned value in line 20 (orange R_{01} in the above equations). Thus the computations are not affected, and the signature is correct. The attacker can conclude that $d_0 = 1$.

Example 5.3.6 Let $d = 2 = 10_2$, and keep the other parameters the same as in Example 5.3.5. Then

$$s = m^d \bmod n = 2^2 \bmod 15 = 4.$$

With Algorithm 5.8, lines 1 and 2 give

$$R_0 = 1, \ R_{00} = 01, \ R_{01} = 00. \quad R_1 = 2, \ R_{10} = 10, \ R_{11} = 00.$$

The intermediate values are

$$
\begin{aligned}
&j = 1 \ d_1 = 1 \\
&\qquad \text{loop line 17 } i = 1 \quad R = 2^\omega R + R_{01} R_1 \bmod n = 0 \\
&\qquad\qquad\qquad\qquad\; i = 0 \quad R = 2^\omega R + R_{00} R_1 \bmod n = 2 \bmod 15 = 2 \\
&\qquad \text{line 20} \qquad R_0 = 2 \ R_{00} = 10, \ R_{01} = 00 \\
&\qquad \text{loop line 22 } i = 1 \quad R = 2^\omega R + R_{11} R_1 \bmod n = 0 \\
&\qquad\qquad\qquad\qquad\; i = 0 \quad R = 2^\omega R + R_{10} R_1 \bmod n = 2 \times 2 \bmod 15 = 4 \\
&\qquad \text{line 25} \qquad R_1 = 4 \ R_{10} = 00, \ R_{11} = 01 \\[6pt]
&j = 0 \ d_0 = 0 \\
&\qquad \text{loop line 6 } \ i = 1 \quad R = 2^\omega R + R_{01} R_1 \bmod n = 0 \\
&\qquad\qquad\qquad\qquad\; i = 0 \quad R = 2^\omega + R_{00} R_1 \bmod n = 8 \\
&\qquad \text{line 9} \qquad R_1 = 8 \ R_{10} = 00, \ R_{11} = 10 \\
&\qquad \text{loop line 11 } i = 1 \quad R = 2^\omega R + R_{01} R_0 \bmod n = 0 \\
&\qquad\qquad\qquad\qquad\; i = 0 \quad R = 2^\omega R + R_{00} R_0 \bmod n = 2 \times 2 \bmod 15 = 4 \\
&\qquad \text{line 14} \qquad R_0 = 4
\end{aligned}
$$

Hence the output is 4.

Suppose the attacker would like to find out what is d_0. They estimate the time for $j = 0$ in the outer loop and $i = 0$ in the loop starting from either line 6 or line 17. Then they inject fault into R_{01} at this time. Suppose the faulty R_{01} has a value 01. The intermediate values will be as follows:

$j = 1$ $d_1 = 1$

      loop line 17 $i = 1$ $R = 0$

                $i = 0$ $R = R_{00} R_1 \bmod n = 2 \bmod 15 = 2$

   line 20 $R_0 = 2$ $R_{00} = 10$, $R_{01} = 00$

   loop line 22 $i = 1$ $R = 0$

                $i = 0$ $R = R_{10} R_1 \bmod n = 2 \times 2 \bmod 15 = 4$

   line 25 $R_1 = 4$ $R_{10} = 00$, $R_{11} = 01$

$j = 0$ $d_0 = 0$

   loop line 6 $i = 1$ $R = 2^\omega R + R_{01} R_1 \bmod n = 0$

                $i = 0$ $R = 2^\omega + R_{00} R_1 \bmod n = 8$

   line 9 $R_1 = 8$ $R_{10} = 00$, $R_{11} = 10$

   loop line 11 $i = 1$ $R = 2^\omega R + R_{01} R_0 \bmod n = 1 \times 4 \bmod 15 = 4$

                $i = 0$ $R = 2^\omega R + R_{00} R_0 \bmod n = 2^2 \times 4 + 2 \times 2 \bmod 15 = 5$

   line 14 $R_0 = 5$

---

where the blue R_{01} is used before the fault injection and the green R_{01} carries the faulty value of R_{01}. Thus the final result will be changed, and the attacker can conclude $d_0 = 0$.

5.3.4.2   Safe Error Attack on the Square and Multiply Algorithm

Before detailing the safe error attack on the square and multiply algorithm, we first consider a fault attack on Algorithm 5.9, where Blakely's method (Algorithm 3.11) is used for computing modular multiplication.

Let $a, b \in \mathbb{Z}_n$ be two integers. Since ℓ_n is the bit length of n, the bit length of a is at most ℓ_n. Recall that $\kappa = \lceil \ell_n / \omega \rceil$. We can store a in κ registers, each containing one a_i and (see also Eq. 3.22)

$$a = \sum_{i=0}^{\kappa-1} a_i (2^\omega)^i. \tag{5.25}$$

We assume the attacker has the knowledge of the correct output for a pair of a and b. And they can rerun the algorithm with the same input, inject fault, and observe the output. Suppose $c = 1$, and a fault is injected during the loop starting from line 3 in the register containing a_{i_0} ($0 \leq i_0 \leq \kappa - 1$), when $i < i_0$. In this case, the fault in a_{i_0} will not affect the output since a_{i_0} is used when i is equal to i_0. On the other hand, if $c = 0$ and a fault is injected in the register containing a_{i_0} during the computation, then the final result will be faulty since the faulty value in a will be returned.

Now, if the attacker does not know the value of c and would like to recover it by fault injection attacks, they can assume that $c = 1$, and the loop in line 3 is executed. Then they inject fault in a_{i_0} at the time when i is less than i_0. Finally, they compare

---

Algorithm 5.9: An algorithm involving computing modular multiplication with Blakely's method

---

Input: n, a, b, c // $n \in \mathbb{Z}$, $n \geq 2$ has bit length ℓ_n; $a, b \in \mathbb{Z}_n$; $c = 0, 1$
Output: $ab \bmod n$ if $c = 1$ and a otherwise
1 **if** $c = 1$ **then**
2     $R = 0$
     // $\kappa = \lceil \ell_n / \omega \rceil$, where ω is the computer's word size
3   **for** $i = \kappa - 1, i \; >= 0, i - -$ **do**
4        $R = 2^{\omega} R + a_i b$
5        $R = R \bmod n$
6     $a = R$
7 **return** a

---

the output with the correct one and recovers the value of c—if the output is correct, $c = 1$; otherwise $c = 0$.

The same attack idea can be applied to the square and multiply algorithm to recover the secret key. With the right-to-left square and multiply algorithm and Blakley's method, the signature $s = m^d \bmod n$ is computed with Algorithm 5.10 (see Algorithms 3.13 and 5.7). Since ℓ_n is the bit length of n, the bit lengths of the variables s and t are at most ℓ_n. We can write

$$s = \sum_{j=0}^{\kappa-1} s_j (2^{\omega})^j, \quad t = \sum_{j=0}^{\kappa-1} t_j (2^{\omega})^j.$$

Then, in Algorithm 5.10, lines 5–9 implement $s = s * t \bmod n$ (line 5 of Algorithm 5.7), and lines 10–14 implement $t = t * t \bmod n$ (line 6 of Algorithm 5.7).

Similar to before, we consider fault injections in the variables in Algorithm 5.10 at a certain time.

Suppose $d_i = 1$, and a fault is injected during the ith iteration of the outer loop and at the time when j is less than j_0 during the loop starting from line 6, in the register containing s_{j_0}, where $0 \leq j_0 \leq \kappa - 1$. The fault in s_{j_0} will not affect the output since s_{j_0} is used when j is equal to j_0 and the value in s is replaced by R in line 9.

Suppose $d_i = 0$, and a fault is injected during the ith iteration of the outer loop in the register containing s_{j_0} ($0 \leq j_0 \leq \kappa - 1$); then the value in s will be changed, and the final result will be different.

From these observations, similarly to the attack on Algorithm 5.9, the attacker first assumes $d_i = 1$ and injects fault in s_{j_0} at the time corresponding to $j < j_0$. If the final result is not changed, the attacker can conclude that $d_i = 1$; otherwise, $d_i = 0$. The attacker can then repeat the attack for different values of i to recover the entire private key.

Algorithm 5.10: RSA signature signing computation with the right-to-left square and multiply algorithm and Blakely's method

---

Input: n, m, d// n is the RSA modulus of bit length ℓ_n; m is the hash value of the message; d is the private key of bit length ℓ_d

Output: $m^d \bmod n$

1 $s = 1$
2 $t = m$
3 **for** $i = 0, i < \ell_d, i + + $ **do**
      // ith bit of d is 1
4   **if** $d_i = 1$ **then**
         // lines 5 - 9 implement $s = s * t \bmod n$
5      $R = 0$
         // $\kappa = \lceil \ell_n / \omega \rceil$, where ω is the computer's word size
6   **for** $j = \kappa - 1, j \geq 0, j - - $ **do**
7        $R = 2^\omega R + s_j t$
8        $R = R \bmod n$
9      $s = R$
      // lines 10 - 14 implement $t = t * t \bmod n$
10   $R = 0$
11   **for** $j = \kappa - 1, j \geq 0, j - - $ **do**
12     $R = 2^\omega R + t_j t$
13     $R = R \bmod n$
14   $t = R$

15 **return** s

Similar techniques can also be applied to attack the left-to-right square and multiply algorithm with Blakely's method. We refer the interested reader to [YJ00].

Example 5.3.7 Let us repeat the computations in Example 5.3.5 with Algorithm 5.10. We have

$$p = 3, \quad q = 5, \quad n = 15, \quad d = 3 = 11_2, \quad m = 2,$$

$$\ell_n = 4, \quad \ell_d = 2, \quad \omega = 2, \quad \kappa = 2.$$

With Algorithm 5.10, lines 1 and 2 give

$$s = 1, \; s_0 = 01, \; s_1 = 00. \quad t = 2, \; t_0 = 10, \; t_1 = 00.$$

The intermediate values during the computation are

$$i = 0 \; d_0 = 1$$

$$
\begin{aligned}
&\text{loop line 6} \quad j = 1 \; R = 2^{\omega}R + s_1 t \bmod n = 0 \\
&\qquad\qquad\quad\; j = 0 \; R = 2^{\omega}R + s_0 t \bmod n = 0 + 2 \bmod 15 = 2 \\
&\text{line 9} \qquad\quad\; s = 2 \; s_0 = 10, \; s_1 = 00 \\
&\text{loop line 11} \; j = 1 \; R = 2^{\omega}R + t_1 t \bmod n = 0 \\
&\qquad\qquad\quad\; j = 0 \; R = 2^{\omega}R + t_0 t \bmod n = 0 + 2 \times 2 \bmod 15 = 4 \\
&\text{line 14} \qquad\; t = 4 \; t_0 = 00, \; t_1 = 01
\end{aligned}
$$

$$i = 1 \; d_1 = 1$$

$$
\begin{aligned}
&\text{loop line 6} \quad j = 1 \; R = 2^{\omega}R + s_1 t \bmod n = 0 \\
&\qquad\qquad\quad\; j = 0 \; R = 2^{\omega}R + s_0 t \bmod n = 0 + 2 \times 4 \bmod 15 = 8 \\
&\text{line 9} \qquad\quad\; s = 8
\end{aligned}
$$

Hence the correct output is 8.

Suppose the attacker would like to find out what is d_0. They make the guess that $d_0 = 1$ and inject faults into s_1 when $i = 0$ for the outer loop and $j = 0$ in the loop starting from line 6. We note that s_1 is used (blue s_1 in the above equations) before $j = 0$ and reassigned value in line 9 (orange s_1 in the above equations). Thus the computations are not affected, and the final result is unchanged. The attacker can conclude that $d_0 = 1$.

Example 5.3.8 Let $d = 2 = 10_2$, and keep the rest of the parameters as in Example 5.3.7. Then

$$s = m^d \bmod n = 2^2 \bmod 15 = 4.$$

With Algorithm 5.10, lines 1 and 2 give

$$s = 1, \; s_0 = 01, \; s_1 = 00. \quad t = 2, \; t_0 = 10, \; t_1 = 00.$$

And the intermediate values are

$$i = 0 \; d_0 = 0$$

$$
\begin{aligned}
&\text{loop line 11} \; j = 1 \; R = 2^{\omega}R + t_1 t \bmod n = 0 \\
&\qquad\qquad\quad\; j = 0 \; R = 2^{\omega}R + t_0 t \bmod n = 0 + 2 \times 2 \bmod 15 = 4 \\
&\text{line 14} \qquad\; t = 4 \; t_0 = 00, \; t_1 = 01
\end{aligned}
$$

$$i = 1 \; d_1 = 1$$

$$
\begin{aligned}
&\text{loop line 6} \quad j = 1 \; R = 2^{\omega}R + s_1 t \bmod n = 0 \\
&\qquad\qquad\quad\; j = 0 \; R = 2^{\omega}R + s_0 t \bmod n = 0 + 1 \times 4 \bmod 15 = 4 \\
&\text{line 9} \qquad\quad\; s = 4
\end{aligned}
$$

Hence the correct output is 4.

Now we consider an attacker who would like to recover the value of d_0. They make the guess that $d_0 = 1$, estimates the time for $i = 0$ in the outer loop and $j = 0$ in the loop starting from line 6, and injects faults into s_1 at this point of time. Since $d_0 = 0$, s is not used in the iteration for $i = 0$. We can assume the fault is injected before the start of the next iteration as the computation time for lines 10–14 is similar to that for lines 5–9.

Suppose the faulty s_1 has a value 01. The intermediate values will be as follows:

$i = 0$ $d_0 = 0$
> loop line 11 $j = 1$ $R = 2^\omega R + t_1 t \bmod n = 0$
> $j = 0$ $R = 2^\omega R + t_0 t \bmod n = 0 + 2 \times 2 \bmod 15 = 4$
> line 14 $t = 4$ $t_0 = 00$, $t_1 = 01$

$i = 1$ $d_1 = 1$
> loop line 6 $j = 1$ $R = 2^\omega R + s_1 t \bmod n = 0 + 1 \times 4 \bmod 15 = 4$
> $j = 0$ $R = 2^\omega R + s_0 t \bmod n = 2^2 \times 4 + 1 \times 4 \bmod 15 = 5$
> line 9 $s = 5$

where the green s_1 is the faulty s_1. The final result is changed, and the attacker can conclude $d_0 = 0$.

5.4 Fault Countermeasures for RSA and RSA Signatures

In this section, we will discuss a few countermeasures for the attacks presented in Sect. 5.3. We keep the same notations as before. p and q are two distinct odd primes, and $n = pq$. $d \in \mathbb{Z}^*_{\varphi(n)}$ is the private key for RSA signatures and $e = d^{-1} \bmod \varphi(n)$. n has bit length ℓ_n. d has bit length ℓ_d with the following binary representation (see Theorem 1.1.1):

$$d = \sum_{i=0}^{\ell_d - 1} d_i 2^i.$$

m denotes the hash value of the message. $s = m^d \bmod n$ is the corresponding signature.

CRT-based implementation of RSA signatures computes

$$s_p := m^{d \bmod (p-1)} \bmod p, \quad s_q := m^{d \bmod (q-1)} \bmod q, \tag{5.26}$$

and s is given by Gauss's algorithm

$$s = s_p y_q q + s_q y_p p \bmod n$$

or by Garner's algorithm

$$s = s_p + ((s_q - s_p) y_p \bmod q) p,$$

where

$$y_q = q^{-1} \bmod p, \quad y_p = p^{-1} \bmod q. \tag{5.27}$$

5.4.1 Shamir's Countermeasure

A simple countermeasure proposed by A. Shamir [Sha97] for the Bellcore attack (Sect. 5.3.1) is to use an *extended modulus*. More specifically, let r be a random ℓ_r-bit prime number. Typically $\ell_r = 32$ [KQ07]. Instead of computing s_p and s_q as given in Eq. 5.26, we compute

$$s_p^* = m^{d \bmod (p-1)(r-1)} \bmod pr, \quad s_q^* = m^{d \bmod (q-1)(r-1)} \bmod qr. \tag{5.28}$$

Then we check if

$$s_p^* \equiv s_q^* \bmod r. \tag{5.29}$$

If yes, the signature s is given by

$$s = s_p^* y_q q + s_q^* y_p p \bmod n. \tag{5.30}$$

Firstly, we note that when there is no fault, by Eq. 5.28,

$$s_p^* \equiv m^{d \bmod (p-1)(r-1)} \bmod p.$$

Let

$$a = d \bmod (p-1)(r-1),$$

then we can write

$$d = a + b(p-1)(r-1)$$

for some integer b. We have

$$d \equiv a \equiv (d \bmod (p-1)(r-1)) \bmod (p-1).$$

By Corollary 1.4.3,

$$s_p^* \equiv m^{d \bmod (p-1)} \bmod p. \tag{5.31}$$

Hence,

$$s_p^* \equiv m^{d \bmod (p-1)} \equiv s_p \bmod p.$$

Similarly,

$$s_q^* \equiv m^{d \bmod (q-1)} \equiv s_q \bmod q.$$

Consequently, s given by Eq. 5.30 satisfies

$$s \equiv s_p^* y_q q \equiv s_p^* \equiv s_p \bmod p, \quad s \equiv s_q^* y_p p \equiv s_q^* \equiv s_q \bmod q$$

and is indeed the signature $m^d \bmod n$.

Furthermore, since r is prime, by a similar argument for Eq. 5.31, we have

$$s_p^* \equiv m^{d \bmod (r-1)} \bmod r, \quad s_q^* \equiv m^{d \bmod (r-1)} \bmod r,$$

which gives

$$s_p^* \equiv s_q^* \bmod r.$$

Suppose the Bellcore attack is to be carried out, and a malicious fault is injected during the computation (Eq. 5.28) of s_p^* or s_q^*, but not both. Without loss of generality, let us assume s_p^* is faulty and s_q^* is computed correctly. Let $s_p^{*\prime}$ denote the faulty s_p^*. The fault will be detected if

$$s_p^{*\prime} \not\equiv s_q^* \bmod r,$$

which means the probability of injecting an undetectable fault is the probability of producing $s_p^{*\prime}$ such that

$$s_p^{*\prime} \equiv s_q^* \bmod r. \tag{5.32}$$

If we assume the fault is injected so that the resulting value of $s_p^{*\prime}$ is random and follows a uniform distribution in $\mathbb{Z}_{pr}$, then the probability for $s_p^{*\prime}$ to satisfy Eq. 5.32 is $1/r$. Thus, with Shamir's countermeasure, the Bellcore attack will be successful with probability $1/r$. When the bit length of r is around 32 bits, this probability is about 2^{-32}.

Example 5.4.1 Let us compute the signature from Example 5.3.1 with Shamir's countermeasure. We have

$$p = 5, \quad q = 7, \quad n = 35, \quad d = 5, \quad m = 6.$$

Suppose $r = 3$. By Eq. 5.28,

$$s_p^* = m^{d \bmod (p-1)(r-1)} \bmod pr = 6^{5 \bmod (4 \times 2)} = 6^5 \bmod 15 = 6,$$

$$s_q^* = m^{d \bmod (q-1)(r-1)} \bmod qr = 6^{5 \bmod (6 \times 2)} = 6^5 \bmod 21 = 6.$$

We can check that

$$s_p^* \equiv s_q^* \equiv 0 \bmod 3.$$

We have shown in Example 3.5.8 that $y_q = 3$ and $y_p = 3$. By Eq. 5.30, the signature is given by

$$s = s_p^* y_q q + s_q^* y_p p \bmod n = 6 \times 3 \times 7 + 6 \times 3 \times 5 \bmod 35 = 6,$$

which agrees with the computations in Example 5.3.1.

Suppose an error occurred during the computation of s_p^* and the faulty value $s_p^{*\prime} = 4$. Then we would have

$$s_p^{*\prime} \not\equiv s_q^* \bmod r.$$

However, in case $s_p^{*\prime} = 9$, we have

$$s_p^{*\prime} \equiv s_q^* \equiv 0 \bmod 3,$$

and the faulty signature will be

$$s' = s_p^{*\prime} y_q q + s_q^* y_p p \bmod n = 9 \times 3 \times 7 + 6 \times 3 \times 5 \bmod 35 = 34.$$

In this case, the attacker can repeat the Bellcore attack by computing

$$q = \gcd(s' - s, n) = \gcd(34 - 6, 35) = \gcd(28, 35) = 7.$$

Example 5.4.2 Let us compute the signature from Example 5.3.2, by Shamir's countermeasure. We have

$$p = 11, \quad q = 13, \quad n = 143, \quad d = 11, \quad m = 2.$$

Suppose $r = 5$. By Eq. 5.28,

$$s_p^* = m^{d \bmod (p-1)(r-1)} \bmod pr = 2^{11} \bmod 55 = 13,$$
$$s_q^* = m^{d \bmod (q-1)(r-1)} \bmod qr = 2^{11} \bmod 65 = 33.$$

We can check that

$$s_p^* \equiv s_q^* \equiv 3 \bmod 5.$$

We have shown in Example 5.3.2 that $y_q = 6$ and $y_p = 6$. By Eq. 5.30, the signature is given by

$$s = s_p^* y_q q + s_q^* y_p p \bmod n = 13 \times 6 \times 13 + 33 \times 6 \times 11 \bmod 143 = 46,$$

which agrees with the computations in Example 5.3.2.

Suppose an error occurred during the computation of s_p^* and the faulty value $s_p^{*\prime} = 10$. Then we would have

$$s_p^{*\prime} \not\equiv s_q^* \bmod r.$$

However, in case $s_p^{*\prime} = 3$, we have

$$s_p^{*\prime} \equiv s_q^* \equiv 3 \bmod 5,$$

and the faulty signature will be

$$s' = s_p^{*\prime} y_q q + s_q^* y_p p \bmod n = 10 \times 6 \times 13 + 33 \times 6 \times 11 \bmod 143 = 98.$$

In this case, the attacker can repeat the Bellcore attack by computing

$$q = \gcd(s' - s, n) = \gcd(98 - 46, 143) = \gcd(52, 143).$$

By the Euclidean algorithm,

$$\begin{aligned}
143 &= 52 \times 2 + 39, &\gcd(52, 143) &= \gcd(52, 39), \\
52 &= 39 \times 1 + 13, &\gcd(52, 39) &= \gcd(39, 13), \\
39 &= 13 \times 3, &q = \gcd(39, 13) &= 13.
\end{aligned}$$

5.4.2 *Infective Countermeasure*

Although Shamir's countermeasure can effectively protect RSA signature computations against the Bellcore attack (Sect. 5.3.1), a simple improved attack is to bypass the check of Eq. 5.29 using an instruction skip.

In this subsection, we will discuss a more sophisticated countermeasure against the Bellcore attack, an *infective countermeasure*, proposed by Sung-Ming et al. [SMKLM02]. The main goal of the countermeasure is to make s_p (Eq. 5.26) faulty if s_q is faulty, hence the name "infective." We have discussed an infective countermeasure for AES in Sect. 5.2.2. We remark that the infective countermeasure was first proposed for RSA signatures.

The same as before, let p and q be distinct odd primes. $n = pq$. d is the private key for RSA signatures. $e = d^{-1} \bmod \varphi(n)$. m is the hash value for the message.

Recall that

$$y_q = q^{-1} \bmod p, \quad y_p = p^{-1} \bmod q.$$

We select a random integer r such that $\gcd(d_r, \varphi(n)) = 1$ and e_r is a small integer, where

$$d_r = d - r,$$

and

$$e_r = d_r^{-1} \bmod \varphi(n). \tag{5.33}$$

Let

$$k_p = \left\lfloor \frac{m}{p} \right\rfloor, \quad k_q = \left\lfloor \frac{m}{q} \right\rfloor.$$

The signature s is then computed using Eqs. 5.34–5.39.

$$s_p = m^{d_r} \bmod p, \tag{5.34}$$

$$\hat{m} = ((s_p^{e_r} \bmod p) + k_p p) \bmod q, \tag{5.35}$$

$$s_q = \hat{m}^{d_r} \bmod q, \tag{5.36}$$

$$s_{dr} = s_p y_q q + s_q y_p p \bmod n, \tag{5.37}$$

$$\widetilde{m} = (s_q^{e_r} \bmod q) + k_q q, \tag{5.38}$$

$$s = s_{dr} \widetilde{m}^r \bmod n. \tag{5.39}$$

In Lemma 5.4.1, we will show that the signature s computed above is indeed equal to the signature given by $m^d \bmod n$.

Lemma 5.4.1

$$s \equiv m^d \bmod n. \tag{5.40}$$

Proof By definition of e_r (Eq. 5.33),

$$d_r e_r \equiv 1 \bmod \varphi(n).$$

Since $\varphi(n) = (p - 1)(q - 1)$, we have

$$d_r e_r \equiv 1 \bmod (p - 1).$$

By Corollary 1.4.3,

$$m^{d_r e_r} \bmod p = m \bmod p.$$

Furthermore,

$$\left\lfloor \frac{m}{p} \right\rfloor p = m - (m \bmod p).$$

Hence,

$$(m^{d_r e_r} \bmod p) + \left\lfloor \frac{m}{p} \right\rfloor p = m. \tag{5.41}$$

By Eqs. 5.34 and 5.35, we have

$$\hat{m} = m \bmod q.$$

By Eq. 5.36,

$$s_q \equiv m^{d_r} \bmod q.$$

Together with Eq. 5.34, it follows from Chinese Remainder Theorem (see Theorem 1.4.7 and Example 1.4.19) that

$$s_{dr} \equiv m^{d_r} \bmod n.$$

Following a similar argument that leads to Eq. 5.41, we can show

$$\widetilde{m} = (s_q^{e_r} \bmod q) + k_q q = m^{d_r e_r} \bmod q + \left\lfloor \frac{m}{q} \right\rfloor q = m.$$

Finally, by Eq. 5.39,

$$s = s_{dr} \widetilde{m}^r \bmod n = m^{d_r} m^r \bmod n = m^{d-r+r} = m^d \bmod n.$$

$\square$

Next, we show that the Bellcore attack cannot succeed if s is calculated using Eqs. 5.34–5.39.

Proposition 5.4.1 *Suppose $p < q$. If s_p is faulty, then s_q is also faulty.*

Proof Let s_p' denote the faulty value of s_p, then $s_p \neq s_p'$. By Corollary 1.4.4,

$$s_p^{e_r} \not\equiv s_p'^{e_r} \bmod p.$$

Since $p < q$,

$$(s_p^{er} \bmod p) \bmod q \neq (s_p'^{er} \bmod p) \bmod q.$$

By Eq. 5.35, $\hat{m}$ is faulty. Thus s_q is also faulty by Eq. 5.36. $\square$

Lemma 5.4.2 *Suppose $p > q$. The cardinality of the set*

$$\left\{(a, b) \mid a, b \in \mathbb{Z}_p, \; a \neq b, \; a \equiv b \bmod q\right\}$$

is given by

$$E := 2(p \bmod q) \left\lfloor \frac{p}{q} \right\rfloor + q \left\lfloor \frac{p}{q} \right\rfloor \left(\left\lfloor \frac{p}{q} \right\rfloor - 1 \right).$$

Proof There are

$$(p \bmod q) \left(\left\lfloor \frac{p}{q} \right\rfloor + 1 \right)$$

many $a \in \mathbb{Z}_p$ such that

$$0 \leq a \bmod q \leq (p \bmod q) - 1.$$

In this case, there are $\left\lfloor \frac{p}{q} \right\rfloor$ of $b \in \mathbb{Z}_p$ such that $b \equiv a \bmod q$.
  There are

$$q - (p \bmod q) \left(\left\lfloor \frac{p}{q} \right\rfloor + 1 \right) = (q - p \bmod q) \left\lfloor \frac{p}{q} \right\rfloor$$

many $a \in \mathbb{Z}_p$ such that

$$p \bmod q \leq a \bmod q \leq q - 1.$$

In this case, there are $\left\lfloor \frac{p}{q} \right\rfloor - 1$ of $b \in \mathbb{Z}_p$ such that $b \equiv a \bmod q$. We have

$$
\begin{aligned}
E &= (p \bmod q) \left(\left\lfloor \frac{p}{q} \right\rfloor + 1 \right) \left\lfloor \frac{p}{q} \right\rfloor + (q - p \bmod q) \left\lfloor \frac{p}{q} \right\rfloor \left(\left\lfloor \frac{p}{q} \right\rfloor - 1 \right) \\
&= (p \bmod q) \left\lfloor \frac{p}{q} \right\rfloor \left\lfloor \frac{p}{q} \right\rfloor + (p \bmod q) \left\lfloor \frac{p}{q} \right\rfloor + q \left\lfloor \frac{p}{q} \right\rfloor \left(\left\lfloor \frac{p}{q} \right\rfloor - 1 \right) \\
&\quad - (p \bmod q) \left\lfloor \frac{p}{q} \right\rfloor \left\lfloor \frac{p}{q} \right\rfloor + (p \bmod q) \left\lfloor \frac{p}{q} \right\rfloor \\
&= 2(p \bmod q) \left\lfloor \frac{p}{q} \right\rfloor + q \left\lfloor \frac{p}{q} \right\rfloor \left(\left\lfloor \frac{p}{q} \right\rfloor - 1 \right).
\end{aligned}
$$

$\square$

Example 5.4.3 Let $p = 7, q = 5$. There are

$$(p \bmod q)\left(\left\lfloor \frac{p}{q} \right\rfloor + 1\right) = 2 \times (1 + 1) = 4$$

many $a \in \mathbb{Z}_7$ such that

$$0 \le a \bmod q \le (p \bmod q) - 1, \quad \text{i.e.,} \quad 0 \le a \bmod 5 \le 1.$$

Those values of a are given by $\{0, 1, 5, 6\}$. In this case, there are

$$\left\lfloor \frac{p}{q} \right\rfloor = \left\lfloor \frac{7}{5} \right\rfloor = 1$$

many $b \in \mathbb{Z}_7$ such that $b \equiv a \bmod q$. In particular, all possible values of (a, b) are given by

$$(0, 5), \quad (5, 0), \quad (1, 6), \quad (6, 1).$$

There are

$$(q - p \bmod q)\left\lfloor \frac{p}{q} \right\rfloor = (5 - 2) \times 1 = 3$$

many $a \in \mathbb{Z}_7$ such that

$$p \bmod q \le a \bmod q \le q - 1, \quad \text{i.e.,} \quad 2 \le a \bmod 5 \le 4.$$

The values of a are given by $\{2, 3, 4\}$. In this case, there are

$$\left\lfloor \frac{p}{q} \right\rfloor - 1 = 0.$$

many $b \in \mathbb{Z}_7$ such that $b \equiv a \bmod q$. For example, there is no other number except for 2 in $\mathbb{Z}_7$ that is congruent to 2 mod 7.

Thus the total number of pairs (a, b) is 4. We can check that

$$E = 2(p \bmod q)\left\lfloor \frac{p}{q} \right\rfloor + q\left\lfloor \frac{p}{q} \right\rfloor\left(\left\lfloor \frac{p}{q} \right\rfloor - 1\right) = 2 \times 2 + 5 \times 0 = 4.$$

Proposition 5.4.2 *Suppose $p > q$. If s_p is faulty, the probability for s_q to be also faulty is*

$$1 - \frac{E}{p(p - 1)}.$$

Proof Let s'_p denote the faulty value of s_p, then $s_p \neq s'_p$. By Corollary 1.4.4,

$$s_p^{e_r} \not\equiv s'^{e_r}_p \mod p.$$

There are $p(p-1)$ distinct pairs (s_p, s'_p). By Eqs. 5.35 and 5.36 and Lemma 5.4.2, there are E possible pairs (s_p, s'_p) that produce the same $\hat{m}$, hence the same s_q.

Thus the probability for s_q to be faulty is

$$1 - \frac{E}{p(p-1)}.$$

$\square$

We note that, in practice, p is large, and p and q are of similar bit lengths. Then E will be small compared to $p(p-1)$.

Example 5.4.4 Let $p = 421, q = 419$, then

$$E = 2(p \bmod q) \left\lfloor \frac{p}{q} \right\rfloor + q \left\lfloor \frac{p}{q} \right\rfloor \left(\left\lfloor \frac{p}{q} \right\rfloor - 1 \right) = 2 \times 2 \times 1 + 419 \times 1 \times 0 = 4$$

and

$$1 - \frac{E}{p(p-1)} = 1 - \frac{4}{421 \times 420} = 0.99998.$$

Proposition 5.4.3 *If s_q is faulty and s_p is computed correctly, the attacker cannot compute $q = \gcd(s'^{e_r}_{dr} - m, n)$ without brute force.*

Proof Suppose s_q is faulty, and s_p is computed correctly. Let $s'_{dr}, \tilde{m}', s'_q$, and s' denote the faulty values of $s_{dr}, \tilde{m}, s_q$, and s, respectively.

To carry out the Bellcore attack, the attacker needs to compute

$$q = \gcd(s'^{e_r}_{dr} - m, n).$$

However, the attacker does not have the knowledge of s'_{dr}. Instead, we can assume that the attacker knows s'. To get s'_{dr}, the attacker needs to compute $\tilde{m}'^r$.

We note that there are $q - 1$ possible values for s'_q. By Corollary 1.4.4 and Eq. 5.38, there are $q - 1$ possible values for $\tilde{m}'$. And by Corollary 1.4.6, there are $q - 1$ possible values for $\tilde{m}'^r \bmod n$. Thus the attacker cannot tell which value in $\mathbb{Z}_n$ is $\tilde{m}'^r \bmod n$ even with the knowledge of m and r because of the unknown $\tilde{m}'$. In conclusion, the attacker needs to brute force all possible values for $\tilde{m}'^r \bmod n$ in $\mathbb{Z}_n$. $\square$

In summary, the Bellcore attack assumes one of s_p and s_q is faulty, but not both. For the infective countermeasure, we have shown:

- When $p < q$, if s_p is faulty, s_q will also be faulty.
- When $p > q$, if s_p is faulty, then s_q has a high probability to be faulty.

- If s_q is faulty and s_p is not faulty, the attacker cannot repeat the attack without brute force.

Example 5.4.5 Let $p = 3$, $q = 5$, and $m = 3$. Then, $n = 15$, $\varphi(n) = 8$. As discussed in Example 3.5.5, $y_p = 2$, $y_q = 2$. Suppose $d = 5$.

To compute the signature with the infective countermeasure, choose $r = 2$, and we have

$$d_r = d - r = 5 - 2 = 3. \quad e_r = 3^{-1} \bmod 8 = 3.$$

$$k_p = \left\lfloor \frac{m}{p} \right\rfloor = \left\lfloor \frac{3}{3} \right\rfloor = 1, \quad k_q = \left\lfloor \frac{m}{q} \right\rfloor = \left\lfloor \frac{3}{5} \right\rfloor = 0.$$

And

$$s_p = m^{d_r} \bmod p = 3^3 \bmod 3 = 0,$$

$$\hat{m} = ((s_p^{e_r} \bmod p) + k_p p) \bmod q = 0 + 3 \bmod 5 = 3,$$

$$s_q = \hat{m}^{d_r} \bmod q = 3^3 \bmod 5 = 27 \bmod 5 = 2,$$

$$s_{dr} = s_p y_q q + s_q y_p p \bmod n = 0 + 2 \times 2 \times 3 \bmod 15 = 12,$$

$$\widetilde{m} = (s_q^{e_r} \bmod q) + k_q q = 2^3 \bmod 5 + 0 = 8 \bmod 5 = 3,$$

$$s = s_{dr} \widetilde{m}^r \bmod n = 12 \times 3^2 \bmod 15 = 108 \bmod 15 = 3.$$

We can verify that

$$s = m^d \bmod n = 3^5 \bmod 15 = 243 \bmod 15 = 3.$$

If s_p is faulty and $s_p' = 1$, then

$$\hat{m}' = ((s_p'^{e_r} \bmod p) + k_p p) \bmod q = (1 + 3) \bmod 5 = 4,$$

$$s_q' = \hat{m}'^{d_r} \bmod q = 4^3 \bmod 5 = 64 \bmod 5 = 4.$$

Thus s_q is also faulty, as has been shown in Proposition 5.4.1.

If s_q is faulty with $s_q' = 1$ and s_p is computed correctly, then

$$s_{dr}' = s_p y_q q + s_q' y_p p \bmod n = 0 + 1 \times 2 \times 3 \bmod 15 = 6.$$

We note that

$$p = \gcd(s_{dr}'^{e_r} - m, n) = \gcd(6^3 - 3, 15) = \gcd(213, 15). \tag{5.42}$$

By the Euclidean algorithm,

$$213 = 15 \times 14 + 3, \; \gcd(213, 15) = \gcd(15, 3),$$
$$15 = 3 \times 5, \qquad p = \gcd(15, 3) = 3.$$

However, the attacker does not have the knowledge of $\widetilde{m}'^{r}$ to get the value of s'_{dr} from s'. Thus from their point of view, any value in $\mathbb{Z}_n = \mathbb{Z}_{15}$ might be $\widetilde{m}'^{r}$. And they cannot compute p as in Eq. 5.42.

Example 5.4.6 Let us compute the signature from Example 5.3.2 with the infective countermeasure. We have

$$p = 11, \; q = 13, \; n = 143, \; m = 2, \; \varphi(n) = 120, \; d = 11, \; y_p = 6, \; y_q = 6.$$

Choose $r = 4$, then

$$d_r = d - r = 11 - 4 = 7.$$

By the extended Euclidean algorithm,

$$120 = 7 \times 17 + 1 \Longrightarrow 1 = 120 - 7 \times 17,$$

hence

$$e_r = d_r^{-1} \bmod \varphi(n) = -17 \bmod 120 = 103.$$

We also have

$$k_p = \left\lfloor \frac{m}{p} \right\rfloor = \left\lfloor \frac{2}{11} \right\rfloor = 0, \quad k_q = \left\lfloor \frac{m}{q} \right\rfloor = \left\lfloor \frac{2}{13} \right\rfloor = 0.$$

And

$$s_p = m^{d_r} \bmod p = 2^7 \bmod 11 = 128 \bmod 11 = 7,$$
$$\hat{m} = ((s_p^{e_r} \bmod p) + k_p p) \bmod q$$
$$= (7^{103} \bmod 11 + 0) \bmod 13 = (7^{103 \bmod 10} \bmod 11) \bmod 13$$
$$= (7^3 \bmod 11) \bmod 13 = (343 \bmod 11) \bmod 13 = 2,$$
$$s_q = \hat{m}^{d_r} \bmod q = 2^7 \bmod 13 = 128 \bmod 13 = 11,$$
$$s_{dr} = s_p y_q q + s_q y_p p \bmod n = 7 \times 6 \times 13 + 11 \times 6 \times 11 \bmod 143 = 128,$$
$$\widetilde{m} = (s_q^{e_r} \bmod q) + k_q q = 11^{103} \bmod 13 + 0$$
$$= 11^{103 \bmod 12} \bmod 13 = 11^7 \bmod 13 = 2,$$
$$s = s_{dr} \widetilde{m}^{r} \bmod n = 128 \times 2^4 \bmod 143 = 2048 \bmod 143 = 46.$$

Suppose s_p is faulty and $s'_p = 2$. Then

$$\hat{m}' = ((s_p'^{e_r} \bmod p) + k_p p) \bmod q = (2^{103} \bmod 11 + 0) \bmod 13 = 2^3 \bmod 11 = 8,$$

$$s'_q = \hat{m}'^{d_r} \bmod q = 8^7 \bmod 13 = 5.$$

Thus s'_q is also faulty, as has been shown in Proposition 5.4.1.

Example 5.4.7 Now let us assume $p = 13$, $q = 11$. Let $d = 11$ and $r = 4$ as in Example 5.4.6. We have

$$n = 143, \quad \varphi(n) = 120, \quad y_p = 6, \quad y_q = 6, \quad d_r = 7, \quad e_r = 103, \quad k_p = 0, \quad k_q = 0.$$

Suppose $m = 12$, then

$$s_p = m^{d_r} \bmod p = 12^7 \bmod 13 = 12,$$
$$\hat{m} = ((s_p^{e_r} \bmod p) + k_p p) \bmod q = (12^{103} \bmod 13 + 0)$$
$$\bmod 11 = (12^7 \bmod 13) \bmod 11 = 12 \bmod 11 = 1,$$
$$s_q = \hat{m}^{d_r} \bmod q = 1^7 \bmod 11 = 1,$$
$$s_{dr} = s_p y_q q + s_q y_p p \bmod n = 12 \times 6 \times 11 + 1 \times 6 \times 13 \bmod 143 = 12,$$
$$\widetilde{m} = (s_q^{e_r} \bmod q) + k_q q = 1^{103} \bmod 11 + 0 = 1,$$
$$s = s_{dr} \widetilde{m}^r \bmod n = 12 \times 1 \bmod 143 = 12.$$

We can check that

$$s = m^d \bmod n = 12^{11} \bmod 143 = 12.$$

Suppose s_p is faulty and $s'_p = 2$. Then

$$\hat{m}' = ((s_p'^{e_r} \bmod p) + k_p p) \bmod q = (2^{103} \bmod 13 + 0) \bmod 13 = 2^7 \bmod 13 = 11,$$

$$s'_q = \hat{m}'^{d_r} \bmod q = 11^7 \bmod 13 = 2.$$

Thus s'_q is also faulty.
    By Lemma 5.4.2,

$$E = 2(p \bmod q)\left\lfloor \frac{p}{q} \right\rfloor + q \left\lfloor \frac{p}{q} \right\rfloor \left(\left\lfloor \frac{p}{q} \right\rfloor - 1\right) = 2 \times 2 + 0 = 4.$$

By Proposition 5.4.2, the probability for s_p to be faulty and s_q to be computed correctly is given by

$$\frac{E}{p(p-1)} = \frac{4}{13 \times (13-1)} = \frac{1}{39} \approx 0.0256.$$

5.4.3   Countermeasure for Attacks on the Square and Multiply Algorithm

In this subsection, we discuss a simple countermeasure proposed in [JPY01] for the attacks discussed in Sect. 5.3.2. It follows a similar idea as Shamir's countermeasure (Sect. 5.4.1) for the Bellcore attack.

First, we choose a small random number r. Compute

$$y = m^d \bmod r, \quad z = m^d \bmod nr.$$

If $z \not\equiv y \bmod r$, we conclude that an error has occurred; otherwise, the signature is given by

$$s = z \bmod n.$$

By Lemma 1.1.1 (6),

$$y = m^d \bmod r, \; z = m^d \bmod nr \implies r|(y - m^d), \; nr|(z - m^d) \implies r|(y - z).$$

If there is no error during the computation, we have $z \equiv y \bmod r$.

Furthermore, we note that the probability of an undetected fault is the probability of

$$z' \equiv y' \bmod r, \tag{5.43}$$

where z' and y' denote the values of z and y when fault is present during the computation. If we assume the fault is random, then the probability of achieving Eq. 5.43 can be approximated by the probability that two random numbers are congruent modulo r, which is $1/r$. If r is an integer of bit length 20, the probability is less than 10^{-6}.

Example 5.4.8 Let us consider the computation from Example 5.3.3. We have

$$p = 3, \quad q = 5, \quad n = 15, \quad d = 3 = d_1 d_0 = 11, \quad m = 2.$$

Following the above countermeasure, suppose $r = 3$, and we have

$$y = m^d \bmod r = 2^3 \bmod 3 = 2, \quad z = m^d \bmod nr = 2^3 \bmod 45 = 8.$$

We can check that

$$z \equiv y \equiv 2 \bmod r.$$

And the signature is given by

$$s = z \bmod n = 8 \bmod 15 = 8.$$

Now if there is a bit flip on the least significant bit of d, d_0, resulting in $d' = 2$, then

$$y' = m^{d'} \bmod r = 2^2 \bmod 3 = 1, \quad z' = m^{d'} \bmod nr = 2^2 \bmod 45 = 4.$$

We have

$$y' \not\equiv z' \bmod r.$$

On the other hand, if the bit flip is on d_1 and we get $d' = 1$, then

$$y' = m^{d'} \bmod r = 2^1 \bmod 3 = 2, \quad z' = m^{d'} \bmod nr = 2^1 \bmod 45 = 2.$$

We have

$$y' \equiv z' \equiv 2 \bmod r.$$

And

$$s' = z' \bmod n = 2 \bmod 15 = 2.$$

In this case, the attack described in Sect. 5.3.2 can be repeated. In particular, the attacker computes

$$\frac{s'}{s} = \frac{2}{8} \bmod 15 = 2^{-2} \bmod 15, \quad m^{2i} = 2^2 \bmod 15 \quad \Longrightarrow \quad \frac{s'}{s} \equiv m^{-2^i} \bmod n.$$

By Eq. 5.21, $d_1 = 1$.

5.4.4 Countermeasures Against the Safe Error Attack

We note that a simple countermeasure exists for the safe error attack presented in Sect. 5.3.4.

We first consider protecting the simple algorithm in Algorithm 5.9. Recall that $a, b \in \mathbb{Z}_n$. ℓ_n is the bit length of n, and the bit lengths of a, b are at most ℓ_n.

$$\kappa = \lceil \ell_n / \omega \rceil,$$

where ω is the word size of the computer. We can store a in κ registers, each containing one a_i and

$$a = \sum_{i=0}^{\kappa-1} a_i (2^\omega)^i.$$

Similarly, we can write b as

$$b = \sum_{i=0}^{\kappa-1} b_i (2^\omega)^i,$$

where each b_i is stored in one register. Then we can modify Algorithm 5.9 to Algorithm 5.11. Suppose $c = 1$ and the fault is in b_{i_0} when $i < i_0$, for some i_0 that satisfies $0 \le i_0 \le \kappa - 1$. Since b_{i_0} is used before the fault happens, the final result will not be affected. Suppose $c = 0$, then a fault in b_{i_0} at any time will not change the final output either. If a fault is injected in a, the output will be faulty no matter what value c takes. Thus, Algorithm 5.11 is not vulnerable to the safe error attack discussed in Sect. 5.3.4.

Algorithm 5.11: Modified Algorithm 5.9 to counter the safe error attack

---

Input: n, a, b, c // $n \in \mathbb{Z}$, $n \ge 2$ has bit length ℓ_n; $a, b \in \mathbb{Z}_n$; $c = 0, 1$
Output: $ab \bmod n$ if $c = 1$ and a otherwise
1 **if** $c = 1$ **then**
2     $R = 0$
    // $\kappa = \lceil \ell_n / \omega \rceil$, where ω is the computer's word size
3   **for** $i = \kappa - 1, i >= 0, i - -$ **do**
4         $R = 2^\omega R + b_i a$
5         $R = R \bmod n$
6     $a = R$
7 **return** a

Similarly, we can change line 18 of Algorithm 5.8 to

$$R = 2^\omega R + R_{1i} R_0.$$

We get Algorithm 5.12. In this case, suppose $d_j = 0$, and a fault is injected in the variable R_{0i_0}, during the jth iteration of the outer loop and at the time $i < i_0$ in the loop starting from line 6, where $0 \le i_0 \le \kappa - 1$. The final output will be faulty because the faulty R_{0i_0} will be used in line 12. If a fault is injected in R_{0i_0} in the loop starting from line 11, since the value in the whole variable R_0 is used in line 12, the fault will propagate to the output.

On the other hand, if $d_j = 1$ and a fault is injected in R_{0i_0} during the jth iteration of the outer loop, specifically at the time $i < i_0$ in the loop starting from line 17,

---

Algorithm 5.12: RSA signature computation with Montgomery powering ladder and Blakely's method (Algorithm 5.8), protected against the safe error attack from Sect. 5.3.4

Input: n, m, d// n is the RSA modulus of bit length ℓ_n; m is the hash value of the message; d is the private key of bit length ℓ_d

Output: $m^d \bmod n$

1 $R_0 = 1$
2 $R_1 = m$
3 **for** $j = \ell_d - 1, j \geq 0, j - -$ **do**
4   **if** $d_j = 0$ **then**
        // lines 5 - 9 implement $R_1 = R_0 R_1 \bmod n$
5         $R = 0$
6   **for** $i = \kappa - 1, i \geq 0, i - -$ **do**
            // $\kappa = \lceil \ell_n / \omega \rceil$, where ω is the word size of the computer
7             $R = 2^\omega R + R_{0i} R_1$
8             $R = R \bmod n$
9         $R_1 = R$
        // lines 10 - 14 implement $R_0 = R_0^2 \bmod n$
10        $R = 0$
11   **for** $i = \kappa - 1, i \geq 0, i - -$ **do**
12           $R = 2^\omega R + R_{0i} R_0$
13           $R = R \bmod n$
14        $R_0 = R$
15   **else**
        // lines 16 - 20 implement $R_0 = R_0 R_1 \bmod n$
16        $R = 0$
17   **for** $i = \kappa - 1, i \geq 0, i - -$ **do**
18           $R = 2^\omega R + R_{1i} R_0$
19           $R = R \bmod n$
20        $R_0 = R$
        // lines 21 - 25 implement $R_1 = R_1^2 \bmod n$
21        $R = 0$
22   **for** $i = \kappa - 1, i \geq 0, i - -$ **do**
23           $R = 2^\omega R + R_{1i} R_1$
24           $R = R \bmod n$
25        $R_1 = R$
26 **return** R_0

---

where $0 \leq i_0 \leq \kappa - 1$, the final output will also be faulty because the faulty R_{0i_0} will be used in line 18. If the fault is injected in the jth iteration of the outer loop in R_{0i_0} in the loop starting from line 22, the fault will stay till the next iteration of the outer loop and affect the output.

If a fault is injected in R_{1i_0} for some i_0, by a similar argument, the signature will always be faulty whether $d_j = 0$ or $d_j = 1$.

Thus Algorithm 5.12 is resistant to the safe error attack discussed in Sect. 5.3.4.1.

Algorithm 5.13: RSA signature signing computation with the right-to-left square and multiply algorithm and Blakely's method (Algorithm 5.10), protected against the safe error attack from Sect. 5.3.4.2

Input: n, m, d // n is the RSA modulus of bit length ℓ_n; m is the hash value of the message; d is the private key of bit length ℓ_d

Output: $m^d \bmod n$

1 $s = 1$
2 $t = m$
3 **for** $i = 0, i < \ell_d, i + + $ **do**
       // ith bit of d is 1
4   **if** $d_i = 1$ **then**
          // lines 5 - 9 implement $s = s * t \bmod n$
5        $R = 0$
          // $\kappa = \lceil \ell_n/\omega \rceil$, where ω is the computer's word size
6   **for** $j = \kappa - 1, j \geq 0, j - - $ **do**
7           $R = 2^\omega R + t_j s$
8           $R = R \bmod n$
9        $s = R$
       // lines 10 - 14 implement $t = t * t \bmod n$
10    $R = 0$
11   **for** $j = \kappa - 1, j \geq 0, j - - $ **do**
12       $R = 2^\omega R + t_j t$
13       $R = R \bmod n$
14    $t = R$

15 **return** s

In the same manner, to protect Algorithm 5.10 against the safe error attack, we can just change line 7 to

$$R = 2^\omega R + t_j s.$$

We get Algorithm 5.13. If $d_i = 1$, a fault during the ith iteration in s_{j_0} ($0 \leq j_0 \leq \kappa - 1$) will affect the result since the faulty s_{j_0} will be used in line 7. If $d_i = 0$, s is not used in the ith iteration, but the faulty s_{j_0} will be used in the next iteration and affect the final output. On the other hand, a fault in t will always propagate to the output since the faulty value will be used in line 12. Thus Algorithm 5.13 is resistant to the safe error attack discussed in Sect. 5.3.4.2.

5.5 Further Reading

Differential fault analysis We have seen the diagonal DFA attack on AES in Sect. 5.1.1.2. Tunstall et. al [TMA11] demonstrated that using this attack and by exploiting the relation between K_{10} (the last round key) and K_9 (the second last round key), the key guesses for K_{10} can be further reduced to 2^{12}. Piret and Quisquater [PQ03] discussed another DFA attack on AES that injects fault to the input of MixColumns in round 9. Phan et al. [PY06] proposed to combine cryptanalysis techniques with DFA to recover the secret key of AES.

DFA attacks on PRESENT implementations can be found in, e.g., [BEG13, WW10, BH15]. A generalization of DFA to SPN ciphers is given in [KHN$^+$19].

Persistent fault analysis (PFA) We discussed the PFA attack on AES in Sect. 5.1.3. PFA can also be applied to other block ciphers, e.g., PRESENT [ZZJ$^+$20] and feistel cipher [CB19]. In [ZZY$^+$19], the authors demonstrated a practical fault injection in the AES Sbox lookup table. In 2020, Xu et al. [XZY$^+$20] discussed PFA attacks in earlier rounds of AES and other SPN ciphers. Notably, AI has also been adopted for PFA to recover the key for AES [COZZ23].

Other fault attack methodologies on symmetric block ciphers There are many other fault attack methods. Here we give more information on a few of them.

Ineffective fault analysis (IFA) was first introduced in [Cla07], where the faults that do not change the intermediate values are exploited. Those faults are called *ineffective faults*. Normally a particular fault model is assumed, e.g., a stuck-at-0 fault model. We note that IFA is dependent on the effect a fault has on the corrupted data. In comparison, the safe error attack (Sect. 5.3.4) does not require a specific fault model, an intermediate value is changed, and the knowledge of whether the faulty value is used or not is exploited.

Statistical ineffective fault attack (SIFA) [DEK$^+$18] combines both SFA (Sect. 5.1.2) and IFA. A nonuniform fault model is assumed, and the attack exploits ineffective faults. More precisely, the dependency between the fault induction being ineffective and the data that is processed is exploited. Different from SFA, SIFA does not require each fault to be successful, but the attack requires repeated plaintext and knowledge of the correct ciphertext (or whether each ciphertext is correct or not). The fault injection is the same as described in Sect. 5.1.2.2. After the attacker obtains a set of ciphertexts, they filter out the correct ones. With each hypothesis of 4 bytes of K_{10}, the attacker can compute a hypothesis of the original byte value s_{00}. Then, statistical methods, such as maximum likelihood as discussed in Sect. 5.1.2.1, can be applied to find the correct key hypothesis. In [DEK$^+$18], the authors provide a detailed theoretical analysis of the number of ciphertexts needed and extensive experimental results.

Collision fault analysis [BK06] injects fault in the earlier rounds of a block cipher implementation. Then the attacker records the faulty ciphertext and finds plaintext that produces the same ciphertext, but without fault. Further analysis using those plaintexts can recover the secret key. If the fault only changes 1 bit or 1 byte of the intermediate value, the attacker can try different plaintexts that only differ at 1 bit or 1 byte.

Algebraic fault analysis (AFA) [CJW10] is similar to DFA. It also exploits differences between correct and faulty ciphertexts. But DFA relies on manual analysis, and AFA expresses cryptographic algorithm in the form of algebraic equations and utilizes SAT solver[1] to recover the key.

Fault sensitivity analysis [LSG+10] exploits the sensitivity of a device to faults. The attack analyzes when a faulty output begins to exhibit some detectable characteristics and utilizes the information to recover the secret key. No knowledge of faulty ciphertext is required for the attack.

Fault attacks on RSA and RSA signatures Shamir's countermeasure (Sect. 5.4.1) for the Bellcore attack (Sect. 5.3.1) was broken in 2002 [ABF+03]. Infective countermeasure (Sect. 5.4.2) for the Bellcore attack was broken in 2006 [YKM06]. There are also various other countermeasures, such as BOS algorithm [BOS03] and Vigilant's algorithm [Vig08].

As mentioned in Sect. 5.3.1, the very first fault attack on cryptographic implementations was proposed in [BDL97] for attacking RSA signatures. In this paper, the authors also discussed an attack aiming at the intermediate values of the square and multiply algorithm to recover the private key. More attacks on the square and multiply algorithm are proposed in, e.g., [Bor06], [SH08].

The first attack on RSA modulus n was proposed in [Sei05], where the goal of the attacker is to corrupt RSA signature verification in a way that there is a high probability that the verification will be successful for signatures created by the attacker using their own private key and message. In more detail, the attack requires the faulty n' to be a prime number known to the attacker such that $\gcd(e, n' - 1) = 1$, where e is the public key of RSA. Then the attacker can compute their private key $d' = e^{-1} \bmod n' - 1$ by the extended Euclidean algorithm and sign their chosen message with their private key. The authors proved that there is a high probability to produce a faulty n' with the above property. Another attack on RSA modulus can be seen in [BCMCC06].

As mentioned in Sect. 5.3, even though no systematic methodologies exist for fault attacks on public key ciphers, the general attack concept can be applied to a different cipher based on a similar intractable problem. For example, the attack on the square and multiply algorithm described in Sect. 5.3.2 can be applied to attack discrete logarithm-based ciphers [BDH+97].

Fault countermeasures for symmetric block ciphers In Sect. 5.2, we have seen two countermeasures for symmetric block ciphers.

Detection-based countermeasures In Sect. 5.2.1, we have discussed encoding-based countermeasures, and we have seen a proposal to use anticodes for the implementation. Similar to the reasoning mentioned in Remark 5.2.1, fault attacks based on knowledge of faulty ciphertext and certain bit-flip fault models can be prevented by encoding-based countermeasures, e.g., IFA, SIFA, and AFA.

---

[1] An SAT solver solves Boolean satisfiability problems. It takes a Boolean logic formula and checks if there is a solution satisfying the formula.

There are also other proposals for different code designs. For example, in [KKT04], the authors proposed a special type of code for hardware countermeasures. The code is defined by

$$C = \left\{ (x, w) \mid x \in \mathbb{F}_2^k, w = (Px)^3 \in \mathbb{F}_2^r \right\},$$

where $H = (P|I)$ is a parity-check matrix for a binary $[n, k]$-code, and $r = n - k$. Akdemir et al. [AWKS12] considered *robust codes*. A binary code of length n is said to be R-robust if

$$\max_{e \neq 0} |C \cap C + e| = R,$$

where

$$C + e = \left\{ c + e \mid c \in C, e \in \mathbb{F}_2^n \right\}.$$

In [GGP09], the authors proposed to use a digest value for the cipher state and update it after each operation. The fault can be detected through the digest values. See also [MSY06] for a comparative study on a few detection-based countermeasures for symmetric block ciphers.

Infective countermeasure Infective countermeasure was first introduced for RSA [SMKLM02] (see Sect. 5.4.2). Then it was adopted for symmetric block cipher in 2012 [GST12], where the authors discussed the implementation for both SPN and Feistel ciphers. In 2014, Tupsamudre et al. broke this countermeasure for AES and proposed an improved version [TBM14] for AES implementations.

As we have seen in Sect. 5.2.2, the infective countermeasure returns a ciphertext in a way that if the attacker does not know the correct ciphertext will not be able to tell if the fault injection was successful or not. However, as shown in [DEK+18], for SIFA, even though the attacker does not know whether the ineffective fault occurred in the target AES round or anywhere else, they can precalculate the probability of faulting the target round and analyze the obtained ciphertext utilizing this probability.

Generally applicable fault countermeasures For fault attack countermeasures, except for those introduced in this chapter, there are also many other techniques. Similarly to SCA countermeasures (Sect. 4.6), we can divide them according to the levels of protection.

Protocol level countermeasure involves designing the usage of cryptographic primitives in a way that certain fault attacks are not possible anymore, e.g., rekeying [MSGR10] or tweak-in-plaintext strategy [BBB+18].

Cryptographic primitive level approaches provide some sort of fault protection directly in the cipher design [BLMR19, BBB+21]. The main advantage is to unburden the implementer from the need to apply additional countermeasures. However, at this point, the fault models covered directly in the design are limited.

Implementation-level countermeasures were the focus of this chapter. We have seen that one common technique is the infective countermeasure, which was discussed in Sect. 5.2.2 for symmetric block ciphers and in Sect. 5.4.2 for RSA. Another common implementation-level countermeasure for both symmetric and asymmetric ciphers is to introduce redundancy, for example, by repeating the computation, e.g., deploying the circuit more than once, single-fault attacks can be detected, or parity check-based countermeasure that allows the detection of faults [KKG03, WKKG04] or by using error-detecting/correcting codes, which was discussed in Sect. 5.2.1 for symmetric block ciphers. Code-based countermeasures for public key cryptosystem can be found in, e.g., [GSK06].

Hardware level countermeasure has been studied for a long time in the smart card industry, for example, using light sensors to detect the chip's opening and voltage/temperature sensors to detect fault injections by voltage glitches or temperature variations [HS13]. A glitch detector can be used against voltage/clock glitching [ZDT+14]. A ring oscillator-based sensor can be utilized for all of these, including EM injection [HBB+16]. On the other hand, there are new ways to induce faults proposed all the time. The main focus in academics is more on countermeasures that aim at managing the effect of fault induction.

Chip package level techniques involve using a special package that prevents the attacker from accessing the chip, for example, packaging that is hard to remove without rendering the chip unusable or packaging with random distribution of connection wires that would be cut during the depackaging process. Also, a layered chip with the memory attached on top of the computation unit provides additional security against FA.

Combined attacks Combined attacks were first proposed in the form of differential behavioral analysis (DBA) which combines DPA with a safe error attack [RM07]. The researchers then followed the idea and proposed attacks on masked implementations of AES [CFGR10, RLK11]. PRESENT implementation was targeted by a combined DFA and a SCADPA-like side-channel method in [PBMB17]. Redundancy-based countermeasure on PRESENT and AES was broken in [SJB+18]. A similar direction was taken in [PNP+20] where different types of countermeasures were evaluated and attacked. A "blind side-channel" (a method where the attacker does not need the value of the cipher output) SIFA was proposed in [APZ21]. A "semi-blind" (no knowledge of the cipher input/output, but ability to repeat the encryption with the same input) combined attack on bit permutation-based ciphers with application to AEAD (authenticated encryption with associated data) schemes was proposed in [HBB22].

Combined countermeasures We have seen encoding-based countermeasures for SCA in Sect. 4.5.1.1 and for FA in Sect. 5.2.1. A proposal of finding optimal codes against both attacks can be found in [BH17]. Various combined countermeasures have also been studied. For example, see [SMG16] for a combined hardware countermeasure based on masking and error-detecting code. [BDF+09] designs a logic design-based solution, and [RDMB+18] discusses both hardware and software countermeasures based on multiparty computation [CD+15]. A sensor-based countermeasure utilizing a ring oscillator with a phase-locked loop was proposed in [RBBC18].

Attacks on post-quantum cryptographic implementations The first practical fault attack on lattice-based key encapsulation schemes was proposed in [RRB$^+$19], targeting the usage of nonce. In [RBRC20] the authors propose several types of attacks, including SCA, fault attacks, and combined attacks on lattice-based schemes. A message-recovery attack on the code-based McEliece algorithm was proposed in [CCD$^+$21]. The attack works by changing the syndrome computation from $\mathbb{F}_2$ to $\mathbb{N}$, making it easy to break the security guarantee of the scheme. In [XIU$^+$21] the authors investigate all the NIST PQC Round 3 KEM candidates w.r.t. fault attacks.

Attacks on neural networks Fault attack techniques have been adopted for attacking neural network implementations recently, with a wide variety of attacker's goals. The very first work, published in 2017, proposes misclassification by bit flips [LWLX17]. Several works followed in this direction [HFK$^+$19, RHF19, RHL$^+$21], proposing more efficient and powerful attacks, mostly utilizing the Rowhammer technique (see Sect. 6.2.1). The same goal was shown to be achievable by instruction skips during the activation function execution [BHJ$^+$18, HBJ$^+$21]. Backdoor/Trojan insertion to do targeted misclassification (a powerful method where the attacker can choose the output class of the model) was proposed in [RHF20, CFZK21, BHOS22]. A model extraction by faults was proposed in [RCYF22, BJH$^+$21]. In such an attack, the adversary tries to learn the model parameters (weights and biases), of a proprietary model.

Chapter 6
Practical Aspects of Physical Attacks

As physical attacks focus on implementations running on real-world devices, there are many practical aspects one needs to consider. When developing attacks and countermeasures, we often work with simplified models of those devices. However, once we move from theoretical assumptions to practice, there might appear deviations stemming from process variations, measurement errors, and various noise sources that are not easy to determine. In this chapter, we will detail practical aspects of side-channel and fault attacks[1] that might be useful when doing experimental evaluations. Apart from that, this chapter will also focus on industrial standards that relate to hardware security.

6.1 Side-Channel Attacks

In the first part of this section, we will explain how information leakage is created by the operation of integrated circuits. In the second part, we will detail the main components of a measurement setup—oscilloscopes and probes.

6.1.1 Origins of Leakage

Current microchips are composed of solid-state metal-oxide-semiconductor field-effect transistors (MOSFETs). There are arrays of positive (NMOS) and negative (PMOS) transistors in each chip, which enable processing digital data composed of 0s and 1s. The reason to combine them in a single circuit is to increase the immunity

---

[1] While we use the term "fault attacks" throughout the book, one can also find the term "fault injection attacks" in the literature, which refers to the same.

© The Author(s), under exclusive license to Springer Nature Switzerland AG 2024

X. Hou, J. Breier, *Cryptography and Embedded Systems Security*,

https://doi.org/10.1007/978-3-031-62205-2_6

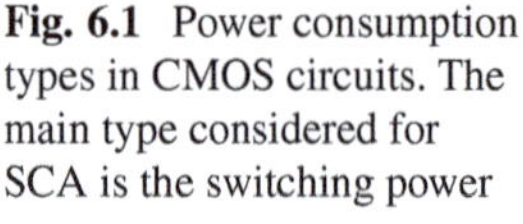

Fig. 6.1 Power consumption types in CMOS circuits. The main type considered for SCA is the switching power

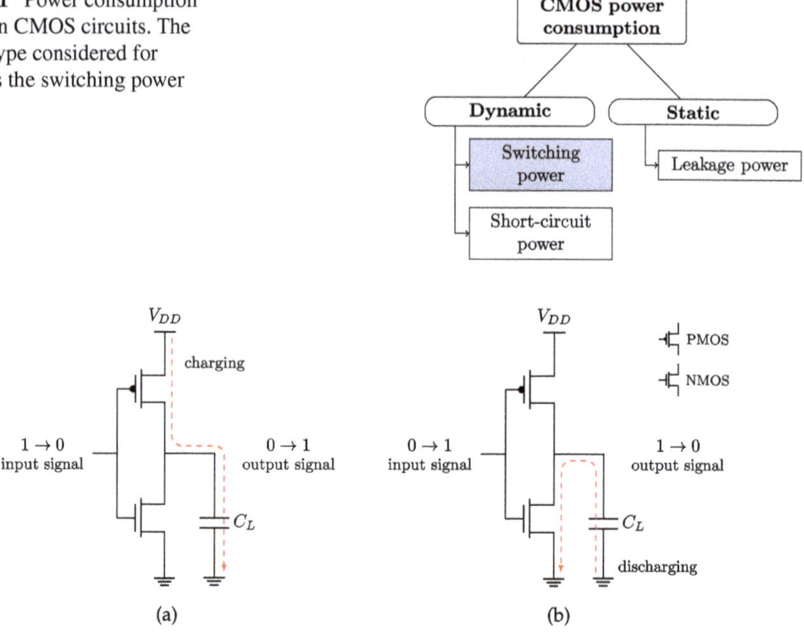

Fig. 6.2 Switching of the CMOS circuit, showing (**a**) the charging path from V_{DD} to C_L and (**b**) the discharging path C_L to GND of the capacitive load

to noise and decrease the static power dissipation, compared to implementing each of these types separately [Cal75]. A circuit consisting of NMOS and PMOS transistors is called a complementary metal oxide semiconductor (CMOS).

The side-channel leakage that comes in the form of electromagnetic leakage of power consumption originates from the physical characteristics of data processing by CMOS-based circuits. Based on these characteristics, leakage models were developed to recover the processed information (see Sect. 4.2.1, part Leakage Models). There are two types of power dissipation in CMOS gates: *static* and *dynamic* (see Fig. 6.1). Static power is consumed even if there is no circuit activity. It is primarily caused by leakage currents that flow when the transistor is in the off state. While this type of power dissipation is not as investigated in the world of SCA as the dynamic one, some works focus on exploiting it [MMR19]. Dynamic power dissipation comes in two forms: short-circuit currents (a short time during the switching of a gate when PMOS and NMOS are conducting simultaneously) and switching power consumption (charge and discharge of the load capacitance).

When considering side channels, the switching power is the most relevant as it directly correlates the processed data with the observable changes in power consumption [Sta10]. This behavior of the CMOS circuit is depicted in Fig. 6.2. Generally, the energy delivery to a CMOS is split into two parts—the charging and the discharging of the load capacitance C_L. During the charging phase, the input gate signal makes a $1 \to 0$ switch, resulting in switching the PMOS transistor on and its NMOS counterpart off. As shown in Fig. 6.2a, in this scenario, the load capacitance C_L is connected to the supply voltage (V_{DD}) via the PMOS transistor, thus allowing the current $I(t)$ to charge C_L. There are two important equations defining the transition from the energy point of view [JAB$^+$03]:

$$E_d = C_L V_{DD}^2,$$

$$E_c = \int_0^\infty I(t)V(t)dt = \frac{1}{2}C_L V_{DD}^2,$$

where E_d is the delivered energy and E_c is the energy stored in the C_L. From these equations, it can be seen that only half of the delivered energy is stored in the capacitor—the PMOS transistor dissipates the other half. Therefore, this power loss during the logic transition can be measured and correlated with the switching activity, resulting in SCA leakage. When the gate signal changes from 0 to 1, the opposite scenario happens—the PMOS transistor is switched off, and NMOS is switched on. The energy E_d stored in C_L is drained to the ground via the NMOS transistor, as can be seen in Fig. 6.2b, thus causing SCA leakage. For more details regarding the power consumption of CMOS circuits, we refer the interested reader to [NYGD22], for example.

It is important to note that there is usually more than one switch during one clock cycle. This is because the input signals to the (multi-input) gate normally do not arrive at the same time, resulting in several switches before the correct output is generated. The output transitions before the stable state are called *glitches*. They are unnecessary for the correct functioning of the circuit and consume a non-negligible amount of dynamic power, ranging between 20% and 70% [SM12]. Glitches are the reason why Boolean masking in hardware, although theoretically secure, can be broken [MPG05]. An approach called *threshold implementation* [NRS11] solves the problem of secure Boolean masking by utilizing multiparty computation and secret sharing.

6.1.2 Measurement Setup

The core of the measurement setup for SCA is the oscilloscope. It can either be connected to the power supply of the DUT for power measurements or can measure electromagnetic (EM) signals through an EM probe.

6.1.2.1 Oscilloscopes

The measurement is normally done with a digital sampling oscilloscope—a device that takes samples of the measured voltage signal over time. The core of such an instrument is an analog-to-digital (ADC) converter, which takes the analog value of

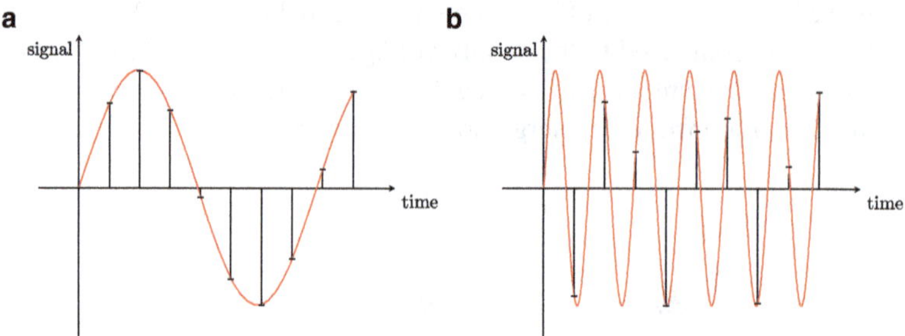

Fig. 6.3 Digital sampling of a continuous signal with ten samples of (**a**) low-frequency signal and (**b**) high-frequency signal

the measured signal (voltage, in our case) at the specified sampling rate and changes it into a digital value. The precision of this value is generally between 8 and 12 bits for midrange oscilloscopes, and the sampling rate ranges from hundreds of mega samples per second (MS/s) to several giga samples per second (GS/s).

When measuring analog signals, such as voltage, with digital devices, it is important to note that we are measuring a continuous value with equipment that *samples* such a value at periodic intervals (which is why we call it a time sample) and stores it in a binary format with limited precision. Therefore, *discretization* is applied twice—first in the time domain and then to the value itself. According to the Nyquist–Shannon sampling theorem [Vai01], the sampling rate of the measurement device should be at least twice the highest frequency component of the measured signal. It is a good rule of thumb to have the oscilloscope sampling rate at least $4\times$ the target device frequency when doing measurements for power analysis attacks. Figure 6.3 shows this phenomenon. The red curve denotes the original analog signal, while the black lines show a sampling of this signal with ten samples over the given time interval. While we can easily reconstruct the original signal if the frequency is low (Fig. 6.3a), it becomes much harder with a high-frequency signal (Fig. 6.3b). The precision of the oscilloscope specifies how many values can the sampled output value take, e.g., an 8-bit ADC would give a range of 256 values, which is sufficient for an SCA attack in most cases.

Another important parameter of the oscilloscope is the analog bandwidth. It is defined as the frequency at which the amplitude measured by the oscilloscope has reduced by 3 dB. To avoid the unnecessary modification of the measured signal, the bandwidth should be at least $3\times$ the target device frequency.

An important task during the acquisition is capturing the correct time window corresponding to the operations we want to measure. In laboratory conditions, it is common to use an artificial trigger signal that indicates the start/end of the encryption. In real-world settings, it is necessary to identify the correct position by examining the captured signal—this is usually done based on the evaluator's expertise.

6.1.2.2 Probes

Near-field electric and magnetic probes are an essential part of the setup when doing electromagnetic side-channel analysis. They can be connected to the oscilloscope in a passive way or with an amplifier. Optionally, a bandpass filter can be used to only pass the relevant frequencies and discard the rest. Several established companies, such as Riscure and Langer, provide probes suitable for EM SCA. Due to the simplicity of the probe design, researchers have also been building their own probes since the early days of SCA [GMO01]. Generally, a coiled copper wire is sufficient, with a coil diameter of at most a few hundred microns. More details on designing near-field probes can be found, for example, in [Siv17].

6.2 Fault Attacks

An interesting aspect of fault attacks is that, unlike with side-channel attacks, the adversary can break the cryptographic security even without the knowledge of the underlying algorithm, for example, by skipping the entire encryption routine by injecting faults in the conditional branches [SWM18].

In this section, we will look into the practical aspects of fault attacks (FAs), such as sample preparation, fault injection techniques and devices, and mechanisms to trigger faults in integrated circuits.

6.2.1 Fault Injection Techniques

In this subsection, we will outline the most popular techniques for FA testing of integrated circuits [BH22].

6.2.1.1 Clock/Voltage Glitching

Clock and voltage glitching techniques are the most accessible in terms of cost as they do not need sophisticated equipment. Initially, they were only performed locally with a device at hand, but with power management techniques such as dynamic voltage and frequency scaling (DVFS), they can also be performed remotely on chips that utilize that technology.

In the case of a voltage glitch, the faults are caused by precise high variations in power supply or by underpowering the device. Power supply variations, or spikes, modify the state of latches of flip-flops, influencing the control and data path logic of the circuit [KJP14]. For example, if the voltage spike happens during memory reading, wrong data may be retrieved. It was also shown that a different shape of the glitch waveform affects the success of the attack [BFP19]. Underpowering, on the

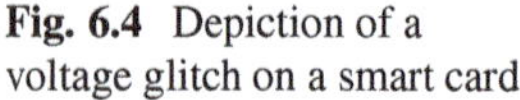

Fig. 6.4 Depiction of a voltage glitch on a smart card

other hand, affects the algorithm continuously and might cause faults throughout the computation. Single faults are possible when the insufficient power supply causes small enough stress so that dysfunctions do not occur immediately after the computation starts and multiple faults do not happen [SGD08]. When the attacker can physically access the target device, voltage glitching is generally easy to implement. It is also the most inexpensive fault injection method as the necessary equipment is wires for connecting to the device and a power source. A local voltage glitch on a smart card is depicted in Fig. 6.4. Voltage glitching attacks were shown to be effective even against security enclaves of Intel [CVM$^+$21] and AMD [BJKS21]. An inexpensive Teensy 4.0 board ($\approx$30 USD) was used for the abovementioned attacks, making them highly practical in terms of equipment cost.

Clock glitch is another technique that can be performed with low-cost equipment. For digital computing devices, it is necessary to synchronize the calculations with either an internal or an external clock. If the clock signal changes, the resulting computation might have a wrong instruction executed or data corrupted. Devices that require an external clock generator can be faulted by supplying a bad clock signal— containing fewer pulses than the normal one [KSV13]. On the other hand, devices that are configured to use an internal clock signal cannot be easily faulted. Clock glitches are generally considered the simplest fault injection method as the attack devices are easy to operate with. For example, clock glitches can be achieved by using low-end field-programmable gate array (FPGA) boards [BGV11, ESH$^+$11].

A relatively new direction in clock/voltage glitching is remote attacks that take advantage of power management systems of modern processors. The security aspects of these systems are rarely considered due to the complexity of devices from the hardware point of view and software executed, cost, and time-to-market constraints [PS19]. CLKSCREW is the first attack in this direction, targeting frequency and voltage manipulation of the Nexus 6 phone, forcing the processor to operate beyond recommended limits [TSS17]. The researchers experimentally injected a one-byte random fault. CLKSCREW can be achieved just by utilizing software control of energy management hardware regulators in the target devices. Similar attacks were also proposed for ARM-based Krait processor [QWLQ19] and Intel SGX [QWL$^+$20]. The main advantage of these attacks is that they are software-based, therefore allowing the threat model to shift from local to remote.

6.2.1.2 Optical Fault Injection

The ionization effect on transistors is a well-known phenomenon, and nowadays it is common to perform failure tolerance testing of integrated circuits. For example, testing robustness and reliability with lasers dates more than half a century back [Hab65]. While there might not be any unexpected effects in standard conditions, there are environments where ionization effects are common and cause unintentional faults, such as the Earth's orbit where satellites are deployed and conditioned to cosmic rays [BSH75]. The first usage of optical fault injection against cryptographic circuits dates back to 2002 when researchers used a flash gun and a laser pointer to set and reset bits in an SRAM [SA02].

The variety of techniques within optical fault injection is vast—from camera flashes to lasers to X-ray beams. It was also shown that with the usage of lasers, one can probe the memory without changing it to check its content [CCT+18], shifting the use case to the realm of side channels.

For security evaluation and certification labs, the method of choice is normally a laser fault injection (LFI), depicted in Fig. 6.5a. Off-the-shelf setups for performing LFI are readily available from companies selling testing equipment. A standard LFI setup consists of the following parts: a laser source, an objective lens, a motorized positioning table, and a controlling device. One can also utilize an infrared ring that allows taking images of the chip from the backside, making the silicon substrate transparent to the camera lens. An example of such an image is in Fig. 6.5b. It is also common to include a digital oscilloscope to precisely check the timing of the laser activation with respect to the cipher execution. While the cost of off-the-shelf LFI testing equipment starts around 100k USD, it was shown that a low-cost setup can be built for around 500 USD [KM20]. Naturally, specialized expertise is required to design and assemble such a setup.

Optical fault injection requires direct access to the chip—from either the front or the backside. That means, in most cases, it is necessary to remove the chip package, by using either mechanical or chemical techniques. There is also an option to use a

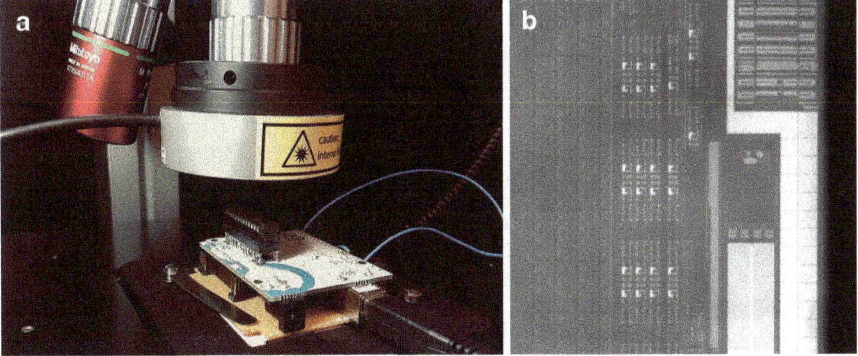

Fig. 6.5 Depiction of (**a**) laser fault injection on an AVR microcontroller mounted on Arduino UNO board and (**b**) zoomed infrared image of the chip

focused ion beam (FIB), but these techniques are generally outside of the budget of a standard testing laboratory, so in this part, we will focus on the two abovementioned methods.

Mechanical techniques are relatively straightforward. They are mostly used for backside decapsulation as the front side of the chip is too sensitive to any physical tampering. They can, for example, involve using inexpensive manual rotary milling machines that grind down the epoxy package. This is recommended mostly for low-cost chips as there is a high risk of overheating or mechanically damaging the die. Another way is to use specialized tools for decapsulation, thinning, and polishing (e.g., Ultra Tec ASAP-1). These tools work in an automated way by slowly milling down the package layers to avoid any damage. Naturally, the main drawback is the cost which typically ranges in tens of thousands of dollars.

Chemical techniques are recommended when the front side of the chip needs to be accessed. In some cases, such as smart cards, acetone is enough to remove the protective plastic (after the outer hard plastic case is removed, e.g., by using a scalpel). When removing the black epoxy package, one might need to use strong acids, such as fuming nitric acid (HNO_3 with a concentration of at least 86%). This typically involves operation in a safe laboratory environment equipped with a fume hood and a proper acid disposal facility. A depiction of such a setup is shown in Fig. 6.6. When using such aggressive acids, there is also a risk of removing the bonding wires using this technique, unless they are either golden or at least gold-plated. More details on decapsulation techniques can be found in [BC16].

When using optical fault injection techniques, it is important to know the absorption depth in silicon as a function of wavelength. This is depicted in Fig. 6.7. The green laser (532 nm) has an absorption depth of $\approx 1.3\ \mu$m; therefore, it can be utilized either for front-side injection (where it can directly access the components)

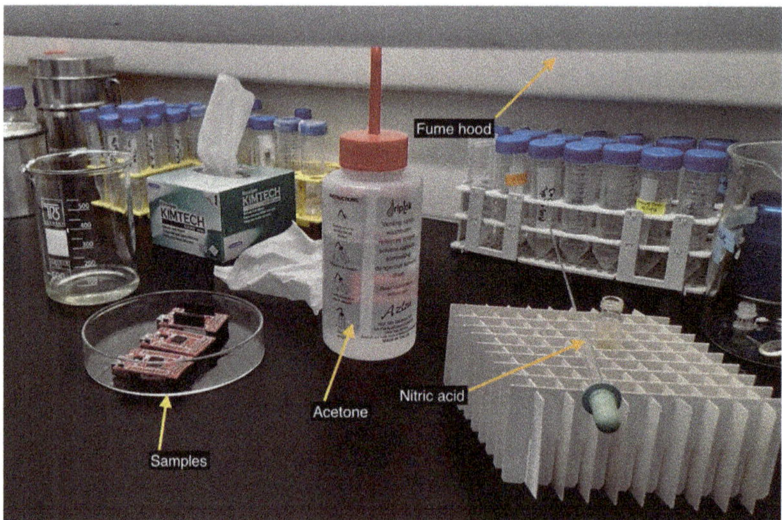

Fig. 6.6 Depiction of a chemical decapsulation by using fuming nitric acid

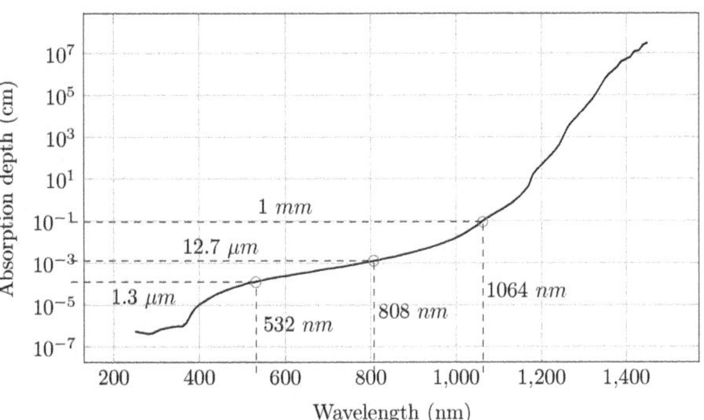

Fig. 6.7 Absorption depth in silicon. The most common laser wavelengths for testing integrated circuits are highlighted—532 nm (green), 808 nm (near-infrared), and 1064 (near-infrared)

or for almost fully removed silicon substrate from the backside. As the latter might damage the chip, it makes sense to use lasers with deeper absorption depth, such as 808 nm or 1064 nm, both from the near-infrared light spectrum. The 1064 nm laser allows a penetration depth up to 1 mm which can often be used even for non-thinned substrate.

There are other fault injection techniques that are related to optical techniques in the way they work. In the area of failure analysis, electron and ion beams have been successfully used to test the reliability of circuits [SA93]. X-ray beams were used to tamper with memories of a microcontroller [ABC+17].

All in all, optical fault injection offers precision and repeatability at a relatively high cost (considering commercial off-the-shelf setups). With specific expertise, it is possible to construct a DIY setup for a much lower price. The main drawback of this technique is the necessity to "see" the chip, which normally requires depackaging and delayering of the chip, making it often impractical outside of laboratory environments. As it is a powerful technique, it is a *de facto* standard for security testing and certification labs which need to consider strong attacker models.

6.2.1.3 Electromagnetic Fault Injection

The electromagnetic fault injection (EMFI) technique is a versatile way to attack chips, allowing targeting both analog and digital blocks. The working principle of EMFI is to generate a changing magnetic field that induces a voltage into the structures of IC surface [DLM20]. In cryptographic circuits, digital logic is used for the algorithm itself, while analog logic controls the clock and random number generators. Below, we discuss the EMFI approaches that can be used to target each of those.

Analog blocks can be targeted by powerful harmonic EM waves. A stable sinusoidal signal can be generated by the attacker at a given frequency that injects a harmonic wave creating a parasitic signal [HHS+11]. This signal can be used to bias the clock behavior or to inject additional power directly and locally into the

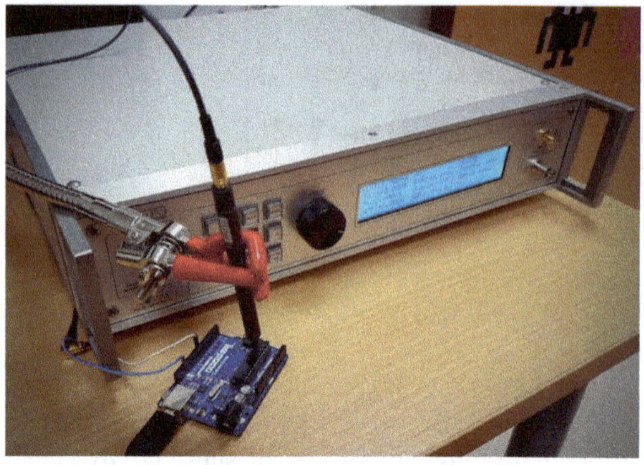

Fig. 6.8 Depiction of a pulsed electromagnetic fault injection on an AVR microcontroller mounted on Arduino UNO board

chip. Harmonic EMFI equipment includes normally a motorized positioning table, a signal generation module, and an oscilloscope.

As digital blocks are clocked, the method of choice is an EM pulse injection during a specified clock cycle [SH07]. When a sharp and sudden EM pulse is injected into the integrated circuit, it can create intense transient currents that change the behavior of logic cells, ultimately causing faults. Standard equipment includes a high-voltage pulse generator and a coil with a ferrite core, serving as an injection probe. Such equipment is depicted in Fig. 6.8.

As the fault analysis methods targeting cryptographic implementation mostly use data faults (bit flips, bit sets/resets, random byte faults, etc.), most of the research is dedicated to pulse fault injection. It is possible to build low-cost EMFI equipment for as low as 50 USD [O'F23]. A more comprehensive ready-to-use device can be bought, for example, from NewAE for $\approx$ 3.3k USD (ChipSHOUTER[2]). While an injector device itself is not enough for a proper testing setup, in [KBJ+22] the authors show how to incorporate ChipSHOUTER in a testbench including XYZ stage and a controller for $\approx$ 7k EUR.[3] If one needs more powerful and precise equipment, Avtech pulse generators can be purchased in a price range between 10k and 20k USD.[4] In that case, a near-field injection probe is needed, which can either be bought for a few hundred USD or manufactured from very inexpensive components. Many resources can be found in the literature on designing and building custom EMFI probes [ORJ+13, Sau13, BKH+19]. The basic building blocks are a ferrite core, a copper wire, and a connector. A generic design of an EMFI probe is depicted in Fig. 6.9.

---

[2] https://www.newae.com/chipshouter

[3] We use currencies stated in original papers, which is why some prices are in USD and some in EUR.

[4] https://www.avtechpulse.com/medium/

Fig. 6.9 A depiction of a generic design of an electromagnetic fault injection probe

Recently, several interesting low-cost custom-built setups were proposed in the literature, capable of performing various attack models:

- *Defeating secure boot* on a multicore 1GHz+ ARM was shown to be practically feasible with just a 350 USD EMFI platform named BADFET [CH17].
- *Bypassing firmware security protection* in various configurations was done by a device called SiliconToaster, a USB-powered EM injector capable of generating 1.2kV of voltage [AH20].
- *Privilege escalation* using a malicious field-replaceable unit (FRU) with a modified mosquito killer spark gap generator was described in [DO22].

From the above, it is evident that EMFI is a popular fault injection technique that is easily accessible due to low cost but at the same time offers a localized and powerful way to defeat secure components on modern chips. Compared to optical fault injection, it does not need direct visibility over the chip and, therefore, leaves out the necessity of cumbersome decapsulation. Moreover, enthusiasts can find many publicly available instructions on how to build a working EMFI setup from easily available off-the-shelf components.

6.2.1.4 Rowhammer Attacks

Rowhammer is a remote fault injection technique that exploits the physical characteristics of DRAM (dynamic random access memory) technology. This attack works by aggressive reading/writing to memory cells adjacent to the target cell, where it causes bit flips [KDK+14]. The attack is made possible by advancing technology which allows shrinking the cells and placing them closer to each other. A smaller cell uses less capacity for charge and therefore provides less tolerance to noise and greater vulnerability to errors [MDB+02]. High cell density further extends this vulnerability by creating electromagnetic coupling effects between them, producing unwanted interactions [KKY+89]. The Rowhammer access patterns are depicted in Fig. 6.10. The *aggressor* row refers to a row that is being hammered by the attacker to flip the bits in the *victim* row. According to [JVDVF+22], three common patterns were shown effective in flipping bits. The *single-sided* pattern uses one aggressor row next to the victim row and the other one far apart. The *double-sided* pattern tightly surrounds the victim row with aggressor rows, increasing the chance of bit flips. Finally, there is an *n-sided* pattern where n refers to $n - 1$ victim rows being hammered by n aggressor rows. The figure shows an example for $n = 4$.

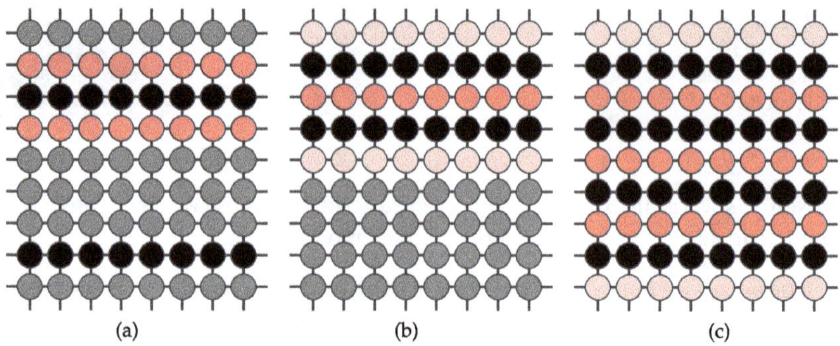

Fig. 6.10 Different ways of spatial arrangement of aggressor rows (black) and target/victim rows (red/pink) in DRAM. (**a**) Single-sided. (**b**) Double-sided. (**c**) 4-sided

As DRAM is the most prevalent technology for nonvolatile memories in modern devices, it is no surprise that the Rowhammer attack was demonstrated on a plethora of targets, ranging from smartphones [VDVFL+16] to cloud environment [ORBG17] to browsers [BRBG16]. Aside from obvious targets such as privilege escalation or cryptanalytic fault attacks, Rowhammer also became popular in the area of hardware attacks on neural networks [TIA+23, YRF20].

There is no need for specialized equipment to perform this attack—it is normally triggered through a software program. While the standard modus operandi is a code execution on the target machine, it was shown that Rowhammer can be realized by sending network packets to the target machine over RDMA-enabled networks [TKA+18].

6.3   Industry Standards

Hardware vulnerability assessment of cryptographic implementations has made its way to industrial standardization. Some products, such as credit cards, need to be evaluated and certified to show they are sufficiently resistant against SCA and FIA. There are two main evaluation frameworks used in the industry: Common Criteria and NIST FIPS 140. We will outline each of them below.

Note A good overview of cybersecurity standards in various industries is maintained by the European Cyber Security Organisation (ECSO) in their *Overview of existing Cybersecurity standards and certification schemes* report [Org17].

A more detailed review of side-channel evaluation standards and methods is given in [ABB+20].

6.3.1 Common Criteria

The Common Criteria for Information Technology Security Evaluation (colloquially known as Common Criteria or CC) is an international standard published in ISO/IEC 15408 [2709] document. It is a general security evaluation framework where users specify their security *functional* and *assurance* requirements in a document called Security Target (ST). ST is defined as an "implementation-dependent statement of security needs for a specific identified Target of Evaluation (TOE)," where TOE is the product that is being certified. ST can conform to one or more Protection Profiles (PPs)—generic documents written by a user or a community for a family of products, such as smart cards, tokens, or firewalls.

The level of security of the evaluated TOE is divided into seven categories—Evaluation Assurance Levels (EALs). While EAL 1 mostly focuses on functional testing with minimum emphasis on security, EAL 7 requires formally verified design and tests. Higher EALs are typically used for military-grade products and require lengthy and expensive evaluation. The CC website[5] lists the accredited labs capable of certifying to a certain EAL and also provides the list of certified products.

The generic steps to be taken before the evaluation can start are as follows:

1. Choosing the National Scheme. CC Certificate Authorizing Schemes were established by 17 countries. Each of them developed their own legislation and norms.
2. Choosing the Target of Evaluation. The TOE and its boundary need to be defined. The TOE can be a part of an IT product, an IT product, a set of an IT product, a technology, or a combination of those.
3. Picking an Evaluation Assurance Level. The evaluation requirements will be based on the EAL. Also, the CC Test Laboratory needs to be certified to evaluate TOEs with the chosen EAL.
4. Choosing the Protection Profile (optional). A suitable PP serves as a guiding document, ensuring that the security features of the TOE align well with the requirements tailored to the category of products to which TOE belongs.
5. Preparing the Security Target. The ST is an implementation-dependent declaration of security needs for the given TOE.
6. Preparing the Evaluation Work Plan. This plan is prepared by the CC Test Laboratory and approved by the Certification Body.

When it comes to SCA, the main area of interest within CC is smart cards. In this context, two documents are used as guidelines for the evaluation, both of them produced through the International Security Certification Initiative (ISCI) and the Joint Interpretation Library (JIL) Hardware Attacks Subgroup (JHAS):

* *Application of Attack Potential to Smart Cards [SI20a]:* The document specifies on how to express the effort required by the attacker to mount a successful attack.

---

[5] https://www.commoncriteriaportal.org

It is related to risk analysis methods and considers the following rating factors: elapsed time, expertise, knowledge of TOE, access to TOE, used equipment, and open samples.

- *Attack Methods for Smart Cards and Similar Devices [SI20b]:* This is a companion document, under limited distribution. It describes the attacks themselves.

The rating method from the listed documents is also adopted in the security evaluation specified by EMVCo, an organization managed by the major payment security players (American Express, Discover, JCB, MasterCard, UnionPay, and Visa). Their aim is to maintain the standardized security level of contact and contactless payment system by managing and evolving the security requirements and related testing processes.

6.3.2   FIPS 140-3

The Federal Information Processing Standard (FIPS) 140-3, Security Requirements for Cryptographic Modules [NIS19] is a document released by the US National Institute of Standards and Technology (NIST). It is applicable to Federal agencies that use cryptographic-based systems. Unlike CC, which specifies an evaluation method and is independent of the underlying algorithms, FIPS 140-3 lists the approved algorithms allowed for usage. The standard specifies six security levels related to physical security, with level 1 stating requirements for protective coating and level 6 requiring countermeasures against differential power/electromagnetic analysis. The product certification is done through the Cryptographic Module Validation Program (CMVP), which is a joint effort between the NIST and the Canadian Centre for Cyber Security.

The FIPS-140 links side-channel evaluation test metrics to the ISO/IEC 17825:2016 standard (with the new version coming in 2024) and the tools and methods to the ISO/IEC 20085-1 and 20085-2 standards.

Appendix A
Proofs

A.1 Matrices

Let R be a commutative ring in this section.

In Definition 1.3.4, we have defined the determinant of a matrix A with coefficients from a commutative ring R. Here we show that the value of $\det(A)$ in Eq. 1.6 does not depend on the choice of i_0 .

Lemma A.1.1 *For any* $0 < i_0 \leq n - 1$

$$\det(A) = \sum_{j=0}^{n-1}(-1)^{i_0+j}a_{i_0 j}\det(A_{i_0 j}) = \sum_{j=0}^{n-1}(-1)^{j}a_{0j}\det(A_{0j}).$$

Proof We prove by induction. For $n = 1$, it is trivially true. For $n = 2$, we can write $A = \begin{pmatrix} a_{00} & a_{01} \\ a_{10} & a_{11} \end{pmatrix}$. Take $i_0 = 0$,

$$\sum_{j=0}^{n-1}(-1)^{i_0+j}a_{i_0 j}\det(A_{i_0 j}) = \sum_{j=0}^{1}(-1)^{i}a_{0j}\det(A_{0j}) = a_{00}a_{11} \quad a_{01}a_{10}.$$

Take $i_0 = 1$,

$$\sum_{j=0}^{n-1}(-1)^{i_0+j}a_{i_0 j}\det(A_{i_0 j}) = \sum_{j=0}^{1}(-1)^{1+j}a_{1j}\det(A_{1j}) = -a_{10}a_{01} + a_{11}a_{00}.$$

Since R is a commutative ring, $a_{00}a_{11} - a_{01}a_{10} = -a_{10}a_{01} + a_{11}a_{00}$, the lemma is true for $n = 2$.

© The Author(s), under exclusive license to Springer Nature Switzerland AG 2024
X. Hou, J. Breier, *Cryptography and Embedded Systems Security*,
https://doi.org/10.1007/978-3-031-62205-2

Now suppose the lemma is true for $n = k$, where $k \geq 2$. In particular, for any $0 \leq j < k$ and $i_0 \neq 0$, we have

$$\det(A_{0j}) = \sum_{\ell=0}^{k-1}(-1)^\ell a_{0j}\det(A_{00,j\ell}) = \sum_{\ell=0}^{k-1}(-1)^{i_0+\ell}a_{i_0 j}\det(A_{0i_0,j\ell}), \qquad (A.1)$$

where $A_{0i,j\ell}$ is obtained from A_{0j} by deleting the ith row and ℓth column. We will show that the lemma is true for $n = k + 1$. Take $i_0 = 0$, we have

$$\sum_{j=0}^{n-1}(-1)^{i_0+j}a_{i_0 j}\det(A_{i_0 j}) = \sum_{j=0}^{k}(-1)^j a_{0j}\det(A_{0j}).$$

Take $i_0 \neq 0$, we have

$$\sum_{j=0}^{n-1}(-1)^{i_0+j}a_{i_0 j}\det(A_{i_0 j}) = \sum_{j=0}^{k}(-1)^{i_0+j}a_{i_0 j}\left(\sum_{\ell=0}^{k-1}(-1)^\ell a_{0\ell}\det(A_{i_0 0,j\ell})\right)$$

$$= \sum_{j=0}^{k}\sum_{\ell=0}^{k-1}(-1)^{i_0+j}a_{i_0 j}(-1)^\ell a_{0j}\det(A_{i_0 0,j\ell})$$

$$= \sum_{j=0}^{k}(-1)^j a_{0j}\left(\sum_{\ell=0}^{k-1}(-1)^{i_0+\ell}a_{i_0 j}\det(A_{i_0 0,j\ell})\right)$$

$$= \sum_{j=0}^{k}(-1)^j a_{0j}\det(A_{0j}),$$

where the last equality follows from Eq. A.1.

By mathematical induction, we have proved the lemma. $\qquad\square$

A.2 Invertible Matrices for the Stochastic Leakage Model

In this section, we will focus on matrices with coefficients from the field of real numbers $\mathbb{R}$. Let n, m be two positive integers.

Definition A.2.1 For any vector $u = (u_0, u_1, \ldots, u_{n-1}) \in \mathbb{R}^n$, the *Euclidean norm* of u, denoted $\|u\|_2$ is defined to be

$$\|u\|_2 = \left(\sum_{i=0}^{n-1}u_i^2\right)^{1/2}.$$

In other words, the Euclidean norm of u is the square root of the scalar product (see Definition 1.3.2) between u and $u^{\top}$:

$$\|u\|_2 = \left(u \cdot u^{\top}\right)^{1/2}. \tag{A.2}$$

Remark A.2.1 It is easy to see that if $\|u\|_2 = 0$, then $u = 0$.

Example A.2.1 Let $u = (1, 2, 5)$, then

$$\|u\|_2 = \sqrt{1 + 2^2 + 5^2} = \sqrt{1 + 4 + 25} = \sqrt{30}.$$

Definition A.2.2 For any two vectors $u, v \in \mathbb{R}^n$, the *Euclidean distance*, denoted $d(u, v)$, is defined to be the Euclidean norm of the vector $u - v$

$$d(u, v) = \|u - v\|_2.$$

Definition A.2.3 The *row rank* (resp. *column rank*) of a matrix A, denoted rank(A) over $\mathbb{R}$ is the maximum number of rows (resp. columns) in A that constitute a set of independent vectors.

The following result is very useful for us. For a proof, see, e.g., [Ber09, Section 2.4].

Theorem A.2.1 *The column rank of a matrix A is equal to its row rank.*

Definition A.2.4 The *rank* of a matrix A, denoted rank(A) is the row rank of A. An $n \times m$ matrix A is said to *have full column rank* if rank(A) $= m$. It is said to *have full row rank* if rank(A) $= n$.

Example A.2.2 Let

$$A = \begin{pmatrix} 1 & 0 & 1 & 1 \\ 0 & 1 & 0 & 1 \\ 1 & 0 & 0 & 1 \end{pmatrix}.$$

We can see that the vectors $\{(1, 0, 1), (0, 1, 0), (1, 0, 0)\}$ are independent but

$$(1, 1, 1) = (1, 0, 1) + (0, 1, 0).$$

Thus A has rank 3. And A has full row rank.

Let A be an $n \times m$ matrix. Take any row vector $u \in \mathbb{R}^n$, we note that uA is a linear combination (see Definition 1.3.9) of rows of A. By Definition 1.3.11, the rows of A are linearly independent if and only if there does not exist a nonzero vector u such that $uA = 0$. Similar results hold for the columns of A. We have proved

Lemma A.2.1 *An $n \times m$ matrix A has full row rank if and only if there does not exist a nonzero vector $\boldsymbol{u} \in \mathbb{R}^n$ such that $\boldsymbol{u}A = \boldsymbol{0}$. A has full column rank if and only if there does not exist a nonzero vector $\boldsymbol{u} \in \mathbb{R}^m$ such that $A\boldsymbol{u}^\top = \boldsymbol{0}$.*

Theorem A.2.2 *An $n \times n$ square matrix A is invertible if and only if $\mathrm{rank}(A) = n$.*

Proof We will provide the proof for the necessity. We refer the readers to [Goc11, Section 3.6] for the proof of the sufficiency.

By Definition 1.3.3, A is invertible if and only if there exists an $n \times n$ matrix B such that $AB = BA = I_n$, where I_n is the $n-$dimensional identity matrix. Suppose A is invertible and $\mathrm{rank}(A) \neq n$. Then by Lemma A.2.1, there exists a nonzero vector $\boldsymbol{u} \in \mathbb{R}^n$ such that $\boldsymbol{u}A = \boldsymbol{0}$. Then we have

$$\boldsymbol{u}AB = \boldsymbol{0}B = \boldsymbol{0} = \boldsymbol{0}I_n,$$

a contradiction. $\square$

Lemma A.2.2 *Let M be an $n \times m$ matrix. The matrix $M^\top M$ is invertible if and only if M has full column rank.*

Proof Let

$$A = M^\top M.$$

Then A is a square matrix of size $m \times m$.

$\implies$ Suppose A is invertible and $\mathrm{rank}(M) \neq m$. By Lemma A.2.1, there exists a nonzero vector $\boldsymbol{u} \in \mathbb{R}^m$ such that $M\boldsymbol{u}^\top = \boldsymbol{0}$. We have

$$A\boldsymbol{u}^\top = M^\top M\boldsymbol{u}^\top = \boldsymbol{0}.$$

By Lemma A.2.1 again, we know that $\mathrm{rank}(A) \neq n$ and according to Theorem A.2.2, A is not invertible. A contradiction.

$\impliedby$ Suppose $\mathrm{rank}(M) = m$ and A is not invertible. By Theorem A.2.2 and Lemma A.2.1, there exists a nonzero vector $\boldsymbol{u} \in \mathbb{R}^m$ such that $A\boldsymbol{u}^\top = \boldsymbol{0}$, which gives (see Eq. A.2 and Remark A.2.1)

$$0 = \boldsymbol{u}M^\top M\boldsymbol{u}^\top = (M\boldsymbol{u}^\top)^\top (M\boldsymbol{u}^\top) = \|M\boldsymbol{u}^\top\|_2 \implies M\boldsymbol{u}^\top = \boldsymbol{0}.$$

By Lemma A.2.1, M does not have full column rank. A contradiction. $\square$

Let us consider the matrix $M_{\boldsymbol{v}}$ from Sect. 4.3.2.2. Suppose we take a collection of m_v different values of $\boldsymbol{v}$ such that the rows of $M_{\boldsymbol{v}}$ are linearly independent, then $M_{\boldsymbol{v}}$ has row rank equal to m_v. By Theorem A.2.1, $M_{\boldsymbol{v}}$ also has column rank m_v. Since $M_{\boldsymbol{v}}$ has m_v columns, by definition, $M_{\boldsymbol{v}}$ has full column rank. It follows from Lemma A.2.2 that the matrix $M_{\boldsymbol{v}}^\top M_{\boldsymbol{v}}$ is invertible.

In particular, with all possible values of v appearing in the rows of M_v, we will have m_v linear independent rows in M_v given by those with Hamming weight 1:

$$(1, 0, 0, \ldots, 0), \quad (0, 1, 0, 0, \ldots, 0), \quad (0, 0, 1, 0, \ldots, 0), \quad \ldots, \quad (0, 0, 0, \ldots, 0, 1).$$

Then M_v has row rank equal to m_v and $M_v^\top M_v$ will be an invertible matrix.

Appendix B
Long Division

In primary school, we learned to do long division for calculating the quotient and remainder of dividing one integer by another integer. For example, to compute

$$1346 = 25 \times q + r,$$

we can write

$$
\begin{array}{r}
53 \\
25 \,\overline{)\,1346} \\
125 \\
\overline{96} \\
75 \\
\overline{21}
\end{array}
$$

and we get $q = 53$, $r = 21$.

Similarly, let us take two polynomials $f(x), g(x) \in F[x]$, where F is a field. We can also compute $f(x)$ divided by $g(x)$ using long division. Let $F = \mathbb{F}_2$. Take

$$f(x) = x^8 + x^4 + x^3 + x + 1 \in \mathbb{F}_2[x],$$

and

$$g(x) = x + 1 \in \mathbb{F}_2[x].$$

We have

© The Author(s), under exclusive license to Springer Nature Switzerland AG 2024
X. Hou, J. Breier, *Cryptography and Embedded Systems Security*,
https://doi.org/10.1007/978-3-031-62205-2

$$
\begin{array}{r}
x^7 + x^6 + x^5 + x^4 + x^2 + x + 1 \\
x+1{\overline{\smash{\big)}\,x^8 + x^4 + x^3 + x + 1}} \\
\underline{x^8 + x^7} \\
x^7 + x^4 + x^3 + x + 1 \\
\underline{x^7 + x^6} \\
x^6 + x^4 + x^3 + x + 1 \\
\underline{x^6 + x^5} \\
x^5 + x^4 + x^3 + x + 1 \\
\underline{x^5 + x^4} \\
x^3 + x + 1 \\
\underline{x^3 + x^2} \\
x^2 + x + 1 \\
\underline{x^2 + x} \\
1
\end{array}
$$

Thus (see Example 1.5.21)

$$
f(x) = (x + 1)(x^7 + x^6 + x^5 + x^4 + x^2 + x + 1) + 1.
$$

Appendix C
DES Sbox

Table C.1 Sboxes in DES (Sect. 3.1.1) round function

15	1	8	14	6	11	3	4	9	7	2	13	12	0	5	10
3	13	4	7	15	2	8	14	12	0	1	10	6	9	11	5
0	14	7	11	10	4	13	1	5	8	12	6	9	3	2	15
13	8	10	1	3	15	4	2	11	6	7	12	0	5	14	9

(a) SB^2_{DES}

10	0	9	14	6	3	15	5	1	13	12	7	11	4	2	8
13	7	0	9	3	4	6	10	2	8	5	14	12	11	15	1
13	6	4	9	8	15	3	0	11	1	2	12	5	10	14	7
1	10	13	0	6	9	8	7	4	15	14	3	11	5	2	12

(b) SB^3_{DES}

7	13	14	3	0	6	9	10	1	2	8	5	11	12	4	15
13	8	11	5	6	15	0	3	4	7	2	12	1	10	14	9
10	6	9	0	12	11	7	13	15	1	3	14	5	2	8	4
3	15	0	6	10	1	13	8	9	4	5	11	12	7	2	14

(c) SB^4_{DES}

2	12	4	1	7	10	11	6	8	5	3	15	13	0	14	9
14	11	2	12	4	7	13	1	5	0	15	10	3	9	8	6
4	2	1	11	10	13	7	8	15	9	12	5	6	3	0	14
11	8	12	7	1	14	2	13	6	15	0	9	10	4	5	3

(continued)

Table C.1 (continued)

(d) SB_{DES}^5

12	1	10	15	9	2	6	8	0	13	3	4	14	7	5	11
10	15	4	2	7	12	9	5	6	1	13	14	0	11	3	8
9	14	15	5	2	8	12	3	7	0	4	10	1	13	11	6
4	3	2	12	9	5	15	10	11	14	1	7	6	0	8	13

(e) SB_{DES}^6

4	11	2	14	15	0	8	13	3	12	9	7	5	10	6	1
13	0	11	7	4	9	1	10	14	3	5	12	2	15	8	6
1	4	11	13	12	3	7	14	10	15	6	8	0	5	9	2
6	11	13	8	1	4	10	7	9	5	0	15	14	2	3	12

(f) SB_{DES}^7

13	2	8	4	6	15	11	1	10	9	3	14	5	0	12	7
1	15	13	8	10	3	7	4	12	5	6	11	0	14	9	2
7	11	4	1	9	12	14	2	0	6	10	13	15	3	5	8
2	1	14	7	4	10	8	13	15	12	9	0	3	5	6	11

(g) SB_{DES}^8

Appendix D
Algebraic Normal Forms for PRESENT Sbox Output Bits

For $i = 1, 2, 3$, define

$$\varphi_i : \mathbb{F}_2^4 \to \mathbb{F}_2$$

$$x \mapsto \text{SB}_{\text{PRESENT}}(x)_i,$$

where $\text{SB}_{\text{PRESENT}}(x)_i$ is the ith bit of $\text{SB}_{\text{PRESENT}}(x)$, the PRESENT Sbox output corresponding to x. In this section, we will compute the algebraic normal forms for φ_i. Similarly to Table 3.13, we construct the table for each φ_i—see Tables D.1, D.2, and D.3.

The coefficients λ are calculated based on Eq. 3.10 and the following equations:

$$\lambda_{0000} = \varphi_i(0000),$$
$$\lambda_{0001} = \varphi_i(0000) + \varphi_i(0001),$$
$$\lambda_{0010} = \varphi_i(0000) + \varphi_i(0010),$$

$$\lambda_{0011} = \varphi_i(0000) + \varphi_i(0010) + \varphi_i(0001) + \varphi_i(0011),$$

$$\lambda_{0100} = \varphi_i(0000) + \varphi_i(0100),$$

$$\lambda_{0101} = \varphi_i(0000) + \varphi_i(0001) + \varphi_i(0100) + \varphi_i(0101),$$

$$\lambda_{0110} = \varphi_i(0000) + \varphi_i(0010) + \varphi_i(0100) + \varphi_i(0110),$$

$$\lambda_{0111} = \sum_{x=0}^{7} \varphi_i(x),$$

$$\lambda_{1000} = \varphi_i(0000) + \varphi_i(1000),$$

$$\lambda_{1001} = \varphi_i(0000) + \varphi_i(0001) + \varphi_i(1000) + \varphi_i(1001),$$

X. Hou, J. Breier, *Cryptography and Embedded Systems Security*,
https://doi.org/10.1007/978-3-031-62205-2

Table D.1 The Boolean function φ_1 takes input x and outputs the 1st bit of $SB_{PRESENT}(x)$. The second last row lists the output of φ_1 for different input values. The last row lists the coefficients (Eq. 3.10) for the algebraic normal form of φ_1

x	0	1	2	3	4	5	6	7	8	9	A	B	C	D	E	F
x_3	0	0	0	0	0	0	0	0	1	1	1	1	1	1	1	1
x_2	0	0	0	0	1	1	1	1	0	0	0	0	1	1	1	1
x_1	0	0	1	1	0	0	1	1	0	0	1	1	0	0	1	1
x_0	0	1	0	1	0	1	0	1	0	1	0	1	0	1	0	1
$SB_{PRESENT}(x)$	C	5	6	B	9	0	A	D	3	E	F	8	4	7	1	2
$\varphi_1(x)$	0	0	1	1	0	0	1	0	1	1	1	0	0	1	0	1
λ_x	0	0	1	0	0	0	0	1	1	0	1	1	1	1	0	0

Table D.2 The Boolean function φ_2 takes input x and outputs the 2nd bit of $SB_{PRESENT}(x)$. The second last row lists the output of φ_2 for different input values. The last row lists the coefficients (Eq. 3.10) for the algebraic normal form of φ_2

x	0	1	2	3	4	5	6	7	8	9	A	B	C	D	E	F
x_3	0	0	0	0	0	0	0	0	1	1	1	1	1	1	1	1
x_2	0	0	0	0	1	1	1	1	0	0	0	0	1	1	1	1
x_1	0	0	1	1	0	0	1	1	0	0	1	1	0	0	1	1
x_0	0	1	0	1	0	1	0	1	0	1	0	1	0	1	0	1
$SB_{PRESENT}(x)$	C	5	6	B	9	0	A	D	3	E	F	8	4	7	1	2
$\varphi_2(x)$	1	1	1	0	0	0	0	1	0	1	1	0	1	1	0	0
λ_x	1	0	0	1	1	0	0	0	1	1	1	1	0	1	0	0

Table D.3 The Boolean function φ_3 takes input x and outputs the 3rd bit of $SB_{PRESENT}(x)$. The second last row lists the output of φ_3 for different input values. The last row lists the coefficients (Eq. 3.10) for the algebraic normal form of φ_3

x	0	1	2	3	4	5	6	7	8	9	A	B	C	D	E	F
x_3	0	0	0	0	0	0	0	0	1	1	1	1	1	1	1	1
x_2	0	0	0	0	1	1	1	1	0	0	0	0	1	1	1	1
x_1	0	0	1	1	0	0	1	1	0	0	1	1	0	0	1	1
x_0	0	1	0	1	0	1	0	1	0	1	0	1	0	1	0	1
$SB_{PRESENT}(x)$	C	5	6	B	9	0	A	D	3	E	F	8	4	7	1	2
$\varphi_3(x)$	1	0	0	1	1	0	1	1	0	1	1	1	0	0	0	0
λ_x	1	1	1	0	0	0	1	1	1	0	0	1	0	1	0	0

$$\lambda_{1010} = \varphi_i(0000) + \varphi_i(0010) + \varphi_i(1000) + \varphi_i(1010),$$

$$\lambda_{1011} = \varphi_i(0) + \varphi_i(1) + \varphi_i(2) + \varphi_i(3) + \varphi_i(8) + \varphi_i(9) + \varphi_i(A) + \varphi_i(B),$$

$$\lambda_{1100} = \varphi_i(0000) + \varphi_i(0100) + \varphi_i(1000) + \varphi_i(1100),$$

$$\lambda_{1101} = \varphi_i(0) + \varphi_i(1) + \varphi_i(4) + \varphi_i(5) + \varphi_i(8) + \varphi_i(9) + \varphi_i(C) + \varphi_i(D),$$

$$\lambda_{1110} = \varphi_i(0) + \varphi_i(2) + \varphi_i(4) + \varphi_i(6) + \varphi_i(8) + \varphi_i(A) + \varphi_i(C) + \varphi_i(E),$$

$$\lambda_{1111} = \sum_{x=0}^{F} \varphi_i(x).$$

By Eq. 3.9, we have

$$
\begin{aligned}
\varphi_1(x) &= \lambda_{0010}x_1 + \lambda_{0111}x_2x_1x_0 + \lambda_{1000}x_3 + \lambda_{1010}x_3x_1 \\
&\quad + \lambda_{1011}x_3x_1x_0 + \lambda_{1100}x_3x_2 + \lambda_{1101}x_3x_2x_0 \\
&= x_1 + x_3 + x_1x_3 + x_2x_3 + x_0x_1x_2 + x_0x_1x_3 + x_0x_2x_3, \\
\varphi_2(x) &= \lambda_{0000} + \lambda_{0011}x_1x_0 + \lambda_{0100}x_2 + \lambda_{1000}x_3 + \lambda_{1001}x_3x_0 \\
&\quad + \lambda_{1010}x_3x_1 + \lambda_{1011}x_3x_1x_0 + \lambda_{1101}x_3x_2x_0 \\
&= 1 + x_2 + x_3 + x_0x_1 + x_0x_3 + x_1x_3 + x_0x_1x_3 + x_0x_2x_3 \\
\varphi_3(x) &= \lambda_{0000} + \lambda_{0001}x_0 + \lambda_{0010}x_1 + \lambda_{0110}x_2x_1 + \lambda_{0111}x_2x_1x_0 \\
&\quad + \lambda_{1000}x_3 + \lambda_{1011}x_3x_1x_0 + \lambda_{1101}x_3x_2x_0 \\
&= 1 + x_0 + x_1 + x_3 + x_1x_2 + x_0x_1x_2 + x_0x_1x_3 + x_0x_2x_3.
\end{aligned}
$$

Appendix E
Encoding-Based Countermeasure for Symmetric Block Ciphers

In Table E.1, we list values in T_{SG}, which are signals for each integer between 00 and $3F$ with Hamming weight 6, computed with the stochastic leakage model obtained in Code-SCA Step 6 from Sect. 4.5.1.1. The sorted version of T_{SG} is shown in Table E.2, where the signals are in ascending order and the words from $\mathbb{F}_2^6$ with Hamming weight 6 are recorded accordingly.

© The Author(s), under exclusive license to Springer Nature Switzerland AG 2024
X. Hou, J. Breier, *Cryptography and Embedded Systems Security*,
https://doi.org/10.1007/978-3-031-62205-2

Table E.1 Table T_{SG}, estimated signals for each integer between 00 and FF with Hamming weight 6, computed with the stochastic leakage model obtained in Code-SCA Step 6 from Sect. 4.5.1.1. The first (resp. second) column contains the hexadecimal (resp. binary) representations of the integers. The last column lists the corresponding estimated signals

3F	00111111	−0.00980
5F	01011111	−0.00976
6F	01101111	−0.01058
77	01110111	−0.01066
7B	01111011	−0.01053
7D	01111101	−0.01059
7E	01111110	−0.00943
9F	10011111	−0.00987
AF	10101111	−0.01069
B7	10110111	−0.01078
BB	10111011	−0.01065
BD	10111101	−0.01071
BE	10111110	−0.00955
CF	11001111	−0.01066
D7	11010111	−0.01074
DB	11011011	−0.01061
DD	11011101	−0.01067
DE	11011110	−0.00951
E7	11100111	−0.01156
EB	11101011	−0.01143
ED	11101101	−0.01149
EE	11101110	−0.01033
F3	11110011	−0.01152
F5	11110101	−0.01158
F6	11110110	−0.01042
F9	11111001	−0.01145
FA	11111010	−0.01029
FC	11111100	−0.01035

Table E.2 Sorted version of T_{SG} from Table E.1 such that the estimated signals (values in the last column) are in ascending order. The hexadecimal (resp. binary) representations of the corresponding integers are in the first (resp. second) column. Words highlighted in blue constitute the chosen binary code with Algorithm 4.5

F5	11110101	−0.01158
E7	11100111	−0.01156
F3	11110011	−0.01152
ED	11101101	−0.01149
F9	11111001	−0.01145
EB	11101011	−0.01143
B7	10110111	−0.01078
D7	11010111	−0.01074
BD	10111101	−0.01071
AF	10101111	−0.01069
DD	11011101	−0.01067
77	01110111	−0.01066
CF	11001111	−0.01066
BB	10111011	−0.01065
DB	11011011	−0.01061
7D	01111101	−0.01059
6F	01101111	−0.01058
7B	01111011	−0.01053
F6	11110110	−0.01042
FC	11111100	−0.01035
EE	11101110	−0.01033
FA	11111010	−0.01029
9F	10011111	−0.00987
3F	00111111	−0.00980
5F	01011111	−0.00976
BE	10111110	−0.00955
DE	11011110	−0.00951
7E	01111110	−0.00943

References

[2709]   ISO/IEC JTC 1/SC 27. ISO/IEC 15408-1: Information technology—Security techniques—Evaluation criteria for IT security—Part 1: Introduction and general model, International Organization for Standardization, 2009.

[ABB+20]   Melissa Azouaoui, Davide Bellizia, Ileana Buhan, Nicolas Debande, Sébastien Duval, Christophe Giraud, Éliane Jaulmes, François Koeune, Elisabeth Oswald, François-Xavier Standaert, et al. A systematic appraisal of side channel evaluation strategies. In *Security Standardisation Research: 6th International Conference, SSR 2020, London, UK, November 30–December 1, 2020, Proceedings 6*, pages 46–66. Springer, 2020.

[ABC+17]   Stéphanie Anceau, Pierre Bleuet, Jessy Clédière, Laurent Maingault, Jean-luc Rainard, and Rémi Tucoulou. Nanofocused x-ray beam to reprogram secure circuits. In *International Conference on Cryptographic Hardware and Embedded Systems*, pages 175–188. Springer, 2017.

[ABF+03]   Christian Aumüller, Peter Bier, Wieland Fischer, Peter Hofreiter, and J-P Seifert. Fault attacks on RSA with crt: Concrete results and practical countermeasures. In *International Workshop on Cryptographic Hardware and Embedded Systems*, pages 260–275. Springer, 2003.

[AFV07]   Frederic Amiel, Benoit Feix, and Karine Villegas. Power analysis for secret recovering and reverse engineering of public key algorithms. In *Selected Areas in Cryptography: 14th International Workshop, SAC 2007, Ottawa, Canada, August 16–17, 2007, Revised Selected Papers 14*, pages 110–125. Springer, 2007.

[AG01]   Mehdi-Laurent Akkar and Christophe Giraud. An implementation of DES and AES, secure against some attacks. In *Cryptographic Hardware and Embedded Systems–CHES 2001: Third International Workshop Paris, France, May 14–16, 2001 Proceedings 3*, pages 309–318. Springer, 2001.

[Age15]   National Security Agency. Commercial National Security Algorithm Suite. https://apps.nsa.gov/iaarchive/programs/iad-initiatives/cnsa-suite.cfm, 2015.

[AGF21]   Rabin Yu Acharya, Fatemeh Ganji, and Domenic Forte. Infoneat: Information theory-based neuroevolution of augmenting topologies for side-channel analysis. *arXiv preprint arXiv:2105.00117*, 2021.

[AH20]   Karim M Abdellatif and Olivier Hériveaux. Silicontoaster: a cheap and programmable em injector for extracting secrets. In *2020 Workshop on Fault Detection and Tolerance in Cryptography (FDTC)*, pages 35–40. IEEE, 2020.

© The Author(s), under exclusive license to Springer Nature Switzerland AG 2024   465
X. Hou, J. Breier, *Cryptography and Embedded Systems Security*,
https://doi.org/10.1007/978-3-031-62205-2

[AK96]   Ross Anderson and Markus Kuhn. Tamper resistance-a cautionary note. In *Proceedings of the second Usenix workshop on electronic commerce*, volume 2, pages 1–11, 1996.

[ANP20]   Alexandre Adomnicai, Zakaria Najm, and Thomas Peyrin. Fixslicing: a new gift representation: fast constant-time implementations of gift and gift-cofb on arm cortex-m. *IACR Transactions on Cryptographic Hardware and Embedded Systems*, pages 402–427, 2020.

[AP20]   Alexandre Adomnicai and Thomas Peyrin. Fixslicing AES-like ciphers: New bitsliced AES speed records on ARM-cortex M and RISC-V. *Cryptology ePrint Archive*, 2020.

[APZ21]   Melissa Azouaoui, Kostas Papagiannopoulos, and Dominik Zürner. Blind side-channel SIFA. In *2021 Design, Automation & Test in Europe Conference & Exhibition (DATE)*, pages 555–560. IEEE, 2021.

[Atm16]   Atmel. AVR Instruction Set Manual. http://ww1.microchip.com/downloads/en/devicedoc/atmel-0856-avr-instruction-set-manual.pdf, 2016.

[AV13]   Kostas Papagiannopoulos Aram Verstegen. Present speed implementation. https://github.com/kostaspap88/PRESENT_speed_implementation, 2013.

[AWKS12] Kahraman D Akdemir, Zhen Wang, Mark Karpovsky, and Berk Sunar. Design of cryptographic devices resilient to fault injection attacks using nonlinear robust codes. *Fault analysis in cryptography*, pages 171–199, 2012.

[BBB$^+$07] Elaine Barker, William Barker, William Burr, William Polk, and Miles Smid. NIST special publication 800-57. *NIST Special publication*, 2007.

[BBB$^+$18] Anubhab Baksi, Shivam Bhasin, Jakub Breier, Mustafa Khairallah, and Thomas Peyrin. Protecting block ciphers against differential fault attacks without re-keying. In *2018 IEEE International Symposium on Hardware Oriented Security and Trust (HOST)*, pages 191–194. IEEE, 2018.

[BBB$^+$21] Anubhab Baksi, Shivam Bhasin, Jakub Breier, Mustafa Khairallah, Thomas Peyrin, Sumanta Sarkar, and Siang Meng Sim. DEFAULT: Cipher Level Resistance Against Differential Fault Attack. In Mehdi Tibouchi and Huaxiong Wang, editors, *Advances in Cryptology—ASIACRYPT 2021*, pages 124–156, Cham, 2021. Springer International Publishing.

[BBB$^+$22] Lejla Batina, Shivam Bhasin, Jakub Breier, Xiaolu Hou, and Dirmanto Jap. On implementation-level security of edge-based machine learning models. In *Security and Artificial Intelligence: A Crossdisciplinary Approach*, pages 335–359. Springer, 2022.

[BBH$^+$20] Shivam Bhasin, Jakub Breier, Xiaolu Hou, Dirmanto Jap, Romain Poussier, and Siang Meng Sim. SITM: See-in-the-middle side-channel assisted middle round differential cryptanalysis on SPN block ciphers. *IACR Transactions on Cryptographic Hardware and Embedded Systems*, pages 95–122, 2020.

[BBJP19] Lejla Batina, Shivam Bhasin, Dirmanto Jap, and Stjepan Picek. {CSI} {NN}: Reverse engineering of neural network architectures through electromagnetic side channel. In *28th USENIX Security Symposium (USENIX Security 19)*, pages 515–532, 2019.

[BC16]   Jakub Breier and Chien-Ning Chen. On determining optimal parameters for testing devices against laser fault attacks. In *2016 International Symposium on Integrated Circuits (ISIC)*, pages 1–4. IEEE, 2016.

[BCDG10] Alexandre Berzati, Cécile Canovas-Dumas, and Louis Goubin. Public key perturbation of randomized RSA implementations. In *Cryptographic Hardware and Embedded Systems, CHES 2010: 12th International Workshop, Santa Barbara, USA, August 17–20, 2010. Proceedings 12*, pages 306–319. Springer, 2010.

[BCG08]   Alexandre Berzati, Cécile Canovas, and Louis Goubin. Perturbating RSA public keys: An improved attack. In *Cryptographic Hardware and Embedded Systems–CHES 2008: 10th International Workshop, Washington, DC, USA, August 10–13, 2008. Proceedings 10*, pages 380–395. Springer, 2008.

[BCMCC06] Eric Brier, Benoît Chevallier-Mames, Mathieu Ciet, and Christophe Clavier. Why one should also secure RSA public key elements. In *Cryptographic Hardware and Embedded Systems-CHES 2006: 8th International Workshop, Yokohama, Japan, October 10–13, 2006. Proceedings 8*, pages 324–338. Springer, 2006.

[BD00] Dan Boneh and Glenn Durfee. Cryptanalysis of RSA with private key d less than n/sup 0.292. *IEEE transactions on Information Theory*, 46(4):1339–1349, 2000.

[BD16] Elaine Barker and Quynh Dang. NIST special publication 800-57 part 1, revision 4. *NIST Special publication*, 2016.

[BDF98] Dan Boneh, Glenn Durfee, and Yair Frankel. An attack on RSA given a small fraction of the private key bits. In *International Conference on the Theory and Application of Cryptology and Information Security*, pages 25–34. Springer, 1998.

[BDF+09] Shivam Bhasin, Jean-Luc Danger, Florent Flament, Tarik Graba, Sylvain Guilley, Yves Mathieu, Maxime Nassar, Laurent Sauvage, and Nidhal Selmane. Combined SCA and DFA countermeasures integrable in a FPGA design flow. In *2009 International Conference on Reconfigurable Computing and FPGAs*, pages 213–218. IEEE, 2009.

[BDH+97] Feng Bao, Robert H Deng, Yongfei Han, A Jeng, A Desai Narasimhalu, and T Ngair. Breaking public key cryptosystems on tamper resistant devices in the presence of transient faults. In *International Workshop on Security Protocols*, pages 115–124. Springer, 1997.

[BDL97] Dan Boneh, Richard A DeMillo, and Richard J Lipton. On the importance of checking cryptographic protocols for faults. In *International conference on the theory and applications of cryptographic techniques*, pages 37–51. Springer, 1997.

[BDPA13] Guido Bertoni, Joan Daemen, Michaël Peeters, and Gilles Van Assche. Keccak. In *Annual international conference on the theory and applications of cryptographic techniques*, pages 313–314. Springer, 2013.

[BDPVA07] Guido Bertoni, Joan Daemen, Michaël Peeters, and Gilles Van Assche. Sponge functions. In *ECRYPT hash workshop*, 2007.

[BECN+06] Hagai Bar-El, Hamid Choukri, David Naccache, Michael Tunstall, and Claire Whelan. The sorcerer's apprentice guide to fault attacks. *Proceedings of the IEEE*, 94(2):370–382, 2006.

[BEG13] Nasour Bagheri, Reza Ebrahimpour, and Navid Ghaedi. New differential fault analysis on present. *EURASIP Journal on Advances in Signal Processing*, 2013:1–10, 2013.

[Bei11] Amos Beimel. Secret-sharing schemes: A survey. In *International conference on coding and cryptology*, pages 11–46. Springer, 2011.

[Ber09] Dennis S Bernstein. *Matrix mathematics: theory, facts, and formulas*. Princeton university press, 2009.

[BFGV12] Josep Balasch, Sebastian Faust, Benedikt Gierlichs, and Ingrid Verbauwhede. Theory and practice of a leakage resilient masking scheme. In *Advances in Cryptology–ASIACRYPT 2012: 18th International Conference on the Theory and Application of Cryptology and Information Security, Beijing, China, December 2–6, 2012. Proceedings 18*, pages 758–775. Springer, 2012.

[BFP19] Claudio Bozzato, Riccardo Focardi, and Francesco Palmarini. Shaping the glitch: optimizing voltage fault injection attacks. *IACR Transactions on Cryptographic Hardware and Embedded Systems*, pages 199–224, 2019.

[BGE+17] Jan Burchard, Mael Gay, Ange-Salomé Messeng Ekossono, Jan Horáček, Bernd Becker, Tobias Schubert, Martin Kreuzer, and Ilia Polian. Autofault: towards automatic construction of algebraic fault attacks. In *2017 Workshop on Fault Diagnosis and Tolerance in Cryptography (FDTC)*, pages 65–72. IEEE, 2017.

[BGK04] Johannes Blömer, Jorge Guajardo, and Volker Krummel. Provably secure masking of AES. In *International workshop on selected areas in cryptography*, pages 69–83. Springer, 2004.

[BGLP13]	Ryad Benadjila, Jian Guo, Victor Lomné, and Thomas Peyrin. Implementing lightweight block ciphers on x86 architectures. In *International Conference on Selected Areas in Cryptography*, pages 324–351. Springer, 2013.

[BGLT04]	Marco Bucci, Michele Guglielmo, Raimondo Luzzi, and Alessandro Trifiletti. A power consumption randomization countermeasure for DPA-resistant cryptographic processors. In *Integrated Circuit and System Design. Power and Timing Modeling, Optimization and Simulation: 14th International Workshop, PATMOS 2004, Santorini, Greece, September 15–17, 2004. Proceedings 14*, pages 481–490. Springer, 2004.

[BGM+03]	Luca Benini, Angelo Galati, Alberto Macii, Enrico Macii, and Massimo Poncino. Energy-efficient data scrambling on memory-processor interfaces. In *Proceedings of the 2003 international symposium on Low power electronics and design*, pages 26–29, 2003.

[BGN+15]	Begül Bilgin, Benedikt Gierlichs, Svetla Nikova, Ventzislav Nikov, and Vincent Rijmen. Trade-offs for threshold implementations illustrated on AES. *IEEE Transactions on Computer-Aided Design of Integrated Circuits and Systems*, 34(7):1188–1200, 2015.

[BGV11]	Josep Balasch, Benedikt Gierlichs, and Ingrid Verbauwhede. An in-depth and black-box characterization of the effects of clock glitches on 8-bit MCUs. In *2011 Workshop on Fault Diagnosis and Tolerance in Cryptography*, pages 105–114. IEEE, 2011.

[BH15]	Jakub Breier and Wei He. Multiple fault attack on PRESENT with a hardware trojan implementation in FPGA. In Gabriel Ghinita and Pedro Peris-Lopez, editors, *2015 International Workshop on Secure Internet of Things, SIoT 2015, Vienna, Austria, September 21–25, 2015*, pages 58–64. IEEE Computer Society, 2015.

[BH17]	Jakub Breier and Xiaolu Hou. Feeding two cats with one bowl: On designing a fault and side-channel resistant software encoding scheme. In *Topics in Cryptology–CT-RSA 2017: The Cryptographers' Track at the RSA Conference 2017, San Francisco, CA, USA, February 14–17, 2017, Proceedings*, pages 77–94. Springer, 2017.

[BH22]	Jakub Breier and Xiaolu Hou. How practical are fault injection attacks, really? *IEEE Access*, 10:113122–113130, 2022.

[BHJ+18]	Jakub Breier, Xiaolu Hou, Dirmanto Jap, Lei Ma, Shivam Bhasin, and Yang Liu. Practical fault attack on deep neural networks. In *Proceedings of the 2018 ACM SIGSAC Conference on Computer and Communications Security*, pages 2204–2206. ACM, 2018.

[BHL18]	Jakub Breier, Xiaolu Hou, and Yang Liu. Fault attacks made easy: Differential fault analysis automation on assembly code. *IACR Transactions on Cryptographic Hardware and Embedded Systems*, pages 96–122, 2018.

[BHL19]	Jakub Breier, Xiaolu Hou, and Yang Liu. On evaluating fault resilient encoding schemes in software. *IEEE Transactions on Dependable and Secure Computing*, 18(3):1065–1079, 2019.

[BHOS22]	Jakub Breier, Xiaolu Hou, Martín Ochoa, and Jesus Solano. Foobar: Fault fooling backdoor attack on neural network training. *IEEE Transactions on Dependable and Secure Computing*, 2022.

[BHT01]	Eric Brier, Helena Handschuh, and Christophe Tymen. Fast primitives for internal data scrambling in tamper resistant hardware. In *Cryptographic Hardware and Embedded Systems–CHES 2001: Third International Workshop Paris, France, May 14–16, 2001 Proceedings 3*, pages 16–27. Springer, 2001.

[BHvW12]	Lejla Batina, Jip Hogenboom, and Jasper GJ van Woudenberg. Getting more from PCA: first results of using principal component analysis for extensive power analysis. In *Topics in Cryptology–CT-RSA 2012: The Cryptographers' Track at the RSA Conference 2012, San Francisco, CA, USA, February 27–March 2, 2012. Proceedings*, pages 383–397. Springer, 2012.

[Bih97]   Eli Biham. A fast new DES implementation in software. In *International Workshop on Fast Software Encryption*, pages 260–272. Springer, 1997.

[BILT04]   Jean-Claude Bajard, Laurent Imbert, Pierre-Yvan Liardet, and Yannick Teglia. Leak resistant arithmetic. In *Cryptographic Hardware and Embedded Systems-CHES 2004: 6th International Workshop Cambridge, MA, USA, August 11–13, 2004. Proceedings 6*, pages 62–75. Springer, 2004.

[BJB18]   Jakub Breier, Dirmanto Jap, and Shivam Bhasin. SCADPA: side-channel assisted differential-plaintext attack on bit permutation based ciphers. In Jan Madsen and Ayse K. Coskun, editors, *2018 Design, Automation & Test in Europe Conference & Exhibition, DATE 2018, Dresden, Germany, March 19–23, 2018*, pages 1129–1134. IEEE, 2018.

[BJH+21]   Jakub Breier, Dirmanto Jap, Xiaolu Hou, Shivam Bhasin, and Yang Liu. Sniff: reverse engineering of neural networks with fault attacks. *IEEE Transactions on Reliability*, 71(4):1527–1539, 2021.

[BJHB19]   Jakub Breier, Dirmanto Jap, Xiaolu Hou, and Shivam Bhasin. On side channel vulnerabilities of bit permutations in cryptographic algorithms. *IEEE Transactions on Information Forensics and Security*, 15:1072–1085, 2019.

[BJHB23]   Jakub Breier, Dirmanto Jap, Xiaolu Hou, and Shivam Bhasin. A desynchronization-based countermeasure against side-channel analysis of neural networks. In *International Symposium on Cyber Security, Cryptology, and Machine Learning*, pages 296–306. Springer, 2023.

[BJKS21]   Robert Buhren, Hans-Niklas Jacob, Thilo Krachenfels, and Jean-Pierre Seifert. One glitch to rule them all: Fault injection attacks against AMD's secure encrypted virtualization. In *Proceedings of the 2021 ACM SIGSAC Conference on Computer and Communications Security*, pages 2875–2889, 2021.

[BJP20]   Shivam Bhasin, Dirmanto Jap, and Stjepan Picek. AES HD dataset—50 000 traces. AISyLab repository, 2020. https://github.com/AISyLab/AES_HD.

[BK06]   Johannes Blömer and Volker Krummel. Fault based collision attacks on AES. In *International Workshop on Fault Diagnosis and Tolerance in Cryptography*, pages 106–120. Springer, 2006.

[BKH+19]   Arthur Beckers, Masahiro Kinugawa, Yuichi Hayashi, Daisuke Fujimoto, Josep Balasch, Benedikt Gierlichs, and Ingrid Verbauwhede. Design considerations for em pulse fault injection. In *International Conference on Smart Card Research and Advanced Applications*, pages 176–192. Springer, 2019.

[BKHL20]   Jakub Breier, Mustafa Khairallah, Xiaolu Hou, and Yang Liu. A countermeasure against statistical ineffective fault analysis. *IEEE Transactions on Circuits and Systems II: Express Briefs*, 67(12):3322–3326, 2020.

[BKL+07]   Andrey Bogdanov, Lars R Knudsen, Gregor Leander, Christof Paar, Axel Poschmann, Matthew JB Robshaw, Yannick Seurin, and Charlotte Vikkelsoe. Present: An ultra-lightweight block cipher. In *International workshop on cryptographic hardware and embedded systems*, pages 450–466. Springer, 2007.

[Bla83]   George R Blakely. A computer algorithm for calculating the product ab modulo m. *IEEE Transactions on Computers*, 100(5):497–500, 1983.

[BLMR19]   Christof Beierle, Gregor Leander, Amir Moradi, and Shahram Rasoolzadeh. Craft: lightweight tweakable block cipher with efficient protection against DFA attacks. *IACR Transactions on Symmetric Cryptology*, 2019(1):5–45, 2019.

[BMV07]   Sanjay Burman, Debdeep Mukhopadhyay, and Kamakoti Veezhinathan. LFSR based stream ciphers are vulnerable to power attacks. In *International Conference on Cryptology in India*, pages 384–392. Springer, 2007.

[Bor06]   Michele Boreale. Attacking right-to-left modular exponentiation with timely random faults. In *Fault Diagnosis and Tolerance in Cryptography: Third International Workshop, FDTC 2006, Yokohama, Japan, October 10, 2006. Proceedings*, pages 24–35. Springer, 2006.

[BOS03] Johannes Blömer, Martin Otto, and Jean-Pierre Seifert. A new CRT-RSA algorithm secure against bellcore attacks. In *Proceedings of the 10th ACM conference on Computer and communications security*, pages 311–320, 2003.

[BP82] HJ Beker and FC Piper. Communications security: a survey of cryptography. *IEE Proceedings A (Physical Science, Measurement and Instrumentation, Management and Education, Reviews)*, 129(6):357–376, 1982.

[BPS+20] Ryad Benadjila, Emmanuel Prouff, Rémi Strullu, Eleonora Cagli, and Cécile Dumas. Deep learning for side-channel analysis and introduction to ASCAD database. *Journal of Cryptographic Engineering*, 10(2):163–188, 2020.

[BPS+21] Ryad Benadjila, Emmanuel Prouff, Rémi Strullu, Eleonora Cagli, and Cécile Dumas. ASCAD SCA database. https://github.com/ANSSI-FR/ASCAD.git, 2021.

[BRBG16] Erik Bosman, Kaveh Razavi, Herbert Bos, and Cristiano Giuffrida. Dedup est machina: Memory deduplication as an advanced exploitation vector. In *2016 IEEE symposium on security and privacy (SP)*, pages 987–1004. IEEE, 2016.

[BS97] Eli Biham and Adi Shamir. Differential fault analysis of secret key cryptosystems. In *Advances in Cryptology–CRYPTO'97: 17th Annual International Cryptology Conference Santa Barbara, California, USA August 17–21, 1997 Proceedings 17*, pages 513–525. Springer, 1997.

[BS08] Bhaskar Biswas and Nicolas Sendrier. Mceliece cryptosystem implementation: Theory and practice. In *International Workshop on Post-Quantum Cryptography*, pages 47–62. Springer, 2008.

[BS12] Eli Biham and Adi Shamir. *Differential cryptanalysis of the data encryption standard*. Springer Science & Business Media, 2012.

[BSH75] Daniel Binder, Edward C Smith, and AB Holman. Satellite anomalies from galactic cosmic rays. *IEEE Transactions on Nuclear Science*, 22(6):2675–2680, 1975.

[BT12] Alessandro Barenghi and Elena Trichina. Fault attacks on stream ciphers. In *Fault Analysis in Cryptography*, pages 239–255. Springer, 2012.

[Buc04] Johannes Buchmann. *Introduction to cryptography*, volume 335. Springer, 2004.

[Cal75] Stephen Calebotta. CMOS, the ideal logic family. *National Semiconductor CMOS Databook, Rev*, 1:2–3, 1975.

[CB19] Andrea Caforio and Subhadeep Banik. A study of persistent fault analysis. In *Security, Privacy, and Applied Cryptography Engineering: 9th International Conference, SPACE 2019, Gandhinagar, India, December 3–7, 2019, Proceedings 9*, pages 13–33. Springer, 2019.

[CCD+21] Pierre-Louis Cayrel, Brice Colombier, Vlad-Florin Drăgoi, Alexandre Menu, and Lilian Bossuet. Message-recovery laser fault injection attack on the classic mceliece cryptosystem. In *Annual International Conference on the Theory and Applications of Cryptographic Techniques*, pages 438–467. Springer, 2021.

[CCT+18] Samuel Chef, Chung Tah Chua, Jing Yun Tay, Yu Wen Siah, Shivam Bhasin, J Breier, and Chee Lip Gan. Descrambling of embedded SRAM using a laser probe. In *2018 IEEE International Symposium on the Physical and Failure Analysis of Integrated Circuits (IPFA)*, pages 1–6. IEEE, 2018.

[CD+15] Ronald Cramer, Ivan Bjerre Damgård, et al. *Secure multiparty computation*. Cambridge University Press, 2015.

[CFGR10] Christophe Clavier, Benoit Feix, Georges Gagnerot, and Mylene Roussellet. Passive and active combined attacks on AES combining fault attacks and side channel analysis. In *2010 Workshop on Fault Diagnosis and Tolerance in Cryptography*, pages 10–19. IEEE, 2010.

[CFZK21] Huili Chen, Cheng Fu, Jishen Zhao, and Farinaz Koushanfar. Proflip: Targeted trojan attack with progressive bit flips. In *Proceedings of the IEEE/CVF International Conference on Computer Vision*, pages 7718–7727, 2021.

[CG16] Claude Carlet and Sylvain Guilley. Complementary dual codes for countermeasures to side-channel attacks. *Adv. Math. Commun.*, 10(1):131–150, 2016.

[CH17] Ang Cui and Rick Housley. BADFET: Defeating Modern Secure Boot Using Second-Order Pulsed Electromagnetic Fault Injection. In *11th USENIX Workshop on Offensive Technologies (WOOT 17)*, 2017.

[Cho22] Charles Q. Choi. IBM Unveils 433-Qubit Osprey Chip. *IEEE Spectrum*, November 2022.

[CJRR99] Suresh Chari, Charanjit S Jutla, Josyula R Rao, and Pankaj Rohatgi. Towards sound approaches to counteract power-analysis attacks. In *Advances in Cryptology—CRYPTO'99: 19th Annual International Cryptology Conference Santa Barbara, California, USA, August 15–19, 1999 Proceedings 19*, pages 398–412. Springer, 1999.

[CJW10] Nicolas T Courtois, Keith Jackson, and David Ware. Fault-algebraic attacks on inner rounds of DES. In *E-Smart'10 Proceedings: The Future of Digital Security Technologies*. Strategies Telecom and Multimedia, 2010.

[CK09] Jean-Sébastien Coron and Ilya Kizhvatov. An efficient method for random delay generation in embedded software. In *Cryptographic Hardware and Embedded Systems-CHES 2009: 11th International Workshop Lausanne, Switzerland, September 6–9, 2009 Proceedings*, pages 156–170. Springer, 2009.

[CK10] Jean-Sébastien Coron and Ilya Kizhvatov. Analysis and improvement of the random delay countermeasure of ches 2009. In *Cryptographic Hardware and Embedded Systems, CHES 2010: 12th International Workshop, Santa Barbara, USA, August 17–20, 2010. Proceedings 12*, pages 95–109. Springer, 2010.

[CK18] Jean-Sébastien Coron and Ilya Kizhvatov. Trace sets with random delays. https://github.com/ikizhvatov/randomdelays-traces.git, 2018.

[Cla07] Christophe Clavier. Secret external encodings do not prevent transient fault analysis. In *Cryptographic Hardware and Embedded Systems-CHES 2007: 9th International Workshop, Vienna, Austria, September 10–13, 2007. Proceedings 9*, pages 181–194. Springer, 2007.

[Cor99] Jean-Sébastien Coron. Resistance against differential power analysis for elliptic curve cryptosystems. In *Cryptographic Hardware and Embedded Systems: First InternationalWorkshop, CHES'99 Worcester, MA, USA, August 12–13, 1999 Proceedings 1*, pages 292–302. Springer, 1999.

[COZZ23] Yukun Cheng, Changhai Ou, Fan Zhang, and Shihui Zheng. DLPFA: Deep learning based persistent fault analysis against block ciphers. *Cryptology ePrint Archive*, 2023.

[CRR03] Suresh Chari, Josyula R Rao, and Pankaj Rohatgi. Template attacks. In *Cryptographic Hardware and Embedded Systems-CHES 2002: 4th International Workshop Redwood Shores, CA, USA, August 13–15, 2002 Revised Papers 4*, pages 13–28. Springer, 2003.

[CT03] Jean-Sébastien Coron and Alexei Tchulkine. A new algorithm for switching from arithmetic to boolean masking. In *International Workshop on Cryptographic Hardware and Embedded Systems*, pages 89–97. Springer, 2003.

[CVM+21] Zitai Chen, Georgios Vasilakis, Kit Murdock, Edward Dean, David Oswald, and Flavio D Garcia. VoltPillager: Hardware-based fault injection attacks against Intel SGX Enclaves using the SVID voltage scaling interface. In *30th USENIX Security Symposium (USENIX Security 21)*, pages 699–716, 2021.

[DAP+22] Anuj Dubey, Afzal Ahmad, Muhammad Adeel Pasha, Rosario Cammarota, and Aydin Aysu. Modulonet: Neural networks meet modular arithmetic for efficient hardware masking. *IACR Transactions on Cryptographic Hardware and Embedded Systems*, pages 506–556, 2022.

[dBLW03] Bert den Boer, Kerstin Lemke, and Guntram Wicke. A DPA attack against the modular reduction within a crt implementation of RSA. In *Cryptographic Hardware and Embedded Systems-CHES 2002: 4th International Workshop Redwood Shores, CA, USA, August 13–15, 2002 Revised Papers 4*, pages 228–243. Springer, 2003.

[DCA20]   Anuj Dubey, Rosario Cammarota, and Aydin Aysu. Maskednet: The first hardware inference engine aiming power side-channel protection. In *2020 IEEE International Symposium on Hardware Oriented Security and Trust (HOST)*, pages 197–208. IEEE, 2020.

[DCRB⁺16]   Thomas De Cnudde, Oscar Reparaz, Begül Bilgin, Svetla Nikova, Ventzislav Nikov, and Vincent Rijmen. Masking AES with shares in hardware. In *International Conference on Cryptographic Hardware and Embedded Systems*, pages 194–212. Springer, 2016.

[DCSA22]   Anuj Dubey, Rosario Cammarota, Vikram Suresh, and Aydin Aysu. Guarding machine learning hardware against physical side-channel attacks. *ACM Journal on Emerging Technologies in Computing Systems (JETC)*, 18(3):1–31, 2022.

[DEK⁺18]   Christoph Dobraunig, Maria Eichlseder, Thomas Korak, Stefan Mangard, Florian Mendel, and Robert Primas. SIFA: exploiting ineffective fault inductions on symmetric cryptography. *IACR Transactions on Cryptographic Hardware and Embedded Systems*, pages 547–572, 2018.

[DLM20]   Mathieu Dumont, Mathieu Lisart, and Philippe Maurine. Modeling and simulating electromagnetic fault injection. *IEEE Transactions on Computer-Aided Design of Integrated Circuits and Systems*, 40(4):680–693, 2020.

[DO22]   Shaked Delarea and Yossi Oren. Practical, low-cost fault injection attacks on personal smart devices. *Applied Sciences*, 12(1):417, 2022.

[DPRS11]   Julien Doget, Emmanuel Prouff, Matthieu Rivain, and François-Xavier Standaert. Univariate side channel attacks and leakage modeling. *Journal of Cryptographic Engineering*, 1:123–144, 2011.

[DR02]   Joan Daemen and Vincent Rijmen. *The design of Rijndael*, volume 2. Springer, 2002.

[Dud14]   Richard M Dudley. *Uniform central limit theorems*, volume 142. Cambridge university press, 2014.

[Dur19]   Rick Durrett. *Probability: theory and examples*, volume 49. Cambridge university press, 2019.

[Dwo15]   Morris Dworkin. SHA-3 Standard: Permutation-Based Hash and Extendable-Output Functions, 2015-08-04 2015.

[DZD⁺18]   A Adam Ding, Liwei Zhang, François Durvaux, François-Xavier Standaert, and Yunsi Fei. Towards sound and optimal leakage detection procedure. In *Smart Card Research and Advanced Applications: 16th International Conference, CARDIS 2017, Lugano, Switzerland, November 13–15, 2017, Revised Selected Papers*, pages 105–122. Springer, 2018.

[EJ96]   Artur Ekert and Richard Jozsa. Quantum computation and shor's factoring algorithm. *Reviews of Modern Physics*, 68(3):733, 1996.

[ESH⁺11]   Sho Endo, Takeshi Sugawara, Naofumi Homma, Takafumi Aoki, and Akashi Satoh. An on-chip glitchy-clock generator for testing fault injection attacks. *Journal of Cryptographic Engineering*, 1(4):265–270, 2011.

[Far70]   PG Farrell. Linear binary anticodes. *Electronics Letters*, 13(6):419–421, 1970.

[FJLT13]   Thomas Fuhr, Éliane Jaulmes, Victor Lomné, and Adrian Thillard. Fault attacks on AES with faulty ciphertexts only. In *2013 Workshop on Fault Diagnosis and Tolerance in Cryptography*, pages 108–118. IEEE, 2013.

[FMP03]   Pierre-Alain Fouque, Gwenaëlle Martinet, and Guillaume Poupard. Attacking unbalanced RSA-crt using spa. In *Cryptographic Hardware and Embedded Systems-CHES 2003: 5th International Workshop, Cologne, Germany, September 8–10, 2003. Proceedings 5*, pages 254–268. Springer, 2003.

[Fou98]   Electronic Frontier Foundation. Cracking DES: Secrets of encryption research, wiretap politics and chip design. https://cryptome.org/jya/cracking-des/cracking-des.htm, 1998.

[FRVD08]   Pierre-Alain Fouque, Denis Réal, Frédéric Valette, and Mhamed Drissi. The carry leakage on the randomized exponent countermeasure. In *International Workshop on Cryptographic Hardware and Embedded Systems*, pages 198–213. Springer, 2008.

[FV03]   Pierre-Alain Fouque and Frédéric Valette. The doubling attack–why upwards is better than downwards. In *Cryptographic Hardware and Embedded Systems-CHES 2003: 5th International Workshop, Cologne, Germany, September 8–10, 2003. Proceedings 5*, pages 269–280. Springer, 2003.

[GBTP08]   Benedikt Gierlichs, Lejla Batina, Pim Tuyls, and Bart Preneel. Mutual information analysis: A generic side-channel distinguisher. In *International Workshop on Cryptographic Hardware and Embedded Systems*, pages 426–442. Springer, 2008.

[GE21]   Craig Gidney and Martin Ekerå. How to factor 2048 bit RSA integers in 8 hours using 20 million noisy qubits. *Quantum*, 5:433, 2021.

[GGJR⁺11] Benjamin Jun Gilbert Goodwill, Josh Jaffe, Pankaj Rohatgi, et al. A testing methodology for side-channel resistance validation. In *NIST non-invasive attack testing workshop*, volume 7, pages 115–136, 2011.

[GGP09]   Laurie Genelle, Christophe Giraud, and Emmanuel Prouff. Securing AES implementation against fault attacks. In *2009 Workshop on Fault Diagnosis and Tolerance in Cryptography (FDTC)*, pages 51–62. IEEE, 2009.

[GHNZ09]   Zheng Gong, Pieter H Hartel, Svetla Nikova, and Bo Zhu. Towards secure and practical MACs for body sensor networks. In *INDOCRYPT*, pages 182–198. Springer, 2009.

[GHO15]   Richard Gilmore, Neil Hanley, and Maire O'Neill. Neural network based attack on a masked implementation of AES. In *2015 IEEE International Symposium on Hardware Oriented Security and Trust (HOST)*, pages 106–111. IEEE, 2015.

[GHP04]   Sylvain Guilley, Philippe Hoogvorst, and Renaud Pacalet. Differential power analysis model and some results. In *Smart Card Research and Advanced Applications VI: IFIP 18th World Computer Congress TC8/WG8. 8 & TC11/WG11. 2 Sixth International Conference on Smart Card Research and Advanced Applications (CARDIS) 22–27 August 2004 Toulouse, France*, pages 127–142. Springer, 2004.

[GJJ22]   Qian Guo, Andreas Johansson, and Thomas Johansson. A key-recovery side-channel attack on classic mceliece implementations. *IACR Transactions on Cryptographic Hardware and Embedded Systems*, pages 800–827, 2022.

[GM11]   Louis Goubin and Ange Martinelli. Protecting AES with shamir's secret sharing scheme. In *Cryptographic Hardware and Embedded Systems–CHES 2011: 13th International Workshop, Nara, Japan, September 28–October 1, 2011. Proceedings 13*, pages 79–94. Springer, 2011.

[GMO01]   Karine Gandolfi, Christophe Mourtel, and Francis Olivier. Electromagnetic analysis: Concrete results. In *Cryptographic Hardware and Embedded Systems–CHES 2001: Third International Workshop Paris, France, May 14–16, 2001 Proceedings 3*, pages 251–261. Springer, 2001.

[GMWM16]   Daniel Gruss, Clémentine Maurice, Klaus Wagner, and Stefan Mangard. Flush+flush: a fast and stealthy cache attack. In *Detection of Intrusions and Malware, and Vulnerability Assessment: 13th International Conference, DIMVA 2016, San Sebastián, Spain, July 7–8, 2016, Proceedings 13*, pages 279–299. Springer, 2016.

[Goc11]   Mark S Gockenbach. *Finite-dimensional linear algebra*. CRC Press, 2011.

[Gou01]   Louis Goubin. A sound method for switching between boolean and arithmetic masking. In *Cryptographic Hardware and Embedded Systems–CHES 2001: Third International Workshop Paris, France, May 14–16, 2001 Proceedings 3*, pages 3–15. Springer, 2001.

[GP99]   Louis Goubin and Jacques Patarin. DES and differential power analysis the "duplication" method. In *Cryptographic Hardware and Embedded Systems: First InternationalWorkshop, CHES'99 Worcester, MA, USA, August 12–13, 1999 Proceedings 1*, pages 158–172. Springer, 1999.

[GSK06]   Gunnar Gaubatz, Berk Sunar, and Mark G Karpovsky. Non-linear residue codes for robust public-key arithmetic. In *Fault Diagnosis and Tolerance in Cryptography:*

Third International Workshop, FDTC 2006, Yokohama, Japan, October 10, 2006. Proceedings, pages 173–184. Springer, 2006.

[GST12] Benedikt Gierlichs, Jörn-Marc Schmidt, and Michael Tunstall. Infective computation and dummy rounds: Fault protection for block ciphers without check-before-output. In *Progress in Cryptology–LATINCRYPT 2012: 2nd International Conference on Cryptology and Information Security in Latin America, Santiago, Chile, October 7–10, 2012. Proceedings 2*, pages 305–321. Springer, 2012.

[Hab65] Donald H Habing. The use of lasers to simulate radiation-induced transients in semiconductor devices and circuits. *IEEE Transactions on Nuclear Science*, 12(5):91–100, 1965.

[HBB⁺16] Wei He, Jakub Breier, Shivam Bhasin, Noriyuki Miura, and Makoto Nagata. Ring oscillator under laser: Potential of PLL-based countermeasure against laser fault injection. In *Fault Diagnosis and Tolerance in Cryptography (FDTC), 2016 Workshop on*, pages 102–113. IEEE, 2016.

[HBB21] Xiaolu Hou, Jakub Breier, and Shivam Bhasin. DNFA: Differential no-fault analysis of bit permutation based ciphers assisted by side-channel. In *2021 Design, Automation & Test in Europe Conference & Exhibition (DATE)*, pages 182–187. IEEE, 2021.

[HBB22] Xiaolu Hou, Jakub Breier, and Shivam Bhasin. SBCMA: Semi-blind combined middle-round attack on bit-permutation ciphers with application to AEAD schemes. *IEEE Transactions on Information Forensics and Security*, 17:3677–3690, 2022.

[HBJ⁺21] Xiaolu Hou, Jakub Breier, Dirmanto Jap, Lei Ma, Shivam Bhasin, and Yang Liu. Physical security of deep learning on edge devices: Comprehensive evaluation of fault injection attack vectors. *Microelectronics Reliability*, 120:114116, 2021.

[HBK23] Xiaolu Hou, Jakub Breier, and Mladen Kovacevic. Another look at side-channel resistant encoding schemes. *IACR Cryptol. ePrint Arch.*, page 1698, 2023.

[HBZL19] Xiaolu Hou, Jakub Breier, Fuyuan Zhang, and Yang Liu. Fully automated differential fault analysis on software implementations of block ciphers. *IACR Transactions on Cryptographic Hardware and Embedded Systems*, pages 1–29, 2019.

[HDD11] Philippe Hoogvorst, Guillaume Duc, and Jean-Luc Danger. Software implementation of dual-rail representation. *COSADE, February*, pages 24–25, 2011.

[Her96] Israel N Herstein. *Abstract algebra*. Prentice Hall, 1996.

[HFK⁺19] Sanghyun Hong, Pietro Frigo, Yiğitcan Kaya, Cristiano Giuffrida, and Tudor Dumitras. Terminal brain damage: Exposing the graceless degradation in deep neural networks under hardware fault attacks. In *28th USENIX Security Symposium (USENIX Security 19)*, pages 497–514, 2019.

[HH11] Ludger Hemme and Lars Hoffmann. Differential fault analysis on the sha1 compression function. In *2011 Workshop on Fault Diagnosis and Tolerance in Cryptography*, pages 54–62. IEEE, 2011.

[HHS⁺11] Yu-ichi Hayashi, Naofumi Homma, Takeshi Sugawara, Takaaki Mizuki, Takafumi Aoki, and Hideaki Sone. Non-invasive EMI-based fault injection attack against cryptographic modules. In *2011 IEEE International Symposium on Electromagnetic Compatibility*, pages 763–767. IEEE, 2011.

[HLMS14] Ronglin Hao, Bao Li, Bingke Ma, and Ling Song. Algebraic fault attack on the sha-256 compression function. *International Journal of Research in Computer Science*, 4(2):1, 2014.

[HOM06] Christoph Herbst, Elisabeth Oswald, and Stefan Mangard. An AES smart card implementation resistant to power analysis attacks. In *International conference on applied cryptography and network security*, pages 239–252. Springer, 2006.

[HPS98] Jeffrey Hoffstein, Jill Pipher, and Joseph H Silverman. NTRU: A ring-based public key cryptosystem. In *International algorithmic number theory symposium*, pages 267–288. Springer, 1998.

[HS13]   Michael Hutter and Jörn-Marc Schmidt. The temperature side channel and heating fault attacks. In *International Conference on Smart Card Research and Advanced Applications*, pages 219–235. Springer, 2013.

[HSP20]   Max Hoffmann, Falk Schellenberg, and Christof Paar. Armory: fully automated and exhaustive fault simulation on arm-m binaries. *IEEE Transactions on Information Forensics and Security*, 16:1058–1073, 2020.

[Hun12]   Thomas W Hungerford. *Algebra*, volume 73. Springer Science & Business Media, 2012.

[HZ12]   Annelie Heuser and Michael Zohner. Intelligent machine homicide: Breaking cryptographic devices using support vector machines. In *Constructive Side-Channel Analysis and Secure Design: Third International Workshop, COSADE 2012, Darmstadt, Germany, May 3–4, 2012. Proceedings 3*, pages 249–264. Springer, 2012.

[JAB+03]   M Rabaey Jan, Chandrakasan Anantha, Nikolic Borivoje, et al. Digital integrated circuits: a design perspective. *Prentice Hall*, 2003.

[Jea16]   Jérémy Jean. TikZ for Cryptographers. https://www.iacr.org/authors/tikz/, 2016.

[JP04]   Jean Jacod and Philip Protter. *Probability essentials*. Springer Science & Business Media, 2004.

[JPY01]   Marc Joye, Pascal Paillier, and Sung-Ming Yen. Secure evaluation of modular functions. In *2001 International Workshop on Cryptology and Network Security*, pages 227–229. Citeseer, 2001.

[JQBD97]   Marc Joye, Jean-Jacques Quisquater, Feng Bao, and Robert H Deng. RSA-type signatures in the presence of transient faults. In *IMA International Conference on Cryptography and Coding*, pages 155–160. Springer, 1997.

[JVDVF+22] Patrick Jattke, Victor Van Der Veen, Pietro Frigo, Stijn Gunter, and Kaveh Razavi. Blacksmith: Scalable rowhammering in the frequency domain. In *2022 IEEE Symposium on Security and Privacy (SP)*, pages 716–734. IEEE, 2022.

[JY03]   Marc Joye and Sung-Ming Yen. The montgomery powering ladder. In *Cryptographic Hardware and Embedded Systems-CHES 2002: 4th International Workshop Redwood Shores, CA, USA, August 13–15, 2002 Revised Papers*, pages 291–302. Springer, 2003.

[KAF+10]   Thorsten Kleinjung, Kazumaro Aoki, Jens Franke, Arjen K Lenstra, Emmanuel Thomé, Joppe W Bos, Pierrick Gaudry, Alexander Kruppa, Peter L Montgomery, Dag Arne Osvik, et al. Factorization of a 768-bit RSA modulus. In *Annual Cryptology Conference*, pages 333–350. Springer, 2010.

[KBJ+22]   Niclas Kühnapfel, Robert Buhren, Hans Niklas Jacob, Thilo Krachenfels, Christian Werling, and Jean-Pierre Seifert. Em-fault it yourself: Building a replicable EMFI setup for desktop and server hardware. In *2022 IEEE Physical Assurance and Inspection of Electronics (PAINE)*, pages 1–7. IEEE, 2022.

[KDB+22]   Satyam Kumar, Vishnu Asutosh Dasu, Anubhab Baksi, Santanu Sarkar, Dirmanto Jap, Jakub Breier, and Shivam Bhasin. Side channel attack on stream ciphers: A three-step approach to state/key recovery. *IACR Transactions Cryptographic Hardware and Embedded. Systems*, 2022(2):166–191, 2022.

[KDK+14]   Yoongu Kim, Ross Daly, Jeremie Kim, Chris Fallin, Ji Hye Lee, Donghyuk Lee, Chris Wilkerson, Konrad Lai, and Onur Mutlu. Flipping bits in memory without accessing them: An experimental study of dram disturbance errors. *ACM SIGARCH Computer Architecture News*, 42(3):361–372, 2014.

[KHN+19]   Mustafa Khairallah, Xiaolu Hou, Zakaria Najm, Jakub Breier, Shivam Bhasin, and Thomas Peyrin. Sok: On DFA vulnerabilities of substitution-permutation networks. In Steven D. Galbraith, Giovanni Russello, Willy Susilo, Dieter Gollmann, Engin Kirda, and Zhenkai Liang, editors, *Proceedings of the 2019 ACM Asia Conference on Computer and Communications Security, AsiaCCS 2019, Auckland, New Zealand, July 09–12, 2019*, pages 403–414. ACM, 2019.

[KJ01]   Paul C Kocher and Joshua M Jaffe. Secure modular exponentiation with leak minimization for smartcards and other cryptosystems, October 2 2001. US Patent 6,298,442.

[KJJ99]   Paul Kocher, Joshua Jaffe, and Benjamin Jun. Differential power analysis. In *Advances in Cryptology—CRYPTO'99: 19th Annual International Cryptology Conference Santa Barbara, California, USA, August 15–19, 1999 Proceedings 19*, pages 388–397. Springer, 1999.

[KJJ10]   Paul C Kocher, Joshua M Jaffe, and Benjamin C Jun. Cryptographic computation using masking to prevent differential power analysis and other attacks, February 23 2010. US Patent 7,668,310.

[KJJR11]   Paul Kocher, Joshua Jaffe, Benjamin Jun, and Pankaj Rohatgi. Introduction to differential power analysis. *Journal of Cryptographic Engineering*, 1:5–27, 2011.

[KJP14]   Raghavan Kumar, Philipp Jovanovic, and Ilia Polian. Precise fault-injections using voltage and temperature manipulation for differential cryptanalysis. In *2014 IEEE 20th International On-Line Testing Symposium (IOLTS)*, pages 43–48. IEEE, 2014.

[KKG03]   Ramesh Karri, Grigori Kuznetsov, and Michael Goessel. Parity-based concurrent error detection of substitution-permutation network block ciphers. In *Cryptographic Hardware and Embedded Systems-CHES 2003: 5th International Workshop, Cologne, Germany, September 8–10, 2003. Proceedings 5*, pages 113–124. Springer, 2003.

[KKT04]   Mark Karpovsky, Konrad J Kulikowski, and Alexander Taubin. Robust protection against fault-injection attacks on smart cards implementing the advanced encryption standard. In *International Conference on Dependable Systems and Networks, 2004*, pages 93–101. IEEE, 2004.

[KKY+89]   Yasuhiro Konishi, Masaki Kumanoya, Hiroyuki Yamasaki, Katsumi Dosaka, and Tsutomu Yoshihara. Analysis of coupling noise between adjacent bit lines in megabit drams. *IEEE Journal of Solid-State Circuits*, 24(1):35–42, 1989.

[KM20]   Martin S Kelly and Keith Mayes. High precision laser fault injection using low-cost components. In *2020 IEEE International Symposium on Hardware Oriented Security and Trust (HOST)*, pages 219–228. IEEE, 2020.

[KMBM17]   Fatma Kahri, Hassen Mestiri, Belgacem Bouallegue, and Mohsen Machhout. Fault attacks resistant architecture for keccak hash function. *International Journal of Advanced Computer Science and Applications*, 8(5), 2017.

[Koç94]   CK Koç. High-speed RSA implementation technical report. *RSA Laboratories, Redwood City*, 1994.

[Koc96]   Paul C Kocher. Timing attacks on implementations of diffie-hellman, RSA, DSS, and other systems. In *Advances in Cryptology—CRYPTO'96: 16th Annual International Cryptology Conference Santa Barbara, California, USA August 18–22, 1996 Proceedings 16*, pages 104–113. Springer, 1996.

[Kos02]   Thomas Koshy. *Elementary number theory with applications*. Academic press, 2002.

[KPH+19]   Jaehun Kim, Stjepan Picek, Annelie Heuser, Shivam Bhasin, and Alan Hanjalic. Make some noise. unleashing the power of convolutional neural networks for profiled side-channel analysis. *IACR Transactions on Cryptographic Hardware and Embedded Systems*, pages 148–179, 2019.

[KPP+22]   Alexandr Alexandrovich Kuznetsov, Oleksandr Volodymyrovych Potii, Nikolay Alexandrovich Poluyanenko, Yurii Ivanovich Gorbenko, and Natalia Kryvinska. *Stream Ciphers in Modern Real-time IT Systems*. Springer, 2022.

[KQ07]   Chong Hee Kim and Jean-Jacques Quisquater. Fault attacks for crt based RSA: New attacks, new results, and new countermeasures. In *IFIP International Workshop on Information Security Theory and Practices*, pages 215–228. Springer, 2007.

[KS09] Emilia Käsper and Peter Schwabe. Faster and timing-attack resistant AES-GCM. In *International Workshop on Cryptographic Hardware and Embedded Systems*, pages 1–17. Springer, 2009.

[KSV13] Duško Karaklajić, Jörn-Marc Schmidt, and Ingrid Verbauwhede. Hardware designer's guide to fault attacks. *IEEE Transactions on Very Large Scale Integration (VLSI) Systems*, 21(12):2295–2306, 2013.

[Kwa00] Matthew Kwan. Reducing the gate count of bitslice DES. *IACR Cryptol. ePrint Arch.*, 2000(51):51, 2000.

[LBM15] Liran Lerman, Gianluca Bontempi, and Olivier Markowitch. A machine learning approach against a masked AES: Reaching the limit of side-channel attacks with a learning model. *Journal of Cryptographic Engineering*, 5:123–139, 2015.

[Len96] Arjen K Lenstra. Memo on RSA signature generation in the presence of faults. Technical report, EPFL, 1996.

[LSG+10] Yang Li, Kazuo Sakiyama, Shigeto Gomisawa, Toshinori Fukunaga, Junko Takahashi, and Kazuo Ohta. Fault sensitivity analysis. In *Cryptographic Hardware and Embedded Systems, CHES 2010: 12th International Workshop, Santa Barbara, USA, August 17–20, 2010. Proceedings 12*, pages 320–334. Springer, 2010.

[LWLX17] Yannan Liu, Lingxiao Wei, Bo Luo, and Qiang Xu. Fault injection attack on deep neural network. In *Proceedings of the 36th International Conference on Computer-Aided Design*, pages 131–138. IEEE, 2017.

[LX04] San Ling and Chaoping Xing. *Coding theory: a first course*. Cambridge University Press, 2004.

[LZC+21] Xiangjun Lu, Chi Zhang, Pei Cao, Dawu Gu, and Haining Lu. Pay attention to raw traces: A deep learning architecture for end-to-end profiling attacks. *IACR Transactions on Cryptographic Hardware and Embedded Systems*, pages 235–274, 2021.

[Mah45] Patrick Mahon. History of hut 8 to December 1941 (1945). *B. Jack Copeland*, page 265, 1945.

[Man03] Stefan Mangard. A simple power-analysis (spa) attack on implementations of the AES key expansion. In *Information Security and Cryptology—ICISC 2002: 5th International Conference Seoul, Korea, November 28–29, 2002 Revised Papers 5*, pages 343–358. Springer, 2003.

[May03] Alexander May. *New RSA vulnerabilities using lattice reduction methods*. PhD thesis, Citeseer, 2003.

[MBFC22] Saurav Maji, Utsav Banerjee, Samuel H Fuller, and Anantha P Chandrakasan. A threshold implementation-based neural network accelerator with power and electromagnetic side-channel countermeasures. *IEEE Journal of Solid-State Circuits*, 2022.

[MDB+02] Jack A Mandelman, Robert H Dennard, Gary B Bronner, John K DeBrosse, Rama Divakaruni, Yujun Li, and Carl J Radens. Challenges and future directions for the scaling of dynamic random-access memory (DRAM). *IBM Journal of Research and Development*, 46(2.3):187–212, 2002.

[MDS99a] Thomas S Messerges, Ezzy A Dabbish, and Robert H Sloan. Investigations of power analysis attacks on smartcards. *Smartcard*, 99:151–161, 1999.

[MDS99b] Thomas S Messerges, Ezzy A Dabbish, and Robert H Sloan. Power analysis attacks of modular exponentiation in smartcards. In *Cryptographic Hardware and Embedded Systems: First InternationalWorkshop, CHES'99 Worcester, MA, USA, August 12–13, 1999 Proceedings 1*, pages 144–157. Springer, 1999.

[Mes00] Thomas S Messerges. Securing the AES finalists against power analysis attacks. In *International Workshop on Fast Software Encryption*, pages 150–164. Springer, 2000.

[MMR19] Thorben Moos, Amir Moradi, and Bastian Richter. Static power side-channel analysis—an investigation of measurement factors. *IEEE Transactions on Very Large Scale Integration (VLSI) Systems*, 28(2):376–389, 2019.

[MMS01a] David May, Henk L Muller, and Nigel P Smart. Non-deterministic processors. In *Information Security and Privacy: 6th Australasian Conference, ACISP 2001 Sydney, Australia, July 11–13, 2001 Proceedings 6*, pages 115–129. Springer, 2001.

[MMS01b] David May, Henk L Muller, and Nigel P Smart. Random register renaming to foil DPA. In *Cryptographic Hardware and Embedded Systems—CHES 2001: Third International Workshop Paris, France, May 14–16, 2001 Proceedings 3*, pages 28–38. Springer, 2001.

[Mon85] Peter L Montgomery. Modular multiplication without trial division. *Mathematics of computation*, 44(170):519–521, 1985.

[Mon87] Peter L Montgomery. Speeding the pollard and elliptic curve methods of factorization. *Mathematics of computation*, 48(177):243–264, 1987.

[MOP08] Stefan Mangard, Elisabeth Oswald, and Thomas Popp. *Power analysis attacks: Revealing the secrets of smart cards*, volume 31. Springer Science & Business Media, 2008.

[MPC00] Lauren May, Lyta Penna, and Andrew Clark. An implementation of bitsliced DES on the pentium MMX TM processor. In *Australasian Conference on Information Security and Privacy*, pages 112–122. Springer, 2000.

[MPG05] Stefan Mangard, Thomas Popp, and Berndt M Gammel. Side-channel leakage of masked CMOS gates. In *Cryptographers' Track at the RSA Conference*, pages 351–365. Springer, 2005.

[MPP16] Houssem Maghrebi, Thibault Portigliatti, and Emmanuel Prouff. Breaking cryptographic implementations using deep learning techniques. In *Security, Privacy, and Applied Cryptography Engineering: 6th International Conference, SPACE 2016, Hyderabad, India, December 14–18, 2016, Proceedings 6*, pages 3–26. Springer, 2016.

[MS77] Florence Jessie MacWilliams and Neil James Alexander Sloane. *The theory of error correcting codes*, volume 16. Elsevier, 1977.

[MS00] Rita Mayer-Sommer. Smartly analyzing the simplicity and the power of simple power analysis on smartcards. In *International Workshop on Cryptographic Hardware and Embedded Systems*, pages 78–92. Springer, 2000.

[MSB16] Houssem Maghrebi, Victor Servant, and Julien Bringer. There is wisdom in harnessing the strengths of your enemy: Customized encoding to thwart side-channel attacks. In *Fast Software Encryption: 23rd International Conference, FSE 2016, Bochum, Germany, March 20–23, 2016, Revised Selected Papers 23*, pages 223–243. Springer, 2016.

[MSGR10] Marcel Medwed, François-Xavier Standaert, Johann Großschädl, and Francesco Regazzoni. Fresh re-keying: Security against side-channel and fault attacks for low-cost devices. In *International Conference on Cryptology in Africa*, pages 279–296. Springer, 2010.

[MSY06] Tal G Malkin, François-Xavier Standaert, and Moti Yung. A comparative cost/security analysis of fault attack countermeasures. In *Fault Diagnosis and Tolerance in Cryptography: Third International Workshop, FDTC 2006, Yokohama, Japan, October 10, 2006. Proceedings*, pages 159–172. Springer, 2006.

[MVOV18] Alfred J Menezes, Paul C Van Oorschot, and Scott A Vanstone. *Handbook of applied cryptography*. CRC press, 2018.

[MWK+22] Catinca Mujdei, Lennert Wouters, Angshuman Karmakar, Arthur Beckers, Jose Maria Bermudo Mera, and Ingrid Verbauwhede. Side-channel analysis of lattice-based post-quantum cryptography: Exploiting polynomial multiplication. *ACM Transactions on Embedded Computing Systems*, 2022.

[MWM21] Thorben Moos, Felix Wegener, and Amir Moradi. Dl-la: Deep learning leakage assessment: A modern roadmap for SCA evaluations. *IACR Transactions on Cryptographic Hardware and Embedded Systems*, pages 552–598, 2021.

[MZMM16] Zdenek Martinasek, Vaclav Zeman, Lukas Malina, and Josef Martinasek. K-nearest neighbors algorithm in profiling power analysis attacks. *Radioengineering*, 25(2):365–382, 2016.

[NIS01] NIST. Federal information processing standards publication (fips) 197. *Advanced Encryption Standard (AES)*, 2001.

[NIS19] NIST. FIPS 140-3: Security Requirements for Cryptographic Modules, National Institute of Standards and Technology. Technical report, Federal Inf. Process. Stds. (NIST FIPS), National Institute of Standards and Technology, Gaithersburg, MD, 2019.

[Nov02] Roman Novak. Spa-based adaptive chosen-ciphertext attack on RSA implementation. In *International Workshop on Public Key Cryptography*, pages 252–262. Springer, 2002.

[NRS11] Svetla Nikova, Vincent Rijmen, and Martin Schläffer. Secure hardware implementation of nonlinear functions in the presence of glitches. *Journal of Cryptology*, 24:292–321, 2011.

[NY21] Yusuke Nozaki and Masaya Yoshikawa. Shuffling countermeasure against power side-channel attack for MLP with software implementation. In *2021 IEEE 4th International Conference on Electronics and Communication Engineering (ICECE)*, pages 39–42. IEEE, 2021.

[NYGD22] Len Luet Ng, Kim Ho Yeap, Magdalene Wan Ching Goh, and Veerendra Dakulagi. Power consumption in CMOS circuits. In *Electromagnetic Field in Advancing Science and Technology*. IntechOpen, 2022.

[O'D14] Ryan O'Donnell. *Analysis of boolean functions*. Cambridge University Press, 2014.

[O'F23] Colin O'Flynn. PicoEMP: A low-cost EMFI platform compared to BBI and voltage fault injection using TDC and external VCC measurements. *Cryptology ePrint Archive*, 2023.

[Ogg] Frédérique Oggier. Lecture notes. https://feog.github.io/. Accessed: 2012-11-30.

[ORBG17] Marco Oliverio, Kaveh Razavi, Herbert Bos, and Cristiano Giuffrida. Secure Page Fusion with VUsion: https://www.vusec.net/projects/VUsion. In *Proceedings of the 26th Symposium on Operating Systems Principles*, pages 531–545, 2017.

[Org17] European Cyber Security Organisation. Overview of existing cybersecurity standards and certification schemes v2, wg1—standardisation, certification, labelling and supply chain management, 2017.

[ORJ$^+$13] Rachid Omarouayache, Jérémy Raoult, Sylvie Jarrix, Laurent Chusseau, and Philippe Maurine. Magnetic microprobe design for em fault attack. In *2013 International Symposium on Electromagnetic Compatibility*, pages 949–954. IEEE, 2013.

[OS05] Elisabeth Oswald and Kai Schramm. An efficient masking scheme for AES software implementations. In *International Workshop on Information Security Applications*, pages 292–305. Springer, 2005.

[Osw] David Oswald. Lecture notes: Hardware and embedded systems security. https://github.com/david-oswald/hwsec_lecture_notes. Accessed: 2012-12-03.

[PBMB17] Sikhar Patranabis, Jakub Breier, Debdeep Mukhopadhyay, and Shivam Bhasin. One plus one is more than two: a practical combination of power and fault analysis attacks on present and present-like block ciphers. In *2017 Workshop on Fault Diagnosis and Tolerance in Cryptography (FDTC)*, pages 25–32. IEEE, 2017.

[PBP21] Guilherme Perin, Ileana Buhan, and Stjepan Picek. Learning when to stop: a mutual information approach to prevent overfitting in profiled side-channel analysis. In *Constructive Side-Channel Analysis and Secure Design: 12th International Workshop, COSADE 2021, Lugano, Switzerland, October 25–27, 2021, Proceedings 12*, pages 53–81. Springer, 2021.

[PCP20] Guilherme Perin, Łukasz Chmielewski, and Stjepan Picek. Strength in numbers: Improving generalization with ensembles in machine learning-based profiled side-channel analysis. *IACR Transactions on Cryptographic Hardware and Embedded Systems*, pages 337–364, 2020.

[PGP⁺19] Ilia Polian, Mael Gay, Tobias Paxian, Matthias Sauer, and Bernd Becker. Automatic construction of fault attacks on cryptographic hardware implementations. *Automated Methods in Cryptographic Fault Analysis*, pages 151–170, 2019.

[PHJ⁺19] Stjepan Picek, Annelie Heuser, Alan Jovic, Shivam Bhasin, and Francesco Regazzoni. The curse of class imbalance and conflicting metrics with machine learning for side-channel evaluations. *IACR Transactions on Cryptographic Hardware and Embedded Systems*, 2019(1):1–29, 2019.

[PM05] Thomas Popp and Stefan Mangard. Masked dual-rail pre-charge logic: DPA-resistance without routing constraints. In *International Workshop on Cryptographic Hardware and Embedded Systems*, pages 172–186. Springer, 2005.

[PMK⁺11] Axel Poschmann, Amir Moradi, Khoongming Khoo, Chu-Wee Lim, Huaxiong Wang, and San Ling. Side-channel resistant crypto for less than 2,300 ge. *Journal of Cryptology*, 24:322–345, 2011.

[PNP⁺20] Athanasios Papadimitriou, Konstantinos Nomikos, Mihalis Psarakis, Ehsan Aerabi, and David Hely. You can detect but you cannot hide: Fault assisted side channel analysis on protected software-based block ciphers. In *2020 IEEE International Symposium on Defect and Fault Tolerance in VLSI and Nanotechnology Systems (DFT)*, pages 1–6. IEEE, 2020.

[PP09] Christof Paar and Jan Pelzl. *Understanding cryptography: a textbook for students and practitioners*. Springer Science & Business Media, 2009.

[PPM17] Robert Primas, Peter Pessl, and Stefan Mangard. Single-trace side-channel attacks on masked lattice-based encryption. In *Cryptographic Hardware and Embedded Systems–CHES 2017: 19th International Conference, Taipei, Taiwan, September 25–28, 2017, Proceedings*, pages 513–533. Springer, 2017.

[PQ03] Gilles Piret and Jean-Jacques Quisquater. A differential fault attack technique against SPN structures, with application to the AES and khazad. In *Cryptographic Hardware and Embedded Systems-CHES 2003: 5th International Workshop, Cologne, Germany, September 8–10, 2003. Proceedings 5*, pages 77–88. Springer, 2003.

[PR13] Emmanuel Prouff and Matthieu Rivain. Masking against side-channel attacks: A formal security proof. In *Annual International Conference on the Theory and Applications of Cryptographic Techniques*, pages 142–159. Springer, 2013.

[Pro13] Emmanuel Prouff. Side channel attacks against block ciphers implementations and countermeasures. *Tutorial presented in CHES*, 2013.

[PS19] Sandro Pinto and Nuno Santos. Demystifying ARM TrustZone: A Comprehensive Survey. *ACM Computing Surveys (CSUR)*, 51(6):1–36, 2019.

[PSKH18] Aesun Park, Kyung-Ah Shim, Namhun Koo, and Dong-Guk Han. Side-channel attacks on post-quantum signature schemes based on multivariate quadratic equations:-rainbow and UOV. *IACR Transactions on Cryptographic Hardware and Embedded Systems*, pages 500–523, 2018.

[PSQ07] Eric Peeters, François-Xavier Standaert, and Jean-Jacques Quisquater. Power and electromagnetic analysis: Improved model, consequences and comparisons. *Integration*, 40(1):52–60, 2007.

[PV13] Konstantinos Papagiannopoulos and Aram Verstegen. Speed and size-optimized implementations of the present cipher for tiny avr devices. In *Radio Frequency Identification: Security and Privacy Issues 9th International Workshop, RFIDsec 2013, Graz, Austria, July 9–11, 2013, Revised Selected Papers 9*, pages 161–175. Springer, 2013.

[PY06] Raphael C W Phan and Sung-Ming Yen. Amplifying side-channel attacks with techniques from block cipher cryptanalysis. In *Smart Card Research and Advanced Applications: 7th IFIP WG 8.8/11.2 International Conference, CARDIS 2006,*

Tarragona, Spain, April 19–21, 2006. Proceedings 7, pages 135–150. Springer, 2006.

[QWL⁺20] Pengfei Qiu, Dongsheng Wang, Yongqiang Lyu, Ruidong Tian, Chunlu Wang, and Gang Qu. Voltjockey: A new dynamic voltage scaling-based fault injection attack on Intel SGX. *IEEE Transactions on Computer-Aided Design of Integrated Circuits and Systems*, 40(6):1130–1143, 2020.

[QWLQ19] Pengfei Qiu, Dongsheng Wang, Yongqiang Lyu, and Gang Qu. Voltjockey: Breaching trustzone by software-controlled voltage manipulation over multi-core frequencies. In *Proceedings of the 2019 ACM SIGSAC Conference on Computer and Communications Security*, pages 195–209, 2019.

[RAL17] Tiago Reis, Diego F Aranha, and Julio López. Present runs fast. In *International Conference on Cryptographic Hardware and Embedded Systems*, pages 644–664. Springer, 2017.

[RBBC18] Prasanna Ravi, Shivam Bhasin, Jakub Breier, and Anupam Chattopadhyay. PPAP and iPPAP: PLL-based protection against physical attacks. In *2018 IEEE Computer Society Annual Symposium on VLSI, ISVLSI 2018, Hong Kong, China, July 8–11, 2018*, pages 620–625. IEEE Computer Society, 2018.

[RBRC20] Prasanna Ravi, Shivam Bhasin, Sujoy Sinha Roy, and Anupam Chattopadhyay. Drop by drop you break the rock-exploiting generic vulnerabilities in lattice-based PKE/KEMs using EM-based physical attacks. *IACR Cryptol. ePrint Arch.*, 2020:549, 2020.

[RCDB22] Prasanna Ravi, Anupam Chattopadhyay, Jan Pieter D'Anvers, and Anubhab Baksi. Side-channel and fault-injection attacks over lattice-based post-quantum schemes (kyber, dilithium): Survey and new results. *ACM Transactions on Embedded Computing Systems*, 2022.

[RCYF22] Adnan Siraj Rakin, Md Hafizul Islam Chowdhuryy, Fan Yao, and Deliang Fan. Deepsteal: Advanced model extractions leveraging efficient weight stealing in memories. In *2022 IEEE Symposium on Security and Privacy (SP)*, pages 1157–1174. IEEE, 2022.

[RDMB⁺18] Oscar Reparaz, Lauren De Meyer, Begül Bilgin, Victor Arribas, Svetla Nikova, Ventzislav Nikov, and Nigel Smart. Capa: the spirit of beaver against physical attacks. In *Advances in Cryptology–CRYPTO 2018: 38th Annual International Cryptology Conference, Santa Barbara, CA, USA, August 19–23, 2018, Proceedings, Part I 38*, pages 121–151. Springer, 2018.

[RGN13] Pablo Rauzy, Sylvain Guilley, and Zakaria Najm. Formally proved security of assembly code against leakage. *IACR Cryptol. ePrint Arch.*, 2013:554, 2013.

[RHF19] Adnan Siraj Rakin, Zhezhi He, and Deliang Fan. Bit-flip attack: Crushing neural network with progressive bit search. In *Proceedings of the IEEE/CVF International Conference on Computer Vision*, pages 1211–1220, 2019.

[RHF20] Adnan Siraj Rakin, Zhezhi He, and Deliang Fan. TBT: Targeted neural network attack with bit trojan. In *Proceedings of the IEEE/CVF Conference on Computer Vision and Pattern Recognition*, pages 13198–13207, 2020.

[RHL⁺21] Adnan Siraj Rakin, Zhezhi He, Jingtao Li, Fan Yao, Chaitali Chakrabarti, and Deliang Fan. T-BFA: Targeted bit-flip adversarial weight attack. *IEEE Transactions on Pattern Analysis and Machine Intelligence*, 44(11):7928–7939, 2021.

[Riv09] Matthieu Rivain. Differential fault analysis on DES middle rounds. In *CHES*, volume 5747, pages 457–469. Springer, 2009.

[RLK11] Thomas Roche, Victor Lomné, and Karim Khalfallah. Combined fault and side-channel attack on protected implementations of AES. In *Smart Card Research and Advanced Applications: 10th IFIP WG 8.8/11.2 International Conference, CARDIS 2011, Leuven, Belgium, September 14–16, 2011, Revised Selected Papers 10*, pages 65–83. Springer, 2011.

[RM07] Bruno Robisson and Pascal Manet. Differential behavioral analysis. In *Cryptographic Hardware and Embedded Systems-CHES 2007: 9th International*

Workshop, Vienna, Austria, September 10–13, 2007. Proceedings 9, pages 413–426. Springer, 2007.

[Ros20] Sheldon M Ross. *Introduction to probability and statistics for engineers and scientists*. Academic press, 2020.

[RP10] Matthieu Rivain and Emmanuel Prouff. Provably secure higher-order masking of AES. In *International Workshop on Cryptographic Hardware and Embedded Systems*, pages 413–427. Springer, 2010.

[RRB+19] Prasanna Ravi, Debapriya Basu Roy, Shivam Bhasin, Anupam Chattopadhyay, and Debdeep Mukhopadhyay. Number "not used" once-practical fault attack on pqm4 implementations of nist candidates. In *Constructive Side-Channel Analysis and Secure Design: 10th International Workshop, COSADE 2019, Darmstadt, Germany, April 3–5, 2019, Proceedings 10*, pages 232–250. Springer, 2019.

[RS09] Mathieu Renauld and François-Xavier Standaert. Algebraic side-channel attacks. In *International Conference on Information Security and Cryptology*, pages 393–410. Springer, 2009.

[RWPP21] Jorai Rijsdijk, Lichao Wu, Guilherme Perin, and Stjepan Picek. Reinforcement learning for hyperparameter tuning in deep learning-based side-channel analysis. *IACR Transactions on Cryptographic Hardware and Embedded Systems*, pages 677–707, 2021.

[RZC+21] Damien Robissout, Gabriel Zaid, Brice Colombier, Lilian Bossuet, and Amaury Habrard. Online performance evaluation of deep learning networks for profiled side-channel analysis. In *Constructive Side-Channel Analysis and Secure Design: 11th International Workshop, COSADE 2020, Lugano, Switzerland, April 1–3, 2020, Revised Selected Papers 11*, pages 200–218. Springer, 2021.

[SA93] Jerry M Soden and Richard E Anderson. Ic failure analysis: techniques and tools for quality reliability improvement. *Proceedings of the IEEE*, 81(5):703–715, 1993.

[SA02] Sergei P Skorobogatov and Ross J Anderson. Optical fault induction attacks. In *International workshop on cryptographic hardware and embedded systems*, pages 2–12. Springer, 2002.

[Sau13] Laurent Sauvage. Electric probes for fault injection attack. In *2013 Asia-Pacific Symposium on Electromagnetic Compatibility (APEMC)*, pages 1–4. IEEE, 2013.

[SBM18] Pascal Sasdrich, René Bock, and Amir Moradi. Threshold implementation in software: Case study of present. In *Constructive Side-Channel Analysis and Secure Design: 9th International Workshop, COSADE 2018, Singapore, April 23–24, 2018, Proceedings 9*, pages 227–244. Springer, 2018.

[SC78] Hiroaki Sakoe and Seibi Chiba. Dynamic programming algorithm optimization for spoken word recognition. *IEEE transactions on acoustics, speech, and signal processing*, 26(1):43–49, 1978.

[Sch00] Bruce Schneier. A self-study course in block-cipher cryptanalysis. *Cryptologia*, 24(1):18–33, 2000.

[Sei05] Jean-Pierre Seifert. On authenticated computing and RSA-based authentication. In *Proceedings of the 12th ACM conference on Computer and communications security*, pages 122–127, 2005.

[SGD08] Nidhal Selmane, Sylvain Guilley, and Jean-Luc Danger. Practical setup time violation attacks on AES. In *2008 Seventh European Dependable Computing Conference*, pages 91–96. IEEE, 2008.

[SH07] Jörn-Marc Schmidt and Michael Hutter. *Optical and EM fault-attacks on CRT-based RSA: Concrete results*. 2007.

[SH08] Jörn-Marc Schmidt and Christoph Herbst. A practical fault attack on square and multiply. In *2008 5th Workshop on Fault Diagnosis and Tolerance in Cryptography*, pages 53–58. IEEE, 2008.

[Sha45] Claude E Shannon. A mathematical theory of cryptography. *Mathematical Theory of Cryptography*, 1945.

[Sha97] A Shamir. Method and apparatus for protecting public key schemes from timing and fault attacks. In *EUROCRYPT'97*, 1997.

[Sha00] Adi Shamir. Protecting smart cards from passive power analysis with detached power supplies. In *Cryptographic Hardware and Embedded Systems–CHES 2000: Second International Workshop Worcester, MA, USA, August 17–18, 2000 Proceedings 2*, pages 71–77. Springer, 2000.

[SHS16] Bodo Selmke, Johann Heyszl, and Georg Sigl. Attack on a DFA protected AES by simultaneous laser fault injections. In *2016 Workshop on Fault Diagnosis and Tolerance in Cryptography (FDTC)*, pages 36–46. IEEE, 2016.

[SI20a] SOG-IS. Application of attack potential to smartcards and similar devices, v3.1, 2020.

[SI20b] SOG-IS. Attack methods for smartcards and similar devices, 2020.

[Sie88] Waclaw Sierpinski. *Elementary Theory of Numbers: Second English Edition (edited by A. Schinzel)*. Elsevier, 1988.

[Siv17] Nimisha Sivaraman. *Design of magnetic probes for near field measurements and the development of algorithms for the prediction of EMC*. PhD thesis, Université Grenoble Alpes, 2017.

[SJB+18] Sayandeep Saha, Dirmanto Jap, Jakub Breier, Shivam Bhasin, Debdeep Mukhopadhyay, and Pallab Dasgupta. Breaking redundancy-based countermeasures with random faults and power side channel. In *2018 Workshop on Fault Diagnosis and Tolerance in Cryptography (FDTC)*, pages 15–22. IEEE, 2018.

[SM12] Pushpa Saini and Rajesh Mehra. A novel technique for glitch and leakage power reduction in CMOS vlsi circuits. *International Journal of Advanced Computer Science and Applications*, 3(10), 2012.

[SM15] Tobias Schneider and Amir Moradi. Leakage assessment methodology: A clear roadmap for side-channel evaluations. In *Cryptographic Hardware and Embedded Systems–CHES 2015: 17th International Workshop, Saint-Malo, France, September 13–16, 2015, Proceedings 17*, pages 495–513. Springer, 2015.

[SMG16] Tobias Schneider, Amir Moradi, and Tim Güneysu. Parti–towards combined hardware countermeasures against side-channel and fault-injection attacks. In *Advances in Cryptology–CRYPTO 2016: 36th Annual International Cryptology Conference, Santa Barbara, CA, USA, August 14–18, 2016, Proceedings, Part II 36*, pages 302–332. Springer, 2016.

[SMKLM02] Yen Sung-Ming, Seungjoo Kim, Seongan Lim, and Sangjae Moon. RSA speedup with residue number system immune against hardware fault cryptanalysis. In *international conference on information security and cryptology*, pages 397–413. Springer, 2002.

[SMR09] Dhiman Saha, Debdeep Mukhopadhyay, and Dipanwita RoyChowdhury. A diagonal fault attack on the advanced encryption standard. *Cryptology ePrint Archive*, 2009.

[SMY09] François-Xavier Standaert, Tal G Malkin, and Moti Yung. A unified framework for the analysis of side-channel key recovery attacks. In *Advances in Cryptology-EUROCRYPT 2009*, pages 443–461. Springer, 2009.

[Sor84] Arthur Sorkin. Lucifer, a cryptographic algorithm. *Cryptologia*, 8(1):22–42, 1984.

[SP06] Kai Schramm and Christof Paar. Higher order masking of the AES. In *Topics in Cryptology–CT-RSA 2006: The Cryptographers' Track at the RSA Conference 2006, San Jose, CA, USA, February 13–17, 2005. Proceedings*, pages 208–225. Springer, 2006.

[SS16] Peter Schwabe and Ko Stoffelen. All the AES you need on cortex-m3 and m4. In *International Conference on Selected Areas in Cryptography*, pages 180–194. Springer, 2016.

[Sta10] François-Xavier Standaert. Introduction to side-channel attacks. *Secure integrated circuits and systems*, pages 27–42, 2010.

[Sti05] Douglas R Stinson. *Cryptography: theory and practice*. Chapman and Hall/CRC, 2005.

[SVK$^+$03] H Saputra, N Vijaykrishnan, M Kandemir, MJ Irwin, and R Brooks. Masking the energy behaviour of encryption algorithms. *IEE Proceedings-Computers and Digital Techniques*, 150(5):274–284, 2003.

[SWM18] Robert Schilling, Mario Werner, and Stefan Mangard. Securing conditional branches in the presence of fault attacks. In *2018 Design, Automation & Test in Europe Conference & Exhibition (DATE)*, pages 1586–1591. IEEE, 2018.

[SWP03] Kai Schramm, Thomas Wollinger, and Christof Paar. A new class of collision attacks and its application to DES. In *Fast Software Encryption: 10th International Workshop, FSE 2003, Lund, Sweden, February 24–26, 2003. Revised Papers 10*, pages 206–222. Springer, 2003.

[TAV02] Kris Tiri, Moonmoon Akmal, and Ingrid Verbauwhede. A dynamic and differential CMOS logic with signal independent power consumption to withstand differential power analysis on smart cards. In *Proceedings of the 28th European solid-state circuits conference*, pages 403–406. IEEE, 2002.

[TBM14] Harshal Tupsamudre, Shikha Bisht, and Debdeep Mukhopadhyay. Destroying fault invariant with randomization: A countermeasure for AES against differential fault attacks. In *Cryptographic Hardware and Embedded Systems–CHES 2014: 16th International Workshop, Busan, South Korea, September 23–26, 2014. Proceedings 16*, pages 93–111. Springer, 2014.

[THM07] Stefan Tillich, Christoph Herbst, and Stefan Mangard. Protecting AES software implementations on 32-bit processors against power analysis. In *Applied Cryptography and Network Security: 5th International Conference, ACNS 2007, Zhuhai, China, June 5–8, 2007. Proceedings 5*, pages 141–157. Springer, 2007.

[TIA$^+$23] M Caner Tol, Saad Islam, Andrew J Adiletta, Berk Sunar, and Ziming Zhang. Don't knock! rowhammer at the backdoor of DNN models. In *2023 53rd Annual IEEE/IFIP International Conference on Dependable Systems and Networks (DSN)*, pages 109–122. IEEE, 2023.

[Tim19] Benjamin Timon. Non-profiled deep learning-based side-channel attacks with sensitivity analysis. *IACR Transactions on Cryptographic Hardware and Embedded Systems*, pages 107–131, 2019.

[TKA$^+$18] Andrei Tatar, Radhesh Krishnan Konoth, Elias Athanasopoulos, Cristiano Giuffrida, Herbert Bos, and Kaveh Razavi. Throwhammer: Rowhammer attacks over the network and defenses. In *2018 USENIX Annual Technical Conference (USENIX ATC 18)*, pages 213–226, 2018.

[TMA11] Michael Tunstall, Debdeep Mukhopadhyay, and Subidh Ali. Differential fault analysis of the advanced encryption standard using a single fault. In *Information Security Theory and Practice. Security and Privacy of Mobile Devices in Wireless Communication: 5th IFIP WG 11.2 International Workshop, WISTP 2011, Heraklion, Crete, Greece, June 1–3, 2011. Proceedings 5*, pages 224–233. Springer, 2011.

[TSS$^+$06] Pim Tuyls, Geert Jan Schrijen, Boris Skoric, Jan Van Geloven, Nynke Verhaegh, and Rob Wolters. Read-proof hardware from protective coatings. In *Ches*, volume 6, pages 369–383. Springer, 2006.

[TSS17] Adrian Tang, Simha Sethumadhavan, and Salvatore Stolfo. CLKSCREW: Exposing the Perils of Security-Oblivious Energy Management. In *26th USENIX Security Symposium (USENIX Security 17)*, pages 1057–1074, 2017.

[TV06] Kris Tiri and Ingrid Verbauwhede. A digital design flow for secure integrated circuits. *IEEE Transactions on Computer-Aided Design of Integrated Circuits and Systems*, 25(7):1197–1208, 2006.

[UXT$^+$22] Rei Ueno, Keita Xagawa, Yutaro Tanaka, Akira Ito, Junko Takahashi, and Naofumi Homma. Curse of re-encryption: A generic power/EM analysis on post-quantum

KEMs. *IACR Transactions on Cryptographic Hardware and Embedded Systems*, pages 296–322, 2022.

[Vai01]   PP Vaidyanathan. Generalizations of the sampling theorem: Seven decades after nyquist. *IEEE Transactions on Circuits and Systems I: Fundamental Theory and Applications*, 48(9):1094–1109, 2001.

[VCGRS13]   Nicolas Veyrat-Charvillon, Benoît Gérard, Mathieu Renauld, and François-Xavier Standaert. An optimal key enumeration algorithm and its application to side-channel attacks. In *Selected Areas in Cryptography: 19th International Conference, SAC 2012, Windsor, ON, Canada, August 15–16, 2012, Revised Selected Papers 19*, pages 390–406. Springer, 2013.

[VCGS14]   Nicolas Veyrat-Charvillon, Benoît Gérard, and François-Xavier Standaert. Soft analytical side-channel attacks. In *Advances in Cryptology–ASIACRYPT 2014: 20th International Conference on the Theory and Application of Cryptology and Information Security, Kaoshiung, Taiwan, ROC, December 7–11, 2014. Proceedings, Part I 20*, pages 282–296. Springer, 2014.

[VDVFL+16] Victor Van Der Veen, Yanick Fratantonio, Martina Lindorfer, Daniel Gruss, Clémentine Maurice, Giovanni Vigna, Herbert Bos, Kaveh Razavi, and Cristiano Giuffrida. Drammer: Deterministic rowhammer attacks on mobile platforms. In *Proceedings of the 2016 ACM SIGSAC conference on computer and communications security*, pages 1675–1689, 2016.

[VEW12]   Camille Vuillaume, Takashi Endo, and Paul Wooderson. RSA key generation: new attacks. In *Constructive Side-Channel Analysis and Secure Design: Third International Workshop, COSADE 2012, Darmstadt, Germany, May 3–4, 2012. Proceedings 3*, pages 105–119. Springer, 2012.

[Vig08]   David Vigilant. RSA with crt: A new cost-effective solution to thwart fault attacks. In *International Workshop on Cryptographic Hardware and Embedded Systems*, pages 130–145. Springer, 2008.

[VW01]   Manfred Von Willich. A technique with an information-theoretic basis for protecting secret data from differential power attacks. In *IMA International Conference on Cryptography and Coding*, pages 44–62. Springer, 2001.

[vWWB11]   Jasper GJ van Woudenberg, Marc F Witteman, and Bram Bakker. Improving differential power analysis by elastic alignment. In *Topics in Cryptology–CT-RSA 2011: The Cryptographers' Track at the RSA Conference 2011, San Francisco, CA, USA, February 14–18, 2011. Proceedings*, pages 104–119. Springer, 2011.

[Wal02]   Colin D Walter. Mist: An efficient, randomized exponentiation algorithm for resisting power analysis. In *Topics in Cryptology—CT-RSA 2002: The Cryptographers' Track at the RSA Conference 2002 San Jose, CA, USA, February 18–22, 2002 Proceedings*, pages 53–66. Springer, 2002.

[Wel47]   Bernard L Welch. The generalization of 'student's' problem when several different population variances are involved. *Biometrika*, 34(1–2):28–35, 1947.

[WHJ+21]   Yoo-Seung Won, Xiaolu Hou, Dirmanto Jap, Jakub Breier, and Shivam Bhasin. Back to the basics: Seamless integration of side-channel pre-processing in deep neural networks. *IEEE Transactions on Information Forensics and Security*, 16:3215–3227, 2021.

[WJB20]   Yoo-Seung Won, Dirmanto Jap, and Shivam Bhasin. Push for more: On comparison of data augmentation and smote with optimised deep learning architecture for side-channel. In *Information Security Applications: 21st International Conference, WISA 2020, Jeju Island, South Korea, August 26–28, 2020, Revised Selected Papers 21*, pages 227–241. Springer, 2020.

[WKKG04]   Kaijie Wu, Ramesh Karri, Grigori Kuznetsov, and Michael Goessel. Low cost concurrent error detection for the advanced encryption standard. In *2004 International Conference on Test*, pages 1242–1248. IEEE, 2004.

[WP20]   Lichao Wu and Stjepan Picek. Remove some noise: On pre-processing of side-channel measurements with autoencoders. *IACR Transactions on Cryptographic Hardware and Embedded Systems*, pages 389–415, 2020.

[WPP22]   Lichao Wu, Guilherme Perin, and Stjepan Picek. I choose you: Automated hyperparameter tuning for deep learning-based side-channel analysis. *IEEE Transactions on Emerging Topics in Computing*, 2022.

[WvWM11]   Marc F Witteman, Jasper GJ van Woudenberg, and Federico Menarini. Defeating RSA multiply-always and message blinding countermeasures. In *Topics in Cryptology–CT-RSA 2011: The Cryptographers' Track at the RSA Conference 2011, San Francisco, CA, USA, February 14–18, 2011. Proceedings*, pages 77–88. Springer, 2011.

[WW10]   Gaoli Wang and Shaohui Wang. Differential fault analysis on present key schedule. In *2010 International Conference on Computational Intelligence and Security*, pages 362–366. IEEE, 2010.

[XIU+21]   Keita Xagawa, Akira Ito, Rei Ueno, Junko Takahashi, and Naofumi Homma. Fault-injection attacks against nist's post-quantum cryptography round 3 KEM candidates. In *International Conference on the Theory and Application of Cryptology and Information Security*, pages 33–61. Springer, 2021.

[XLZ+18]   Sen Xu, Xiangjun Lu, Kaiyu Zhang, Yang Li, Lei Wang, Weijia Wang, Haihua Gu, Zheng Guo, Junrong Liu, and Dawu Gu. Similar operation template attack on RSA-crt as a case study. *Science China Information Sciences*, 61:1–17, 2018.

[XZY+20]   Guorui Xu, Fan Zhang, Bolin Yang, Xinjie Zhao, Wei He, and Kui Ren. Pushing the limit of PFA: enhanced persistent fault analysis on block ciphers. *IEEE Transactions on Computer-Aided Design of Integrated Circuits and Systems*, 40(6):1102–1116, 2020.

[Yeh14]   James J Yeh. *Real analysis: theory of measure and integration*. World Scientific Publishing Company, 2014.

[YJ00]   Sung-Ming Yen and Marc Joye. Checking before output may not be enough against fault-based cryptanalysis. *IEEE Transactions on computers*, 49(9):967–970, 2000.

[YKM06]   Sung-Ming Yen, Dongryeol Kim, and SangJae Moon. Cryptanalysis of two protocols for RSA with crt based on fault infection. In *International Workshop on Fault Diagnosis and Tolerance in Cryptography*, pages 53–61. Springer, 2006.

[YMY+20]   Honggang Yu, Haocheng Ma, Kaichen Yang, Yiqiang Zhao, and Yier Jin. Deepem: Deep neural networks model recovery through em side-channel information leakage. In *2020 IEEE International Symposium on Hardware Oriented Security and Trust (HOST)*, pages 209–218. IEEE, 2020.

[YRF20]   Fan Yao, Adnan Siraj Rakin, and Deliang Fan. {DeepHammer}: Depleting the intelligence of deep neural networks through targeted chain of bit flips. In *29th USENIX Security Symposium (USENIX Security 20)*, pages 1463–1480, 2020.

[ZBHV20]   Gabriel Zaid, Lilian Bossuet, Amaury Habrard, and Alexandre Venelli. Methodology for efficient CNN architectures in profiling attacks. *IACR Transactions on Cryptographic Hardware and Embedded Systems*, pages 1–36, 2020.

[ZDT+14]   Loic Zussa, Amine Dehbaoui, Karim Tobich, Jean-Max Dutertre, Philippe Maurine, Ludovic Guillaume-Sage, Jessy Clediere, and Assia Tria. Efficiency of a glitch detector against electromagnetic fault injection. In *2014 Design, Automation & Test in Europe Conference & Exhibition (DATE)*, pages 1–6. IEEE, 2014.

[ZLZ+18]   Fan Zhang, Xiaoxuan Lou, Xinjie Zhao, Shivam Bhasin, Wei He, Ruyi Ding, Samiya Qureshi, and Kui Ren. Persistent fault analysis on block ciphers. *IACR Transactions on Cryptographic Hardware and Embedded Systems*, pages 150–172, 2018.

[ZZJ+20]   Fan Zhang, Yiran Zhang, Huilong Jiang, Xiang Zhu, Shivam Bhasin, Xinjie Zhao, Zhe Liu, Dawu Gu, and Kui Ren. Persistent fault attack in practice. *IACR Transactions on Cryptographic Hardware and Embedded Systems*, pages 172–195, 2020.

[ZZY+19]   Yiran Zhang, Fan Zhang, Bolin Yang, Guorui Xu, Bin Shao, Xinjie Zhao, and Kui Ren. Persistent fault injection in FPGA via BRAM modification. In *2019 IEEE Conference on Dependable and Secure Computing (DSC)*, pages 1–6. IEEE, 2019.

Index

Symbols
$\mathcal{M}_{n \times n}(R)$, 23
$\mathcal{R}^d$, 67
$\varphi(n)$, 38
n-bit binary string, 27
$\mathbb{Z}_n$, 34
$\mathbb{Z}_n^*$, 38

A
Advanced Encryption Standard (AES),
139
AES T-tables, 158
Affine cipher, 109

B
Bellcore attack, 392
Binary code, 57
  anticode, 65
  binary (n, M)−code, 57
  binary (n, M, d)−code, 58
  binary $[n, k, d]$− linear code, 61
  codeword, 57
  dimension, 61
  dual code, 62
  error correcting, 59
  error-detecting, 58
  generator matrix, 63
  length, 57
  linear, 60
  maximum distance, 65
  (minimum) distance, 58
  n−repetition code, 61
  parity-check code, 62
  parity-check matrix, 63
  size, 57
Bit, 20
Bitwise AND, 27
Bitwise OR, 27
Bitwise XOR, 27
Blakely's method, 182
Boolean function, 159
  algebraic normal form, 160
  indicator function, 160
  truth table, 159
Borel set, 67
Byte, 27

C
Caesar cipher, 108
CBC mode, 127
Chinese Remainder Theorem, 44
Correlation coefficient, 81
Cryptographic primitives, 102
Cryptosystem, 104
  block cipher, 105
    block length, 132
    Feistel cipher, 133
    key length, 132
    key schedule, 132
    master key, 132
    round function, 132
    Sbox, 133
    SPN cipher, 133
  computationally secure, 107
  perfectly secure, 107, 124
  secure in practice, 107
  stream cipher, 105

9 783031 622045